FOURTH EDITION

VISUALIZING

ENVIRONMENTAL SCIENCE

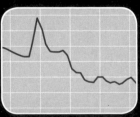

VISUALIZING
ENVIRONMENTAL SCIENCE

FOURTH EDITION

David M. Hassenzahl
Mary Catherine Hager
Linda R. Berg

WILEY

Credits

EXECUTIVE VP AND PUBLISHER Kaye Pace
SENIOR EDITOR Rachel Falk
DIRECTOR OF DEVELOPMENT Barbara Heaney
MANAGER, PRODUCT DEVELOPMENT Nancy Perry
ASSISTANT CONTENT EDITOR Lauren Samuelson
WILEY VISUALIZING PROJECT EDITOR Brian B. Baker
WILEY VISUALIZING SENIOR EDITORIAL ASSISTANT
    Tiara Kelly
ASSOCIATE DIRECTOR OF MARKETING Jeffrey
    Rucker

MARKETING MANAGER Carrie Ayers
CONTENT MANAGER Juanita Thompson
SENIOR PRODUCTION EDITOR Trish McFadden
SENIOR MEDIA EDITOR Linda Muriello
SENIOR MEDIA SPECIALIST Svetlana Barskaya
CREATIVE DIRECTOR Harry Nolan
COVER DESIGNER Harry Nolan
PHOTO EDITOR Sheena Goldstein

COVER CREDITS: Main Image: © Siegfried Layda/Photographers Choice/Getty Images
          Bottom left: © Matthias Kulka/Corbis
          Bottom second from left: © WDG Photo/Shutterstock
          Bottom center: © Mint Images-Frans Lanting/Getty Images
          Bottom right: © Peter Adams/The Image Bank/Getty Images

Back Cover image: Schlesinger, W. H. *Biogeochemistry: An Analysis of Global Change*, 2nd edition.
          Academic Press, San Diego (1997) and is based on several sources.

This book was set in Baskerville by codeMantra, and printed and bound by Quad/Graphics. The cover was printed by Quad/Graphics.

ISBN: 9781118169834
BRV ISBN: 9781118176863

How Is Wiley Visualizing Different?

Wiley Visualizing differs from competing textbooks by uniquely combining two powerful elements: visual pedagogy integrated with a comprehensive text, and the inclusion of interactive multimedia through "the *WileyPLUS.*" Together these elements deliver rigorous content using methods that engage students with the material. Each key concept is supported by supporting data, images, text, and diagrams that are crafted to maximize student learning.

(1) Visual Pedagogy. Wiley Visualizing is based on decades of research on the use of visuals in learning.[1] Using the cognitive theory of multimedia learning, which is backed up by hundreds of empirical research studies, Wiley's authors select relevant visualizations and graphical displays of information for their texts that specifically support students' thinking and learning, then organize the updated content to integrate the new knowledge with prior knowledge. Visuals and text are conceived and planned together in ways that clarify and reinforce major concepts while allowing students to understand the details. This commitment to distinctive and consistent visual pedagogy sets Wiley Visualizing apart from other textbooks.

(2) Interactive Multimedia. Wiley Visualizing is based on the understanding that learning is an active process of knowledge construction. *Visualizing Environmental Science, Fourth Edition* is therefore tightly integrated with multimedia activities provided in *WileyPLUS*.

Wiley Visualizing is designed as a natural extension of how we learn

Visuals, comprehensive text, and learning aids are integrated to display facts, concepts, processes, and principles more effectively than words alone can. To understand why the Wiley Visualizing approach is effective, it is first helpful to understand how we learn.

1. Our brain processes information using two channels: visual and verbal. Our *working memory* holds information that our minds process as we learn. In working memory we begin to make sense of words and pictures, and build verbal and visual models of the information.

2. When the verbal and visual models of corresponding information are connected in working memory, we form more comprehensive, or integrated, mental models.

3. When we link these integrated mental models to our prior knowledge, which is stored in our *long-term* memory, we build even stronger mental models. When an integrated mental model is formed and stored in long-term memory, real learning begins.

The effort our brains put forth to make sense of instructional information is called *cognitive load*. There are two kinds of cognitive load: productive cognitive load, such as when we're engaged in learning or exert positive effort to create mental models; and unproductive cognitive load, which occurs when the brain is trying to make sense of needlessly complex content or when information is not presented well. The learning process can be impaired when the amount of information to be processed exceeds the capacity of working memory. Well-designed visuals and text with effective pedagogical guidance can reduce the unproductive cognitive load in our working memory.

---

[1] Mayer, R.E. (Ed.) 2005. *The Cambridge Handbook of Multimedia Learning.* New York: Cambridge University Press.

Wiley Visualizing is designed for engaging and effective learning

The visuals and text in *Visualizing Environmental Science, Fourth Edition* are specially integrated to present complex processes in clear steps and with compelling representations, organize related pieces of information, and integrate related information sources. This approach, along with the use of interactive multimedia, minimizes unproductive cognitive load and helps students engage with the content. When students are engaged, they are storing information in long-term memory, and thinking critically about both new information and their previous beliefs. This leads to better thinking, greater knowledge, and ultimately to academic success.

Research shows that well-designed visuals, integrated with comprehensive text, can improve the efficiency with which a learner processes information. In this regard, SEG Research, an independent research firm, conducted a national, multisite study evaluating the effectiveness of Wiley Visualizing. Its findings indicate that students using Wiley Visualizing products (both print and multimedia) were more engaged in the course, exhibited greater retention throughout the course, and made significantly greater gains in content area knowledge and skills, as compared to students in similar classes that did not use Wiley Visualizing.[2]

The use of *WileyPLUS* can also increase learning. According to a white paper titled "Leveraging Blended Learning for More Effective Course Management and Enhanced Student Outcomes" by Peggy Wyllie of Evince Market Research & Communications,[3] studies show that effective use of online resources can improve learning outcomes. Pairing supportive online resources with face-to-face instruction can help students to learn and reflect on material, and deploying multimodal learning methods can help students to engage with the material and retain their acquired knowledge. *WileyPLUS* provides students with an environment that stimulates active learning and enables them to optimize the time they spend on their coursework. Continual assessment/remediation is also key to helping students stay on track. The *WileyPLUS* system facilitates instructors' course planning, organization, and delivery and provides a range of flexible tools for easy design and deployment of activities and tracking of student progress for each learning objective.

Figure 5.5: Energy flow through a food chain Textual elements have been physically integrated with the visual elements. This eliminates split attention (dividing our attention between several sources of different information). The arrows visually display processes, easing the way we recognize relationships.

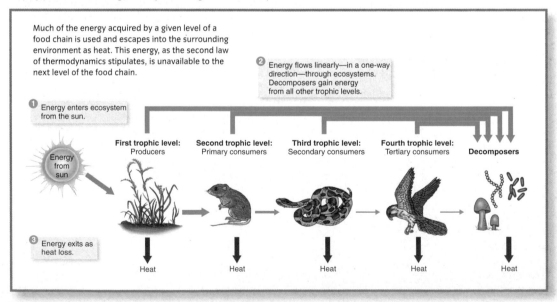

[2]SEG Research. 2009. "Improving Student-Learning with Graphically-Enhanced Textbooks: A Study of the Effectiveness of the Wiley Visualizing Series." Available online at www.segmeasurement.com.
[3]Peggy Wyllie. 2009. "Leveraging Blended Learning for More Effective Course Management and Enhanced Student Outcomes." Available online at http://catalog.wileyplus.com./about/instructors/whitepaper.html.

How Are the Wiley Visualizing Chapters Organized?

Student engagement requires more than just providing visuals, text, and interactivity—it entails motivating students to learn. Student engagement can be behavioral, cognitive, social, and/or emotional. It is easy to get bored or lose focus when presented with large amounts of information, and it is easy to lose motivation when the relevance of the information is unclear. Wiley Visualizing reorganizes course content into manageable learning objectives and relates it to everyday life.

The content in Wiley Visualizing is organized into learning modules. Each module has a clear instructional objective, one or more examples, and an opportunity for assessment. These modules are the building blocks of Wiley Visualizing.

Each Wiley Visualizing chapter engages students from the start

Chapter opening text and visuals introduce the subject and connect the student with the material that follows.

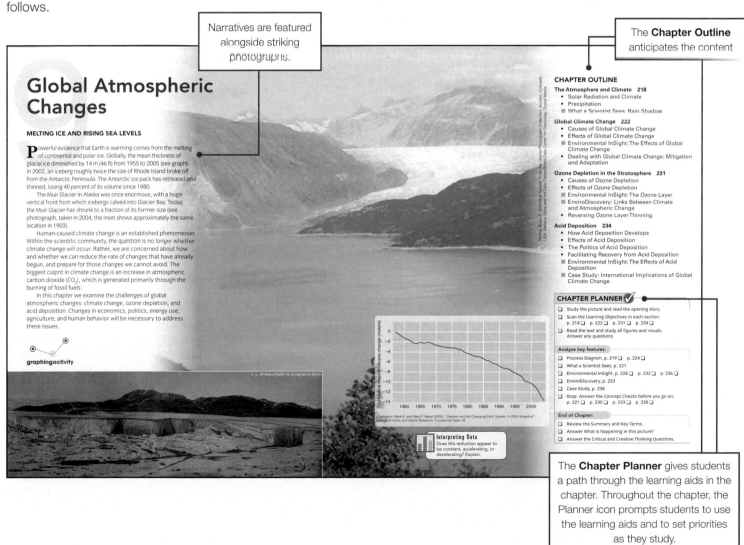

Narratives are featured alongside striking photographs.

The **Chapter Outline** anticipates the content

Global Atmospheric Changes

MELTING ICE AND RISING SEA LEVELS

Powerful evidence that Earth is warming comes from the melting of continental and polar ice. Globally, the mean thickness of glacial ice diminished by 14 m (46 ft) from 1955 to 2005 (see graph). In 2002, an iceberg roughly twice the size of Rhode Island broke off from the Antarctic Peninsula. The Antarctic ice pack has retreated and thinned, losing 40 percent of its volume since 1980.

The Muir Glacier in Alaska was once enormous, with a huge vertical front from which icebergs calved into Glacier Bay. Today, the Muir Glacier has shrunk to a fraction of its former size (see photograph, taken in 2004; the inset shows approximately the same location in 1903).

Human-caused climate change is an established phenomenon. Within the scientific community, the question is no longer whether climate change will occur. Rather, we are concerned about how and whether we can reduce the rate of changes that have already begun, and prepare for those changes we cannot avoid. The biggest culprit in climate change is an increase in atmospheric carbon dioxide (CO_2), which is generated primarily through the burning of fossil fuels.

In this chapter we examine the challenges of global atmospheric changes: climate change, ozone depletion, and acid deposition. Changes in economics, politics, energy use, agriculture, and human behavior will be necessary to address these issues.

graphingactivity

Interpreting Data
Does this reduction appear to be constant, accelerating, or decelerating? Explain.

CHAPTER OUTLINE

The Atmosphere and Climate 218
- Solar Radiation and Climate
- Precipitation
- What a Scientist Sees: Rain Shadow

Global Climate Change 222
- Causes of Global Climate Change
- Effects of Global Climate Change
- Environmental InSight: The Effects of Global Climate Change
- Dealing with Global Climate Change: Mitigation and Adaptation

Ozone Depletion in the Stratosphere 231
- Causes of Ozone Depletion
- Effects of Ozone Depletion
- Environmental InSight: The Ozone Layer
- EnviroDiscovery: Links Between Climate and Atmospheric Change
- Reversing Ozone Layer Thinning

Acid Deposition 234
- How Acid Deposition Develops
- Effects of Acid Deposition
- The Politics of Acid Deposition
- Facilitating Recovery from Acid Deposition
- Environmental InSight: The Effects of Acid Deposition
- Case Study: International Implications of Global Climate Change

CHAPTER PLANNER ✓

- ☐ Study the picture and read the opening story.
- ☐ Scan the Learning Objectives in each section: p. 218 ☐ p. 222 ☐ p. 231 ☐ p. 234 ☐
- ☐ Read the text and study all figures and visuals. Answer any questions.

Analyze key features:
- ☐ Process Diagram, p. 219 ☐ p. 224 ☐
- ☐ What a Scientist Sees, p. 221 ☐
- ☐ Environmental InSight, p. 228 ☐ p. 232 ☐ p. 236 ☐
- ☐ EnviroDiscovery, p. 233 ☐
- ☐ Case Study, p. 238 ☐
- ☐ Stop: Answer the Concept Checks before you go on: p. 221 ☐ p. 230 ☐ p. 233 ☐ p. 238 ☐

End of Chapter:
- ☐ Review the Summary and Key Terms.
- ☐ Answer What is happening in this picture?
- ☐ Answer the Critical and Creative Thinking Questions.

The **Chapter Planner** gives students a path through the learning aids in the chapter. Throughout the chapter, the Planner icon prompts students to use the learning aids and to set priorities as they study.

What Is the Organization of This Book?

We begin *Visualizing Environmental Science 4e* with an introduction of the environmental dilemmas we face in our world today, emphasizing particularly how unchecked population growth and economic inequity complicate our ability to solve these problems. We stress that solutions rest in understanding the science underlying these problems. They also require creativity and diligence at all levels, from individual commitment to international cooperation. Indeed, a key theme integrated throughout the fourth edition is the local to global scales of environmental science. We offer concrete suggestions that students can adopt to make their own difference in solving environmental problems, and we explain the complications that arise when solutions are tackled on a local, regional, national, or global scale. New to the fourth edition are "Sustainable Citizen" questions that compel students to consider how global issues are addressed where they live.

Yet *Visualizing Environmental Science 4e* is not simply a checklist of "to do" items to save the planet. In the context of an engaging visual presentation, we offer solid discussions of such critical environmental concepts as sustainability, conservation and preservation, and risk analysis. We weave the threads of these concepts throughout our treatment of ecological principles and their application to various ecosystems, the impacts of human population change, and the problems associated with our use of the world's resources. We particularly instruct students in the importance of ecosystem services to a functioning world, and the threats that restrict our planet's ability to provide such services.

This text is intended to provide introductory content primarily for nonscience undergraduate students. The accessible format of *Visualizing Environmental Science 4e*, coupled with our assumption that students have little prior knowledge of environmental sciences, allows students to easily make the transition from jumping-off points in the early chapters to the more complex concepts they encounter later. With its interdisciplinary presentation, which mirrors the nature of environmental science itself, this book is appropriate for use in one-semester and one-quarter environmental science courses offered by a variety of departments, including environmental studies and sciences, biology, ecology, agriculture, earth sciences, and geography.

Visualizing Environmental Science 4e is organized around the premise that humans are inextricably linked to the world's environmental dilemmas. Understanding how different parts of Earth's systems change, and how those changes affect other parts and systems, prepares us to make better choices as we deal with environmental problems we encounter everyday in the media and our lives.

- Chapters 1 through 4 establish the groundwork for understanding the environmental issues we face, how environmental sustainability and human values play a critical role in addressing these issues, how the environmental movement developed over time and was shaped by economics, and how environmental threats from many sources create health hazards that must be evaluated.

- Chapters 5, 6, and 7 present the intricacies of ecological concepts in a human-dominated world, including energy flow and the cycling of matter through ecosystems, and the various ways that species interact and divide resources. Gaining familiarity with these concepts allows students to better appreciate the variety of terrestrial and aquatic ecosystems that are then introduced, and to develop a richer understanding of the implications of human population change for the environment.

- The remaining 11 chapters deal with the world's resources as we use them today and as we assess their availability and impacts for the future. These issues cover a broad spectrum, including the sources and effects of air pollution, climate and global atmospheric change, freshwater resources, causes and effects of water pollution, the ocean and fisheries, mineral and soil resources, land resources, agriculture and food resources, biological resources, solid and hazardous waste, and nonrenewable and renewable energy resources. Recognizing the importance of the global ocean to environmental issues, we are particularly pleased to dedicate an entire chapter to a discussion of ocean processes and resources.

New to this Edition

In this edition, the authors have significantly expanded graphical representations of information. This includes additional graphs and tables throughout the text, but with a particular emphasis on the start of each chapter. Each chapter now opens with a representative story that includes prose, compelling imagery, and a visual display of information. A challenging question about the story engages the reader in the chapter's topic.

To actively encourage students to synthesize and apply content, "Sustainable Citizen" questions at the end of each chapter challenge students to examine how their own practices and beliefs might affect the local and global environment, or be applied to implementing solutions to environmental problems. A few new examples of new material in this edition include:

- A new chapter opener focusing on the 50th anniversary of Rachel Carson's *Silent Spring.*

- A new map on human population trends that includes data on arable land, global population density changes, and comparisons in population density trends in developed and developing regions.

- A case study on coping with conflict minerals, emphasizing coltan mining in the Democratic Republic of the Congo.

- New data on worldwide energy trends, including solar thermal and wind electrical generation and biomass ethanol consumption.

- Additional graphs associated with many photos throughout the text as well as with all chapter openers to reinforce the importance of data and graphing in environmental science.

- Wiley has partnered with Smart Science® to create new **Graphing Activities** for WileyPLUS which allow students to work through the scientific process by collecting real data and graphing it and then thinking critically about what that data demonstrates. Students are able to see the data represented on various graph types, and to compare different data sets against one another. Each graphing activity module is linked to the appropriate chapter and includes questions that allow instructors to test students' understanding of the activities.

Smart Science®

Finally, recognizing the educational value of integrating text with graphics and imagery, we have focused on improving the quality of process diagrams and have continued to revise our art program, layout, and design to provide students with a visually stunning, content-rich, image-based learning experience.

Guided Chapter Tour

Scientific Literacy and Data Analysis

Students are given an overview of the basics of environmental science and presented with many opportunities to interact with real-world data and real-world situations. As students learn about environmental science in this context, they are developing the critical thinking skills they can use to apply in making daily environmentally conscious decisions for the rest of their lives.

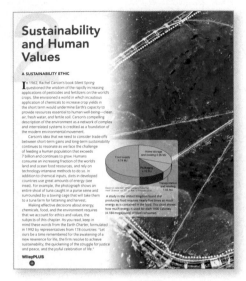

Chapter Openers are visually appealing, and include graphs which help students link real-world data to the photos presented.

Case Studies use a critical-thinking approach to walk students through the particular characteristics defining a real-life environmental challenge. The case studies present a problem and in some cases describe attempted solutions. Students then explore the information necessary to appreciate the significance of the featured example.

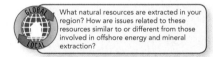

Global-Local questions help students apply environmental science conditions in distant locations to the places where they live.

Think Critically questions let students analyze the material and develop insights into essential concepts.

EnviroDiscovery essays explore an area or topic of relevance. Students synthesize the material for greater understanding.

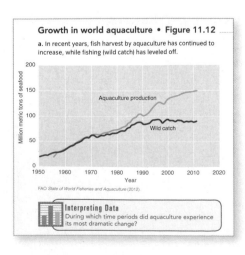

Interpreting Data questions help students evaluate graphs, figures, and data sets.

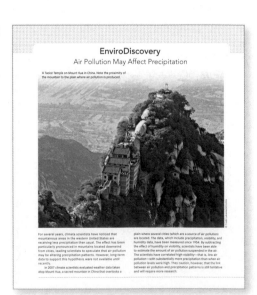

Visuals

Our unique visual design engages students, and keeps them interested in the content.

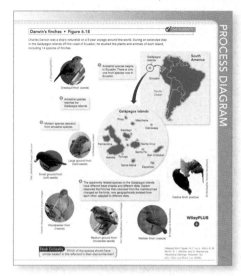

Process Diagrams provide in-depth coverage of processes correlated with clear, step-by-step narrative, enabling students to grasp important topics with less effort.

Interactive Process Diagrams provide additional visual examples and descriptive narrative of the diagrams that appear in the text. These diagrams allow students to Build the Process interactively to be sure they fully understand the process. Look for them in *WileyPLUS* when you see the WileyPLUS icon.

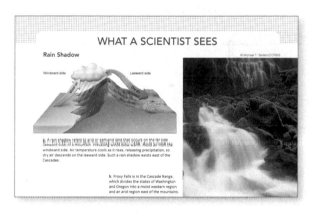

What a Scientist Sees highlights a concept or phenomenon that would stand out to a professional in the field. Photos and figures are used to compare how a nonscientist and a scientist see the issues, and students apply their observational skills to answer questions.

Margin Glossary identifies key terms for students within the margin of the text so they can easily identify which terms to pay close attention to.

enhanced greenhouse effect
Additional atmospheric warming produced as human activities increase atmospheric concentrations of greenhouse gases.

radiative forcing
For greenhouse gases, the capacity to retain heat in Earth's atmosphere.

Environmental InSight features are multipart visual sections that focus on a key concept or topic in the chapter, exploring it in detail or in a broader context using a combination of photos, figures, and data.

WileyPLUS **Icon** indicates when there is additional visual content such as videos, animations, and interactivities available in *WileyPLUS*.

WileyPLUS

Guided Chapter Tour

Student-Focused Pedagogy

Chapter pedagogy helps students navigate through the chapter, and clearly understand what they need to focus on to succeed.

CHAPTER OUTLINE

The Importance of Water 244
- Properties of Water
- The Hydrologic Cycle and Our Supply of Fresh Water

Water Resource Problems 246
- Aquifer Depletion
- Overdrawing of Surface Waters
- Salinization of Irrigated Soil
- Global Water Issues

Water Management 254
- Dams and Reservoirs: Managing the Columbia River
- Water Conservation
- Environmental InSight: Water Conservation

Water Pollution 258
- Types of Water Pollution
- What a Scientist Sees: Oligotrophic and Eutrophic Lakes
- Sources of Water Pollution
- Groundwater Pollution

Improving Water Quality 264
- Purification of Drinking Water
- Municipal Sewage Treatment
- Controlling Water Pollution
- Case Study: China's Three Gorges Dam

Chapter Outline at the beginning of each chapter tells students what they will encounter in the chapter

CHAPTER PLANNER ✓

- ❑ Study the picture and read the opening story.
- ❑ Scan the Learning Objectives in each section:
  p. 244 ❑ p. 246 ❑ p. 254 ❑ p. 258 ❑ p. 264 ❑
- ❑ Read the text and study all figures and visuals. Answer any questions.

Analyze key features

- ❑ National Geographic Map, p. 252
- ❑ Environmental InSight, p. 257
- ❑ What a Scientist Sees, p. 260
- ❑ Process Diagram, p. 264 ❑ p. 265 ❑
- ❑ Case Study, p. 269
- ❑ Stop: Answer the Concept Checks before you go on:
  p. 246 ❑ p. 254 ❑ p. 256 ❑ p. 263 ❑ p. 268 ❑

End of Chapter

- ❑ Review the Summary and Key Terms.
- ❑ Answer What is happening in this picture?
- ❑ Answer the Critical and Creative Thinking Questions.

Chapter Planner at the beginning of each chapter shows students which key features they should pay close attention to.

LEARNING OBJECTIVES

1. **Define** *toxicology* and *epidemiology.*
2. **Explain** why public water supplies are monitored for fecal coliform bacteria despite the fact that most strains of *E. coli* do not cause disease.
3. **Describe** the link between environmental changes and emerging diseases, such as swine flu.

Learning Objectives at the start of each section indicate in behavioral terms the concepts that students are expected to master while reading the section.

CONCEPT CHECK

1. **What** is the difference between toxicology and epidemiology?
2. **Why** is the fecal coliform test performed on public drinking water supplies?
3. **How** is the incidence of swine flu related to human activities that alter the environment?

Concept Check questions at the end of each section allow students to test their comprehension of the learning objectives.

Assessment

Wiley Visualizing offers students ample practice material for assessing their understanding of each study objective. Students know exactly what they are getting out of each study session through immediate feedback and coaching.

The **Summary** revisits each learning objective, with relevant accompanying images taken from the chapter; these visual clues reinforce important elements.

Critical and Creative Thinking Questions challenge students to think more broadly about chapter concepts. The level of these questions ranges from simple to advanced; they encourage students to think critically and develop an analytical understanding of the ideas discussed in the chapter.

Sustainable Citizen questions give students a chance to think about issues that are relevant to their everyday lives, so they can give a better sense of how they can personally relate to and be involved in the science they have learned in the chapter.

What is happening in this picture? presents an uncaptioned photograph that is relevant to a chapter topic and illustrates a situation students are not likely to have encountered previously. The photograph is paired with questions that ask the students to describe and explain what they can observe in the photo based on what they have learned.

WileyPLUS

WileyPLUS is designed for different learning styles, different levels of proficiency, and different levels of preparation. All students are unique, and *WileyPLUS* empowers them to take advantage of their individual strengths. Students receive timely access to resources that address their demonstrated needs, and get immediate feedback and remediation when needed.

Read, Study, & Practice

WileyPLUS offers many opportunities for student self-assessment linked to the relevant portions of the text. Students can take control of their own learning and practice until they master the material.

- **E-book:** *WileyPLUS* integrates the entire digital textbook with the most effective instructor and student resources to fit every learning style.

- **Audio Glossary and Flashcards** are available on the student companion site and in *WileyPLUS*. They provide students with easy access to digital flashcards so they can test themselves on key terms.

- **Videos:** Students think critically and solve the problems of real-life situations with a rich collection of over 75 videos from a variety of sources. Each video is linked to the text sections, and questions allow students to solve problems online.

- **Virtual Field Trip Videos:** Students go to nine different places around the world on Virtual Field Trips and gain a better understanding of the environment and our impact on it. Through these video-based field trips students gain virtual on-the-ground experience using their *WileyPLUS* course.

- **Graphing Activities** are new to this edition. They allow students to work through the scientific process by collecting real data, graphing it, and then thinking critically about what that data demonstrates. Students are able to see the data represented on various graph types, and to compare different data sets against one another. Each graphing activity module is linked to the appropriate chapter and includes questions that allow instructors to test students' understanding of the activities. Graphing activity icons on the chapter openers indicate when there is a module that corresponds with that chapter.

graphingactivity

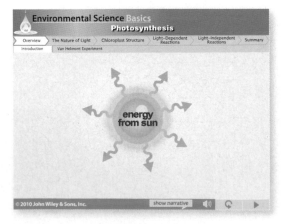

- **Environmental Science Basics:** Driven by instructor feedback for the most important topics for students to understand about environmental science, Environmental Science Basics provides a suite of animated concepts and tutorials to give students a solid grounding in the key basic environmental concepts. Concepts ranging from global climate change to sustainable agriculture are presented across 21 modules in easy-to-understand language.

- **Practice Quizzes** created by Brian Mooney, Johnson & Wales University, are available on the student companion site as well as in *WileyPLUS*. These quizzes include questions about each section of the book and offer students immediate feedback so they can gauge their own understanding of the chapter material.

Assignments and Gradebook

WileyPLUS includes pre-created assignments, which instructors can edit, in addition to tools that allow them to create their own assignment materials. *WileyPLUS* simplifies and automates such tasks as assessing student performance, making assignments, scoring student work, keeping grades, and more.

- **Learning Objectives**: All content in *WileyPLUS,* including media and questions, are mapped to Learning Objectives that correspond to each section of the text. Instructors can sort by learning objective to get the exact content they need for their students.

- **Test Bank** questions are available for instructors to create assignments for students that will be automatically graded. There are approximately 75 multiple choice, true/false, and essay questions available for each chapter.

- **Pre- and Post-Lecture Questions** are available to help assess student performance before and after they come to class.

- **Video Questions** are available for every video that appears in the course, so instructors can test their students' comprehension of the video content.

- **Graphing Questions** are available for each graphing activity module, so instructors can assess their students' understanding of the graphing activity content.

- **Gradebook** reports show all the assignment students have completed or attempted to date.

Prepare and Present

- *WileyPLUS* allow instructors to easily add and manage presentation materials for student reference or use in class.

- **QuickStart** includes ready-to-use question assignments and presentations for instructors to optimize their time.

- Course materials, including lecture and image **PowerPoints**, video and animation **Lecture Launcher PowerPoints**, and **Wiley's Visual Library for Biology**, help you personalize lessons and optimize your time.

WileyPLUS empowers you with the tools and resources you need to make your teaching even more effective.

To learn more about *WileyPLUS* or to request a test drive, visit **www.wileyplus.com**

Wiley Visualizing Site

The Wiley Visualizing site hosts a wealth of information for instructors using Wiley Visualizing, including ways to maximize the visual approach in the classroom and a white paper titled 'How Visuals Can Help Students Learn' by Matt Leavitt, instructional design consultant. Visit Wiley Visualizing at **www.wiley.com/college/visualizing**

Instructor's Manual
(Available in *WileyPLUS* and on the book companion site)

Answers to the Think Critically, Interpreting Data, Concept Check, Global–Local and Critical and Creative Thinking questions that appear in the printed text are available in an Instructor's Manual created by Chuck McKinney at Oakland City University.

Also available for each chapter is an In-class Activities instructor's manual.

Test Bank
(Available in *WileyPLUS* and on the book companion site)

Many visuals from the textbook are also included in the Test Bank by Keith Hench at Kirkwood Community College. The Test Bank has approximately 1300 test items, with at least 25% of them incorporating visuals from the book. The test items include multiple choice, true/false, text entry, and essay questions which test a variety of comprehension levels. The test bank is available online in MS Word files, as a computerized Test Bank, and within *WileyPLUS*. The easy-to-use test-generation program fully supports graphics, print tests, student answer sheets, and answer keys. The software's advanced features allow you to produce an exam to your exact specifications.

Lecture Launchers
(Available in *WileyPLUS*)

Each video available in the *WileyPLUS* course is accompanied by a Lecture Launcher PowerPoint to facilitate in-class use.

Lecture PowerPoints, Image PowerPoints and JPGs
(Available in *WileyPLUS* and on the book companion site)

A complete set of highly visual PowerPoint presentations—one per chapter—by Erica Kipp, Pace University, and Janet Wolkenstein, Hudson Valley Community College, is available online and in *WileyPLUS* to enhance classroom presentations. Tailored to the text's topical coverage and learning objectives, these presentations are designed to convey key text concepts, illustrated by embedded text art.

All photographs, figures, maps, and tables from the text are available as jpgs and PowerPoints, and can be used as you wish in the classroom. These electronic files allow you to easily incorporate images into your own PowerPoint presentations as you choose, or to create your own handouts.

Biology Visual Library
(Available in *WileyPLUS* and on the book companion site)

All illustrations from the text are contained in the Visual Library and can be viewed with or without labels. Also included in the Visual Library are the illustrations from Wiley's other biology and environmental science titles.

Clicker Questions
(Available in *WileyPLUS* and on the book companion site)

Clicker questions, written by Julie Weinert at Southern Illinois University Carbondale, are available in MS Word files, and can be converted to appropriate clicker formats upon request.

Wiley Faculty Network

The Wiley Faculty Network (WFN) is a global community of faculty, connected by a passion for teaching and a drive to learn, share, and collaborate. Their mission is to promote the effective use of technology and enrich the teaching experience. Connect with the Wiley Faculty Network to collaborate with your colleagues, find a mentor, attend virtual and live events, and view a wealth of resources all designed to help you grow as an educator. Visit the Wiley Faculty Network at www.wherefacultyconnect.com.

ALSO AVAILABLE

Environmental Science: Active Learning Laboratories and Applied Problem Sets, 2e by Travis Wagner and Robert Sanford, both of the University of Southern Maine, presents specific labs that use natural and social science concepts and encourages a "hands-on" approach to evaluating the impacts from the environmental/human interface. The laboratory and **homework activities are designed to be low cost and to reflect a sustainable approach** in both practice and theory. *Environmental Science: Active Learning Laboratories and Applied Problem Sets, 2e* is available as a stand-alone, in a package with, or customized with *Visualizing Environmental Science 4e*. Contact your Wiley representative for more information.

Climate Change: What the Science Tells Us by Charles Fletcher discusses the most recent research focusing on the causes and effects of climate change and offers strategies to help learners understand why and how scientists have come to this conclusion. This book can be packaged or customized with *Visualizing Environmental Science, 4e*.

Wiley Custom Select

Wiley Custom Select gives you the freedom to build your course materials exactly the way you want them. Offer your students a cost-efficient alternative to traditional texts. In a simple three-step process, create a solution containing the content you want, in the sequence you want, delivered how you want. Visit Wiley Custom Select at http://customselect.wiley.com.

Book Companion Site

All instructor resources (the Test Bank, Instructor's Manual, PowerPoint presentations, and all textbook illustrations and photos in jpg format) are housed on the book companion site (www.wiley.com/college/berg). Student resources include self quizzes and flashcards.

How Has Wiley Visualizing Been Shaped by Contributors?

Wiley Visualizing and the *WileyPlus* learning envrironment would not have come about without lots of people, each of whom played a part in sharing their research and contributing to this new approach.

Academic research consultants

Richard Mayer, Professor of Psychology, UC Santa Barbara. *Mayer's Cognitive Theory of Multimedia Learning* provided the basis on which we designed our program. He continues to provide guidance to our author and editorial teams on how to develop and implement strong, pedagogically effective visuals and use them in the classroom.

Jan L. Plass, Professor of Educational Communication and Technology in the Steinhardt School of Culture, Education, and Human Development at New York University. Plass co-directs the NYU Games for Learning Institute and is the founding director of the CREATE Consortium for Research and Evaluation of Advanced Technology in Education.

Matthew Leavitt, Instructional Design Consultant, advises the Visualizing team on the effective design and use of visuals in instruction and has made virtual and live presentations to university faculty around the country regarding effective design and use of instructional visuals.

Independent research studies

SEG Research, an independent research and assessment firm, conducted a national, multisite effectiveness study of students enrolled in entry-level college Psychology and Geology courses. The study was designed to evaluate the effectiveness of Wiley Visualizing. You can view the full research paper at www.wiley.com/college/visualizing/huffman/efficacy.html.

Instructor and student contributions

Throughout the process of developing the concept of guided visual pedagogy for Wiley's Visualizing, we benefited from the comments and constructive criticism provided by the instructors and colleagues listed below. We offer our sincere appreciation to these individuals for their helpful reviews and general feedback:

Reviewers, Focus Group Participants, and Survey Respondents

James Abbott, *Temple University*
Melissa Acevedo, *Westchester Community College*
Shiva Achet, *Roosevelt University*
Denise Addorisio, *Westchester Community College*
Dave Alan, *University of Phoenix*
Sue Allen-Long, *Indiana University Purdue*
Robert Amey, *Bridgewater State College*
Nancy Bain, *Ohio University*
Corinne Balducci, *Westchester Community College*
Steve Barnhart, *Middlesex County Community College*
Stefan Becker, *University of Washington—Oshkosh*
Callan Bentley, *Northern Virginia Community College*
Valerie Bergeron, *Delaware Technical & Community College*
Andrew Berns, *Milwaukee Area Technical College*
Gregory Bishop, *Orange Coast College*
Rebecca Boger, *Brooklyn College*
Scott Brame, *Clemson University*
Joan Brandt, *Central Piedmont Community College*
Richard Brinn, *Florida International University*
Jim Bruno, *University of Phoenix*
William Chamberlin, *Fullerton College*

Oiyin Pauline Chow, *Harrisburg Area Community College*
Laurie Corey, *Westchester Community College*
Ozeas Costas, *Ohio State University at Mansfield*
Christopher Di Leonardo, *Foothill College*
Dani Ducharme, *Waubonsee Community College*
Mark Eastman, *Diablo Valley College*
Ben Elman, *Baruch College*
Staussa Ervin, *Tarrant County College*
Michael Farabee, *Estrella Mountain Community College*
Laurie Flaherty, *Eastern Washington University*
Susan Fuhr, *Maryville College*
Peter Galvin, *Indiana University at Southeast*
Andrew Getzfeld, *New Jersey City University*
Janet Gingold, *Prince George's Community College*
Donald Glassman, *Des Moines Area Community College*
Richard Goode, *Porterville College*
Peggy Green, *Broward Community College*
Stelian Grigoras, *Northwood University*
Paul Grogger, *University of Colorado*
Michael Hackett, *Westchester Community College*
Duane Hampton, *Western Michigan University*

R. Laurence Davis, *Northeastern Cave Conservancy, Inc.*
JodyLee Estrada Duek, *Pima Community College*
Catherine M. Etter, *Cape Cod Community College*
Brad C. Fiero, *Pima Community College*
Michael Freake, *Lee University*
Jennifer Frick-Ruppert, *Brevard College*
Todd G. Fritch, *Northeastern University*
Marcia L. Gillette, *Indiana University, Kokomo*
Arthur Goldsmith, *Hallandale High*
Cliff Gottlieb, *Shasta College*
Peggy Green, *Broward Community College*
Stelian Grigoras, *Northwood University*
Syed E. Hasan, *University of Missouri—Kansas City*
Carol Hoban, *Kennesaw State University*
Guang Jin, *Illinois State University*
Dawn Keller, *Hawkeye Community College*
Martin Kelly, *Genesee Community College*
David Kitchen, *University of Richmond*
Paul Kramer, *Farmingdale State College*
Meredith Gooding Lassiter, *Wiona State University*
Ernesto Lasso de la Vega, *Edison College*
Madelyn E. Logan, *North Shore Community College*
Linda Lyon, *Frostburg State University*
Timothy F. Lyon, *Ball State University*
Robert S. Mahoney, *Johnson & Wales University at Florida*
Heidi Marcum, *Baylor University*
Matthew H. McConeghy, *Johnson & Wales University*
Rick McDaniel, *Henderson State University*
Brian Mooney, *Johnson & Wales University at North Carolina*
Jacob Napieralski, *University of Michigan, Dearborn*
Renee Nerish, *Mercer County Community College*
Leslie Nesbitt, *Niagara University*
Ken Nolte, *Shasta College*
Natalie Osterhoudt, *Broward Community College*

Barry Perlmutter, *Community College of Southern Nevada*
Neal Phillip, *Bronx Community College*
Thomas E. Pliske, *Florida International University*
Katherine Prater, *Texas Wesleyan University*
Uma Ramakrishnan, *Juniata College*
Sabine Rech, *San Jose State University*
Shamili A. Sandiford, *College of DuPage*
Thomas Sasek, *University of Louisiana at Monroe*
Howie Scher, *University of Rochester*
Nan Schmidt, *Pima Community College*
Richard B. Schultz, *Elmhurst College*
Richard Shaker, *University of Wisconsin, Milwaukee*
Charles Shorten, *West Chester University*
Jerry Skinner, *Keystone College*
Roy Sofield, *Chattanooga State Technical Community College*
Bo Sosnicki, *Florida Community College at Jacksonville*
Ravi Srinivas, *University of St. Thomas*
David Steffy, *Jacksonville State University*
Andrew Suarez, *University of Illinois*
Charles Venuto, *Brevard Community College, Cocoa Campus*
Margaret E. Vorndam, *Colorado State University Pueblo*
Laura J. Vosejpka, *Northwood University*
Maud M. Walsh, *Louisiana State University*
John F. Weishampel, *University of Central Florida*
Karen Wellner, *Arizona State University*
Arlene Westhoven, *Ferris State University*
Susan M. Whitehead, *Becker College*
John Wielichowski, *Milwaukee Area Technical College*
Hernan Aubert, *Pima Community College*
Keith Hench, *Kirkwood Community College*
Dawn Keller, *Hawkeye Community College*
Dale Lambert, *Tarrant Community College*
Ashok Malik, *Evergreen Valley College.*
Janice Padula, *Clinton College*

Reviewers of the Fourth Edition

David Bass, *University of Central Oklahoma*
Greta Bolin, *University of North Texas*
Arielle Burlett, *Chatham University*
Wilbert Butler, *Tallahassee Community College*
Katie Chenu, *Seattle Central Community College*
Arielle Conti, *American University*
David Cronin, *Cleveland State University*
Michael Draney, *University of Wisconsin- Green Bay*
James Patrick Dunn, *Grand Valley State University*
Rus Higley, *Highline Community College*
Mariana Lecknew, *American Military University*
Kirt Leuschner, *College of the Desert*
Kamau Mbuthia, *Bowling Green State University*

Chuck McKinney, *Oakland City University*
Brian Mooney, *Johnson & Wales University*
Jessica Mooney, *Chatham University*
Joy Perry, *University of Wisconsin- Fox Valley*
Neal Phillip, *Bronx Community College*
Ellison Robinson, *Midlands Technical College*
Pamela Scheffler, *Hawaii Community College*
Julie Stoughton, *University of Nevada Reno*
Zachary Taylor, *Willamette University*
Ruthanne Thompson, *University of North Texas*
Janet Wolkenstein, *Hudson Valley Community College*
David Wyatt, *Sacramento City College*

Special Thanks

We are extremely grateful to the many members of the editorial and production staff at John Wiley and Sons who guided us through the challenging steps of developing this book. Their tireless enthusiasm, professional assistance, and endless patience smoothed the path as we found our way. We thank in particular Senior Editor Rachel Falk, who expertly launched and directed the revision; Lauren Samuelson, Assistant Editor, for coordinating the development and revision process; Jeffrey Rucker, Executive Marketing Manager, and Clay Stone, Marketing Manager, for a superior marketing effort; and Chloe Moffett, Editorial Assistant, for her constant attention to detail. Thanks also to Bonnie Roth, Senior Product Designer, for insightful work in developing our *WileyPLUS* course as well as the other media components. We also thank Trish McFadden, Senior Production Editor, for expertly helping us through the production process. Thanks to Kathy Naylor, who managed and helped develop our art program. We thank Senior photo editors MaryAnn Price and Sheena Goldstein for their unflagging, always swift work in researching and obtaining many of our text images. We thank Kristine Carney for the beautiful new interior design and for her constant attention to page layout, as well as Harry Nolan and Wendy Lai for our stunning new cover. Thank you to Kaye Pace, Vice President and Executive Publisher; and Anne Smith, Vice President and Executive Publisher; for providing guidance and support to the rest of the team throughout the revision.

About the Authors

David M. Hassenzahl is the Founding Dean of the School of Sustainability and the Environment at Chatham University. An internationally recognized scholar of sustainability and risk analysis, his research focuses on incorporating scientific information and expertise into public decision. He holds a B.A. in Environmental Science and Paleontology from the University of California at Berkeley, and a Ph.D. from Princeton University's Woodrow Wilson School. His efforts in climate change education have been supported by the National Science Foundation, and recognition of his work includes the Society for Risk Analysis Outstanding Educator Award and the UNLV Foundation Distinguished Teaching Award. Dr. Hassenzahl is a Senior Fellow of the National Council for Science and the Environment, and a founding member of the Association of Environmental Studies and Sciences. Prior to his academic career, Dr. Hassenzahl worked in the private sector as an environmental manager, and as an inspector for the (San Francisco) Bay Area Air Quality Management District.

Mary Catherine Hager is a professional science writer and editor specializing in life and earth sciences. She received a double-major B.A. in environmental science and biology from the University of Virginia and an M.S. in zoology from the University of Georgia. Ms. Hager worked as an editor for an environmental consulting firm and as a senior editor for a scientific reference publisher. For the past 20 years, she has written and edited for environmental science, biology, and ecology textbooks for target audiences ranging from middle school to college. Additionally, she has published articles in environmental trade magazines and edited federal and state reports addressing wetlands conservation issues. Her writing and editing pursuits are a natural outcome of her scientific training and curiosity, coupled with her love of reading and effective communication.

Linda R. Berg is an award-winning teacher and textbook author. She received a B.S. in science education, an M.S. in botany, and a Ph.D. in plant physiology from the University of Maryland. Dr. Berg taught at the University of Maryland—College Park for 17 years and at St. Petersburg College in Florida for 8 years. She has taught introductory courses in environmental science, biology, and botany to thousands of students and has received numerous teaching and service awards. Dr. Berg is also the recipient of many national and regional awards, including the National Science Teachers Association Award for Innovations in College Science Teaching, the Nation's Capital Area Disabled Student Services Award, and the Washington Academy of Sciences Award in University Science Teaching. During her career as a professional science writer, Dr. Berg has authored or co-authored numerous editions of several leading college science textbooks. Her writing reflects her teaching style and love of science.

Contents in Brief

1   The Environmental Challenges We Face   2

2   Sustainability and Human Values   26

3   Environmental History, Politics, and Economics   48

4   Risk Analysis and Environmental Health Hazards   72

5   How Ecosystems Work   96

6   Ecosystems and Evolution   126

7   Human Population Change and the Environment   158

8   Air and Air Pollution   190

9   Global Atmospheric Changes   216

10   Freshwater Resources and Water Pollution   242

11 The Ocean and Fisheries 272

12 Mineral and Soil Resources 296

13 Land Resources 320

14 Agriculture and Food Resources 348

15 Biodiversity and Conservation 372

16 Solid and Hazardous Waste 396

17 Nonrenewable Energy Resources 418

18 Renewable Energy Resources 442

Graphing Appendix 466
Glossary 475
Index 480

Contents

1 The Environmental Challenges We Face   2

Human Impacts on the Environment   4
- ENVIRODISCOVERY: Green Roofs   8

Sustainability and the Environment   12

Environmental Science   16

How We Handle Environmental Problems   20
- ENVIRODISCOVERY:
  Getting Past NIMBY   22
- CASE STUDY:
  The New Orleans Disaster   23

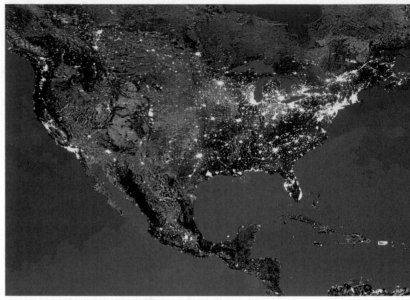

Earth Imaging/Stone/Getty Images

Cristina Redinger-Libolt/Botanica/Getty Images, Inc.

2 Sustainability and Human Values   26

Human Use of the Earth   28

Human Values and Environmental Problems   31

Environmental Justice   35

An Overall Plan for Sustainable Living   36
- CASE STUDY:
  The Loess Plateau in China   44

3 Environmental History, Politics, and Economics 48

Conservation and Preservation of Resources 50

Environmental History 51

■ ENVIRODISCOVERY:
Environmental Literacy 58

Environmental Legislation 59

Environmental Economics 62

■ CASE STUDY:
Tradable Permits and Acid Rain 68

Greg Dale/NG Image Collection

Wolfgang Flamish/zefa/©Corbis

4 Risk Analysis and Environmental Health Hazards 72

A Perspective on Risks 74

Environmental Health Hazards 77

Movement and Fate of Toxicants 81

Determining Health Effects of Pollutants 85

■ ENVIRODISCOVERY:
Smoking: A Significant Risk 88

The Precautionary Principle 90

■ CASE STUDY:
Endocrine Disrupters 92

5 How Ecosystems Work 96

What Is Ecology? 98

The Flow of Energy Through Ecosystems 100

The Cycling of Matter in Ecosystems 106

Ecological Niches 113
- WHAT A SCIENTIST SEES:
  Resource Partitioning 115

Interactions Among Organisms 116
- ENVIRODISCOVERY:
  Bee Colonies Under Threat 118
- CASE STUDY:
  Global Climate Change: How Does
  It Affect the Carbon Cycle? 122

© Philip James Corwin/CORBIS

Sarah Leen/NG Image Collection

6 Ecosystems and Evolution 126

Earth's Major Biomes 128
- ENVIRODISCOVERY:
  Using Goats to Fight Fires 138

Aquatic Ecosystems 142
- WHAT A SCIENTIST SEES:
  Zonation in a Large Lake 143

**Population Responses to Changing
Conditions over Time: Evolution** 147

**Community Responses to Changing
Conditions over Time: Succession** 151
- CASE STUDY:
  Wildfires 154

7 Human Population Change and the Environment   158

Population Ecology   160

Human Population Patterns   165

Demographics of Countries   170

Stabilizing World Population   174

    ■ WHAT A SCIENTIST SEES:
      Education and Fertility   179

    ■ ENVIRODISCOVERY:
      Microcredit Programs   180

Population and Urbanization   181

    ■ CASE STUDY:
      Urban Planning in Curitiba, Brazil   186

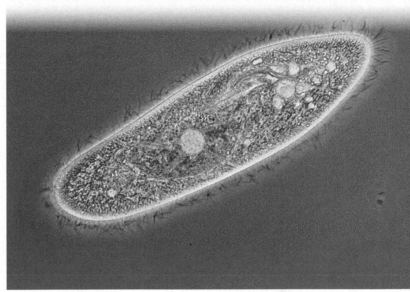

Michael Abbey/Science Source

8 Air and Air Pollution   190

The Atmosphere   192

Types and Sources of Air Pollution   196

    ■ WHAT A SCIENTIST SEES:
      Air Pollution from Volcanoes   199

Effects of Air Pollution   201

    ■ ENVIRODISCOVERY: Air Pollution May
      Affect Precipitation   203

Controlling Air Pollutants   206

Indoor Air Pollution   209

    ■ CASE STUDY:
      Curbing Air Pollution in Chattanooga   212

Kenneth Garrett/NG Image Collection

9 Global Atmospheric Changes 216

The Atmosphere and Climate 218

■ WHAT A SCIENTIST SEES:
Rain Shadow 221

Global Climate Change 222

Ozone Depletion in the Stratosphere 231

■ ENVIRODISCOVERY:
Links Between Climate
and Atmospheric Change 233

Acid Deposition 234

■ CASE STUDY:
International Implications
of Global Climate Change 238

© America/Alamy

Courtesy U. S. Dept. of Energy

10 Freshwater Resources and Water Pollution 242

The Importance of Water 244

Water Resource Problems 246

Water Management 254

Water Pollution 258

■ WHAT A SCIENTIST SEES:
Oligotrophic and Eutrophic Lakes 260

Improving Water Quality 264

■ CASE STUDY:
China's Three Gorges Dam 269

11 The Ocean and Fisheries 272

The Global Ocean 274

Major Ocean Life Zones 278
- ENVIRODISCOVERY:
  Otters in Trouble 282

Human Impacts on the Ocean 284
- WHAT A SCIENTIST SEES:
  Modern Commercial Fishing Methods 286
- WHAT A SCIENTIST SEES:
  Ocean Warming and Coral Bleaching 289

Addressing Ocean Problems 291
- CASE STUDY:
  The Dead Zone in the Gulf of Mexico 293

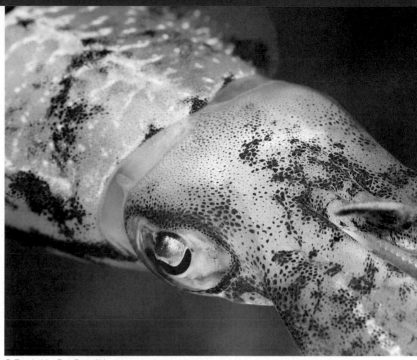

© Travis VanDenBerg/Alamy

JAMES L. AMOS/NG Image Collection

12 Mineral and Soil Resources 296

Plate Tectonics and the Rock Cycle 298

Economic Geology: Useful Minerals 302

Environmental Implications of Mineral Use 306
- ENVIRODISCOVERY:
  Not-so-Precious Gold 307

Soil Properties and Processes 309
- WHAT A SCIENTIST SEES:
  Soil Profile 310

Soil Problems and Conservation 312
- CASE STUDY:
  Coping with "Conflict Minerals" 317

13 Land Resources 320

Land Use in the United States 322

Forests 324
- ◼ WHAT A SCIENTIST SEES:
  Harvesting Trees 327
- ◼ ENVIRODISCOVERY:
  Ecologically Certified Wood 328

Rangelands 333

National Parks and Wilderness Areas 336

Conservation of Land Resources 341
- ◼ CASE STUDY:
  The Tongass Debate over Clear-Cutting 344

Neale Clark/Robert Harding World Images/Getty Images

14 Agriculture and Food Resources 348

World Food Problems 350

The Principal Types of Agriculture 353

Challenges of Agriculture 355

Solutions to Agricultural Problems 360
- ◼ ENVIRODISCOVERY:
  A New Weapon against Locust Swarms 361

Controlling Agricultural Pests 364
- ◼ WHAT A SCIENTIST SEES:
  Pesticide Use and New Pest Species 366
- ◼ CASE STUDY:
  Organic Agriculture 368

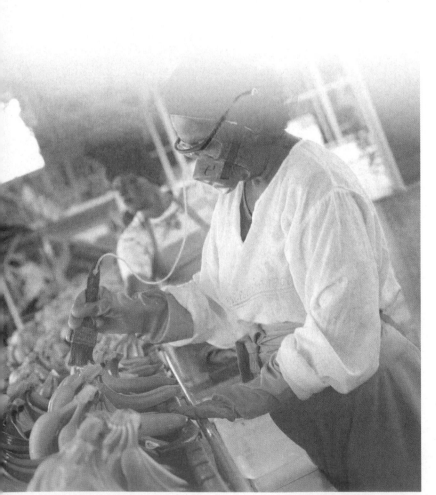

Mayela Lopez/AFP/Getty Images

15 Biodiversity and Conservation   372

Species Richness and Biological Diversity   374

Endangered and Extinct Species   378

- ENVIRODISCOVERY:
  Is Your Coffee Bird Friendly®?   380
- WHAT A SCIENTIST SEES:
  Where Is Declining Biological
  Diversity the Most Serious?   381

Conservation Biology   386

Conservation Policies and Laws   390

- CASE STUDY:
  The Challenges of Protecting
  Rare Species   393

Joel Sartore/NG Image Collection

© Greg Smith/© Corbis

16 Solid and Hazardous Waste   396

Solid Waste   398

- WHAT A SCIENTIST SEES:
  Sanitary Landfills   401
- ENVIRODISCOVERY:
  The U.S.–China Recycling Connection   404

Reducing Solid Waste   404

Hazardous Waste   409

- ENVIRODISCOVERY:
  Handling Nanotechnology Safely   410

Managing Hazardous Waste   412

- CASE STUDY:
  High-Tech Waste   415

17 Nonrenewable Energy Resources 418

Energy Consumption 420

Coal 421

Oil and Natural Gas 423

Nuclear Energy 430

   ■ WHAT A SCIENTIST SEES:
    Yucca Mountain 437

   ■ ENVIRODISCOVERY:
    A Nuclear Waste Nightmare 438

   ■ CASE STUDY:
    The Arctic National Wildlife Refuge 439

Natalie B. Fobes/NG Image Collection

Mark Thiessen/NG Image Collection

18 Renewable Energy Resources 442

Direct Solar Energy 444

   ■ WHAT A SCIENTIST SEES:
    Photovoltaic Cells 446

Indirect Solar Energy 450

Other Renewable Energy Sources 456

Energy Solutions: Conservation and Efficiency 458

   ■ ENVIRODISCOVERY: Netting the
    Benefits of Home Energy Production 461

   ■ CASE STUDY: Green Architecture 463

Graphing Appendix 466

Glossary 475

Index 480

InSight Features

These multipart visual presentations focus on a key concept or topic in the chapter.

Chapter 1
Population Growth and Poverty • Environmental Exploitation

Chapter 2
A Plan for Sustainable Living

Chapter 3
Economics and the Environment

Chapter 4
Bioaccumulation and Biomagnification

Chapter 5
Symbiotic Relationships

Chapter 6
How Climate Shapes Terrestrial Biomes • Evidence for Evolution

Chapter 7
Demographics of Countries

Chapter 8
The Atmosphere

Chapter 9
The Effects of Global Climate Change • The Ozone Layer • The Effects of Acid Deposition

Chapter 10
Water Conservation

Chapter 11
Ocean Currents • Human Impacts on the Ocean

Chapter 12
Soil Conservation

Chapter 13
Tropical Deforestation • National Parks

Chapter 14
World Hunger • Impacts of Industrialized Agriculture

Chapter 15
Threats to Biodiversity • Efforts to Conserve Species

Chapter 16
Recycling in the United States

Chapter 17
The *Exxon Valdez* and *Deepwater Horizon* Oil Spills

Chapter 18
Wind Energy

Process Diagrams

These series or combinations of figures and photos describe and depict a complex process.

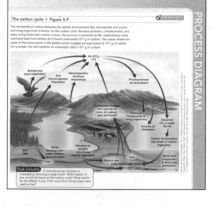

Chapter 1
The Scientific Method • Addressing Environmental Problems

Chapter 2
Cascading Responses of Increased Carbon Dioxide Through the Environment

Chapter 3
Environmental Impact Statements

Chapter 4
Four Steps for Risk Assessment

Chapter 5
Energy Flow Through a Food Chain • Food Web at the Edge of an Eastern U.S. Deciduous Forest • The Carbon Cycle • The Hydrologic Cycle • The Nitrogen Cycle • The Sulfur Cycle • The Phosphorus Cycle

Chapter 6
Darwin's Finches • Primary Succession on Glacial Moraine • Secondary Succession on an Abandoned Field in North Carolina

Chapter 8
The Coriolis Effect

Chapter 9
Fate of Solar Radiation that Reaches Earth • Enhanced Greenhouse Effect

Chapter 10
Treatment of Water For Municipal Use • Primary and Secondary Sewage Treatment

Chapter 11
El Niño–Southern Oscillation (ENSO)

Chapter 12
The Rock Cycle

Chapter 13
Role of Forests in the Hydrologic Cycle

Chapter 14
Energy Inputs in Industrialized Agriculture • Genetic Engineering

Chapter 16
Mass Burn, Waste-to-Energy Incinerator • Integrated Waste Management

Chapter 17
Petroleum Refining • Nuclear Fission • Pressurized Water Reactor

Chapter 18
Active Solar Water Heating

WileyPLUS

WileyPLUS is a research-based online environment for effective teaching and learning.

WileyPLUS builds students' confidence because it takes the guesswork out of studying by providing students with a clear roadmap:

- **what to do**
- **how to do it**
- **if they did it right**

It offers interactive resources along with a complete digital textbook that help students learn more. With *WileyPLUS*, students take more initiative so you'll have greater impact on their achievement in the classroom and beyond.

Now available for

Bb

Blackboard

For more information, visit www.wileyplus.com

WileyPLUS

ALL THE HELP, RESOURCES, AND PERSONAL SUPPORT YOU AND YOUR STUDENTS NEED!
www.wileyplus.com/resources

1st DAY OF CLASS ...AND BEYOND!

2-Minute Tutorials and all of the resources you and your students need to get started

WileyPLUS Student Partner Program

Student support from an experienced student user

Wiley Faculty Network

Collaborate with your colleagues, find a mentor, attend virtual and live events, and view resources
www.WhereFacultyConnect.com

WileyPLUS Quick Start

Pre-loaded, ready-to-use assignments and presentations created by subject matter experts

Technical Support 24/7
FAQs, online chat, and phone support
www.wileyplus.com/support

© Courtney Keating/ iStockphoto

Your *WileyPLUS* Account Manager, providing personal training and support

FOURTH EDITION

VISUALIZING

ENVIRONMENTAL SCIENCE

1 The Environmental Challenges We Face

A WORLD IN CRISIS

Over three billion years ago, just before the first life forms arose, Earth's surface and climate were inhospitable by modern standards, but contained abundant raw materials. As early life forms evolved, they began to use the sun's energy to exploit Earth's resources. Eventually they shaped the landscape, altered the global climate, and modified the chemical makeup of the ocean and soils. These changes, along with shifts in Earth's orbit and the sun's energy output, created conditions in which the millions of species that now inhabit our planet arose and adapted.

Today, the human species is the most significant agent of environmental change on our planet. Our intellectual capacity has even made it possible for us to venture into space, allowing us a view of the uniqueness of our planet in the solar system (see photograph). However, our burgeoning population is overwhelming Earth's regenerative capacity. We transform forests, prairies, and deserts to meet our needs and desires, and we consume ever increasing amounts of Earth's abundant but finite resources—rich topsoil, clean water, and breathable air. We are eradicating thousands upon thousands of unique species as we destroy or alter their habitats. In 2009, a total of 5566 species were classified as endangered worldwide (see insert). Our activities now affect processes, including climate and nutrient cycles, from the local to the global level.

This book introduces the major environmental impacts that humans have on Earth, and considers ways to reduce those impacts that undermine economic, social, and environmental well-being. Most importantly, it explains why we must minimize human impact on our planet. We can't afford to ignore the environment because our lives, as well as those of future generations, depend on it. As a wise proverb says, "We have not inherited the world from our ancestors, we have only borrowed it from our children."

WileyPLUS

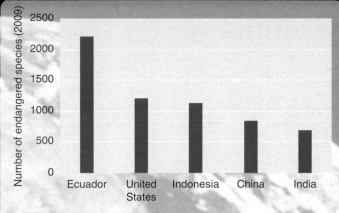

Number of endangered species (2009)

Based on data from International Union for Conservation of Nature 2009,
Red List of Threatened Species.

CHAPTER OUTLINE

Human Impacts on the Environment 4
- The Gap Between Rich and Poor Countries
- Environmental InSight: Population Growth and Poverty
- Population, Resources, and the Environment
- EnviroDiscovery: Green Roofs

Sustainability and the Environment 12
- Environmental InSight: Environmental Exploitation

Environmental Science 17
- The Goals of Environmental Science
- Science as a Process

How We Handle Environmental Problems 20
- EnviroDiscovery: Getting Past NIMBY
- Case Study: The New Orleans Disaster

CHAPTER PLANNER ✓

- ❏ Study the picture and read the opening story
- ❏ Scan the Learning Objectives in each section:
  p. 4 ❏   p. 12 ❏   p. 17 ❏   p. 20 ❏
- ❏ Read the text and study all figures and visuals. Answer any questions.

Analyze key features

- ❏ Environmental InSight, p. 5 ❏   p.13 ❏
- ❏ EnviroDiscovery, p. 8 ❏   p. 22 ❏
- ❏ National Geographic Map, p. 14
- ❏ Process Diagram, p. 19 ❏   p. 21 ❏
- ❏ Case Study, p. 23
- ❏ Stop: Answer the Concept Checks before you go on:
  p. 11 ❏   p. 12 ❏   p. 20 ❏   p. 21 ❏

End of Chapter

- ❏ Review the Summary and Key Terms.
- ❏ Answer What is happening in this picture?
- ❏ Answer the Critical and Creative Thinking Questions.

NASA/NG Image Collection

Human Impacts on the Environment

LEARNING OBJECTIVES

1. **Distinguish** among highly developed countries, moderately developed countries, and less developed countries.

2. **Relate** human population size to natural resources and resource consumption.

3. **Describe** the three factors that are most important in determining human impact on the environment.

The satellite photograph in **Figure 1.1a** is a portrait of about 450 million people. The tiny specks of light represent cities, and the great metropolitan areas, such as New York City along the northeastern seacoast, are ablaze with light. This represents the most significant factor impacting the health of Earth's environment: a large and growing human population.

In 2011 the human population as a whole passed 7 billion individuals. Not only is this number incomprehensibly large, but our population has grown this large in a very brief span of time. In 1960 the human population was only 3 billion (**Figure 1.1b**). By 1975 there were 4 billion people, and by 1987 there were 5 billion. The more than 7 billion people who currently inhabit our planet consume vast quantities of food and water, occupy or farm much of the most productive land, use a great deal of energy and raw materials, and produce much waste.

Despite most countries' involvement with family planning, population growth rates don't change overnight. Several billion people will be added to the world in the 21st century, so even if we remain concerned about population and even if our solutions are very effective, the coming decades may very well see many problems.

On a global level, nearly one in four people lives in extreme **poverty** (**Figure 1.1c**). By one measure, living in poverty is defined as having a per person income of less than $2 per day, expressed in U.S. dollars adjusted for purchasing power. About 3.3 billion people—nearly half of the world's population—currently live

poverty A condition in which people are unable to meet their basic needs for food, clothing, shelter, education, or health.

highly developed countries Countries with complex industrialized bases, low rates of population growth, and high per person incomes.

at this level of poverty. Poverty is associated with a short life expectancy, illiteracy, and inadequate access to health services, safe water, and balanced nutrition.

The world population may stabilize toward the end of the 21st century, given the family planning efforts that are currently under way. Population experts at the Population Reference Bureau have noticed a decrease in the fertility rate worldwide to a current average of 2.5 children per woman, and the fertility rate is projected to continue to decline in coming decades.

The fertility rate varies from country to country, from 1.7 in highly developed countries to 4.5 in some of the least developed countries. Population experts have made various projections for the world population at the end of the 21st century, from about 7.7 billion to 10.6 billion, depending primarily on how fast the fertility rate decreases.

No one knows whether Earth can support so many people indefinitely. Finding ways for it to do so represents one of the greatest challenges of our times. Among the tasks to be accomplished is feeding a world population considerably larger than today's without destroying the biological communities that support life on our planet. The quality of life available to our children and grandchildren will depend to a large extent on our ability to develop a sustainable system of agriculture to feed the world's people.

A factor as important as population size is a population's level of **consumption**, which is the human use of material and energy. Consumption is intimately connected to a country's **economic growth**, the expansion in output of a nation's goods and services. The world's economy is growing at an enormous rate, yet this growth is unevenly distributed across the nations of the world.

The Gap Between Rich and Poor Countries

Generally speaking, countries are divided into rich (the "haves") and poor (the "have-nots"). Rich countries are known as **highly developed countries**. The United States, Canada, most of Europe, and Japan, which represent about

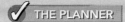
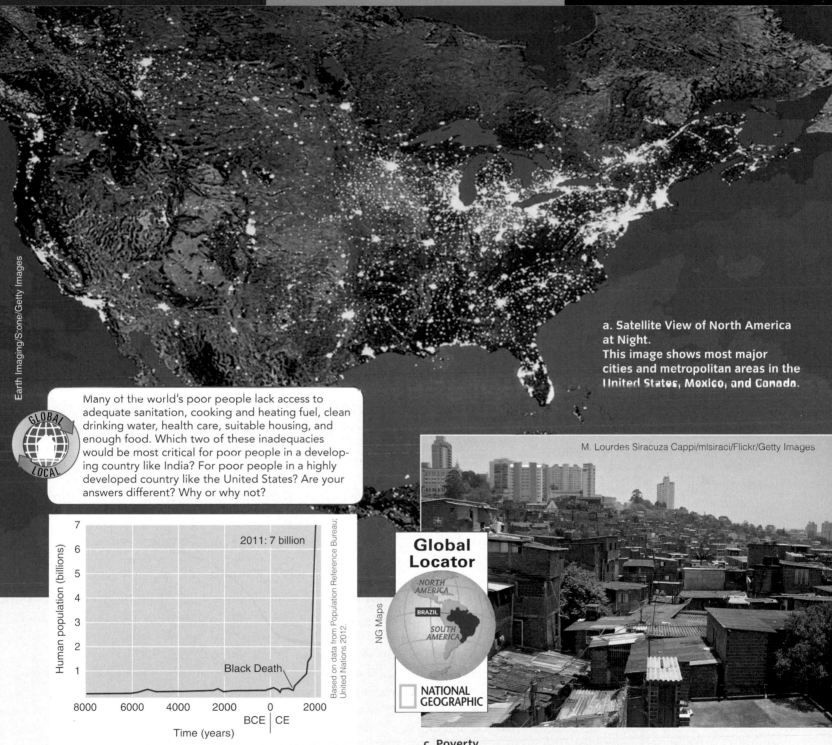

Earth Imaging/Stone/Getty Images

a. Satellite View of North America at Night.
This image shows most major cities and metropolitan areas in the United States, Mexico, and Canada.

Many of the world's poor people lack access to adequate sanitation, cooking and heating fuel, clean drinking water, health care, suitable housing, and enough food. Which two of these inadequacies would be most critical for poor people in a developing country like India? For poor people in a highly developed country like the United States? Are your answers different? Why or why not?

M. Lourdes Siracuza Cappi/mlsiraci/Flickr/Getty Images

2011: 7 billion

Human population (billions)

Black Death

8000   6000   4000   2000   0   2000

BCE | CE

Time (years)

Based on data from Population Reference Bureau; United Nations 2012.

NG Maps

Global Locator

NORTH AMERICA

BRAZIL

SOUTH AMERICA

NATIONAL GEOGRAPHIC

b. Human Population Growth.
It took thousands of years for the human population to reach 1 billion (in 1800). In 2011, Earth's human population surpassed 7 billion. (Black Death refers to a devastating disease, probably bubonic plague, that decimated Europe and Asia in the 14th century.)

c. Poverty.
Despite a growing national economy, many residents of Brazil continue to live in deep poverty in favelas (slums) like this one near Sao Paulo. Note the luxury hotel in the background.

a. A typical Japanese family, from Tokyo, with their possessions. People in highly developed countries consume a disproportionate share of natural resources.

b. A typical Mexican family, from Guadalajara, with their possessions. Economic development in this moderately developed country has allowed many people to enjoy a middle-class lifestyle. Other Mexicans live in poverty, however.

18 percent of the world's population, are highly developed countries (**Figure 1.2a**).

Poor countries, in which about 82 percent of the world's population live, fall into two subcategories: moderately developed and less developed. Turkey, South Africa, Thailand, and Mexico are examples of **moderately developed countries** (MDCs) (**Figure 1.2b**). People living in MDCs have fewer opportunities for income, education, and health care than people living in highly developed countries.

Examples of **less developed countries** (LDCs) include Haiti, Bangladesh, Rwanda, Laos, Ethiopia, and Mali (**Figure 1.2c**). Cheap, unskilled labor is abundant in LDCs, but capital for investment is scarce. To improve their economic conditions, many LDCs must borrow money from banks in highly developed countries. Most economies of LDCs are agriculturally based, often relying on only one or a few crops. As a result, crop failure or a low world market value for that crop is catastrophic to the economy. Hunger, disease, and illiteracy are common in LDCs.

moderately developed countries Countries with medium levels of industrialization and per person incomes lower than those of highly developed countries.

less developed countries Countries with low levels of industrialization, very high rates of population growth, very high infant mortality rates, and very low per person incomes relative to highly developed countries.

Population, Resources, and the Environment

Inhabitants of the United States and other highly developed countries consume many more resources per person than do citizens of developing countries. This high rate of resource consumption affects the environment at least as much as the rapid population growth that is occurring in other parts of the world. China and India, the world's most populous countries, include many of the world's poorest people, a growing middle class, and a few of the world's wealthiest people. Both countries have growing populations and expanding economies.

We can make two useful generalizations about the relationships among population growth, consumption of natural resources, and environmental degradation. First, the amount of resources essential to an individual's survival is small, but rapid population growth (often found in developing countries) tends to overwhelm and deplete a country's soils, forests, and other natural resources. Second, in highly developed nations, individual

c. A typical family from Kouakourou, Mali, with all their possessions. The rapidly increasing number of people in less developed countries overwhelms their natural resources, even though individual resource requirements may be low.

Peter Menzel

demands on natural resources are far greater than the requirements for mere survival. To satisfy their desires as well as their basic needs, many people in more affluent nations deplete resources and degrade the global environment through increased consumption of energy (through such uses as heating, transportation, and manufacturing), material goods (such as cars, televisions, and cellular phones), and agricultural products (including food, animal feed, and wood products).

Types of Resources When examining the effects of population on the environment, it is important to distinguish between nonrenewable and renewable natural resources. **Nonrenewable resources** include minerals (such as silicon, iron, and copper) and fossil fuels (coal, oil, and natural gas). Natural processes do not replenish nonrenewable resources within a reasonable duration on the human timescale. Fossil fuels, for example, take millions of years to form.

In addition to a nation's population and its level of resource use, several other factors affect the way nonrenewable resources are used—including how efficiently the resource is extracted and processed and how much of it is required or consumed. Nonetheless, the inescapable

> **nonrenewable resources** Natural resources that are present in limited supplies and are depleted as they are used.

fact is that Earth has a finite supply of nonrenewable resources that sooner or later will be exhausted. In time, technological advances may help find or develop substitutes for nonrenewable resources. Slowing the rate of population growth and resource consumption will help us buy time to develop such alternatives.

Some examples of **renewable resources** are trees, fishes, fertile agricultural soil, and fresh water. Nature replaces these resources fairly rapidly, on a scale of days to decades. Forests, fisheries, and agricultural land are particularly important renewable resources in developing countries because they provide food. Indeed, many people in developing countries are subsistence farmers who harvest just enough food for their families to survive.

> **renewable resources** Resources that are replaced by natural processes and that can be used forever, provided they are not overexploited in the short term.

Rapid population growth can cause renewable resources to be overexploited. For example, large numbers of poor people must grow crops on land—such as mountain slopes or tropical rain forests—that is poorly suited for farming. Although this practice may provide a short-term solution to the need for food, it does not work in the long run because when these lands are cleared for farming, their agricultural productivity declines rapidly and severe environmental deterioration occurs. Renewable resources, then, are *potentially* renewable. They must be used in a manner that allows natural processes time to replace or replenish them.

The effects of population growth on natural resources are particularly critical in developing countries. The economic growth of developing countries is frequently tied to the exploitation of their natural resources, often for export to highly developed countries. Developing countries are faced with the difficult choice of exploiting natural resources to provide for their expanding populations in the short term (that is, to pay for food or to cover debts) or conserving those resources for future generations. It is instructive to note that the economic growth and development of the United States and of other highly developed nations came about through the exploitation—and in some cases the destruction—of their resources. Continued growth and development in highly developed countries now relies significantly on the importation of these resources from less developed countries.

EnviroDiscovery
Green Roofs

Green roof
City Hall is one of many buildings in Chicago with a living green roof.

A roof that is completely or partially covered with vegetation and soil is known as a *green roof*. Also called *eco-roofs*, green roofs can provide several environmental benefits. For one thing, the plants and soil are effective insulators, reducing heating costs in winter and cooling costs in summer. The rooftop mini-ecosystem filters pollutants out of rainwater and reduces the amount of stormwater draining into sewers. In urban areas, green roofs provide wildlife habitat, even on the tops of tall buildings. A city with multiple green roofs provides "stepping stones" of habitat that enable migrating birds and insects to pass unharmed through the city. Green roofs can also be used to grow vegetable and fruit crops and to provide an outdoor refuge for people living or working in the building. Green roofs allow urban systems to more closely resemble the natural systems they have replaced.

Green roofs may be added to existing buildings, but it is often easier and less expensive to install them in new buildings. Modern green roofs, which are designed to support the additional weight of soil and plants, consist of several layers that hold the soil in place, stop plant roots from growing through the rooftop, and drain excess water, thereby preventing leaks. Currently, Chicago, Illinois, is the U.S. city with the largest total area of green roofs (see photograph). One of the largest individual green roofs in the United States is on the Ford Motor Company's Plant in Dearborn, Michigan.

Poverty is tied to the effects of population pressures on natural resources and the environment. Poor people in developing countries find themselves trapped in a vicious cycle of poverty. They use environmental resources for short-term gain (that is, to survive), but this exploitation degrades the resources and diminishes long-term prospects of economic development.

Population Size and Resource Consumption

Resource issues are clearly related to population size: At a given level of consumption, a larger population consumes more resources and causes more environmental damage than does a smaller population. However, not all people consume the same amounts of resources. There is great variability in consumption patterns across different populations both within a single country and among the different countries of the world.

Variation in consumption is associated with economic status, geography (especially whether people live in rural, suburban, or urban areas), culture, and other social and personal factors. A resident of a city who walks to work, rarely eats meat, owns few belongings, and has a small, well-insulated home may consume a fraction of the resources as compared to a resident of nearby suburbs.

Consumption is both an economic and a social act. Consumption provides the consumer with a sense of identity as well as status among peers. The media, including the advertising industry, promote consumption as a way to achieve happiness. We are encouraged to spend, to consume.

People in highly developed countries can be extravagant and wasteful consumers; their use of resources is greatly out of proportion to their numbers. A single child born in a highly developed country such as the United States causes a greater impact on the environment and on resource depletion than perhaps 20 children born in a developing country. Many natural resources are needed to provide the automobiles, air conditioners, disposable diapers, cell phones, DVD players, computers, clothes, newspapers, athletic shoes, furniture, books, and other "comforts" of life in highly developed nations. Thus, the disproportionately large consumption of resources by the United States and other highly developed countries affects natural resources and the environment as much as or more than the population explosion in the developing world.

Highly developed nations represent less than 20 percent of the world's population, yet they consume significantly more than half of its resources. According to the Worldwatch Institute, highly developed countries account for the lion's share of total resources consumed:

- 86 percent of aluminum used
- 76 percent of timber harvested
- 68 percent of energy produced
- 61 percent of meat eaten
- 42 percent of the fresh water consumed

These nations also generate 75 percent of the world's pollution and waste (**Figure 1.3**).

Consumption • Figure 1.3

American consumption is actively promoted in Times Square advertisements. Highly developed nations, such as the United States, consume more than 50 percent of the world's resources, produce 75 percent of its pollution and waste, and represent only 18 percent of its total population.

Raga Jose Fuste/Prisma/SuperStock

Ecological footprints Each person has an **ecological footprint**, an amount of productive land, fresh water, and ocean required on a continuous basis to supply that person with food, wood, energy, water, housing, clothing, transportation, and waste disposal. In the *Living Planet Report 2008*, scientists calculated that Earth has about 11.4 billion hectares (28.2 billion acres) of productive land and water. If we divide this area by the global human population, it indicates that each person is allotted about 1.8 hectares (4.3 acres). However, the average global ecological footprint is currently about 2.7 hectares (6.7 acres) per person, which means we have an *ecological overshoot*—we have depleted our allotment. We can see the short-term results around us—forest destruction, degradation of croplands, loss of biological diversity, declining ocean fisheries, local water shortages, and increasing pollution. The long-term outlook, if we do not seriously address our consumption of natural resources, is potentially disastrous.

The developing nation of India is the world's second largest country in terms of population, so even though its per capita footprint is low, the country's total footprint is high (**Figure 1.4**). In France, the per capita ecological footprint is high at 4.9 hectares (12.1 acres), but its footprint as a country is relatively low, at 298.1 million hectares (736.6 million acres). The United States, which has the world's third largest population, has a per capita ecological footprint of 9.4 hectares (23.3 acres); the U.S. footprint as a country is a whopping 2810 million hectares (6943 million acres). If all people in the world had the same lifestyle and level of consumption as the average American, and assuming no changes in technology, we would need four additional planets the size of Earth to support us all.

As developing nations increase their economic growth and improve their standard of living, more and more people in those countries purchase consumer goods. By the early 2000s, more new cars were sold annually in Asia than in North America and western Europe combined. These new consumers may not consume at the high level of the average consumer in a highly developed nation, but their consumption has increasingly adverse effects on the environment. For example, air pollution from traffic in urban centers in developing countries is bad and getting worse every year. Millions of dollars are lost to health problems caused by air pollution in these cities.

Population, consumption, and environmental impact When you turn on the tap to brush your teeth in the morning, you probably do not think about where the water comes from or about the environmental consequences of removing it from a river or the ground. All the materials that make up the products we use every day come from Earth, and these materials eventually are returned to Earth, mainly in sanitary landfills.

Such human impacts on the environment are difficult to assess. One way to estimate them is to consider the three factors most important in determining environmental impact (I):

- The number of people (P).

- The affluence per person, which is a measure of the consumption, or amount of resources used per person (A).

- The environmental effects (resources needed and wastes produced) of the technologies used to obtain and consume the resources (T).

This method of assessment is usually referred to as the IPAT equation: $I = P \times A \times T$.

Biologist **Paul R. Ehrlich** and physicist **John P. Holdren** first proposed the IPAT model in the 1970s. It shows the mathematical relationship between environmental impacts and the forces that drive them. To determine the environmental impact of carbon dioxide (CO_2) emissions from motor vehicles, for example, multiply the population by the number of cars per person (affluence or consumption per person) by the average annual CO_2 emissions per year (technological impact). This model demonstrates that although improving motor vehicle efficiency and developing cleaner technologies will reduce pollution and environmental degradation, a larger reduction will result if population and per person consumption are also controlled.

The three factors in the IPAT equation are always changing in relation to each other. Generally, this equation tells us that as population and affluence increase, environmental impacts will increase as well. However, technological improvements can reduce impacts. For example, flat-screen televisions contain less materials and require less energy to produce, transport, and operate than did equivalent-sized televisions produced two decades ago.

Ecological footprints • Figure 1.4

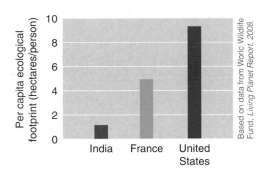

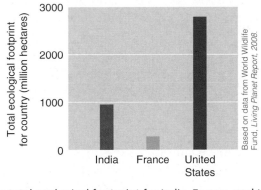

a. The average ecological footprint of a person living in India, France, or the United States. For example, the average Indian requires 0.9 hectare (2.2 acres) of productive land and ocean to meet his or her resource requirements.

b. The total ecological footprint for India, France, and the United States. Notice that India, although having a low per capita ecological footprint, has a relatively large total footprint as a country because of its large population. If everyone in the world had the same level of consumption as the average American, it would take the resources and area of 5 Earths.

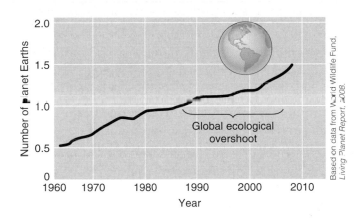

c. Earth's ecological footprint has been increasing over time. By 2008, humans were using the equivalent of 1.5 Earths, a situation that is not sustainable.

Calculate your individual ecological footprint online. (Search for "ecological footprint.") Are you living sustainably? Suggest two things that you could do to lower your ecological footprint.

The IPAT equation helps to identify what we don't know or understand about consumption and its environmental impact. For example, which kinds of consumption have the greatest destructive impact on the environment? How can we alter the activities of these environmentally disruptive consumption patterns? What combination of technological advances and behavioral changes can create simultaneous improvements in economic, environmental, and social conditions? It will take years to address such questions, but the answers should help decision makers in business and government formulate policies that will alter consumption patterns in an environmentally responsible way. The ultimate goal should be to make consumption sustainable so that humanity's current practices do not compromise the ability of future generations to use and enjoy the riches of our planet.

To summarize, as human numbers and consumption increase worldwide, so does humanity's impact on Earth, posing new challenges to us all. Success in achieving sustainability in population size and consumption will require the cooperation of all the world's peoples.

CONCEPT CHECK

1. **How** do highly developed countries, moderately developed countries, and less developed countries differ regarding population growth and per person incomes?

2. **How** is human population growth related to natural resource depletion and environmental degradation?

3. **What** can the three factors of the IPAT equation tell us about measuring and reducing harmful environmental impacts?

Sustainability and the Environment

LEARNING OBJECTIVES

1. **Define** *sustainability*.
2. **Identify** human behaviors that threaten environmental sustainability.

Sustainability is an organizing principle for this text. *Sustainability* is achieved when the environment can function indefinitely without going into a decline from the stresses that human society imposes on natural systems (such as fertile soil, water, and air) (**Figure 1.5**). Sustainability applies at many levels, including the individual, communal, regional, national, and global levels.

Sustainability is based in part on the following ideas:

> **sustainability**
> The ability to meet humanity's current needs without compromising the ability of future generations to meet their needs.

- We must think simultaneously about economic, social and environmental well-being.

- We must consider the effects of our actions on the health and well being of the natural environment, including all living things.

- Earth's resources are not present in infinite supply. We must live within limits that let renewable resources such as fresh water regenerate for future needs.

- We must understand all the costs to the environment and to society of products we consume.

- All of Earth's inhabitants share a responsibility for living sustainably.

Many environmental experts think that human society is not operating sustainably because of the following human behaviors (**Figures 1.6** and **1.7** [the National Geographic map]):

- We are using nonrenewable resources such as fossil fuels as if they were present in unlimited supplies.

- We are using renewable resources such as fresh water and forests faster than they are replenished naturally.

- We are polluting the environment—the land, rivers, ocean, and atmosphere—with toxins as if the capacity of the environment to absorb them were limitless.

- Our numbers continue to grow, despite Earth's finite ability to feed us and to absorb our wastes.

- Our activities disrupt the ability of natural processes to regenerate; this happens from the local to the global scale.

If left unchecked, these activities may threaten the life-support systems of Earth to the extent that recovery is impossible. Our first goal should be to critically evaluate which changes our society is willing to make.

At first glance, issues of sustainability may seem simple. The solutions are more complex and challenging, in part because of various interacting ecological, societal, and economic factors. Our incomplete scientific understanding of how the environment works and how human choices affect the environment is a major reason that sustainability is difficult to achieve. Even for established environmental problems, political and social controversy often prevents widespread acceptance that an environmental threat is real.

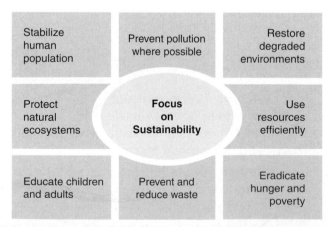

Focus on sustainability • Figure 1.5

Environmental sustainability requires a long-term perspective to promote economic, social, and environmental well-being, such as the goals shown here.

CONCEPT CHECK **STOP**

1. **What** is sustainability?
2. **Which** human behaviors threaten sustainability?

Environmental InSight

a. A worker prepares frozen giant blue fin tuna for auction at a the Tsukiji market in Tokyo, Japan. These increasingly rare fish are caught around the globe.

Keith Douglas/All Canada Photos/SuperStock, Inc.

© AHowden/Japan Stock Photography/Alamy

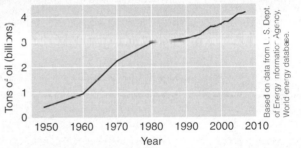

Based on data from U. S. Dept. of Energy Information Agency, World energy database.

Annual World Oil Consumption, 1950 to 2007. This increase is unsustainable.

 Interpreting Data
By approximately how much did annual world oil consumption increase in the past 20 years? Is the increase sustainable? Why or why not?

b. This section of forest in British Columbia, Canada, has been clear cut, leaving it vulnerable to erosion.

Tom Bonaventure/Photographer's Choice/Getty Images, Inc.

WileyPLUS
➕

c. Oil Refinery at Grangemouth, United Kingdom.
Both highly developed and developing countries depend largely on oil for economic development.

Global environmental issues • Figure 1.7

These issues occur locally at so many places around the planet that they are global in scope.

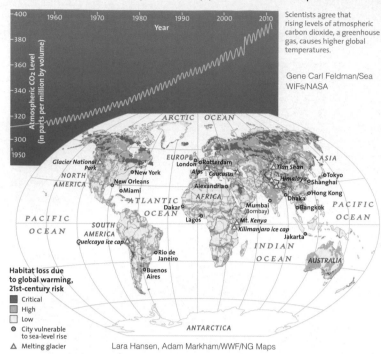

Scientists agree that rising levels of atmospheric carbon dioxide, a greenhouse gas, causes higher global temperatures.

Gene Carl Feldman/Sea WIFs/NASA

Habitat loss due to global warming, 21st-century risk
- Critical
- High
- Low
- ○ City vulnerable to sea-level rise
- △ Melting glacier

Lara Hansen, Adam Markham/WWF/NG Maps

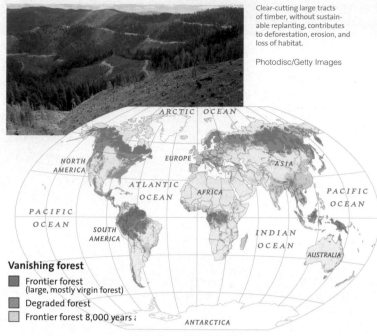

Clear-cutting large tracts of timber, without sustainable replanting, contributes to deforestation, erosion, and loss of habitat.

Photodisc/Getty Images

Vanishing forest
- Frontier forest (large, mostly virgin forest)
- Degraded forest
- Frontier forest 8,000 years ٔ

Global Forest Watch/WRI/NG Maps

▲ GLOBAL WARMING

Temperatures across the world are increasing at a rate not seen at any other time in the last 10,000 years. Although climate variationis a natural phenomenon, human activities that release carbon dioxide and other greenhouse gases into the atmosphere—industrial processes, fossil fuel consumption, deforestation, and land use change—are the primary cause of this warming trend. Scientists predict that if this trend continues, one-third of plant and ani-mal habitats will be dramatically altered and more than one million species will be threatened with extinction in the next 50 years. And even small increases in global temperatures can melt glaciers and polar ice sheets, raising sea levels and flooding coastal cities and towns.

▲ DEFORESTATION

Of the 13 million hectares (32 million acres) of forest lost each year, mostly to make room for agriculture, more than half are in South America and Africa, where many of the world's tropical rain forests and terrestrial plant and animal species can be found. Loss of habitat in such species-rich areas takes a toll on the world's biodiversity. Deforested areas also release, instead of absorb, carbon dioxide into the atmosphere, contribut-ing to global climate change. Deforestation can also affect local climates by reducing evaporative cooling, leading to decreased rainfall and higher temperatures.

Linked to pollution, black band disease kills a coral head near the island of Curaçao.

Caroline Rogers/USGS

Decline in fish catches, post-1970's
- Cod and cod-like fish (haddock, hake)
- Flatfish (flounder, sole)
- Perch-like fish (grouper, snapper)
- ○ Diseased coral reef

NOAA/NMFS/FigureEP-WCMC/NG Maps

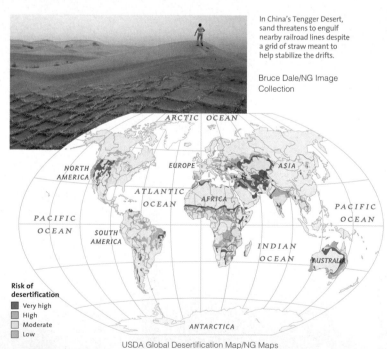

In China's Tengger Desert, sand threatens to engulf nearby railroad lines despite a grid of straw meant to help stabilize the drifts.

Bruce Dale/NG Image Collection

Risk of desertification
- Very high
- High
- Moderate
- Low

USDA Global Desertification Map/NG Maps

▲ THREATENED OCEANS

Oceans cover more than two-thirds of the Earth's surface and are home to at least half of the world's biodiversity, yet they are the least understood ecosystems. The combined stresses of overfishing, pollution, increased carbon dioxide emissions, global climate change, and coastal development are having a serious impact on the health of oceans and ocean species. Over 70% of the world's fish species are depleted or nearing depletion, and 50% of coral reefs worldwide are threatened by human activities.

▲ DESERTIFICATION

Climate variability and human activities, such as grazing and conversion of natural areas to agricultural use, are leading causes of desertification, the degradation of land in arid, semiarid, and dry subhumid areas. The environmental consequences of desertification are great—loss of topsoil, increased soil salinity, damaged vegetation, regional climate change, and a decline in biodiversity. Equally critical are the social consequences—more than 2 billion people live in and make a living off these dryland areas, covering about 41% of Earth's surface.

ASIA
Projected minimum area of sea ice, 2050
North Pole
2000
1950
NORTH AMERICA
EUROPE

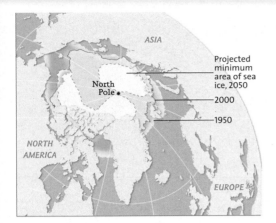

▲ **POLAR ICE CAP**
Over the last 50 years, the extent of polar sea ice has noticeably decreased. Since 1970 alone, an area larger than Norway, Sweden, and Denmark combined has melted. This trend is predicted to accelerate as temperatures rise in the Arctic and across the globe.

National/Naval Ice Center/NG Maps

GeoBytes

ACIDIFYING OCEANS
Oceans are absorbing an unprecedented 20 to 25 million tonnes (22 to 28 millions tons) of carbon dioxide each day, increasing the water's acidity.

ENDANGERED REEFS
Some 95% of coral reefs in Southeast Asia have been destroyed or are threatened.

RECORD TEMPERATURES
The 1990s were the warmest decade on record in the last century.

WARMING ARCTIC
While the world as a whole has warmed nearly 0.6°C (1°F) over the last hundred years, parts of the Arctic have warmed 4 to 5 times as much in only the last 50 years.

DISAPPEARING RAIN FORESTS
Scientists predict that the world's rain forests will disappear within the next one hundred years if the current rate of deforestation continues.

OIL POLLUTION
Nearly 1.3 million tonnes (1.4 million tons) of oil seep into the world's oceans each year from the combined sources of natural seepage, extraction, transportation, and consumption.

ACCIDENTAL DROWNINGS
Entanglement in fishing gear is one of the greatest threats to marine mammals.

A FAREWELL TO FROGS?
Worldwide, almost half of the 5,700 named amphibian species are in decline.

With the growth of scientific record keeping, observation, modeling, and analysis, our understanding of Earth's environment is improving. Yet even as we deepen our insight into environmental processes, we are changing what we are studying. At no other time in history have humans altered their environment with such speed and force. Nothing occurs in isolation, and stress in one area has impacts elsewhere. Our agricultural and fishing practices, industrial processes, extraction of resources, and transportation methods are leading to extinctions, destroying habitats, devastating fish stocks, disturbing the soil, and polluting the oceans and the air. As a result, biodiversity is declining, global temperatures are rising, polar ice is shrinking, and the ozone layer continues to thin.

▼ **POLLUTION**
No corner of the earth is immune to pollution, be it in the air, soil, or water. Concentrations of pollution can be found in the industrialcenters of North America, Europe, and, increasingly, Asia—and areas downwind or downstream from them. Shipping routes are sources of pollution, from oil spills to garbage dumpings.

NORTH AMERICA
Toronto
New York
Los Angeles
Mexico
ATLANTIC OCEAN
PACIFIC OCEAN
SOUTH AMERICA
São Paulo
Buenos Aires
London
Paris
EUROPE
Moscow
ASIA
Beijing
Tokyo
Cairo
AFRICA
Lagos
Mumbai (Bombay)
INDIAN OCEAN
Jakarta
AUSTRALIA
Sydney
PACIFIC OCEAN
ANTARCTICA
Environmental stress factor
Accident
Industrial
Oil rig explosion
Oil spill
Acid rain
Deforestation
Temperate forest
Tropical forest
Desertification
Pollution from shipping

NG Maps

NATIONAL GEOGRAPHIC

▶ **OZONE DEPLETION**
First noted in the mid-1980s, the springtime "ozone hole" over the Antarctic continues to grow. With sustained efforts to restrict chlorofluorocarbons (CFCs) and other ozone-depleting chemicals, scientists have begun to see what they hope is a leveling off in the rate of depletion. Stratospheric ozone shields the Earth from the sun's ultraviolet radiation. Thinning of this protective layer puts people at risk for skin cancer and cataracts. It can also have devastating effects on the Earth's biological functions.

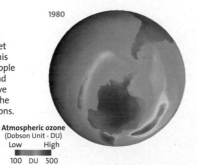

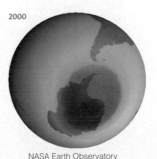

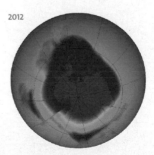

1980
2000
2012
Atmospheric ozone
(Dobson Unit - DU)
Low
High
100 DU 500

NASA Earth Observatory

Environmental Science

LEARNING OBJECTIVES

1. **Define** *environmental science*.
2. **Outline** the steps of the scientific method.

Environmental science is an interdisciplinary field that combines information from many disciplines, such as biology, geography, chemistry, geology, physics, economics, sociology (particularly *demography*, the study of populations), cultural anthropology, natural resource management, agriculture, engineering, law, politics, and ethics.

Ecology, the discipline of biology that studies the interrelationships between organisms and their environment, is a basic tool of environmental science. **Atmospheric science** is a branch of environmental science that includes the study of weather and climate, greenhouse gases, and other airborne pollutants. **Environmental chemistry** examines chemicals in the environment, including air, soil, and water pollution (**Figure 1.8**). **Geosciences**—for example, environmental geology and physical geography—study a wide range of environmental topics, such as

environmental science The interdisciplinary study of humanity's relationship with other organisms and the physical environment.

soil erosion, groundwater use, ocean pollution, and climate. Scientists in these subdisciplines not only evaluate environmental quality but also develop ways to restore damaged environments.

The Goals of Environmental Science

Environmental scientists try to establish general principles about how the natural world functions. They use these principles to develop viable solutions to environmental problems—solutions that are based as much as possible on scientific knowledge.

Environmental problems are generally complex, however, and scientific understanding of them is often less complete than we would like. Environmental scientists are often called on to reach scientific consensus before the data are complete. As a result, they often cannot give precise answers and so instead make recommendations based on what they think is most likely to occur.

Many of the environmental problems considered in this book require urgent attention. Yet environmental science is not simply a "doom and gloom" listing of problems, coupled with predictions of a bleak future. To the contrary, the focus of environmental science, and our

Environmental research chemist • Figure 1.8

This scientist is performing an experiment to purify air pollution gases. The chemical reaction he is studying converts toxic pollutants into nontoxic chemicals.

© PHOTOTAKE Inc./Alamy

focus as individuals and as world citizens, is on identifying, understanding, and solving problems that we as a society have generated. It is encouraging that individuals, businesses, and governments are already doing a great deal, although more must be done to address the problems of today's world.

Science as a Process

The key to successfully solving any environmental problem is rigorous scientific evaluation. It is important to understand clearly just what science is, as well as what it is not. Most people think of **science** as a body of knowledge—a collection of facts about the natural world. However, science is also a dynamic *process*, a systematic way to investigate the natural world. Science seeks to reduce the apparent complexity of our world to general principles,

which are then used to make predictions, solve problems, or provide new insights.

Scientists collect objective **data** (singular, *datum*), the information with which science works. Data are collected through observation and experimentation and then analyzed or interpreted (**Figure 1.9**). Scientific conclusions are inferred from the available data and are not based on faith, emotion, or intuition. Scientists publish their findings in scientific journals, and other scientists examine and critique their work. A requirement of science is repeatability—that is, observations and experiments must produce consistent data when they are repeated. Scrutiny by other scientists reveals any inconsistencies in results or interpretation. These errors are discussed openly, and ways to eliminate them are developed.

Science is an ongoing enterprise, and scientific concepts must be reevaluated in light of newly discovered

Data collection • Figure 1.9

A researcher observes genetically modified rice plants. Photographed at Cornell University, New York. (Inset) A scientist records his data on a computer in the laboratory.

© Guy Croft SciTech/Alamy

Josheph Kahn/The New York Times /Redux Pictures

data. Thus, scientists never claim to know the final answer about anything because scientific understanding changes.

This is not to say, however, that all ideas are equally valid. Rather, the scientific processes evaluate and then reject ideas that are inconsistent with theory and data. Thus, despite many uncertainties, science provides usable insights on many aspects of environmental change and management. While science tells us what *is* and what *can be*, it cannot tell us what *should be*. Questions about what should be are in the realm of religion, ethics, policy, and philosophy. Once we have used these approaches to decide on our priorities and preferences, science is the most useful tool available to help us achieve them. Science aims to discover and better understand the general principles that govern the operation of the natural world.

The Scientific Method The established processes that scientists use to answer questions or solve problems are collectively called the **scientific method**. Although there are many variations of the scientific method, it basically involves five steps:

1. Recognize a question or an unexplained phenomenon in the natural world.

2. Develop a *hypothesis*, or the expected answer to the question.

3. Design and perform an experiment to test the hypothesis.

4. Analyze and interpret the data to reach a conclusion.

5. Share new knowledge with the scientific community.

Although the scientific method is often portrayed as a linear sequence of events, science is rarely as straightforward or tidy as the scientific method implies (**Figure 1.10**). Good science involves creativity, not only in recognizing questions and developing hypotheses but also in designing experiments. Because scientists try to expand our current knowledge, their work is in the realm of the unknown. Many creative ideas end up as dead ends, and there are often temporary setbacks or reversals of direction as scientific knowledge progresses. Scientific knowledge often expands haphazardly, with the "big picture" emerging slowly from confusing and sometimes contradictory details.

Scientific discoveries are often incorrectly portrayed in the media as "new facts" that have just come to light. At a later time, additional "new facts" that question the validity of the original study are reported. If you were to read the scientific

> **scientific method**
> The way a scientist approaches a problem, by formulating a hypothesis and then testing it.

papers on which such media reports are based, however, you would find that all the scientists involved probably made very tentative conclusions based on their data. Science progresses from uncertainty to less uncertainty, not from certainty to greater certainty. Thus, science is self-correcting over time, despite the fact that it never "proves" anything.

The Importance of Prediction Scientists formulate **hypotheses** (plural of hypothesis) based on what they think to be true based on prior scientific work. A hypothesis is a statement of that expectation. A hypothesis is useful if it can be falsified (shown to be wrong) and tested. A hypothesis is a prediction that can be subjected to experimentation. When an experiment refutes a prediction, scientists must carefully recheck the entire experiment.

If the prediction is still refuted, then the hypothesis must be rejected. The more verifiable predictions a hypothesis makes, the more valid that hypothesis is.

Each of the many factors that influence a process is called a **variable**. To evaluate alternative hypotheses about a specific variable, it is necessary to hold all other variables constant so that they are not misleading or confusing. To test a hypothesis about a variable, we carry out two forms of the experiment in parallel. In the **experimental group**, the chosen variable is altered in a known way. In the **control group**, that variable isn't altered. In all other respects, the experimental group and the control group are the same. We then ask, "What is the difference, if any, between the outcomes for the two groups?" Any difference must be due to the influence of the variable we changed because all other variables remained the same. Much of the challenge of science lies in designing control groups and in successfully isolating a single variable from all other variables.

Theories A **theory** is an integrated explanation of numerous hypotheses, each of which is supported by a large body of observations and experiments. A theory condenses and simplifies many data that previously appeared to be unrelated. Because a theory demonstrates the relationships among different data, it simplifies and clarifies our understanding of the natural world. A good theory grows as additional information becomes known. It predicts new data and suggests new relationships among a range of natural phenomena.

Theories are the solid ground of science, the explanations of which we are most sure. This definition contrasts sharply with the general public's use of the word *theory*, which implies lack of knowledge or a guess. In this

The scientific method • Figure 1.10

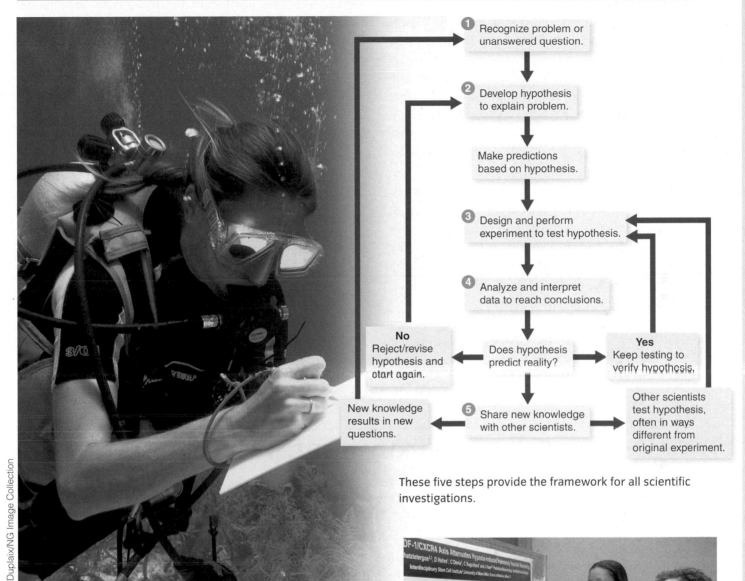

1 Recognize problem or unanswered question.

2 Develop hypothesis to explain problem.

Make predictions based on hypothesis.

3 Design and perform experiment to test hypothesis.

4 Analyze and interpret data to reach conclusions.

No Reject/revise hypothesis and start again.

Does hypothesis predict reality?

Yes Keep testing to verify hypothesis.

New knowledge results in new questions.

5 Share new knowledge with other scientists.

Other scientists test hypothesis, often in ways different from original experiment.

These five steps provide the framework for all scientific investigations.

Nicole Duplaix/NG Image Collection

3 A field scientist makes observations critical to understanding damage to coral reefs from global climate change. Photographed at Turneffe Atoll, Belize.

© Jeff Greenberg/Alamy

5 Many scientists present their research during poster sessions at scientific meetings. This allows their work to be critically assessed by others in the scientific community.

Think Critically What is the relationship between a hypothesis and an experiment?

book, the word *theory* is always used in its scientific sense, to refer to a broadly conceived, logically coherent, and well-supported explanation.

Unfortunately, many questions that are most important to environmental scientists cannot be formulated as testable hypotheses. For example, we cannot design an experiment to test the hypothesis that if we double the carbon dioxide concentration of the atmosphere, global average temperatures will increase. Consequently, much of environmental science requires that we apply our best understanding of theory and data to arrive at conclusions about what we expect will happen.

Despite the fact that theories are generally accepted, there is no absolute truth in science, only varying degrees of uncertainty. Science is continually evolving as new evidence comes to light, and therefore its conclusions are always provisional or uncertain. It is always possible that the results of a future experiment will contradict a prevailing theory and show at least one aspect of it to be false.

Uncertainty, however, does not mean that scientific conclusions are invalid. For example, overwhelming evidence links cigarette smoking and incidence of lung cancer. We can't state with absolute certainty which smokers will be diagnosed with lung cancer, but this uncertainty does not mean that there is no correlation between smoking and lung cancer. On the basis of the available evidence, we say that people who smoke have an increased risk of developing lung cancer.

In conclusion, the aim of science is to increase human comprehension by explaining the processes and events of nature. Scientists work under the assumption that all phenomena in the natural world have natural causes, and they formulate theories to explain these phenomena. The process of science as a human endeavor has shaped the world we live in and transformed our views of the universe and how it works.

CONCEPT CHECK **STOP**

1. **What** is environmental science? What are some of the disciplines involved in environmental science?

2. **What** are the five steps of the scientific method? Why is each important?

How We Handle Environmental Problems

LEARNING OBJECTIVES

1. **List** and briefly describe the five stages of solving environmental problems.

Before examining the environmental problems discussed in the remaining chapters of this book, let's consider the elements that contribute to solving those problems. How, for example, can we handle water pollution in a river (**Figure 1.11**)? At what point are conclusions regarded as certain enough to warrant action? Who makes the decisions, and what are the trade-offs? Viewed simply, there

Monitoring water pollution • Figure 1.11

This pollution control officer is measuring the oxygen level in the Severn River near Shrewsbury, England. When dissolved oxygen levels are high, pollution levels (of sewage, fertilizer, and such) are low.

Ben Osborne/Stone/Getty Images

Addressing environmental problems • Figure 1.12

These five steps provide a framework for addressing environmental problems.

① Scientific assessment:
Problem is defined, hypotheses are tested, and models are constructed to show how present situation developed and to predict future course of events.

Example:
Scientists find higher-than-normal levels of bacteria are threatening a lake's native fish and determine the cause is human-produced pollution.

② Risk analysis:
Potential effects of various interventions—including doing nothing—are analyzed to determine risks associated with each particular course of action.

If no action is taken, fishing resources—a major source of income in the region—will be harmed. If pollution is reduced appreciably, fishery will recover.

③ Public engagement:
Changing public attitudes involves explaining the problem, presenting available alternatives for action, and revealing probable risks, results, and costs of each choice.

Public is informed of the ramifications—in this case, loss of income—if problem is not addressed.

⑤ Long-term environmental management:
Results of any action taken should be carefully monitored to see the environmental problem is being addressed.

Water quality in lake is tested frequently, and fish populations are monitored to ensure they do not decline.

④ Political considerations:
Elected officials, often at urging of their constituencies, implement a course of action based on scientific evidence as well as economic and social considerations.

Elected officials, supported by the public, pass legislation to protect lake and develop lake cleanup plan.

Think Critically Despite having a framework for addressing environmental problems, many problems are either incorrectly addressed or not addressed adequately. Offer at least one possible reason for such failures.

---

are five stages in addressing an environmental problem (**Figure 1.12**):

1. Scientific assessment
2. Risk analysis
3. Public engagement
4. Political considerations
5. Long-term environmental management

These five stages represent an ideal approach to systematically addressing environmental problems. In real life, seeking solutions to environmental problems is rarely so neat and tidy, particularly when the problem is exceedingly complex, of regional or global scale, or has high costs and unclear benefits for the money invested. Quite often, the public becomes aware of a problem, which triggers discussion of remediation before the problem is clearly identified and scientifically assessed.

CONCEPT CHECK STOP

1. **What** are the five steps used to solve an environmental problem?

EnviroDiscovery
Getting Past NIMBY

Our highly industrialized, high-consumption economy produces substantial amounts of waste, many of them dangerous and long-lived. Keeping these wastes where they are generated can create significant threats to human and environmental health and safety. Consequently, waste producers constantly seek locations for permanent disposal. Unfortunately, any disposal site exposes some people to threats from either the facility or associated transportation.

When people hear that a power plant, an incinerator, or a hazardous waste disposal site may be situated nearby, residents often react negatively. Their objections are often referred to as the NIMBY ("not in my back yard") response. In most situations, the people who would be exposed to the new threat are not the people who stand to gain from them. In other cases, people who are labeled as NIMBYs object to being excluded from the decision-making process. Developers and public planners often fail to engage people living in low-income urban areas, older suburban areas, or rural areas to help make decisions that affect their neighborhoods.

Exacerbating the NIMBY response is the failure by companies and government to develop processes for listening and responding to public concerns. The experts they bring in are seen as part of the problem and are distrusted by local residents. When experts are not trusted, people don't believe their analyses, no matter how scientifically valid. Experts, who typically do not have training in effective communication, interpret this distrust as ignorance or emotion. Resentment and conflict follow.

Consider the disposal of radioactive waste from nuclear power plants. There is broad agreement that the best long-term solution is to safely isolate radioactive waste, preferably deep underground, for thousands of years. However, rather than explore a range of possible disposal sites, the U.S. government, backed by the nuclear energy industry, committed in 1982 to explore only a single disposal site, Yucca Mountain in Nevada. It then spent the next three decades studying only that site. As Nevada became more politically powerful, its residents objected to the process, which was often interpreted as an "antiscientific" NIMBY attitude. Only recently (in 2010) did a new process begin, one that incorporates broad perspectives and stakeholders in a national conversation.

Most people agree that our generation has the responsibility to dispose of wastes we generate. Failure to find appropriate long-term solutions can result in more dangerous short-term solutions, or illegal and unsafe dumping. For existing wastes and technologies, then, planners should use approaches that look for socially, economically, and environmentally sound solutions— that is, for sustainable solutions.

The constant recurrence of the NIMBY phenomenon suggests that we should also consider a more sophisticated, systems-based approach. Life-cycle assessment takes a systems perspective on technological threats. Rather than ask "how do we safely dispose of wastes," life-cycle assessment considers how to change processes and materials in a way that minimizes waste production. In many cases, life-cycle assessment leads to innovations that save money, require less energy, and produce fewer wastes.

Not in my backyard

Steam rises from two of the cooling towers of a nuclear power plant. All nuclear power plants store highly radioactive spent fuel on site because there is currently no place to safely dispose of it.

The New Orleans Disaster

Hurricane Katrina, which hit the north-central Gulf Coast in August 2005, was one of the most devastating storms in U.S. history. It produced a storm surge that caused severe damage to New Orleans as well as to other coastal cities and towns in the region. The high waters caused levees and canals to fail, flooding 80 percent of New Orleans and many nearby neighborhoods.

Most people are aware of the catastrophic loss of life and property caused by Katrina. Here we focus on how humans altered the geography and geology of the New Orleans area in ways that exacerbated the storm damage.

Over the years, engineers constructed a system of canals to aid navigation and a system of levees to control flooding because the city is at or below sea level. The canals allowed salt water to intrude and kill the freshwater marsh vegetation. The levees prevented the deposition of sediments that remain behind after floodwaters subside. (The sediments are now deposited in the Gulf of Mexico.) Under natural conditions, these sediments replenish and maintain the delta, building up coastal wetlands.

As the city has grown, new development has taken place on wetlands—bayous, waterways, and marshes—that were drained and filled in. Before their destruction, these coastal wetlands provided some protection against flooding from storm surges. We are not implying that had Louisiana's wetlands been intact, New Orleans would not have suffered any damage from a hurricane of Katrina's magnitude. However, had these wetlands been largely unaltered, they would have moderated the storm's damage by absorbing much of the water from the storm surge.

Another reason that Katrina devastated New Orleans is that the city has been subsiding (sinking) for many years, primarily because New Orleans is built on unconsolidated sediment (no bedrock underneath). Many wetlands scientists also attribute this subsidence to the extraction of the area's rich supply of underground natural resources—groundwater, oil, and natural gas. As these resources are removed, the land compacts, lowering the city. New Orleans and nearby coastal areas are subsiding an average of 6 mm each year (see image). At the same time, the sea level has been rising an average of 1 mm to 2.5 mm per year due to human-induced changes in climate.

UPI Photo/IKONOS/NewsCom

Satellite image of flooding in New Orleans following Hurricane Katrina

Along the left (west) side is a levee from Lake Ponchartrain (top) that failed so that water inundated the New Orleans area east of the levee. Areas on the far left top remained dry. Part of the Mississippi River is shown at lower center.

WileyPLUS
+

Global Locator

NG Maps

NEW ORLEANS

NATIONAL GEOGRAPHIC

Summary

1 Human Impacts on the Environment 4

1. **Highly developed countries** are countries that have complex industrialized bases, low rates of population growth, and high per person incomes. **Moderately developed countries** are developing countries that have medium levels of industrialization and average per person incomes lower than those of highly developed countries. **Less developed countries (LDCs)** are developing countries with low levels of industrialization, very high rates of population growth, very high infant mortality rates, and very low per person incomes (relative to highly developed countries). **Poverty**, which is common in LDCs, is a condition in which people are unable to meet their basic needs for food, clothing, shelter, education, or health.

2. The increasing global population is placing stresses on the environment, as humans consume ever-increasing quantities of food and water, use more energy and raw materials, and produce enormous amounts of waste and pollution. **Nonrenewable resources** are natural resources that are present in limited supplies and are depleted as they are used. **Renewable resources** are resources that natural processes replace and that therefore can be used forever, provided that they are not exploited in the short term.

3. The forces that drive environmental impact can be modeled by the IPAT equation, $I = P \times A \times T$. Environmental impact (I) has three factors: the number of people (P); the affluence per person (A), which is a measure of the consumption, or amount of resources used per person; and the environmental effect of the technologies used to obtain and consume those resources (T).

2 Sustainability and the Environment 12

1. **Sustainability** is the ability to meet humanity's current needs without compromising the ability of future generations to meet their needs. Sustainability is achieved when the environment can function indefinitely without going into a decline from the stresses that human society imposes on natural systems. Taking a sustainability perspective requires that we think simultaneously about economic, social, and environmental well-being.

2. Human behaviors that threaten environmental sustainability include overuse of renewable and nonrenewable resources, pollution, and population growth.

Stabilize human population	Prevent pollution where possible	Restore degraded environments
Protect natural ecosystems	Focus on Sustainability	Use resources efficiently
Educate children and adults	Prevent and reduce waste	Eradicate hunger and poverty

3 Environmental Science 17

1. **Environmental science** is the interdisciplinary study of humanity's relationship with other organisms and the nonliving physical environment. Environmental science encompasses many problems involving human numbers, Earth's natural resources, and environmental pollution. While science always includes some degree of uncertainty, it nevertheless provides useful information for many environment-related decisions.

2. The **scientific method** is the way a scientist approaches a problem, by formulating a **hypothesis** and then testing it by means of an experiment. (1) A scientist recognizes and states the problem or unanswered question. (2) The scientist develops a hypothesis, or an educated guess, to explain the problem. (3) An experiment is designed and performed to test the hypothesis. (4) **Data**, the results obtained from the experiment, are analyzed and interpreted to reach a conclusion. (5) The conclusion is shared with the scientific community.

4 How We Handle Environmental Problems 20

1. Addressing environmental problems ideally requires five stages. (1) Scientific assessment involves identifying a potential environmental problem and collecting data to construct a model. (2) Risk analysis evaluates the potential effects of intervention. (3) Public engagement occurs when the results of scientific assessment and risk analysis are placed in the public arena. (4) In political considerations, elected or appointed officials implement a particular risk-management strategy. (5) Long-term environmental management monitors the effects of the action taken.

Key Terms

- environmental science 16
- highly developed countries 4
- less developed countries 6
- moderately developed countries 6
- nonrenewable resources 7
- poverty 4
- renewable resources 7
- scientific method 18
- sustainability 12

What is happening in this picture?

- What are these people protesting?

- How is this opposition an example of NIMBY?

- What sorts of political processes and scientific information might make it easier to find a long-term solution to nuclear waste management?

© David H. Wells/CORBIS

Critical and Creative Thinking Questions

1. Provide arguments for and against the following statement: "Population growth in developing countries is of much more concern than is population growth in highly developed countries."

2. Why is population growth often linked to excess resource extraction and consumption?

3. Explain why a country with the world's largest level of consumption may not have the largest population.

4. Explain how population, affluence, and technology interact in complex ways.

5. Do you think our current worldwide population growth and economic growth are sustainable? Why or why not?

6. Give at least two examples of things that you can do as an individual to promote sustainability.

7. How does the field of environmental science involve science? economics? politics?

8. Your throat feels scratchy, and you think you're coming down with a cold. You take a couple of vitamin C pills and feel better. You conclude that vitamin C helps prevent colds. Is your conclusion valid from a scientific standpoint? Why or why not?

9. People want scientists to give them precise, definitive answers to environmental problems. Explain why this is not possible, and why science is, nonetheless, useful for managing environmental problems.

10. Examine the graph, which shows an estimate of the discrepancy between the wealth of the world's poorest countries and that of the richest countries.

a. How has the distribution of wealth changed from the 1800s to the present? How would you explain this difference?

b. Based on the trend evident in this graph, predict what the graph might look like in 100 years.

c. Some economists think that our current path of economic growth is unsustainable. Do the data in this graph support or refute this idea? Explain your answer.

Sustainable Citizen Question

11. Making effective personal and professional decisions requires access to appropriate, high-quality information. In this chapter, we have described how the scientific process can inform decisions. Where would you go for information to make decisions about the environment? How would you rank the following as trusted sources? Which of these do you rely on?

- Friends and family
- Blogs and wiki sites
- News media (Web sites, newspapers, magazines, television, radio)
- Scientific journals (see image)
- Textbooks
- Political leaders

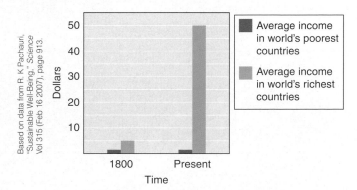

Based on data from R. K. Pachauri, "Sustainable Well-Being," *Science* Vol 315 (Feb 16 2007), page 913.

Legend:
- Average income in world's poorest countries
- Average income in world's richest countries

(Graph: Dollars vs Time, 1800 and Present)

Rural Nevada and Climate Change: Vulnerability, Beliefs, and Risk Perception

Ahmad Saleh Safi,[1,*] William James Smith, Jr.,[2] and Zhnongwei Liu[3]

In this article, we present the results of a study investigating the influence of vulnerability to climate change as a function of physical vulnerability, sensitivity, and adaptive capacity on climate change risk perception. In 2008/2009, we surveyed Nevada ranchers and farmers to assess their climate change-related beliefs, and risk perceptions, political orientations, and socioeconomic characteristics. Ranchers' and farmers' sensitivity to climate change was measured through estimating the proportion of their household income originating from highly scarce water-dependent agriculture to the total income. Adaptive capacity was measured as a combination of the Social Status Index and the Poverty Index. Utilizing water availability and use, and population distribution GIS databases; we assessed water resource vulnerability in Nevada by zip code as an indicator of physical vulnerability to climate change. We performed correlation tests and multiple regression analyses to examine the impact of vulnerability and its three distinct components on risk perception. We find that vulnerability is not a significant determinant of risk perception. Physical vulnerability alone also does not impact risk perception. Both sensitivity and adaptive capacity increase risk perception. While age is not a significant determinant of it, gender plays an important role in shaping risk perception. Yet, general beliefs such as political orientations and climate change-specific beliefs such as be-lieving in the anthropogenic... climate change are... affecting the... observed

Safi, Ahmad Saleh; William James Smith, Jr., and Zhnongwei Lui, "Rural Nevada and Climate Change: Vulnerability, Beliefs, and Risk Perception," *Risk Analysis*, Volume 32, Issue 6, June, 2012: 1041-1059.

THE PLANNER

Sustainability and Human Values

A SUSTAINABILITY ETHIC

In 1962, Rachel Carson's book *Silent Spring* questioned the wisdom of the rapidly increasing applications of pesticides and fertilizers on the world's crops. She envisioned a world in which incautious application of chemicals to increase crop yields in the short term would undermine Earth's capacity to provide resources essential to human well-being—clean air, fresh water, and fertile soil. Carson's compelling description of the environment as a network of complex and interrelated systems is credited as a foundation of the modern environmental movement.

Carson's idea that we need to consider trade-offs between short-term gains and long-term *sustainability* continues to resonate as we face the challenge of feeding a human population that exceeds 7 billion and continues to grow. Humans consume an increasing fraction of the world's land and ocean food resources, and rely on technology-intensive methods to do so. In addition to chemical inputs, diets in developed countries use great amounts of energy (see inset). For example, the photograph shows an entire shoal of tuna caught in a purse seine and surrounded by a towing cage that will take them to a tuna farm for fattening and harvest.

Making effective decisions about energy, chemicals, food, and the environment requires that we account for ethics and values, the subjects of this chapter. As you read, keep in mind these words from the Earth Charter, formulated in 1992 by representatives from 178 countries: "Let ours be a time remembered for the awakening of a new reverence for life, the firm resolve to achieve sustainability, the quickening of the struggle for justice and peace, and the joyful celebration of life."

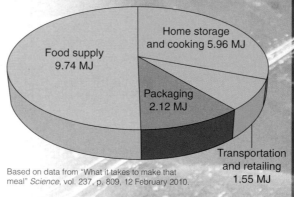

Based on data from "What it takes to make that meal" *Science*, vol. 237, p. 809, 12 February 2010.

A study in the United Kingdom found that producing food requires nearly five times as much energy as is contained in the food. This chart shows how much energy is used for each 1000 Calories (4.184 megajoules) of food consumed.

WileyPLUS

CHAPTER OUTLINE

Human Use of the Earth   28
- Sustainable Consumption

Human Values and Environmental Problems   31
- Worldviews

Environmental Justice   35
- Environmental Justice and Ethical Issues

An Overall Plan for Sustainable Living   36
- ■ Environmental InSight: A Plan for Sustainable Living
- Recommendation 1: Eliminate Poverty and Stabilize the Human Population
- Recommendation 2: Protect and Restore Earth's Resources
- Recommendation 3: Provide Adequate Food for All People
- Recommendation 4: Mitigate Climate Change
- Recommendation 5: Design Sustainable Cities
- ■ Case Study: The Loess Plateau in China

© Gavin Newman/Alamy

CHAPTER PLANNER ✓

- ❑ Study the picture and read the opening story
- ❑ Scan the Learning Objectives in each section:
  p. 28 ❑   p. 31 ❑   p. 35 ❑   p. 36 ❑
- ❑ Read the text and study all figures and visuals. Answer any questions.

Analyze key features

- ❑ Environmental InSight, p. 37
- ❑ Process Diagram, p. 42
- ❑ Case Study, p. 44
- ❑ Stop: Answer the Concept Checks before you go on:
  p. 30 ❑   p. 34 ❑   p. 36 ❑   p. 43 ❑

End of Chapter

- ❑ Review the Summary and Key Terms.
- ❑ Answer What is happening in this picture?
- ❑ Answer the Critical and Creative Thinking Questions.

Human Use of the Earth

LEARNING OBJECTIVES

1. **Define** *sustainable development.*
2. **Outline** some of the complexities associated with the concept of sustainable consumption.
3. **Contrast** voluntary simplicity and technological progress.

E nvironmental sustainability is a concept that people have discussed for many years. *Our Common Future*, the 1987 report of the U.N. World Commission on Environment and Development, presented the closely related concept of **sustainable development** (**Figure 2.1**). The authors of *Our Common Future* pointed out that sustainable development includes meeting the needs of the world's poor. The report also linked the environment's ability to meet present and future needs to the state of technology and social organization existing at a given time and in a given place. The number of people, their degree of affluence (that is, their level of consumption), and their choices of technology all interact to produce the total effect of a given society, or of society at large, on the sustainability of the environment.

> **sustainable development**
> Economic growth that meets the needs of the present without compromising the ability of future generations to meet their needs.

Even using the best technologies imaginable, Earth's productivity still has limits, and our use of it can't be expanded indefinitely. *Sustainable development can occur only within the limits of the environment.* To live within these limits, population growth must be held at a level that we can sustain, and we must identify ways to maintain our high standard of living while consuming far fewer resources. The world does not contain nearly enough resources to sustain everyone at the level of consumption that is enjoyed in the United States, Europe, and Japan. Suitable strategies, however, do exist to reduce these levels of consumption without concurrently reducing the real quality of life.

Sustainable Consumption

As you saw in Chapter 1, pollution and degradation of the environment are exacerbated as individuals in a population consume larger amounts of resources. People living in highly developed nations typically consume disproportionately large shares of Earth's resources and contribute

Sustainable development • Figure 2.1

Three factors—environmentally sound decisions, economically viable decisions, and socially equitable decisions—interact to promote sustainable development.

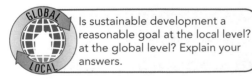

Is sustainable development a reasonable goal at the local level? at the global level? Explain your answers.

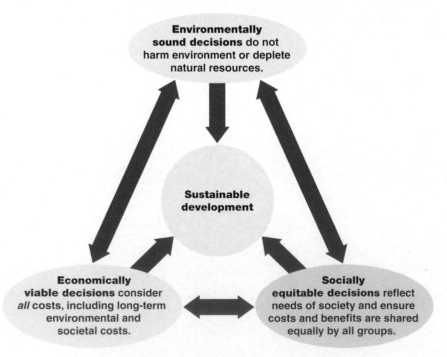

The challenge of eradicating poverty • Figure 2.2

a. A homeless man eats lunch at a mission in Seattle, Washington. Seattle's homeless population is estimated at 8000.

b. Garbage litters the yard around an orphanage in Kolkata (formerly called Calcutta) India.

Global Locator

INDIA

NG Maps

NATIONAL GEOGRAPHIC

© Dan Lamont/Corbis

Johnny Haglund/Lonely Planet Images/Getty Images, Inc.

disproportionately to environmental degradation. *Lifestyle* is interpreted broadly to include goods and services bought for food, clothing, housing, travel, recreation, and entertainment. In evaluating consumption, all aspects of the production, use, and disposal of these goods and services are taken into account, including environmental costs. Such an analysis provides a sense of what it means to consume sustainably versus unsustainably.

Sustainable consumption, like sustainable development, forces us to address whether our present actions undermine the long-term ability of the environment to meet the needs of future generations. Factors that affect sustainable consumption include population, economic activities, technology choices, social values, and government policies.

At the global level, sustainable consumption requires the eradication of poverty (**Figure 2.2**). Indeed, the term *sustainability* was originally adopted to acknowledge that developing countries should not be expected to avoid or reduce environmental damage when to do so hindered socially equitable economic development. This in turn requires that poor people *increase* their consumption of certain

> **sustainable consumption**
> The use of goods and services that satisfy basic human needs and improve the quality of life but that also minimize resource use.

essential resources. For their increased consumption to be sustainable, however, the consumption patterns of people in highly developed countries must change.

Widespread adoption of sustainable consumption will not be easy. It will require major changes in the consumption patterns and lifestyles of most people in highly developed countries. Some examples of promoting sustainable consumption include switching from motor vehicles to public transport and bicycles and developing durable, repairable, recyclable products.

An increasing number of people in the United States and other highly developed nations have embraced a type of sustainable consumption known as **voluntary simplicity**, which recognizes that individual happiness and quality of life are not necessarily linked to the accumulation of material goods. People who embrace voluntary simplicity recognize that a person's values and character define that individual more than how many things he or she owns. This belief requires a change in behavior as people purchase and use fewer items than they might have formerly. It is a commitment at the individual level to saving the planet for future generations.

Light bulb options • Figure 2.3

The incandescent light bulb has been used in the United States for over 130 years with little change between the original made by Thomas Edison (**a**) and the modern bulb (**b**). However, compact fluorescent (**c**) and light-emitting diode (**d**) technologies provide equivalent lighting with a fraction of the energy consumption.

SSPL/Hulton Archive/Getty Images   ElementalImaging/iStockphoto   CGinspiration/iStockphoto   GIPhotoStock/Science Source Images

a   b   c   d

One example of voluntary simplicity is car sharing. Car-sharing programs, which are designed for people who use a car occasionally, offer an economical alternative to individual ownership. Car sharing may also reduce the numbers of cars manufactured. Studies show that most car sharers drive significantly fewer miles than they did before they joined the program.

As people adopt new lifestyles, they must be educated so that they understand the reasons for changing practices that may be highly ingrained or traditional. Formal education and informal education are both important in bringing about change and in contributing to sustainable consumption. If people understand the way the natural world functions, they can appreciate their own place in it and value sustainable actions.

While some individuals choose sustainable consumption and voluntary simplicity, many people do not. Many equate these ideas with unnecessary sacrifice, and object to the idea of compulsory reductions in consumption. However, many scientists and population experts increasingly advocate a shift to sustainable consumption now, before it is forced on us by an environmentally degraded, resource-depleted world.

Continued technological progress represents a promising opportunity for maintaining high standards of living while using fewer resources. One good example is the transition from incandescent light bulbs to compact fluorescent light bulbs and light-emitting diodes (LEDs) (**Figure 2.3**). These shifts are driven by both policy and economics. In Australia, for example, traditional incandescent bulbs have been banned, and some policy makers in the United States promote similar policies. In many places, electricity providers promote the adoption of lower-energy light bulbs by households. And many businesses have made the change on their own, finding that the higher cost of the new bulbs is offset by reduced energy costs within the first year or two. Any long-term involvement in the condition of the world must start with individuals—our values, attitudes, and practices. Each of us makes a difference, and it is ultimately our collective activities that make the world what it is.

CONCEPT CHECK

1. **What** is sustainable development?

2. **What** is sustainable consumption? How is it linked to a reduction in world poverty?

3. **How** do voluntary simplicity and technological progress contribute to sustainable consumption?

Human Values and Environmental Problems

LEARNING OBJECTIVES

1. **Define** *environmental ethics.*
2. **Discuss** distinguishing features of the Western and deep ecology worldviews.

We now shift our attention to the views of different individuals and societies and how those views affect our ability to understand and solve sustainability problems. **Ethics** is the branch of philosophy that is derived through the logical application of human **values**. These values are the principles that an individual or a society considers important or worthwhile. Values are not static entities but change as societal, cultural, political, and economic priorities change. Ethics helps us determine which forms of conduct are morally acceptable and unacceptable, right and wrong. Ethics plays a role in any types of human activities that involve intelligent judgment and voluntary action. Whenever alternative, conflicting values occur, ethics helps us choose which value is better, or worthier, than other values.

Environmental ethics examines moral values to determine how humans should relate to the natural environment. Environmental ethicists consider such issues as what role we should play in determining the fate of Earth's resources, including other species, or how we might develop an environmental ethic that is acceptable in the short term for us as individuals and also in the long term for our species and the planet. These issues and others like them are difficult intellectual questions that involve political, economic, societal, and individual trade-offs.

Environmental ethics considers not only the rights of people living today, both individually and collectively, but also the rights of future generations (**Figure 2.4**). This aspect of environmental ethics is critical because the impacts of today's activities and technologies are changing the environment. In some cases these impacts may be

environmental ethics A field of applied ethics that considers the moral basis of environmental responsibility.

Richard Nowitz/NG Images

Tomorrow's generation • Figure 2.4

The choices made today will determine whether future generations, such as these students from Bailey Elementary School in Falls Church, Virginia, will inherit a sustainable world.

felt for hundreds or even thousands of years. Addressing issues of environmental ethics puts us in a better position to use science and technology for long-term environmental sustainability.

Worldviews

Each of us has a particular **worldview**—that is, a personal perspective based on a collection of our basic values that helps us make sense of the world, understand our place and purpose in it, and determine right and wrong behaviors. These worldviews lead to behaviors and lifestyles that may or may not be compatible with environmental sustainability.

Two extreme **environmental worldviews** are the Western worldview and the deep ecology worldview. These two worldviews, admittedly broad generalizations, are at nearly opposite ends of a spectrum of worldviews relevant to global sustainability problems, and they approach environmental responsibility in radically different ways.

The traditional **Western worldview**, also known as the *expansionist worldview*, is human centered and utilitarian. It mirrors the beliefs of the 19th-century **frontier attitude**, a desire to conquer and exploit nature as quickly as possible (**Figure 2.5**). The Western worldview

also advocates the inherent rights of individuals, accumulation of wealth, and unlimited consumption of goods and services to provide material comforts. According to the Western worldview, humans have a primary obligation to humans and are therefore responsible for managing natural resources to benefit human society. Thus, any concerns about the environment are derived from human interests.

The Western worldview is for many an entrenched belief. It is buttressed by the observation that in most highly developed countries, expanded use of natural resources has historically been closely associated with improvements in quality of life. The **deep ecology worldview** is a diverse set of viewpoints that dates from the 1970s and is based on the work of **Arne Naess**, a Norwegian philosopher,

Western worldview • Figure 2.5

a. Logging operations in 1884. This huge logjam occurred on the St. Croix River near Taylors Falls, Minnesota.

© Minnesota Historical Image Collection/Corbis

Global Locator

NORTH AMERICA

BRAZIL

SOUTH AMERICA

NG Maps

NATIONAL GEOGRAPHIC

b. The Western worldview in operation today. These logs were cut from plantations of non-native eucalyptus tress, which have replaced 30 million hectares (75 million acres) of tropical rain forest in Brazil's Atlantic forest.

AFP/GETTY IMAGES/Newscom

and others, including ecologist **Bill Devall** and philosopher **George Sessions**. The principles of deep ecology, as expressed by Naess in *Ecology, Community and Lifestyle* (1989), include:

1. Both human and nonhuman life have **intrinsic value** (**Figure 2.6**). The value of nonhuman life forms is independent of the usefulness they may have for narrow human purposes.

2. Richness and diversity of life forms contribute to the flourishing of human and nonhuman life on Earth.

3. Humans have no right to reduce this richness and diversity except to satisfy vital needs.

4. Present human interference with the nonhuman world is excessive, and the situation is rapidly worsening.

5. The flourishing of human life and cultures is compatible with a substantial decrease in the human population. The flourishing of nonhuman life requires such a decrease.

6. Improving human well-being requires economic, technological, and ideological changes.

7. The ideological change is mainly that high quality of life need not be synonymous with high levels of consumption.

8. Those who subscribe to the foregoing points have an obligation to participate in the attempt to implement the necessary changes.

For many people in highly developed countries, the deep ecology worldview represents a radical shift

Philosophers recognize two kinds of value, instrumental and intrinsic • Figure 2.6

a. According to the deep ecology worldview (right side of triangle), organisms have intrinsic value—that is, they are valued for their own sake, not for the goods and services they provide.

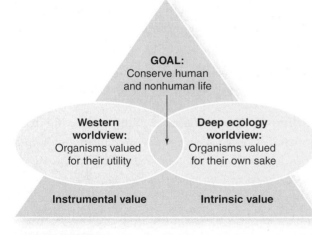

GOAL:
Conserve human
and nonhuman life

Western worldview:
Organisms valued
for their utility

Deep ecology worldview:
Organisms valued
for their own sake

Instrumental value   **Intrinsic value**

Think Critically Why is there an overlapping goal between these two extreme worldviews?

b. A tree trunk has grown around the head of Buddha at Wat Mahathat in Thailand, symbolizing the oneness of Buddha with nature. Buddhists practice the stilling of human desires, the reduction of consumption, and the contemplation of nature. Like Buddhism, many of the world's other religions espouse the intrinsic value of living things.

Cristina Redinger-Libolt/Botanica/Getty Images, Inc.

Priit Vesilind/NG Image Collection

Embracing deep ecology • Figure 2.7

At one time or another, most of us yearn for the simpler life that the tenets of deep ecology advocate. However, there are far too many people and there is far too little land for us all to embrace this lifestyle. Photographed on Gotland Island, Sweden.

in how humans relate to the environment. The deep ecology worldview stresses that all forms of life have the right to exist and that humans are not different or separate from other organisms. Humans have an obligation to themselves and to all other organisms living on Earth. The deep ecology worldview advocates sharply curbing human population growth. It does not advocate returning to a society free of today's technological advances but instead proposes a significant rethinking of our use of current technologies and alternatives. It asks individuals and societies to share an inner spirituality connected to the natural world.

Most people today do not fully embrace either the Western worldview or the deep ecology worldview. The Western worldview is **anthropocentric** and emphasizes the importance of humans as the overriding concern in the grand scheme of things. In contrast, the deep ecology worldview is **biocentric** and views humans as one species among others. The planet's natural resources could not support its more than 7 billion humans if each consumed the high level of goods and services sanctioned by the Western worldview. On the other hand, the world as envisioned by the deep ecology worldview could support only a fraction of the existing human population (**Figure 2.7**).

These worldviews, while not practical for widespread adoption, are useful to keep in mind as you examine various environmental issues in later chapters. In the meantime, you should think about your own worldview and discuss it with others—whose worldviews will probably be different from your own.

As you study this book, consider the following questions:

What is your worldview? Is it closer to a Western worldview or a deep ecology worldview?

What are the short-term and long-term consequences of your worldview for economic, social, and environmental well-being?

In what ways could you maintain or improve your own quality of life while consuming fewer resources?

According to **Robert Cahn**, a 20th-century environmental journalist: *The main ingredients of an environmental ethic are caring about the planet and all of its inhabitants, allowing unselfishness to control the immediate self-interest that harms others, and living each day so as to leave the lightest possible footprints on the planet.* We must develop and incorporate into our culture a long-lasting, environmentally sensitive worldview if the environment is to be sustainable for us, for other living organisms, which are linked to us through a long evolutionary history, and for future generations of both human beings and other life forms.

CONCEPT CHECK 🛑 STOP

1. **What** is environmental ethics?
2. **What** assumptions underlie the Western worldview? the deep ecology worldview?

Environmental Justice

LEARNING OBJECTIVE

1. Define *environmental justice.*

In the early 1970s, the Board of National Ministries of the American Baptist Churches coined the term *eco-justice* to link social and environmental ethics. At the local level, eco-justice encompasses environmental inequities faced by low-income minority communities. Many studies indicate that low-income communities and/or communities of color are more likely than others to have chemical plants, hazardous waste facilities, sanitary landfills, sewage treatment plants, and incinerators (**Figure 2.8**). A 1990 study at Clark Atlanta University, for example, found that six of Houston's eight incinerators were located in predominantly black neighborhoods. Such communities often have limited involvement in the political process and may not even be aware of their exposure to increased levels of pollutants.

Because people in low-income communities frequently lack access to sufficient health care, they may not be treated adequately for exposure to environmental contaminants. The high incidence of asthma in many minority communities, for example, may be caused or exacerbated by exposure to environmental pollutants. Many studies have examined how environmental pollutants interact with other socioeconomic factors to cause health problems. It is challenging to show to what extent a polluted environment is responsible for the disproportionate health problems of poor and minority communities. For example, a 1997 study of residents in San Francisco's polluted Bayview Hunters Point area found that hospitalization rates for chronic illnesses were the highest in the state. Researchers failed, however, to link these illnesses to an increased exposure to toxic pollutants, in part because an illness caused by exposure to pollution is clinically identical to the same illness that is caused by factors other than pollution.

In addition to their increased exposure to pollution, low-income communities may not receive equal benefits from federal cleanup programs. Several studies have reported that toxic waste sites in white communities were cleaned up faster and more thoroughly than those in Latino and African-American communities.

Cases in the 1990s like the Houston incinerators and the Bayview Hunters Point illnesses led then President Clinton to sign an executive order requiring that all

A children's playground overlooks a pulp mill • Figure 2.8

Poor minority neighborhoods often have the most polluted and degraded environments. Photographed in Kingsport, Tennessee.

Jenny Hager/The Image Works

federal agencies consider environmental justice as they make planning decisions. Nonetheless, environmental injustice continues; throughout the world, the poor tend to bear greater environmental burdens than the wealthy. For example, in oil-rich Nigeria, where per capita income hovers around $2000 per year, people who live around oil extraction and refining equipment are exposed to elevated levels of air and water pollution on a regular basis. Nigerians often face more environmental downsides and fewer economic benefits from oil extraction than do wealthy people elsewhere in the world.

> **environmental justice** The right of every citizen to adequate protection from environmental hazards.

Environmental Justice and Ethical Issues

There is an increasing awareness that environmental decisions such as where to locate a hazardous waste landfill have important ethical dimensions. The most basic ethical dilemma centers on the rights of the poor and disenfranchised versus the rights of the rich and powerful. Whose rights should have priority in these decisions? Is it ethically just if environmental burdens and benefits are not equally shared?

The challenge is to find and adopt solutions that respect all individuals, including those yet to be born.

Environmental justice is a fundamental human right in an ethical society. Although we may never completely eliminate past environmental injustices, we have a moral imperative to prevent them today so that their negative effects do not disproportionately affect any particular segment of society.

In response to these concerns, a growing environmental justice movement has emerged at the grassroots level as a strong motivator for change. Advocates are calling for special efforts to clean up hazardous sites in low-income neighborhoods, from inner-city streets to Native American reservations. On an international level, advocates of environmental justice point out that industrialized countries are obligated to help less-developed countries cope with climate change. These countries often suffer disproportionately from the problems caused by climate change, while it is the fossil fuel consumption in highly developed countries that is largely responsible for the changing climate.

CONCEPT CHECK 🛑 STOP

1. **What** is environmental justice, and which communities are exposed to a disproportionate share of environmental hazards?

An Overall Plan for Sustainable Living

LEARNING OBJECTIVES

1. **Relate** poverty and population growth to carrying capacity and global sustainability.
2. **Discuss** problems related to loss of forests and declining biological diversity.
3. **Describe** the extent of food insecurity.
4. **Define** *enhanced greenhouse effect* and explain how stabilizing climate is related to energy use.
5. **Describe** at least two problems in cities in the developing world.

There is no shortage of suggestions for ways to address the world's many environmental problems. We have organized this section around the five recommendations for sustainable living presented in the 2006 book *Plan B 2.0: Rescuing a Planet Under Stress and a Civilization in Trouble*

by Lester R. Brown. If we as individuals and collectively as governments were to focus our efforts and financial support on Brown's plan, we think the quality of human life would be much improved. Brown's five recommendations for sustainable living are:

1. Eliminate poverty and stabilize the human population.
2. Protect and restore Earth's resources.
3. Provide adequate food for all people.
4. Mitigate climate change.
5. Design sustainable cities.

Seriously addressing these recommendations offers hope for the kind of future we want for our children and grandchildren (**Figure 2.9**).

Barry Iverson/Time Life Pictures/Getty Images

Tim Laman/NG Image Collection

Eco Images/Universal Images Group/Getty Images

Family Planning in Egypt.
A woman at a health clinic in Egypt learns about family planning and birth control.

Recommendation 1: Eliminate Poverty and Stabilize the Human Population.

Restoration in Indonesia.
Mangrove trees are planted at low tide to help restore a coastal estuary.

Recommendation 2: Protect and Restore Earth's Resources.

Feeding the World's People.
Farmers in Cameroon harvest potatoes.

Recommendation 3: Provide Adequate Food for All People.

Richard B. Levine/NewsCom

© Iain Masterton/Alamy

Energy Neutral Construction.
Photovoltaic panels cover a south facing wall of the Solaire, a building in Battery Park City, New York. Increasingly, energy is generated at buildings where it is used.

Recommendation 4: Mitigate Climate Change.

WileyPLUS

Bicycle Rack in Amsterdam.
Residents in the Netherlands ride bicycles an average of 573 mi (917 km) per year.

Recommendation 5: Design Sustainable Cities.

Recommendation 1: Eliminate Poverty and Stabilize the Human Population

The ultimate goal of economic development is to make it possible for humans throughout the world to enjoy long, healthy lives. A serious complication lies in the fact that the distribution of the world's resources is unequal. Residents of the United States are collectively the wealthiest people who have ever existed, with the highest standard of living (shared with a few other rich countries). The United States, with fewer than 5 percent of the world's people, controls about 25 percent of the world's economy but depends on other nations for this prosperity. Yet we often seem unaware of this relationship and tend to underestimate our effects on the environment that supports us.

Failing to confront the problem of poverty around the world makes it impossible to attain global sustainability. For example, most people would find it unacceptable that about 24,000 infants and children under age 5 die each day (2008 data from U.N. Children's Fund). Most of these deaths could have been prevented through access to adequate food and basic medical techniques and supplies. For us to allow so many to go hungry and to live in poverty threatens the global ecosystem that sustains us all. Everyone must have a reasonable share of Earth's productivity. As U.S. President **Franklin Delano Roosevelt** said in his second inaugural address in 1937, "The test of our progress is not whether we add more to the abundance of those who have much; it is whether we provide enough for those who have too little."

Raising the standard of living for poor countries requires the universal education of children and the elimination of illiteracy (**Figure 2.10**). Improving the status of women is crucial because women are often disproportionately disadvantaged in poor countries. In many developing countries, women have few rights and little legal ability to protect their property, their rights to their children, and their income.

We have entered an era of global trade, within which we must establish guidelines for national, corporate, and individual behaviors. For example, the flow of money from developing countries to highly developed countries has exceeded the flow in the other direction for many years. Former West German Chancellor **Willy Brandt** termed this phenomenon "a blood transfusion from the sick to the healthy." A world that values social justice and environmental sustainability must reverse this flow. Debts from the poorest countries should be forgiven more readily than they are now, and international development assistance should be enhanced.

Population growth rates are generally highest where poverty is most intense. If we pay consistent attention to overpopulation and devote the resources necessary to make family planning available for everyone, the human population will stabilize. If we do not continue to

Global Locator

CAMBODIA

NG Maps

NATIONAL GEOGRAPHIC

Children at work • Figure 2.10

These girls are not at school because they are employed as weavers at looms in a workshop. Photographed in Cambodia.

Justin Guariglia/NG Image Collection

emphasize family planning measures, we simply will not achieve population stability.

To stay within Earth's carrying capacity, we must reach and sustain a stable population and reduce excessive consumption. These goals must be coupled with educational programs everywhere, so that people understand that Earth's carrying capacity is not unlimited. There is no hope for a peaceful world without overall population stability, and there is no hope for regional economic sustainability without regional population stability.

> **carrying capacity**
> The maximum population that can be sustained by a given environment or by the world as a whole.

Recommendation 2: Protect and Restore Earth's Resources

To build a sustainable society, we must preserve the natural systems that support us. The conservation of nonrenewable resources, such as oil and minerals, is obvious, although discoveries of new supplies of nonrenewable resources sometimes give the illusion that they are inexhaustible. Renewable resources such as forests, biodiversity, soils, fresh water, and fisheries must be used in ways that ensure their long-term productivity. Their capacity for renewal must be understood and respected. However, renewable resources have been badly damaged over the past 200 years. Until environmental sustainability becomes a part of economic calculations, susceptible natural resources will continue to be consumed unsustainably, driven by short-term economics.

The World's Forests Many of the world's forests are being cut, burned, or seriously altered at a frightening rate. In many parts of the developed world, old-growth forests are rare and becoming more so. Much of England's forests were cut down centuries ago to build ships and produce charcoal for fuel. As North America was colonized, forests were first harvested along the eastern seaboard. With expansion of the United States and Canada, forests on the west coast were exploited for construction, paper making, and fuel.

More recently, deforestation in developing countries, has contributed to climate change and degradation of soils. Tropical forests are particularly threatened by overexploitation. Many products—hardwoods; foods such as beef, bananas, coffee, and tea; and medicines—come to the industrialized world from the tropics. As trees are destroyed, only a small fraction of them are replanted.

The pressure of rapid population growth and widespread poverty also harms the world's forests. In many developing countries, forests have traditionally served as a "safety valve" for the poor, who, by consuming small tracts of forest on a one-time basis and moving on, find a source of food, shelter, and clothing. But now the numbers of people in developing countries are too great for their forests to support.

Tropical rain forests—biologically the world's richest terrestrial areas—have been reduced to less than half their original area. Methods of forest clearing that were suitable when population levels were lower and forests had time to recover from temporary disturbances simply do not work any longer. Forests, if managed carefully, can be a renewable resource. However, unsustainable use of forest resources occurs when more trees are harvested than are replaced, or when areas are replanted with low-biodiversity commercial tree plantations.

Loss of Biodiversity We have a clear interest in protecting Earth's **biological diversity** and managing it sustainably because we obtain

> **biological diversity** The number and variety of Earth's organisms.

from living organisms all our food, most medicines, many building and clothing materials, biomass for energy, and numerous other products. In addition, organisms and the natural environment provide an array of **ecosystem services** without which we would not survive. These services include the protection of watersheds and soils, the development of fertile agricultural lands, the determination of both local climate and global climate, and the maintenance of habitats for animals and plants.

Over the next few decades, we can expect human activities to cause the rate of extinction to increase to perhaps hundreds of species a day. How big a loss is this? Unfortunately, we still have limited knowledge about the world's biological diversity. An estimated five-sixths of all species have not yet been scientifically described. Some 80 percent of the species of plants, animals, fungi, and microorganisms on which we depend are found in developing countries. How will these relatively poor countries sustainably manage and conserve these precious resources? Biological diversity is an intrinsically local problem, and each nation must address it for the sake of its own people's future, as well as for the world at large. Like most other challenges of sustainable development, biological diversity can be addressed adequately only if we provide international assistance where needed, including help in training scientists and engineers from developing countries.

Biological diversity and human cultural diversity are intertwined: They are, in fact, two sides of the same coin. **Cultural diversity** is Earth's variety of human communities, each with its individual languages, traditions, and identities (**Figure 2.11**). Cultural diversity enriches the collective human experience. For that reason, the U.N. Educational, Scientific, and Cultural Organization supports the protection of minorities in the context of cultural diversity.

Recommendation 3:
Provide Adequate Food for All People

Globally, more than 800 million people lack access to the food needed for healthy, productive lives. This estimate, according to a 2011 report by the U.N. Food and Agriculture Organization, includes a high percentage of children. Children are particularly susceptible to food deficiencies because their brains and bodies cannot develop properly without adequate nutrition.

Most malnourished people live in rural areas of the poorest developing nations. The link between poverty and **food insecurity** is inescapable. Since 2002, price increases have contributed to food insecurity worldwide (**Figure 2.12**). In addition, food prices have become more volatile, changing from week to week or day to day. Such uncertainty is particularly challenging for the world's poorest people, who spend disproportionate percentages of their income on food.

Improving agriculture is one of the highest priorities for achieving global sustainability. In general, grain production per person has kept pace with human population growth over the past 50 years. However, expanded agricultural productivity has taken place at high environmental costs. Moreover, the global population continues to expand, putting additional pressure on food production.

Much of Earth's agriculturally suitable land is either already under cultivation or covered by development such as roads and buildings. One way to increase the productivity of agricultural land is through **multicropping**, or growing more than one crop per year. For example,

food insecurity
The condition in which people live with chronic hunger and malnutrition.

Humans are part of the web of life • Figure 2.11

Portrait of a Yanomami father and son in Roraima State, Brazil. The Yanomami are Brazil's last large Stone Age tribe. Intrusion into isolated areas such as the Amazon Basin threatens both biological diversity and the cultures of indigenous people who have lived in harmony with nature for hundreds of generations.

Michael Nichols/NG ImageCollection

Food volatility since 2002 • Figure 2.12

From 1975 until the end of the last century, global food prices were mostly in decline. However, beginning in 2002, food prices have been both increasing and less predictable. Food price index is the price of food as compared to the price in 2002, adjusted for inflation.

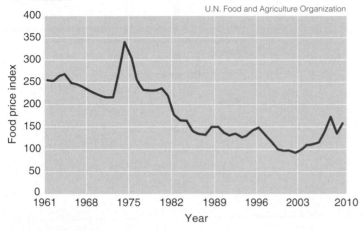

U.N. Food and Agriculture Organization

Interpreting Data

In the early 1970s, global energy prices increased dramatically. What sort of relationship does this graph suggest between prices of food and energy at that time?

Damage to soil resources • Figure 2.13

Erosion is a serious form of soil degradation. Careful stewardship of the land prevents such damage. Photographed in West Pokot, Kenya.

© Alexandra E. Jones; Ecoscene/Corbis

winter wheat and summer soybean crops are grown in some areas of the United States. However, multicropping can be accomplished only in regions where water supplies are adequate for irrigation. Also, care must be taken to prevent a decline in soil fertility from such intensive use and avoid further harm from chemical and energy inputs.

The negative environmental effects of agriculture, including loss of soil fertility, soil erosion, aquifer depletion, and soil, water, and air pollution, must be brought under control (**Figure 2.13**). Many strategies exist to retard the loss of topsoil, conserve water, conserve energy, and reduce the use of agricultural chemicals. For example, in **conservation tillage**, residues from previous crops are left in the soil, partially covering it and helping to hold topsoil in place.

We must develop sustainable agricultural systems that provide improved dietary standards, such as the inclusion of high-quality protein in diets in developing countries. China's expanding use of **aquaculture** is an example of efficient protein production. The carp that are raised in Chinese aquaculture are efficient at converting food into high-quality protein. In China, fish production by aquaculture now exceeds poultry production. However, aquaculture, like all other human endeavors, has negative environmental effects that must be addressed for it to be sustainable on a large scale.

Recommendation 4: Mitigate Climate Change

A widely discussed human effect on the environment is climate change caused by the **enhanced greenhouse effect**. Both highly developed and developing countries contribute to major increases in CO_2 in the atmosphere, as well as to the increasing amounts of methane, nitrous oxide, tropospheric ozone, and CFCs. The most important greenhouse gas, CO_2, is produced when we burn fossil fuels—coal, oil, and natural gas.

> **enhanced greenhouse effect** The additional warming produced by increased levels of gases that absorb infrared radiation.

Although Earth's climate has been relatively stable during the past 10,000 years, human activities are causing it to change. The average global temperature increased by over 1° Celsius during the past century; more than half of that warming occurred since 1975. Precipitation

Cascading responses of increased carbon dioxide through the environment • Figure 2.14

✓ THE PLANNER

1 Most people know that an increase in atmospheric CO_2 leads to global warming, but this phenomenon is far from a simple cause-and-effect relationship. Increasing CO_2 may cause a cascade of interacting responses throughout the Earth system.

Coral bleaching

2 Effects of increased atmospheric CO_2 on the ocean.

Peter Scoones/Science Source

Increase in atmospheric CO_2

Climate warming

Increase in dissolved CO_2 in ocean

Changes in precipitation patterns

Changes in plant growth

Melting of arctic tundra

Changes in ocean chemistry (more acidic)

Changes in plant community composition

Release of methane (CH_4)

Harm to corals and animals with shells

3 Effects of increased atmospheric CO_2 on land plants and animals.

Changes in animal community composition

4 Positive feedbacks, such as release of methane from melting tundra, accelerates climate change

Changes in ocean food web dynamics

Changes in terrestrial food web dynamics

Increased extinctions

Increased extinctions

Think Critically Where do human activities fit into this diagram?

---

patterns have shifted in many places. Climate scientists generally agree that Earth's climate will continue to change rapidly during the 21st century.

These changes will likely have serious effects because Earth's organisms, as well as modern society, have evolved and successfully adapted to conditions as they are. Keeping in mind that the change from the last ice age to the present was accompanied by an increase in global temperature of 5° Celsius puts the consequences of the present change, the most rapid of the past 10,000 years, into perspective.

We often say that an increase in atmospheric CO_2 leads to climate warming, and this is true. However, the increase in CO_2, like other human impacts, is not

a simple cause-and-effect relationship but instead a cascade of interacting responses that ripple through the environment (**Figure 2.14**). We cannot begin to predict how these changes will affect humans or other organisms.

Stabilizing the climate requires a comprehensive energy plan to include phasing out fossil fuels in favor of renewable energy (such as solar and wind power), increasing energy conservation, and improving energy efficiency. In addition, stopping and reversing the destruction of rain forests is critical to storing carbon in trees. Many national and local governments as well as corporations, colleges and universities, and environmentally aware individuals are setting goals to cut

Recommendation 5: Design Sustainable Cities

At the beginning of the Industrial Revolution, in approximately 1800, only 3 percent of the world's people lived in cities and 97 percent were rural, living on farms or in small towns. In the two centuries since then, population distribution has changed radically—toward the cities. More people live in Mexico City today than were living in all the cities of the world 200 years ago. This is a staggering difference in the way people live. Over 50 percent of the world's population now lives in cities, and the percentage continues to grow. In industrialized countries such as the United States and Canada, almost 80 percent of people live in cities.

City planners around the world are trying a variety of approaches to make cities more livable. Many cities are developing urban transportation systems to reduce the use of cars and the problems associated with them, such as congested roads, large areas devoted to parking, and air pollution. Urban transportation ranges from mass transit subways and light rail systems to pedestrian and bicycle pathways.

Investing in urban transportation in ways other than building more highways encourages commuters to use forms of transportation other than automobiles. To encourage mass transit, some cities also tax people using highways into and out of cities during business hours. When a city is built around people instead of cars—such as establishing parks and open spaces instead of highways and parking lots—urban residents gain an improved quality of life. Air pollution, including the emission of climate-warming CO_2, is substantially reduced.

Water scarcity is a major issue for many cities of the world. Some city planners think that innovative approaches must be adopted where water resources are scarce. These approaches would replace the traditional one-time water use that involves water purification before use, treatment of sewerage and industrial wastes after use, and then discharge of the treated water. For example, certain places, such as cities like Singapore, recycle some of their wastewater after it has been treated.

Effectively dealing with the problems in squatter settlements is an urgent need. Evicting squatters does not address the underlying problem of poverty. Instead,

© Peter Treanor/Alamy

Squatter settlement • Figure 2.15

Manila, in the Philippines, is a city of contrasts, with gleaming modern skyscrapers and abjectly poor squatter settlements.

cities should incorporate some sort of plan for the eventual improvement of squatter settlements (**Figure 2.15**). Providing basic services—such as clean water to drink, transportation (so people can find gainful employment), and garbage pickup—would help improve the quality of life for the poorest of the poor.

CONCEPT CHECK STOP

1. **What** is the global extent of poverty?
2. **What** are two ecosystem services provided by natural resources such as forests and biological diversity?
3. **What** is food insecurity?
4. **How** is stabilizing climate related to energy use? Deforestation?
5. **What** are two serious problems in urban environments?

The Loess Plateau in China

The Loess Plateau covers about 640,000 km² (247,000 mi²) in east-central China. This area is named for **loess**, the fine-grained, silty soil deposited there by windstorms following the retreat of ice age glaciers. (*Loess*, pronounced "luss" in the United States, is derived from a German word meaning "loose.") Loess is a fine-grained sedimentary deposit found in many areas of the world. However, it is thickest and most extensive in China. The Loess Plateau covers much of the North China Plain and, to the west, the hilly basin of the Yellow River. It averages 75 m (250 ft) thick.

The loess, which is thick and fertile, was at one time an important resource for China. It provided a fertile agricultural soil that fed millions of people. Chinese people also dug homes in the loess; these homes were cool in summer and warm in winter, although they were prone to collapse from earthquakes.

Loess is easily eroded by wind and water, particularly when vegetation is removed from the surface. The Loess Plateau is semiarid, so water is often in short supply. Lack of water, in combination with centuries of deforestation and overgrazing, turned much of the Loess Plateau into a nonproductive desert.

In 1994 the Loess Plateau Watershed Rehabilitation Project was established to reclaim the land from encroaching desert. Portions of the Loess Plateau were reforested (see photograph). Chinese people living in the area were educated about the causes of land degradation and encouraged to keep their livestock in pens instead of allowing the animals to roam freely and overgraze the land. As portions of the Loess Plateau have slowly recovered, it is turning green again, and less silt is washing into the Yellow River.

Global Locator

NG Maps

CHINA

NATIONAL GEOGRAPHIC

Jim Richardson/NG Image Collection

Tree seedlings have been planted in small earthworks as part of a plan to reduce erosion and restore the hills in the Loess Plateau.

Summary

1 Human Use of the Earth 28

1. **Sustainable development** is economic growth that meets the needs of the present without compromising the ability of future generations to meet their own needs. Environmentally sound decisions, economically viable decisions, and socially equitable decisions interact to promote sustainable development.

2. **Sustainable consumption** is the use of goods and services that satisfy basic human needs and improve the quality of life but also minimize the use of resources so they are available for future use.

3. **Voluntary simplicity** recognizes that individual happiness and quality of life are not necessarily linked to the accumulation of material goods. Technological progress, whether driven by policy or economics, can contribute to a high quality of life while putting fewer demands on Earth's resources.

2 Human Values and Environmental Problems 31

1. **Environmental ethics** is a field of applied ethics that considers the moral basis of environmental responsibility and how far this responsibility extends. Environmental ethicists consider how humans should relate to the natural environment.

2. **An environmental worldview** is a worldview that helps us make sense of how the environment works, our place in the environment, and right and wrong environmental behaviors. The **Western worldview** is an understanding of our place in the world based on human superiority and dominance over nature, the unrestricted use of natural resources, and increased economic growth to manage an expanding industrial base. The **deep ecology worldview** is an understanding of our place in the world based on harmony with nature, a spiritual respect for life, and the belief that humans and all other species have equal worth.

3 Environmental Justice 35

1. **Environmental justice** is the right of every citizen, regardless of age, race, gender, social class, or other factor, to adequate protection from environmental hazards. Environmental justice is a fundamental human right in an ethical society. A growing environmental justice movement has emerged at the grassroots level. Globally, environmental justice includes promoting economic development without imposing disproportionate environmental risks.

4 An Overall Plan for Sustainable Living 36

1. Failing to confront the problem of poverty makes it impossible to attain global sustainability. To stay within Earth's **carrying capacity**, the maximum population that can be sustained indefinitely, it will be necessary to reach a stable population and reduce excessive consumption.

© Minnesota Historical Image/Corbis

2. The world's forests are being cut, burned, and seriously altered for timber and other products that the global economy requires. Also, rapid population growth and poverty are putting pressure on forests. **Biological diversity**, the number and variety of Earth's organisms, is declining at an alarming rate. Humans are part of Earth's web of life and are entirely dependent on that web for survival.

3. **Food insecurity** is the condition in which people live with chronic hunger and malnutrition. Globally, more than 800 million people lack access to the food needed for healthy, productive lives.

4. The **enhanced greenhouse effect** is the additional warming produced by increased levels of gases that absorb infrared radiation. An increase in atmospheric CO_2, mostly produced when fossil fuels are burned and rain forests are destroyed, leads to climate warming. To stabilize climate, we must phase out fossil fuels in favor of renewable energy, increased energy conservation, and improved energy efficiency, and reduce or reverse deforestation.

5. The air in cities in the developing world is badly polluted with exhaust from motor vehicles. Illegal squatter settlements proliferate in cities; the poorest inhabitants build dwellings using whatever materials they can scavenge. Squatter settlements have the worst water, sewage, and solid waste problems.

Key Terms

- biological diversity 39
- carrying capacity 39
- deep ecology worldview 32
- enhanced greenhouse effect 41

- environmental ethics 31
- environmental justice 36
- environmental worldview 32
- food insecurity 40

- sustainable consumption 29
- sustainable development 28
- Western worldview 32

What is happening in this picture?

- These fishermen are pulling up a net of jellyfish, which have proliferated, harming local fish populations. Suggest a possible reason that jellyfish swarms have become so common.

- Given that pollution and climate change are being blamed for the increase in jellyfish, propose a plan to correct the problem. Will your plan be a quick fix, or will it take many years to address? Why?

- Where would you put "Proliferating jellyfish swarms" in Figure 2.14?

© Junji Kurokawa/AP/Wide World Photos

Critical and Creative Thinking Questions

1. Development is sometimes equated with economic growth. Explain the difference between sustainable development and development as an indicator of economic growth, using the figure shown to the right.

2. How are sustainable consumption and voluntary simplicity related?

3. How do the three factors shown in the figure interact to promote sustainable development?

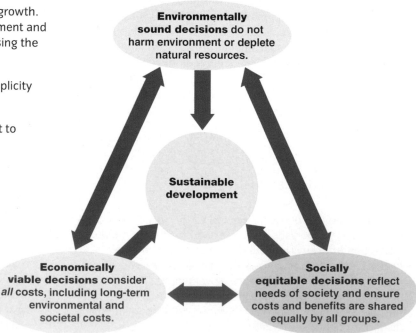

Environmentally sound decisions do not harm environment or deplete natural resources.

Sustainable development

Economically viable decisions consider *all* costs, including long-term environmental and societal costs.

Socially equitable decisions reflect needs of society and ensure costs and benefits are shared equally by all groups.

4. State whether each of the following statements reflects the Western worldview, the deep ecology worldview, or both. Explain your answers.

   a. Species exist to be used by humans.

   b. All organisms, humans included, are interconnected and interdependent.

   c. There is a unity between humans and nature.

   d. Humans are a superior species capable of dominating other organisms.

   e. Humans should protect the environment.

   f. Nature should be used, not preserved.

   g. Economic growth will help Earth manage an expanding human population.

   h. Humans have the right to modify the environment to benefit society.

   i. All forms of life are intrinsically valuable and therefore have the right to exist.

5. What social groups generally suffer the most from environmental pollution and degradation? What social groups generally benefit from this situation?

6. Why is human population control an important part of global sustainability?

7. How is forest destruction related to declining biological diversity?

8. What is food insecurity? How does food insecurity affect the environment?

9. Discuss two ways to make cities more sustainable.

10–12. The graphs below show a computer simulation by the U.S. National Climate Assessment. In **(a)**, the level of atmospheric CO_2 is projected for the 21st century. As a result of increasing levels of CO_2 in the atmosphere, more CO_2 dissolves in ocean water, where it forms carbonic acid. In **(b)** we can see that the increasing acidity dissolves and weakens coral skeletons, which are composed of calcium carbonate. (Values in **a** and **b** are midrange projections.)

10. Why could rising CO_2 levels in the atmosphere be catastrophic to corals and other shell-forming organisms?

11. How do these graphs relate to Figure 2.14?

Sustainable Citizen Question

12. How might the loss of corals and shell-forming organisms impact you? Others in your community? Do all of Earth's people share equally in impacts from and responsibility for ocean acidization? Explain.

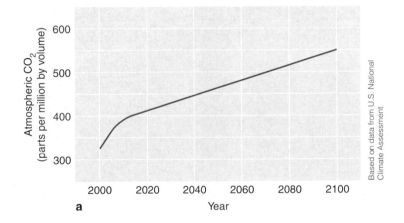

a   Year

Based on data from U.S. National Climate Assessment

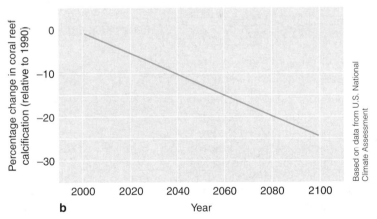

b   Year

Based on data from U.S. National Climate Assessment

THE PLANNER

Environmental History, Politics, and Economics

RENEWABLE ENERGY POLICY CHALLENGES

For the past 20 years, governments have struggled to develop climate change policies. Both the costs of reducing climate change and the effects of not doing so are huge, highly uncertain, and spread out over time, space, and people. And while the scientific community agrees that human-caused climate change is happening and will worsen, many people in the United States, among them influential policy makers, remain deeply skeptical.

Among the biggest issues in the climate debate is how to shift to alternative energy sources. Fossil fuels—coal, oil, and natural gas—are by far the largest source of the greenhouse gases that are changing our climate. Since 2011, while the costs of coal and oil are high and rising, the cost of natural gas has dropped. For these and other reasons, the United States produces less energy from renewable resources than do several other countries (see inset).

Finding suitable locations for alternatives can be a daunting policy challenge. A solar installation can require large amounts of space. In Germany, a town trying to maintain its medieval character is concerned that a proposed solar project will damage its tourism industry. Some California lawmakers object to solar panels in a wilderness area, while in Nevada, concerns about endangered desert tortoise habitat may limit installation site options. Elsewhere, aesthetic, noise, and environmental concerns threaten the launch of potential wind farm projects (see photograph). This chapter explores how environmental policy making requires attention to ethics, economics, culture, and politics as well as to science.

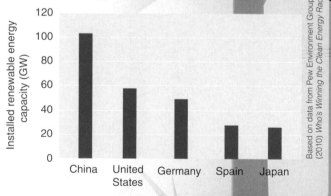

Renewable Energy in Select Countries, 2010.

Installed renewable energy capacity (GW) — China, United States, Germany, Spain, Japan

Based on data from Pew Environment Group (2010) *Who's Winning the Clean Energy Race?*

Think Critically What historical, political, or geographic factors contribute to this difference in renewable energy consumption?

Richard Ellis/age fotostock/SuperStock, Inc.

CHAPTER OUTLINE

Conservation and Preservation of Resources   50

Environmental History   51
- Protecting Forests
- Establishing National Parks and Monuments
- Conservation in the Mid-20th Century
- The Environmental Movement
- ■ EnviroDiscovery: Environmental Literacy

Environmental Legislation   59
- Environmental Regulations
- Accomplishments of Environmental Legislation

Environmental Economics   62
- National Income Accounts and the Environment
- ■ Environmental InSight: Economics and the Environment
- An Economist's View of Pollution
- Economic Strategies for Pollution Control
- ■ Case Study: Tradable Permits and Acid Rain

CHAPTER PLANNER ✓

- ❏ Study the picture and read the opening story
- ❏ Scan the Learning Objectives in each section:
  p. 50 ❏   p. 51 ❏   p. 59 ❏   p. 62 ❏
- ❏ Read the text and study all figures and visuals. Answer any questions.

Analyze key features

- ❏ EnviroDiscovery, p. 58
- ❏ Process Diagram, p. 59
- ❏ Environmental InSight, p. 63
- ❏ Case Study, p. 68
- ❏ Stop: Answer the Concept Checks before you go on:
  p. 50 ❏   p. 56 ❏   p. 61 ❏   p. 67 ❏

End of Chapter

- ❏ Review the Summary and Key Terms.
- ❏ Answer What is happening in this picture?
- ❏ Answer the Critical and Creative Thinking Questions.

Conservation and Preservation of Resources

LEARNING OBJECTIVE

1. **Define** *conservation* and *preservation*, and distinguish between them.

Resources are any part of the natural environment used to promote the welfare of people or other species. Examples of resources include air, water, soil, forests, minerals, and wildlife. **Conservation** is the sensible and careful management of natural resources. Humans have practiced conservation of natural resources for thousands of years. More than 3000 years ago, the Phoenicians terraced hilly farmland to prevent soil erosion. More than 2000 years ago, the Greeks practiced crop rotation to maintain yields on farmlands, and the Romans practiced irrigation.

Other cultures around the world developed similar methods. Modern agriculture continues to develop conservation techniques (**Figure 3.1a**). In addition to agriculture, targets of conservation include energy, water, mineral, forest, fishery, and other resources. Conservation methods can be technological—such as low-flow shower heads—or behavioral—such as shorter showers.

In contrast, **preservation** involves setting aside undisturbed areas, maintaining them in a pristine state, and protecting them from human activities that might alter their "natural" state (**Figure 3.1b**). The decision to preserve places can be controversial: Resources in undisturbed places often have substantial economic value, while the value of nature in a preserved state is difficult to quantify.

Both conservation and preservation became pressing concerns in the early 20th century. At that time, expanding industrialization, coupled with enormous growth in the human population, began to increase pressure on the world's supply of natural resources. As the global population continues to grow, both conservation and preservation will contribute to sustainability. They will help ensure that future generations will have access to essential resources.

CONCEPT CHECK STOP

1. **What** is conservation? preservation? How do they differ?

Conservation and preservation • Figure 3.1

a. Plowing and planting fields in curves that conform to the natural contours of the land conserves soil by reducing erosion.

b. The Eduardo Avaroa Andean Fauna National Reserve in Bolivia preserves flamingos and other wildlife populations and their habitats, such as this one on Laguna Colorada.

USDA/NG Image Collection

© David Noton Photography/Alamy

Environmental History

LEARNING OBJECTIVES

1. **Briefly outline** the environmental history of the United States.

2. **Describe** the contributions of the following people to our perspective on the environment: John James Audubon, Henry David Thoreau, George Perkins Marsh, Theodore Roosevelt, Gifford Pinchot, John Muir, Franklin Roosevelt, Aldo Leopold, Wallace Stegner, Rachel Carson, Paul Ehrlich, Julian Simon, and Wangari Maathai.

3. **Distinguish** between utilitarian conservationists and biocentric preservationists.

4. **Explain** how a systems perspective helps us understand human impacts on the environment.

From the establishment of the first permanent English colony at Jamestown, Virginia, in 1607, the first two centuries of U.S. history were a time of widespread environmental destruction. European settlers exploited land, timber, wildlife, rich soil, clean water, and other resources that had been used sustainably by native peoples for thousands of years. The settlers did not recognize that the bountiful natural resources of North America would one day become scarce. During the 1700s and most of the 1800s, many Americans had a **frontier attitude**, a desire to conquer nature and put its resources to use in the most lucrative manner possible.

Two characteristics of European settlers and their descendants drove this unsustainable resource use: rapid population growth and high per person consumption. European settlements tended to be more densely populated than were those of natives, and settlers accumulated more permanent material goods (houses, roads, wagons, furniture, tools, and clothing).

Protecting Forests

The great forests of the Northeast were cut down within a few generations of European settlement, and, shortly after the Civil War in the 1860s, loggers began deforesting the Midwest at an alarming rate. Within 40 years, they had deforested an area the size of Europe, stripping Minnesota, Michigan, and Wisconsin of virgin forest. By 1897 the sawmills of Michigan had processed 160 billion board

feet of white pine, leaving less than 6 billion board feet standing in the whole state.

During the 19th century, many U.S. naturalists began to voice concerns about conserving natural resources. **John James Audubon** (1785–1851) painted lifelike portraits of birds and other animals in their natural surroundings that aroused widespread public interest in the wildlife of North America (**Figure 3.2**). **Henry David Thoreau** (1817–1862),

Tanagers • Figure 3.2

This portrayal is one of 500 engravings in Audubon's classic, *The Birds of America*, completed in 1844. Shown are two male Louisiana tanagers (also called western tanagers, top) and male and female scarlet tanagers (bottom).

Courtesy Library of Congress

a prominent U.S. writer, lived for 2 years on the shore of Walden Pond near Concord, Massachusetts. There he observed nature and contemplated how people could simplify their lives to live in harmony with the natural world. **George Perkins Marsh** (1801–1882) was a farmer, linguist, and diplomat at various times during his life. Today he is most remembered for his book *Man and Nature*, published in 1864, which provided one of the first discussions of humans as agents of global environmental change.

> **utilitarian conservationist**
> A person who values natural resources because of their usefulness to humans but uses them sensibly and carefully.

In 1875 a group of public-minded citizens formed the American Forestry Association with the intent of influencing public opinion against the wholesale destruction of America's forests. Sixteen years later, in 1891, the **Forest Reserve Act** (which was part of the General Land Law Revision Act) gave the U.S. president the authority to establish forest reserves on public (federally owned) land. Benjamin Harrison (1833–1901), Grover Cleveland (1837–1908), William McKinley (1843–1901), and Theodore Roosevelt (1858–1919) used this law to put a total of 17.4 million hectares (43 million acres) of forest, primarily in the West, out of the reach of loggers.

In 1907 angry Northwest congressmen pushed through a bill stating that national forests could no longer be created by the president but would require an act of Congress. Roosevelt signed the bill into law but only after designating 21 new national forests that totaled 6.5 million hectares (16 million acres).

Roosevelt appointed **Gifford Pinchot** (1865–1946) the first head of the U.S. Forest Service. Both Roosevelt and Pinchot were **utilitarian conservationists** who viewed forests in terms of their usefulness to people—such as in providing jobs and renewable resources. Pinchot supported expanding the nation's forest reserves and managing them scientifically (for instance, harvesting trees only at the rate at which they regrow). Today, national forests are managed for multiple uses, from biological habitats to recreation to timber harvest to cattle grazing.

Establishing National Parks and Monuments

Congress established the world's first national park in 1872, after a party of Montana explorers reported on the natural beauty of the canyon and falls of the Yellowstone River. Yellowstone National Park now includes parts of Idaho, Montana, and Wyoming. In 1890 the **Yosemite National Park Bill** established the Yosemite and Sequoia national parks in California, largely in response to the efforts of a single man, naturalist and writer **John Muir** (1838–1914) (**Figure 3.3**). Muir, who as a child

President Theodore Roosevelt (*left*) and John Muir • Figure 3.3

This photo was taken on Glacier Point above Yosemite Valley, California.

Bettman/Corbis Images

Hetch Hetchy Valley in Yosemite • Figure 3.4

Some environmental battles involving the protection of national parks were lost. John Muir's Sierra Club fought with the city of San Francisco over its efforts to dam a river and form a reservoir in the beautiful Hetch Hetchy Valley, which lay within Yosemite National Park. In 1913 Congress approved the dam. The State of California is considering restoring Hetch Hetchy, at an estimated cost as high as $10 billion. Hetch Hetchy Valley before (a) and after (b) the dam was built.

emigrated from Scotland with his family, was a **biocentric preservationist**. Muir also founded the *Sierra Club*, a national conservation organization that is still active on a range of environmental issues.

In 1906 Congress passed the **Antiquities Act**, which authorized the president to set aside sites that had scientific, historic, or prehistoric importance. By 1916 there were 16 national parks and 21 national monuments, under the loose management of the U.S. Army. Today there are 58 national parks and 74 national monuments under the management of the National Park Service.

> **biocentric preservationist**
> A person who believes in protecting nature from human interference because all forms of life deserve respect and consideration.

Controversy over preservation battles, such as the Hetch Hetchy Valley conflict, generated a strong sentiment that the nation should better protect its national parks (**Figure 3.4**). In 1916 Congress created the National Park Service to manage the national parks and monuments for the enjoyment of the public, "without impairment." It was this clause that gave a different outcome to another battle, fought in the 1950s between conservationists and dam builders over the construction of a dam within Dinosaur National Monument. Preservationists convinced decision makers that to fill the canyon with 400 feet of water

Everett Collection/Newscom

Aldo Leopold • Figure 3.5

Leopold's *A Sand County Almanac* is widely considered an environmental classic.

would "impair" it. This victory for conservation established the "use without impairment" clause as the firm backbone of legal protection afforded our national parks and monuments.

Conservation in the Mid-20th Century

During the Great Depression, the federal government financed many conservation projects to provide jobs for the unemployed. During his administration, **Franklin Roosevelt** (1882–1945) established the Civilian Conservation Corps, which employed 500,000 young men to plant trees, make paths and roads in national parks and forests, build dams to control flooding, and perform other activities that protected natural resources.

During the droughts of the 1930s, windstorms carried away much of the topsoil in parts of the Great Plains, forcing many farmers to abandon their farms and search for work elsewhere. The *American Dust Bowl* alerted the United States to the need for soil conservation, and President Roosevelt formed the Soil Conservation Service in 1935.

Aldo Leopold (1886–1948), a wildlife biologist and environmental visionary, greatly influenced the conservation movement of the mid- to late 20th century (**Figure 3.5**). His textbook *Game Management*, published in 1933, supported the passage of a 1937 act in which new taxes on sporting weapons and ammunition funded wildlife management and research. Leopold also wrote about humanity's relationship with nature and the need to conserve wilderness areas in *A Sand County Almanac*, published in 1949. Leopold argued for a land ethic and the sacrifices that such an ethic requires.

Leopold profoundly influenced many American thinkers and writers, including **Wallace Stegner** (1909–1993), who penned his famous "Wilderness Essay" in 1962. Stegner's essay helped create support for the passage of the *Wilderness Act* of 1964. Stegner wrote:

> *Something will have gone out of us as a people if we ever let the remaining wilderness be destroyed; if we permit the last virgin forests to be turned into comic books and plastic cigarette cases; if we drive the few remaining members of the wild species into zoos or to extinction; if we pollute the last clean air and dirty the last clean streams and push our paved roads through the last of the silence, so that never again will Americans be free in their own country from the noise, the exhausts, the stinks of human and automotive waste . . .*
>
> *We simply need that wild country available to us, even if we never do more than drive to its edge and look in. For it can be a means of reassuring ourselves of our sanity as creatures, a part of the geography of hope.*

During the 1960s, public concern about pollution and resource quality increased, in large part due to the work of marine biologist **Rachel Carson** (1907–1964). Carson wrote about interrelationships among living organisms, including humans, and the natural environment (**Figure 3.6**).

In her most famous work, *Silent Spring*, published in 1962, Carson wrote against the indiscriminate use of pesticides:

> *Pesticide sprays, dusts, and aerosols are now applied almost universally to farms, gardens, forests, and homes—nonselective chemicals that have the power to kill every insect, the "good" and the "bad," to still the song of birds and the leaping of fish in the streams, to coat the leaves with a deadly film, and to linger on in soil—all this though the intended target may be only a few weeds or insects. Can anyone believe it is possible to lay down such a barrage of poisons on the surface of the earth without making it unfit for all life? They should not be called "insecticides," but "biocides."*

Silent Spring heightened public awareness and concern about the dangers of using DDT and other pesticides, including poisoning birds and other wildlife

Rachel Carson • Figure 3.6

Carson's book *Silent Spring* heralded the beginning of the environmental movement.

and contaminating human food supplies. Ultimately, the book led to restrictions on the use of certain pesticides. Around this time, the media increasingly covered environmental incidents, such as hundreds of deaths in New York City from air pollution (1963), closed beaches and fish kills in Lake Erie from water pollution (1965), and detergent foam in a Pennsylvania creek (1966).

Rachel Carson's approach to the environment emphasized the value of taking a **systems perspective**. A systems perspective acknowledges that changes or activities in one place can impact environmental conditions in distant places or in the future. Further, these changes can be difficult to predict and may not be recognized until after significant or irreversible damage has been done. Carson's example of a systems perspective was that pesticides intended to improve crop yields could also kill other organisms. Since *Silent Spring* was published, pesticides have been found around the globe, including in the fatty tissues of polar bears, penguins, and deep-sea fishes.

> **systems perspective** A perspective that considers not just immediate or intended effects of activities, but all of the impacts of those activities in other places or at other times.

Incorporating a systems perspective into environmental management can be a challenge. It is often difficult to predict the long-term or long-distance impacts of an activity. Consequently, management strategies that focus on one aspect of the environment can have unintended consequences. For example, in the 1990s California decided to require that MTBE (methyl tert-butyl ether) be added to gasoline to make it burn more cleanly, thereby improving air quality. However, shortly after MTBE was introduced, it began appearing as a contaminant in groundwater (MTBE is toxic). Similarly, there is a constant debate about whether wastes should be buried, thereby taking up space and sometimes leaching to groundwater, or incinerated, resulting in a variety of toxic air pollutants.

However, a systems perspective can also present better solutions to some environmental problems. For example, rather than spraying pesticides, farmers can eliminate nesting spots for pests or import wasps that eat the pests. Likewise, rather than decide between burying or burning wastes, we can consider strategies to minimize or reuse wastes.

In 1968, when the population of Earth was "only" 3.5 billion people, ecologist **Paul Ehrlich** published *The Population Bomb*. In it he described the stress that such a huge number of people impose on Earth's life support system, including global depletion of fertile soil, groundwater, and other living organisms. Ehrlich's book raised the public's awareness of the dangers of overpopulation and triggered debates about how to deal effectively with population issues.

Ehrlich's critics, in particular **Julian Simon** (1932–1998), countered that technological advances outpace the negative impacts of population growth. A decade into the 21st century, both sides of this issue have strong advocates. Ehrlich continues to point out water, climate, agriculture, and other global stresses, while many economists counter that the collapse Ehrlich predicted has not occurred.

The Environmental Movement

Until 1970 the voice of **environmentalists**, people concerned about the environment, was heard in the United States primarily through societies such as the Sierra Club and the National Wildlife Federation. There was no generally perceived **environmental movement** until the spring of 1970, when **Gaylord Nelson**, former senator of Wisconsin, urged Harvard graduate student **Denis Hayes** to organize the first nationally celebrated Earth Day. This event awakened U.S. environmental consciousness to population growth, overuse of resources, and pollution and degradation of the environment. On Earth

Earth Day 1990 in Washington, DC
• Figure 3.7

Day 1970, an estimated 20 million people in the United States planted trees, cleaned roadsides and riverbanks, and marched in parades to support improvements in resource conservation and environmental quality.

In the years that followed the first Earth Day, environmental awareness and the belief that individual actions could repair the damage humans were doing to Earth became a pervasive popular movement. Musicians and other celebrities popularized environmental concerns. Many of the world's religions—such as Christianity, Judaism, Islam, Hinduism, Buddhism, Taoism, Shintoism, Confucianism, and Jainism—embraced environmental themes such as protecting endangered species and controlling global climate change.

By Earth Day 1990, the movement had spread around the world, signaling the rapid growth in environmental consciousness. An estimated 200 million people in 141 nations demonstrated to increase public awareness of the importance of individual efforts ("Think globally, act locally") (**Figure 3.7**). The theme of Earth Day 2000, "Clean Energy Now," reflected the dangers of global climate change and what individuals and communities could do: Replace fossil fuel energy sources with solar electricity, wind power, and the like. However, by 2000 many environmental activists had begun to think that the individual actions Earth Day espouses, while collectively important, are not as important as pressuring governments and large corporations to make environmentally friendly decisions. Among the most important people in the global environmental movement, **Wangari Maathai**

Wangari Maathai • Figure 3.8

Wangari Maathi was awarded the Noble Peace Prize for her efforts advancing sustainability in her native Kenya and worldwide.

(1940–1911) established the Greenbelt Movement in Kenya (**Figure 3.8**). Maathai organized women in rural areas, showing that they could simultaneously improve their social, economic, and environmental conditions—that is, the sustainability of their communities. For her efforts, Maathai was awarded the Nobel Peace Prize in 2004. In 2012 global environmental concern was expressed as thousands of people from around the world converged on Rio de Janeiro, Brazil, to pressure world leaders to take action on a variety of global environmental problems. **Figure 3.9** shows a timeline of selected environmental events since Earth Day 1970.

CONCEPT CHECK STOP

1. **How** did public perception of the environment evolve during the 20th century?

2. **What** did Rachel Carson contribute to the environmental movement?

3. **What** distinguishes utilitarian conservationists from biocentric preservationists?

4. **How** can a systems perspective improve environmental management?

Timeline of selected environmental events, from 1970 to the present • Figure 3.9

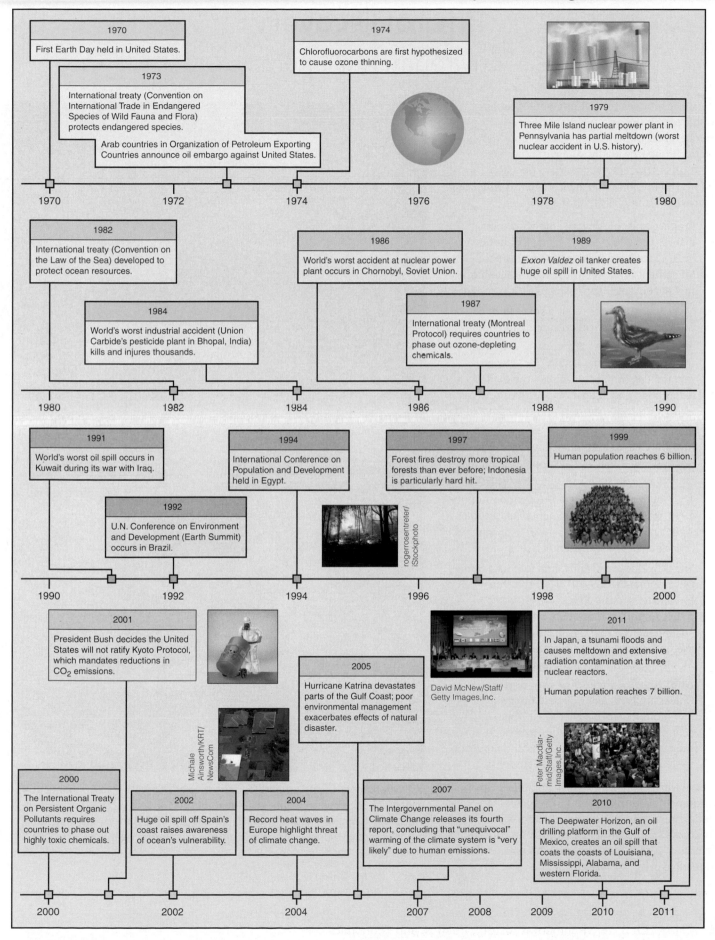

1970
First Earth Day held in United States.

1974
Chlorofluorocarbons are first hypothesized to cause ozone thinning.

1973
International treaty (Convention on International Trade in Endangered Species of Wild Fauna and Flora) protects endangered species.

Arab countries in Organization of Petroleum Exporting Countries announce oil embargo against United States.

1979
Three Mile Island nuclear power plant in Pennsylvania has partial meltdown (worst nuclear accident in U.S. history).

1970   1972   1974   1976   1978   1980

1982
International treaty (Convention on the Law of the Sea) developed to protect ocean resources.

1986
World's worst accident at nuclear power plant occurs in Chornobyl, Soviet Union.

1989
Exxon Valdez oil tanker creates huge oil spill in United States.

1984
World's worst industrial accident (Union Carbide's pesticide plant in Bhopal, India) kills and injures thousands.

1987
International treaty (Montreal Protocol) requires countries to phase out ozone-depleting chemicals.

1980   1982   1984   1986   1988   1990

1991
World's worst oil spill occurs in Kuwait during its war with Iraq.

1994
International Conference on Population and Development held in Egypt.

1997
Forest fires destroy more tropical forests than ever before; Indonesia is particularly hard hit.

1999
Human population reaches 6 billion.

1992
U.N. Conference on Environment and Development (Earth Summit) occurs in Brazil.

rogerosentreter/iStockphoto

1990   1992   1994   1996   1998   2000

2001
President Bush decides the United States will not ratify Kyoto Protocol, which mandates reductions in CO_2 emissions.

2005
Hurricane Katrina devastates parts of the Gulf Coast; poor environmental management exacerbates effects of natural disaster.

David McNew/Staff/Getty Images,Inc.

2011
In Japan, a tsunami floods and causes meltdown and extensive radiation contamination at three nuclear reactors.

Human population reaches 7 billion.

Michale Ainsworth/KRT/NewsCom

Peter Macdiar-mid/Staff/Getty Images,Inc.

2000
The International Treaty on Persistent Organic Pollutants requires countries to phase out highly toxic chemicals.

2002
Huge oil spill off Spain's coast raises awareness of ocean's vulnerability.

2004
Record heat waves in Europe highlight threat of climate change.

2007
The Intergovernmental Panel on Climate Change releases its fourth report, concluding that "unequivocal" warming of the climate system is "very likely" due to human emissions.

2010
The Deepwater Horizon, an oil drilling platform in the Gulf of Mexico, creates an oil spill that coats the coasts of Louisiana, Mississippi, Alabama, and western Florida.

2000   2002   2004   2007   2008   2009   2010   2011

EnviroDiscovery
Environmental Literacy

Because responses to environmental problems depend on the public's awareness and understanding of the issues and the underlying scientific concepts involved, environmental education is critical to appropriate decision making. The emphasis on environmental education has grown dramatically over the years:

- Prepared by coursework at their schools, almost 100,000 U.S. high school students in 2011 took the College Board Advanced Placement exam in Environmental Science, a test accepted by approximately 1700 colleges.

- More than 30 states require some form of environmental education in primary and secondary schools.

- In 2012, the National Council for Science and the Environment launched the Climate Adaptation and Mitigation e-Learning portal to provide curricular resources for climate educators.

- In 2009, the Association of Environmental Studies and Sciences was formed to support college and university faculty and students.

- The National Environmental Education Act of 1990 requires the Environmental Protection Agency to increase public awareness and knowledge of environmental issues.

- The U.N. Decade of Education for Sustainable Development (2005–2014) is dedicated to improving basic education, including public understanding about environmental sustainability. Programs focus on major themes, such as water, climate change, biodiversity, and disaster prevention.

- As of 2012 the American College and University President's Climate Commitment had more than 670 signatories. These schools agree to take actions to reduce their greenhouse gas emissions and require sustainability education for all students.

The North American Association for Environmental Education has issued guidelines for educators to help them select materials such as textbooks and films that are based on sound scientific evidence and that present a balanced perspective on environmental problems.

a. Elementary school children around the world, such as these students testing tap water in Shanghai, learn about the environment through direct experimentation.

b. Environmentalist and primatologist Jane Goodall meets with Connecticut middle school students involved in the Roots and Shoots program, a youth-based environmental action organization that Goodall started. The program includes tens of thousands of members and chapters in nearly 100 countries.

Environmental Legislation

LEARNING OBJECTIVES

1. **Explain** why the National Environmental Policy Act is the cornerstone of U.S. environmental law.

2. **Describe** how environmental impact statements provide powerful protection of the environment.

3. **Explain** the Environmental Protection Agency's role in environmental policy.

W ell-publicized ecological disasters, such as the 1969 oil spill off the coast of Santa Barbara, California, and overwhelming public support for the Earth Day movement, led to the **National Environmental Policy Act (NEPA)** of 1970. The Environmental Protection Agency (EPA) was created in July of the same year. A key provision of NEPA requires the federal government to consider the environmental impacts of proposed federal actions, such as financing highway or dam construction, when making decisions about that action. NEPA provides the basis for developing detailed **environmental impact statements (EISs)** to accompany every federal recommendation or proposal for legislation. An EIS is a document that describes the nature and purpose of the proposal, its short- and long-term environmental impacts, and possible alternatives that would create fewer adverse effects. NEPA also requires solicitation of public comments when preparing an EIS, which generally provides a broader perspective on the proposal and its likely effects.

NEPA established the **Council on Environmental Quality** to monitor the required EISs and report directly to the president. Because this council had no enforcement powers, NEPA was originally considered innocuous, more a statement of good intentions than a regulatory policy. During the next few years, however, environmental activists took people, corporations, and the federal government to court to challenge their EISs or use them to block proposed development. The courts decreed that EISs had to thoroughly analyze the environmental consequences of anticipated projects on soil, water, and endangered species and that EISs be made available to the public (**Figure 3.10**). These rulings put sharp teeth

Environmental impact statements • Figure 3.10

✓ THE PLANNER

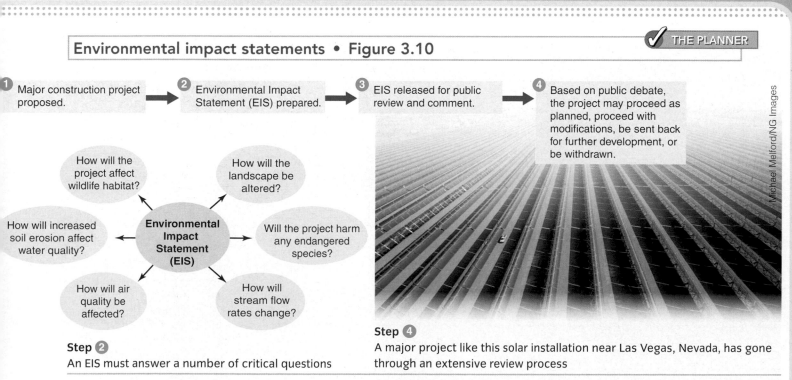

1. Major construction project proposed.

2. Environmental Impact Statement (EIS) prepared.

3. EIS released for public review and comment.

4. Based on public debate, the project may proceed as planned, proceed with modifications, be sent back for further development, or be withdrawn.

Environmental Impact Statement (EIS)

- How will the project affect wildlife habitat?
- How will the landscape be altered?
- How will increased soil erosion affect water quality?
- Will the project harm any endangered species?
- How will air quality be affected?
- How will stream flow rates change?

Step 2
An EIS must answer a number of critical questions

Step 4
A major project like this solar installation near Las Vegas, Nevada, has gone through an extensive review process

Michael Melford/NG Images

PROCESS DIAGRAM

into NEPA—particularly the provision for public scrutiny, which places intense pressure on federal agencies to respect EIS findings.

NEPA revolutionized environmental protection in the United States. Federal agencies manage federal highway construction, flood and erosion control, military projects, and many other public works. They oversee nearly one-third of the land in the United States. Federal holdings include fossil fuel and mineral reserves, millions of hectares of public grazing land, and public forests. Since 1970 few federal activities have been initiated without some sort of environmental review. NEPA has also influenced environmental legislation in at least 36 states and in many other countries.

Although almost everyone agrees that NEPA has successfully reduced adverse environmental impacts of federal activities and projects, it has its critics. Some environmentalists complain that EISs are sometimes incomplete or that reports are ignored when decisions are made. Other critics think the EISs delay important projects ("paralysis by analysis") because the documents are too involved, take too long to prepare, and are often the targets of lawsuits.

Environmental Regulations

In the United States, Congress has passed federal laws including the Clean Air Act, the Clean Water Act, and the Resource Conservation and Recovery Act to serve as the foundation of environmental management. Most of these laws have been around in some form since the middle of the 20th century, but new laws and amendments in the early 1970s created the basis for environmental law as we know it today.

The Environmental Protection Agency (EPA), part of the executive branch of the federal government, is responsible for translating each law's language into specific regulations. Before the EPA can enforce new regulations, several rounds of public comments allow affected parties to present their views. The EPA is required to respond to all of these comments. The Office of Management and Budget then assesses the anticipated environmental impacts of each new regulation. Implementation and enforcement often fall to state governments, which must send the EPA details for achieving the goals of the new regulations.

Accomplishments of Environmental Legislation

WileyPLUS ⊕

During the period since Earth Day 1970, Congress has updated or passed almost 40 major environmental laws that address a wide range of issues, such as endangered species, clean water, clean air, energy conservation, hazardous wastes, and pesticides. This tough interlocking mesh of laws has greatly improved environmental quality.

Despite imperfections, environmental legislation has had overall positive effects. Since 1970,

- Fifteen national parks have been established (**Figure 3.11**), and the National Wilderness Preservation

Joshua Tree National Park, California • Figure 3.11

Formerly a national monument, Joshua Tree was declared a national park in 1994.

System now totals more than 44 million hectares (109 million acres).

- Millions of hectares of farmland particularly vulnerable to erosion have been withdrawn from production, reducing soil erosion by more than 60 percent.

- Many endangered species are recovering, and the American alligator, California gray whale, and bald eagle have recovered enough to be removed from the endangered species list. (However, dozens of other species, such as the manatee and Kemp's ridley sea turtle, have suffered further declines or extinction since 1970.)

Although we still have a long way to go, pollution control efforts through legislation have been particularly successful. According to the EPA's *2008 Report on the Environment*:

- Emissions of six important air pollutants have dropped by more than 25 percent since 1990. (Carbon dioxide emissions, however, have continued to rise.)

- Since 1990, levels of wet sulfate, a major component of acid rain, have dropped by 20 to 35 percent.

- In 2007 almost 90 percent of the U.S. population got its drinking water from community water systems with no violations of EPA standards, up from around 75 percent in 1993 (**Figure 3.12**).

- In 2008 45 percent of municipal solid waste generated in the United States was combusted for energy recovery or recovered for composting or recycling, up from 6 percent in the 1960s.

- By 2007 the EPA considered human exposures to contamination to be under control at 93 percent of the 1968 listed hazardous waste sites.

In the 1960s and 1970s, pollution was often obvious—witness the Cuyahoga River in Cleveland, Ohio, which burst into flames from the oily pollutants on its surface several times. Legislators, the media, and the public typically perceive things like burning rivers as serious threats that require immediate attention, without regard to the cost. As the effects of global climate change become more obvious, public pressure to develop policies to reduce greenhouse gas emissions have grown. Recognizing the high costs of historical legislation and the power of

Water treatment plant • Figure 3.12

The water supply for a town or city is treated before use so it is safe to drink. Photographed in Sarasota, Florida.

Tampa Bay Times/ZUMAPRESS.com

markets to drive innovative solutions, policy makers increasingly look to economics as part of the solution to environmental problems.

CONCEPT CHECK	

1. **Why** is the National Environmental Policy Act the cornerstone of U.S. environmental law?

2. **What** are environmental impact statements?

3. **What** is the EPA's role in environmental regulation?

Environmental Economics

LEARNING OBJECTIVES

1. **Explain** how economics is related to natural capital. Make sure you include sources and sinks.

2. **Give** two reasons why national income accounts are incomplete estimates of national economic performance.

3. **Distinguish** among the following economic terms: marginal cost of pollution, marginal cost of pollution abatement, and optimum amount of pollution.

4. **Describe** various incentive-based regulatory approaches, including environmental taxes and tradable permits.

Economics is the study of how people use their limited resources, given the assumption that each person strives to satisfy unlimited wants. Economists try to understand the consequences of the ways in which people, businesses, and governments allocate their resources. Seen through an economist's eyes, the world is one large marketplace, where resources are allocated to a variety of uses, and where goods—a car, a pair of shoes, a barrel of oil—and

services—a haircut, a museum tour, an education—are consumed and paid for. In a free market, supply and demand determine the price of a good (**Figure 3.13**). If something in great demand is in short supply, its price will be high. High prices encourage suppliers to produce more of a good or service, as long as the selling price is equal to the cost of producing the good or service. This interaction of demand, supply, price, and cost underlies much of what happens in the U.S. economy, from the price of a hamburger to the cycles of economic expansion (increase in economic activity) and recession (slowdown in economic activity).

Economies depend on the natural environment as *sources* for raw materials and *sinks* for waste products (**Figure 3.14**). Both sources and sinks contribute to **natural capital**. Under the assumptions of economics, the environment has value when it provides natural capital for

> **natural capital**
> Earth's resources and processes that sustain living organisms, including humans; includes minerals, forests, soils, water, clean air, wildlife, and fisheries.

The Hibernia oil platform on the Grand Banks in the Atlantic Ocean • Figure 3.13

When demand for crude oil goes up, economists expect that more will be pumped, and the price of a barrel of crude oil will increase.

Global Locator

GRAND BANKS

NG Maps

NATIONAL GEOGRAPHIC

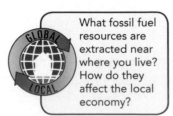

GLOBAL LOCAL

What fossil fuel resources are extracted near where you live? How do they affect the local economy?

Randy Olson/NG Image Collection

human production and consumption. Resource degradation and pollution represent the overuse of natural capital. *Resource degradation* is the overuse of sources, and *pollution* is the overuse of sinks; both threaten our long-term economic future.

National Income Accounts and the Environment

Much of our economic well-being flows from natural capital—such as land, rivers, the ocean, oil, timber, and the air we breathe—rather than human-made assets.

Ideally, for the purposes of economic and environmental planning, **national income accounts** should include natural resource depletion and environmental degradation. Two measures used in national income accounting are *gross domestic product* (*GDP*) and *net domestic product* (*NDP*). Both GDP and NDP provide estimates of national economic performance that are used to make important policy decisions.

> **national income accounts** Measures of the total income of a nation's goods and services for a given year.

Unfortunately, current national income accounting practices provide an incomplete or inaccurate measure of income because they do not incorporate environmental factors. Two important conceptual problems exist with the way national income accounts currently handle the economic use of natural resources and the environment: natural resource depletion and the costs and benefits of pollution control. Better accounting for environmental quality would help address whether for any given activity the benefits (both economic and environmental) exceed the costs.

Other methods have been suggested that take a sustainability perspective on national income accounting by including measures of social and environmental well-being. For example, the **genuine progress indicator** (**GPI**) includes human development and natural capital depletion.

Natural Resource Depletion If a manufacturing firm produces some product (output) but in the process wears out a portion of its plant and equipment, the firm's output is counted as part of GDP, but the depreciation of capital is subtracted in the calculation of NDP. Thus NDP

Environmental InSight

Economics and the environment
- ### Figure 3.14

THE PLANNER

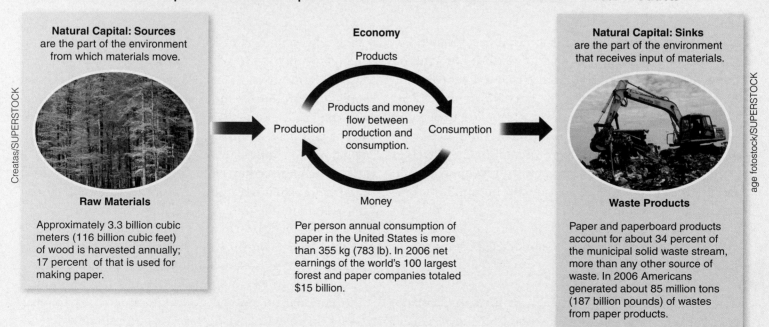

Economies Depend on Natural Capital Sources for Raw Materials and Sinks for Waste Products

Natural Capital: Sources are the part of the environment from which materials move.

Raw Materials

Approximately 3.3 billion cubic meters (116 billion cubic feet) of wood is harvested annually; 17 percent of that is used for making paper.

Economy

Products

Products and money flow between production and consumption.

Production → Consumption

Money

Per person annual consumption of paper in the United States is more than 355 kg (783 lb). In 2006 net earnings of the world's 100 largest forest and paper companies totaled $15 billion.

Natural Capital: Sinks are the part of the environment that receives input of materials.

Waste Products

Paper and paperboard products account for about 34 percent of the municipal solid waste stream, more than any other source of waste. In 2006 Americans generated about 85 million tons (187 billion pounds) of wastes from paper products.

Pollution cleanup and GDP • Figure 3.15

Air emissions from chemical plants like this one in Cleveland, Ohio (a) are far lower than they would be absent air quality regulations. The costs associated with controlling air pollution from industry, transportation, and other sectors are far smaller than are the resulting benefits to health, property, and the environment (b). The improved air quality is not included in the calculation of GDP.

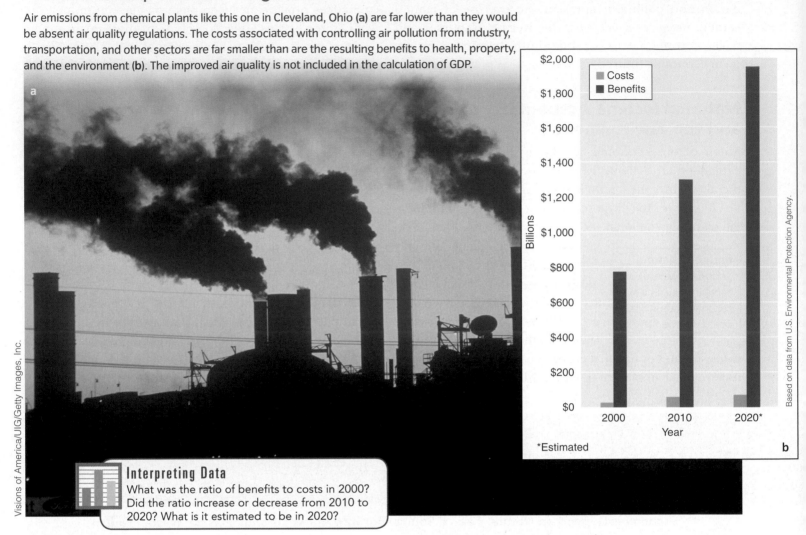

Visions of America/UIG/Getty Images, Inc.

Based on data from U.S. Environmental Protection Agency.

*Estimated

Interpreting Data
What was the ratio of benefits to costs in 2000? Did the ratio increase or decrease from 2010 to 2020? What is it estimated to be in 2020?

is a measure of the net production of the economy, after a deduction for used-up capital. In contrast, when an oil company drains oil from an underground field, the value of the oil produced is counted as part of the nation's GDP, but no offsetting deduction to NDP is made to account for the fact that nonrenewable resources were used up.

In principle, the draining of the oil field is a type of depreciation, and the oil company's net product should be accordingly reduced. The same point applies to any other natural resource that is depleted in the process of production. Natural capital is a very large part of a country's economic wealth, and we should treat it the same as human-made capital.

The Costs and Benefits of Pollution Control

Imagine that a company has the following choices: It can produce $100 million worth of products and, at the same time, release toxic emissions into the air. Alternatively, if the company uses 10 percent of its workers to properly dispose of its wastes, it avoids polluting but gets only $90 million of products. Under current national income accounting rules, if the firm chooses to pollute rather than not to pollute, it will make a larger contribution to GDP ($100 million rather than $90 million) because the national income accounts attach no explicit value to clean air. In an ideal accounting system, the economic cost of environmental degradation is subtracted in the calculation of a firm's contribution to GDP, and activities that improve the environment—because they provide real economic benefits—are added to GDP (**Figure 3.15**).

Incorporating resource depletion and pollution into national income accounting is important because GDP and related statistics are used continually in policy analyses. An increasing number of economists, government planners, and scientists support replacing GDP and NDP with a more comprehensive measure of national income accounting that includes estimates of both depletion of natural capital and the environmental cost of economic activities (**Figure 3.16**).

An Economist's View of Pollution

An important aspect of the operation of a free-market system is that the person consuming a product should pay for all the cost of producing it. However, production or consumption of a product often has an **external cost**.

> **external cost**
> A harmful environmental or social cost that is borne by people not directly involved in selling or buying a product.

A product's market price does not usually reflect an external cost—that is, the buyer or seller doesn't pay for all of the costs associated with production. As a result, a market system with externalities generally does not operate in the most efficient way.

Consider the following example of an external cost. If an industry makes a product and, in so doing, also releases a pollutant into the environment, the product is bought at a price that reflects the cost of inputs such as labor, energy, buildings, and raw materials. However, the price does not reflect costs from when pollutants damage the environment, which is the external cost of the product. (One common external cost of many products is air pollution released when fossil fuels are burned to transport manufacturing components or finished goods.)

Because this environmental damage is not included in the product's price and because the consumer may not know that the pollution exists or that it harms the environment, the cost of the pollution has no impact on the consumer's decision to buy the product. As a result, consumers of the product may buy more of it than they would if its true cost, including the cost of pollution, were reflected in the selling price.

The failure to add the price of environmental damage to the cost of products generates a market force that encourages pollution. From the perspective of economics, then, one of the causes of the world's pollution problem is the failure to include external costs in the prices of goods.

We now examine industrial pollution from an economist's viewpoint, as a policymaking failure. Keep in mind,

Waterfall in Great Smoky Mountain National Park • Figure 3.16

Resources removed from pristine areas such as this one would not be counted as a loss in standard national income accounting. Photographed in North Carolina.

Tim Fitzharris/NG Image Collection

however, that lessons about the economics of industrial pollution also apply to other environmental issues (such as resource degradation) where harm to the environment is a consequence of economic activity.

How Much Pollution Is Acceptable? Economics entails trying to get the most goods or services from limited resources. Consequently, economic solutions to environmental harm view that harm as a loss of resources, a loss that can be assessed in monetary terms. For example, if a coal mining operation pollutes a well that was previously used for drinking water, the economic loss would be equal to the cost of providing clean drinking water to the people who previously drank from the well.

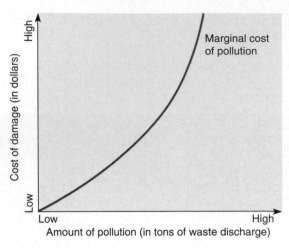

Marginal cost of pollution • Figure 3.17

At low pollution levels, the environment may absorb the damage, so that the marginal cost of one added unit of pollution is near zero. As the level of pollution rises, the cost in terms of human health and a damaged environment increases sharply. At very high levels of pollution, the cost soars.

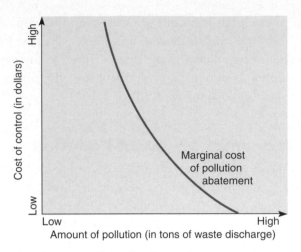

Marginal cost of pollution abatement • Figure 3.18

At high pollution levels, the marginal cost of eliminating one unit of pollution is low. As more and more pollution is eliminated from the environment, the cost of removing each additional (marginal) unit of pollution increases.

How can policy makers decide when it is better to stop polluting the well, clean the water in the well, or import water from another location? Economists analyze the marginal costs of environmental quality and of other goods to answer such questions. A **marginal cost** is the additional cost associated with one more unit of something.

The trade-off between protecting environmental quality and producing more goods involves balancing marginal costs of two kinds: (1) the external cost, in terms of environmental damage, of more pollution (the marginal cost of pollution) and (2) the cost, in terms of giving up goods, of eliminating pollution (the marginal cost of pollution abatement).

Determining the **marginal cost of pollution** involves assessing the risks associated with the pollution—for example, damage to health, property, or agriculture. (See Chapter 4 for a discussion of risk assessment.) Once the risk is known, it must be monetized. This means that injuries, deaths, loss of species, and other damages must be assigned dollar values.

> **marginal cost of pollution** The added cost of an additional unit of pollution.

Let's consider a simple example involving the marginal cost of sulfur dioxide, a type of air pollution produced during the combustion of fuels containing sulfur. Sulfur dioxide is removed from the atmosphere as acid rain, which causes damage to the environment, particularly aquatic ecosystems. Economists add up the harm of each additional unit of pollution—in this example, each ton of sulfur dioxide added to the atmosphere. As the total amount of pollution increases, the harm of each additional unit usually also increases, and as a result, the curve showing the marginal cost of pollution slopes upward, as in **Figure 3.17**.

The **marginal cost of pollution abatement** tends to rise as the level of pollution declines, as shown in **Figure 3.18**. It is relatively inexpensive to reduce automobile exhaust emissions by half, but costly devices are required to reduce the remaining emissions by half again. For this reason, the curve showing the marginal cost of pollution abatement slopes downward.

In **Figure 3.19**, the two marginal-cost curves from Figures 3.17 and 3.18 are plotted together on one graph, called a **cost–benefit diagram**. Economists use this diagram to identify the point at which the marginal

> **marginal cost of pollution abatement** The added cost of reducing one unit of a given type of pollution.
>
> **cost–benefit diagram** A diagram that helps policy makers make decisions about costs of a particular action and benefits that would occur if that action were implemented.

cost of pollution equals the marginal cost of abatement—that is, the point where the two curves intersect. As far as economics is concerned, this point represents an

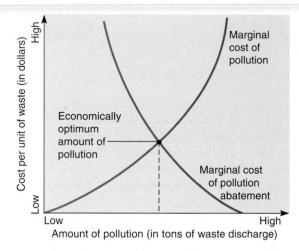

Cost–benefit diagram • Figure 3.19

Economists identify the optimum amount of pollution as the amount at which the marginal cost of pollution equals the marginal cost of pollution abatement (the point at which the two curves intersect). If more pollution than the optimum is allowed, the social cost is unacceptably high. If less than the optimum amount of pollution is allowed, the pollution abatement cost is unacceptably high.

optimum amount of pollution. At this optimum, the cost to society of having less pollution is offset by the benefits to society of the activity creating the pollution.

> **optimum amount of pollution** The amount of pollution that is economically most desirable.

There are two major objections to the economist's concept of optimum pollution. First, it is difficult to determine the true cost of environmental damage caused by pollution. The web of relationships within the environment is extremely intricate and may be more vulnerable to pollution damage than is initially obvious, sometimes with disastrous results. When cost estimates are highly uncertain, economics may lead to poor decisions. Second, many people find the notion of putting prices on lives, species, and wilderness to be unethical.

Economic Strategies for Pollution Control

> **command and control regulation** Pollution control laws that work by setting limits on levels of pollution.

Command and control regulations and incentive-based regulations are two ways that governments control pollution. To date, most pollution control efforts in the United States have involved **command and control regulation**. Sometimes command and control laws require use of a specific pollution control method, such as the use of catalytic converters in cars to decrease polluting exhaust emissions. In other cases, a quantitative goal is set. For example, the Clean Air Act Amendments of 1990 established a goal of a 60 percent reduction in nitrogen oxide emissions in passenger cars by the year 2003. Usually, all polluters must comply with the same rules and regulations, regardless of their particular circumstances.

Economists are concerned that command and control regulations can have excessively high costs. They argue that using economic tools can achieve the same environmental benefits at lower cost. Consequently, most economists, whether progressive or conservative, prefer **incentive-based regulation** over command and control regulation. Ideally, incentive-based regulation forces producers to internalize external cost, thereby achieving the optimum amount of pollution.

> **incentive-based regulation** Pollution control laws that work by establishing emission targets and providing industries with incentives to reduce emissions.

The two most common incentive-based regulatory approaches are environmental taxes and tradable permits. **Environmental taxes** are designed to be equal to the externality caused by a polluter. Unfortunately, this amount can be highly uncertain and so is often very difficult to set. **Tradable permit** approaches, also known as *cap and trade*, set an allowable amount of pollution and then let different companies buy and sell the right to release that pollution. Companies or individuals who can easily reduce their emissions sell some of their pollution rights to those who cannot.

CONCEPT CHECK 🛑 STOP

1. **What** is natural capital? How is economics related to natural capital?

2. **Why** are national income accounts incomplete estimates of total national economic performance?

3. **How** are marginal cost of pollution, marginal cost of pollution abatement, and optimum amount of pollution related?

4. **How** do command and control regulation and incentive-based regulation differ regarding pollution control?

Tradable Permits and Acid Rain

Many international policy experts believe that a cap and trade system is the most promising approach to managing the problem of climate change. They argue that setting a global (or nation-by-nation) cap on greenhouse gases would encourage people to find innovative and inexpensive ways to reduce emissions. However, incentive-based environmental regulations remain less familiar than command and control regulations. The example of tradable sulfur emissions permits to reduce the effects of acid rain demonstrates how effective the approach can be.

When coal containing sulfur is burned, sulfur dioxide is created and released, causing acid rain (see Chapter 8). Through the 1970s and 1980s, Environmental Protection Agency (EPA) regulations reduced sulfur emissions primarily by mandating command and control solutions. This meant that many large coal-burning power plants had to install specific, and often very expensive, equipment. By the late 1980s, these facilities knew of less expensive options for reducing sulfur emissions but had no incentive to adopt them.

Consequently, the Clean Air Act Amendments of 1990 allowed the EPA to limit the amount of sulfur that could be emitted, with a smaller amount allowed each year, and then sell the rights to these emissions. Each year, the EPA allows the Chicago Board of Trade to auction permits to emit sulfur; companies may then buy and sell these permits as needed during the year. Industries quickly adopted a variety of technologies, such as removing sulfur before burning coal, and met the EPA's sulfur reduction goals ahead of schedule and at a lower-than-expected cost.

Tradable permits have not worked as well in all cases. Attempts to reduce water pollution have had mixed results, especially when more than one pollutant is involved. Grandfathering, or exempting older facilities, has undermined other efforts. And in the sulfur case, the EPA was able to establish clear goals and accurately measure emissions, both of which may prove a challenge for a greenhouse gas cap and trade system. Nonetheless, the success of tradable sulfur emissions suggests that incentive-based regulation has a promising future.

Coal-burning power plants in the United States, such as this one in West Virginia, emit sulfur that causes acid rain. Until the 1990s, the EPA mandated emission control technology. Since the 1990s, companies have had more flexibility in how to reduce their emissions.

Skip Brown/NG Image Collection

These buildings in Ottawa, Ontario, Canada, have been damaged by acid rain. This is an example of an externality caused in part by sulfur emissions from coal-burning power plants in the United States.

Ted Spiegel/NG Image Collection

Summary

1 Conservation and Preservation of Resources 50

1. **Conservation** is the sensible and careful management of natural **resources**, such as air, water, soil, forests, minerals, and wildlife. **Preservation** involves setting aside undisturbed areas, maintaining them in a pristine state, and protecting them from human activities.

USDA/NG Image Collection

2 Environmental History 51

1. The first two centuries of U.S. history were a time of widespread environmental destruction. During the 1700s and early 1800s, most Americans had a desire to conquer and exploit nature as quickly as possible. During the 19th century, many U.S. naturalists became concerned about conserving natural resources. The earliest conservation legislation revolved around protecting land—forests, parks, and monuments. By the late 20th century, environmental awareness had become a pervasive popular movement.

2. **John James Audubon**'s art aroused widespread interest in the wildlife of North America. **Henry David Thoreau** wrote about living in harmony with the natural world. **George Perkins Marsh** wrote about humans as agents of global environmental change. **Theodore Roosevelt** appointed **Gifford Pinchot** as the first head of the U.S. Forest Service. Pinchot supported expanding the nation's forest reserves and managing forests scientifically. The Yosemite and Sequoia national parks were established largely in response to the efforts of naturalist **John Muir**. **Franklin Roosevelt** established the Civilian Conservation Corps and the Soil Conservation Service. In *A Sand County Almanac*, **Aldo Leopold** wrote about humanity's relationship with nature. **Wallace Stegner** helped create support for the passage of the Wilderness Act of 1964. **Rachel Carson** published *Silent Spring*, alerting the public about the dangers of uncontrolled pesticide use. **Paul Ehrlich** published *The Population Bomb*, which raised the public's awareness of the dangers of overpopulation. **Julian Simon**, taking an economist's perspective, challenged Ehrlich's concerns about growth. **Wangari Maathai** was awarded the Nobel Peace Prize in 2004 for demonstrating that social, economic, and environmental well-being can be improved simultaneously.

3. A **utilitarian conservationist** is a person who values natural resources because of their usefulness to humans but uses them sensibly and carefully. A **biocentric preservationist** is a person who believes in protecting nature because all forms of life deserve respect and consideration.

4. A **systems perspective** considers not just immediate or intended effects of activities, but all of the impacts of those activities in other places or at other times. Finding pesticides sprayed on farms in the central United States in animals at the north and south poles demonstrates the importance of a systems perspective.

Todd Gipstein/NG Image Collection

3 Environmental Legislation 59

1. Since 1970 the federal government has addressed many environmental problems. The **National Environmental Policy Act (NEPA)** of 1970 established the **Council on Environmental Quality** to monitor required **environmental impact statements (EISs)** and report directly to the president.

2. By requiring EISs that are open to public scrutiny, NEPA initiated serious environmental protection in the United States. NEPA allows citizen suits, in which private citizens take violators, whether they are private industries or government-owned facilities, to court for noncompliance.

3. The United States Congress passes environmental legislation such as the Clean Air Act. The EPA is tasked with turning these laws into environmental regulation. The EPA either directly enforces the law or transfers authority to individual states.

Greg Dale/NG Image Collection

4 **Environmental Economics 62**

1. **Economics** is the study of how people use their limited resources to try to satisfy their unlimited wants. Economies depend on the natural environment as sources for raw materials and sinks for waste products. Both sources and sinks contribute to **natural capital**, which is Earth's resources and processes that sustain living organisms, including

humans. Natural capital includes minerals, forests, soils, water, clean air, wildlife, and fisheries.

2. **National income accounts** are measures of the total income of a nation's goods and services for a given year. An **external cost** is a harmful environmental or social cost that is borne by people not directly involved in buying or selling a product. National income accounts are incomplete estimates of national economic performance because they do not include both natural resource depletion and the environmental costs of economic activities. Many economists, government planners, and scientists support more comprehensive income accounting that includes these estimates.

3. From an economic point of view, the appropriate amount of pollution is a trade-off between harm to the environment and inhibition of development. The **marginal cost of pollution** is the added cost of an additional unit of pollution. The **marginal cost of pollution abatement** is the added cost of reducing one unit of a given type of pollution. Economists think the use of resources for pollution abatement should increase only until the cost of abatement equals the cost of the pollution damage. This results in the **optimum amount of pollution**—the amount of pollution that is economically most desirable.

4. **Incentive-based regulations** take advantage of economic markets to reduce environmental damage. **Environmental taxes** require polluters to pay an amount equal to the harm they cause. **Tradable permit** systems limit the total amount of a pollutant that can be released, allowing people to buy and sell rights to emit and reduce emissions as inexpensively as possible.

Key Terms

- biocentric preservationist 53
- command and control regulation 67
- cost–benefit diagram 66
- external cost 65
- incentive-based regulation 67

- marginal cost of pollution 66
- marginal cost of pollution abatement 66
- national income accounts 63
- natural capital 62

- optimum amount of pollution 67
- systems perspective 55
- utilitarian conservationist 52

MANAN VATSYAYANA/Stringer/AFP/Getty Images

What is happening in this picture?

- This photo was taken in 2010. What event is taking place?

- Note the ages of the individuals in this photo. Do people's attitudes toward the environment change as they grow older? How and why?

Research how Earth Day was celebrated last year. What issues did people focus on?

Critical and Creative Thinking Questions

1. Is a ban on logging in a national park an example of conservation or preservation? Explain.

2. Explain why policy making for renewable energy projects requires attention to ethics, economics, culture, and politics as well as to science.

3. Describe how writers influenced environmental history in the 19th and 20th centuries.

4. List at least three issues that would be included in a national income account that incorporates issues of sustainability.

5. Explain why minimizing environmental damage to a large resource like the Mississippi River requires that we take a systems perspective.

6. The National Environmental Policy Act (NEPA) is sometimes called the "Magna Carta of environmental law." What is meant by such a comparison?

7. How would a utilitarian conservationist's perspective on recreation in national parks differ from that of a biocentric preservationist?

8. How would an economist approach the problem of climate change?

9. In the graph shown below, is the amount of pollution indicated by the vertical dashed line more or less than the economically optimum amount of pollution? Explain your answer.

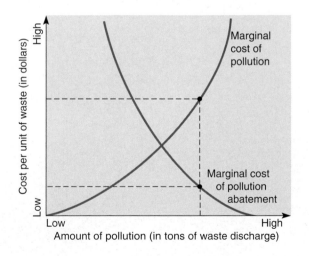

10. If you were an economist examining the previous graph, would you recommend increasing or decreasing pollution abatement measures? Why?

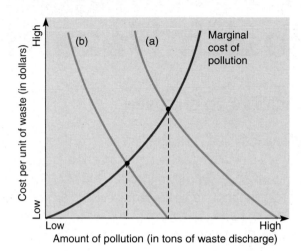

11. The graph above shows two curves, labeled a and b, that represent marginal cost of pollution abatement. In this hypothetical situation, technological innovations were developed between 2002 and 2012 that lowered the abatement cost. Which curve corresponds to 2002 and which to 2012? Explain your answer.

Sustainable Citizen Question

12. If you were a member of Congress, what legislation would you introduce to deal with each of the following problems?

- Toxins from a major sanitary landfill are polluting your state's groundwater.

- Acid rain from a coal-burning power plant in a nearby state is harming the trees in your state. Loggers and foresters are upset.

- There is a high incidence of cancer in the area of your state where heavy industry is concentrated.

Risk Analysis and Environmental Health Hazards

PESTICIDES AND CHILDREN

Historically, analyses of environmental threats to human health have assumed that all people are alike. However, it is increasingly clear that many factors, including genetics, medical conditions, and diet can lead to individuals having different responses to the same exposures. Age is another such factor, and evidence shows that pesticides can be a greater threat to children than to adults for two reasons.

First, children often face greater *exposure*, from playing in contaminated areas, putting their hands and other objects into their mouths or accidentally consuming unsecured pesticides. Second, children's developing bodies can exhibit greater *response* from a given amount of pesticide than do less sensitive adults. Pesticides are common environmental health threats, and they have a range of effects, including cancers and mental or physical disabilities.

Research indicates that pesticide exposure can affect the development of intelligence and motor skills in young children. A study published in *Environmental Health Perspectives* compared two groups of rural Yaqui Indian preschoolers in Mexico, where pesticides are often used on crops for export (see photograph). These two nearly identical groups differed mainly in their exposure to pesticides: One group lived in a farming community where pesticides were used frequently, and the other lived in an area where pesticides were rarely used. When asked to draw a person, most of the 17 children from the low-pesticide area drew recognizable stick figures (see part **a** of inset), whereas most of the 34 children from the high-pesticide area drew meaningless lines and circles (see part **b**). Additional tests of simple mental and physical skills revealed similar striking differences between the two groups of children.

WileyPLUS

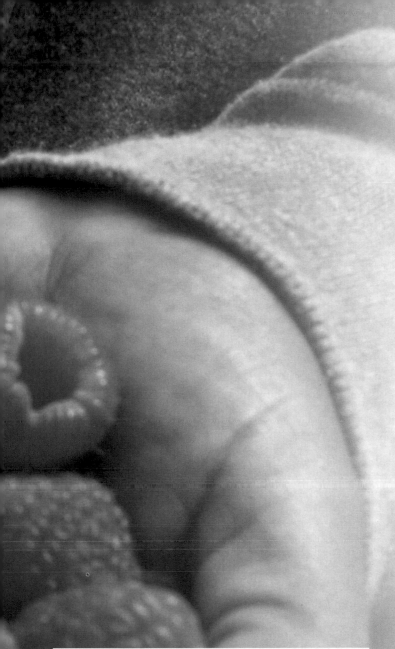

CHAPTER OUTLINE

A Perspective on Risks 74

Environmental Health Hazards 77
- Disease-Causing Agents in the Environment
- Environmental Changes and Emerging Diseases

Movement and Fate of Toxicants 81
- Environmental InSight: Bioaccumulation and Biomagnification
- Mobility in the Environment
- The Global Ban of Persistent Organic Pollutants

Determining Health Effects of Pollutants 85
- Cancer-Causing Substances
- Risk Assessment of Chemical Mixtures
- Children and Chemical Exposure
- EnviroDiscovery: Smoking: A Significant Risk

The Precautionary Principle 90
- Case Study: Endocrine Disrupters

a   b

E. A. Guilette, M. M. Meza, M. G. Aguilar, A. D. Soto, and I. I Garcia, "An anthropological approach to the evaluation of preschool children exposed to pesticides in Mexico." *Environmental Health Perspectives* (May 1998).

Kevin G. Hall/MCT/Getty Images. Inc.

CHAPTER PLANNER ✓

- ❏ Study the picture and read the opening story.
- ❏ Scan the Learning Objectives in each section:
  p. 74 ❏   p. 77 ❏   p. 81 ❏   p. 85 ❏   p. 90 ❏
- ❏ Read the text and study all figures and visuals. Answer any questions.

Analyze key features

- ❏ Process Diagram, p. 75
- ❏ Environmental InSight, p. 82
- ❏ EnviroDiscovery, p. 88
- ❏ Case Study, p. 92
- ❏ Stop: Answer the Concept Checks before you go on:
  p. 76 ❏   p. 81 ❏   p. 84 ❏   p. 89 ❏   p. 92 ❏

End of Chapter

- ❏ Review the Summary and Key Terms.
- ❏ Answer What is happening in this picture?
- ❏ Answer the Critical and Creative Thinking Questions.

A Perspective on Risks

LEARNING OBJECTIVES

1. **Define** *risk* and *risk assessment*.
2. **Explain** how risk assessment helps us manage potential health threats.

Threats to our health, particularly from toxic chemicals in the environment, make big news. Many of these stories are more sensational than factual. Human health in highly developed counties is generally better today than at any previous time in our history, although life expectancy in some of the poorest U.S. counties has begun to decline.

This does not mean that you should ignore chemicals that humans introduce into the environment. Nor should you discount all the stories that the news media sometimes sensationalize, since they can identify serious health threats that we can manage only if we are aware of them. Exposure to lead, organic pesticides, paints, and other chemicals was much higher several decades ago than they are now. Awareness led to both voluntary and legally mandated reductions in the threats we now face.

Risk is inherent in all our actions and in everything in our environment. All of us take risks every day of our lives. Walking on stairs involves a small risk, but a risk nonetheless because sometimes people die from falls on stairs. Using household appliances leads to some risk of

risk The probability of harm (such as injury, disease, death, or environmental damage) occurring under certain circumstances.

electrocution when wires are faulty or people operate appliances unsafely. Driving or riding in a car or flying in a jet has risks that are easier for most of us to recognize. Yet few of us hesitate to get in a car or board a plane because of the associated risk. In order to successfully manage risks, we must have a sense of their causes, likelihoods, and effects (**Figure 4.1**).

Each of us uses intuition, habit, and experience to make many decisions regarding risk every day. However, environmental and health risks often affect many individuals, and the best choices cannot always be made on an intuitive or routine level. **Risk management** is the process of identifying, assessing, and reducing risks.

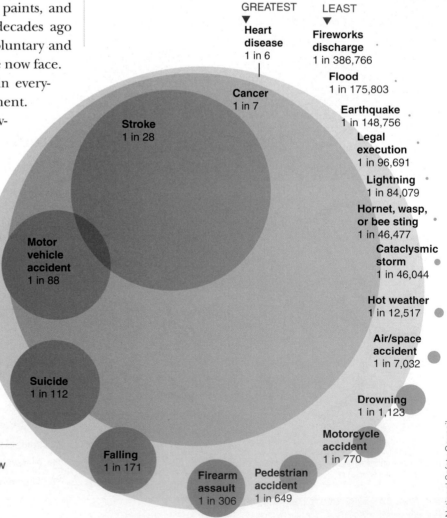

Lifetime probability of death by selected causes • Figure 4.1

These 2011 data are for U.S. residents. Note that few of these risks apply to everyone. For example, only motorcyclists can die in motorcycle accidents.

GREATEST ▼
LEAST ▼

Heart disease 1 in 6
Fireworks discharge 1 in 386,766
Flood 1 in 175,803
Cancer 1 in 7
Earthquake 1 in 148,756
Stroke 1 in 28
Legal execution 1 in 96,691
Lightning 1 in 84,079
Hornet, wasp, or bee sting 1 in 46,477
Cataclysmic storm 1 in 46,044
Motor vehicle accident 1 in 88
Hot weather 1 in 12,517
Air/space accident 1 in 7,032
Suicide 1 in 112
Drowning 1 in 1,123
Motorcycle accident 1 in 770
Falling 1 in 171
Firearm assault 1 in 306
Pedestrian accident 1 in 649

National Safety Council

Four steps for risk assessment • Figure 4.2

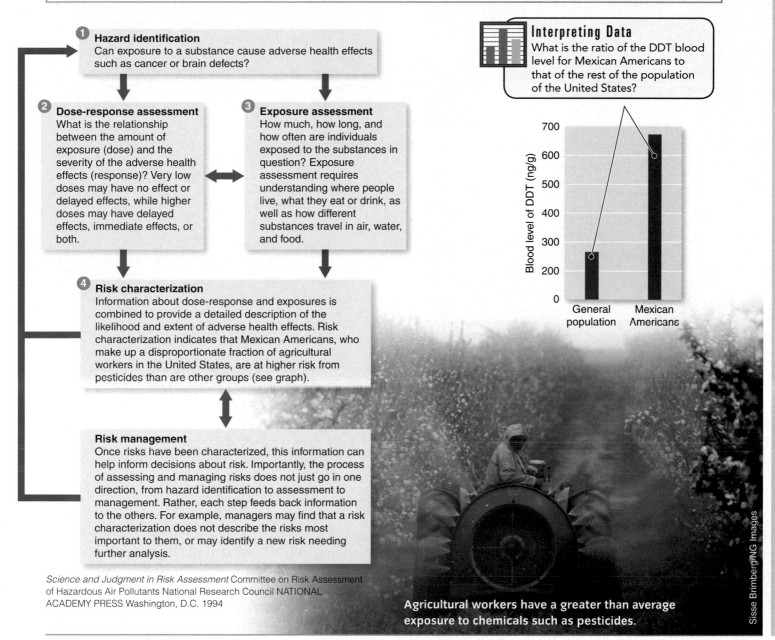

① Hazard identification
Can exposure to a substance cause adverse health effects such as cancer or brain defects?

② Dose-response assessment
What is the relationship between the amount of exposure (dose) and the severity of the adverse health effects (response)? Very low doses may have no effect or delayed effects, while higher doses may have delayed effects, immediate effects, or both.

③ Exposure assessment
How much, how long, and how often are individuals exposed to the substances in question? Exposure assessment requires understanding where people live, what they eat or drink, as well as how different substances travel in air, water, and food.

④ Risk characterization
Information about dose-response and exposures is combined to provide a detailed description of the likelihood and extent of adverse health effects. Risk characterization indicates that Mexican Americans, who make up a disproportionate fraction of agricultural workers in the United States, are at higher risk from pesticides than are other groups (see graph).

Risk management
Once risks have been characterized, this information can help inform decisions about risk. Importantly, the process of assessing and managing risks does not just go in one direction, from hazard identification to assessment to management. Rather, each step feeds back information to the others. For example, managers may find that a risk characterization does not describe the risks most important to them, or may identify a new risk needing further analysis.

Science and Judgment in Risk Assessment Committee on Risk Assessment of Hazardous Air Pollutants National Research Council NATIONAL ACADEMY PRESS Washington, D.C. 1994

Interpreting Data
What is the ratio of the DDT blood level for Mexican Americans to that of the rest of the population of the United States?

Blood level of DDT (ng/g) — General population / Mexican Americans

Agricultural workers have a greater than average exposure to chemicals such as pesticides.

Sisse Brimberg/NG Images

The four steps involved in **risk assessment** for adverse health effects are summarized in **Figure 4.2**. For example, we might learn that a chemical is a hazard—that is, it can harm human health. Next, we assess what sort of harm the chemical can do, how potent it is, and how much of it people are exposed to. We use these data to characterize the risk—that is, how much risk it poses, to which people, and how it can be reduced. Finally, we can use this characterization to guide our management choices—that is, what we do (if anything) to reduce the risk. Once

risk assessment
The quantitative and qualitative characterization of risks so that they can be compared, contrasted, and managed.

a risk assessment is performed, its results are combined with relevant political, social, and economic considerations to determine how we can best avoid, reduce, or eliminate a particular risk and, if so, what we should do. This evaluation includes the development and implementation of laws to regulate hazardous substances.

Risk assessment estimates the probability that an event will occur and lets us set priorities and manage risks in an appropriate way. As an example, consider a person who smokes a pack of cigarettes a day and drinks well water

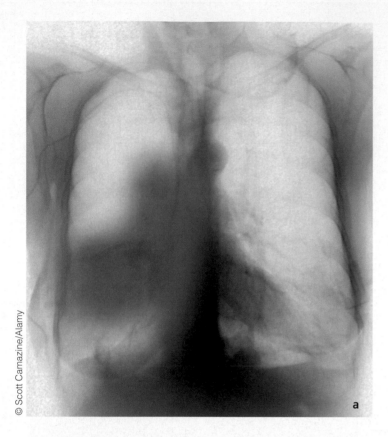

a

Lung cancer • Figure 4.3

a. Cancer was diagnosed in the right lung of this 73-year-old woman (shown by the red areas) after years of heavy smoking, a high-risk behavior. **b.** Data suggest that smoking is on the decline among younger people, although the poor and some minority groups have disproportionately high smoking rates. Percentage of U.S. students (grades 9–12) who smoked cigarettes on at least 1 or 2 of the past 30 days (CDCP 2011).

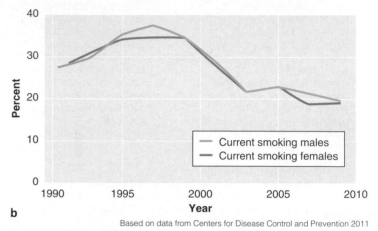

b

Based on data from Centers for Disease Control and Prevention 2011

Interpreting Data
In what year did smoking peak among males? Among females?

containing traces of the cancer-causing chemical trichloroethylene (in amounts permitted under Environmental Protection Agency [EPA] limits). Without knowledge of risk assessment, this person might buy bottled water in an attempt to reduce his or her chances of getting cancer. Based on risk assessment calculations, the annual risk of death from smoking is 0.00059, or 5.9×10^{-4}, whereas the annual risk from drinking water with EPA-accepted levels of trichloroethylene is 0.000000002, or 2.0×10^{-9}. This means that this person is almost *300,000 times* more likely to get cancer from smoking than to get it from ingesting such low levels of trichloroethylene (**Figure 4.3a**). Knowing this, the person in our example would, we hope, stop smoking.

A dilemma that makes risk management a challenge is that risk experts and nonexperts often disagree about which risks are most worrisome. Experts are often surprised when people seem to be far more concerned about small risks, such as those from exposure to small doses of a chemical, than about large risks, like those associated with obesity and smoking. We know, for example, that the average life expectancy of smokers is more than eight years less than that of nonsmokers, and almost one-third of all smokers die from diseases created or exacerbated by tobacco smoke.

There are several explanations for this type of thinking. One is that most of the decisions we make about risks are based on habit and culture, not analysis. Indeed, it would take far too much time and effort to apply analysis to all risk decisions. Fortunately, culture can shift peoples' habits over time. For example, the number of adolescents in the United States who smoke has dropped over the past two decades (**Figure 4.3b**).

Several factors determine which risks we are concerned about. One is trust in institutions—when people believe that business and government are managing risks, they are less concerned about them. Another is that we find some risks—such as dying from cancer—to be more dreadful than others—such as car accidents. We worry more about risks that we dread, even when analysis suggests they are less common. Similarly, we are more concerned about risks that are unfamiliar and those that we don't feel we can control.

CONCEPT CHECK STOP

1. **What** are risk and risk assessment?

2. **What** are the four steps of risk assessment?

Environmental Health Hazards

LEARNING OBJECTIVES

1. **Define** *toxicology* and *epidemiology*.
2. **Explain** why public water supplies are monitored for fecal coliform bacteria despite the fact that most strains of *E. coli* do not cause disease.
3. **Describe** the link between environmental changes and emerging diseases, such as swine flu.

The human body is exposed to many kinds of chemicals in the environment. Both natural and synthetic chemicals are in the air we breathe, the water we drink, and the food we eat. All chemicals, even "safe" chemicals such as sodium chloride (table salt), are toxic if exposure is high enough. For example, a 1-year-old child will die from ingesting about 2 tablespoons of table salt; table salt is also harmful to people with heart or kidney disease. Chemicals with adverse effects are known as **toxicants**.

Toxicology is one of two main approaches we use to understand threats to human health. Toxicologists (a) study the effects of toxicants on living organisms, (or parts of organisms, such as cells in a test tube), (b) evaluate the mechanisms that cause toxicity, and (c) develop ways to prevent or minimize adverse effects. (Developing appropriate handling or exposure guidelines for specific toxicants is one of these ways.)

Epidemiology involves studying how chemicals (toxicants), biological agents (disease), and physical hazards (accidents, radiation) affect the health of human populations. Epidemiologists study large groups of people and investigate a range of possible causes and types of diseases and injuries.

The effects of toxicants following exposure can be immediate (*acute toxicity*) or prolonged (*chronic toxicity*). Symptoms of **acute toxicity** range from dizziness and nausea to death. Acute toxicity occurs immediately to several days following a single exposure. In comparison, **chronic toxicity** generally produces damage following long-term, low-level exposure to a toxicant. Toxicologists know far less about chronic toxicity than they do about acute toxicity, partly because the symptoms of chronic toxicity often

> **toxicology** The study of toxicants, chemicals with adverse effects on health.
>
> **epidemiology** The study of the effects of chemical, biological, and physical agents on the health of human populations.
>
> **acute toxicity** Adverse effects that occur within a short period after high-level exposure to a toxicant.
>
> **chronic toxicity** Adverse effects that occur after a long period of low-level exposure to a toxicant.

mimic those of other chronic diseases associated with risky lifestyle patterns, poor nutrition, and aging. Also, it is difficult to isolate a causative agent from among the multiple toxicants we are routinely exposed to.

Disease-Causing Agents in the Environment

Disease-causing agents are *infectious* organisms, such as bacteria, viruses, protozoa, and parasitic worms that cause diseases. Typhoid, cholera, bacterial dysentery, polio, and infectious hepatitis are some of the most common bacterial or viral diseases that are transmissible through contaminated food and water. Diseases such as these are considered environmental health hazards. We do not discuss other human diseases, such as acquired immunodeficiency syndrome (AIDS), that are not transmissible through the environment.

The vulnerability of water supplies to waterborne disease-causing agents was dramatically demonstrated in 2000, when the first waterborne outbreak in North America of a deadly strain of *Escherichia coli* occurred in Ontario, Canada. Several people were killed, and several thousand became sick. Prior to this outbreak, this deadly *E. coli* strain had been transmitted almost exclusively through contaminated food.

The largest outbreak of a waterborne disease ever recorded in the United States occurred in 1993, when a microorganism (*Cryptosporidium*) contaminated the water supply in the greater Milwaukee area. About 370,000 people developed diarrhea. These and similar outbreaks raise concerns about the safety of our drinking water. Because sewage-contaminated water is an environmental threat to public health, periodic tests are made for the presence of sewage in our drinking water supplies. The best indicator of sewage-contaminated water is the presence of the common intestinal bacterium *E. coli* because it doesn't appear in the environment except from human and animal feces. Tests such as those for *E. coli* are used to indicate the possible presence of various disease-causing agents

Some human diseases transmitted by polluted water • Table 4.1

Disease	Type of organism	Symptoms
Cholera	Bacterium	Severe diarrhea, vomiting; fluid loss of as much as 20 quarts per day causes cramps and collapse
Dysentery	Bacterium	Infection of the colon causes painful diarrhea with mucus and blood in the stools; abdominal pain
Enteritis	Bacterium	Inflammation of the small intestine causes general discomfort, loss of appetite, abdominal cramps, and diarrhea
Typhoid	Bacterium	Early symptoms include headache, loss of energy, fever; later, a pink rash appears, along with (sometimes) hemorrhaging in the intestines
Infectious hepatitis	Virus	Inflammation of liver causes jaundice, fever, headache, nausea, vomiting, severe loss of appetite, muscle aches, and general discomfort
Poliomyelitis	Virus	Early symptoms include sore throat, fever, diarrhea, and aching in limbs and back; when infection spreads to spinal cord, paralysis and atrophy of muscles occur
Cryptosporidiosis	Protozoon	Diarrhea and cramps that last up to 22 days
Amoebic dysentery	Protozoon	Infection of the colon causes painful diarrhea with mucus and blood in the stools; abdominal pain
Schistosomiasis	Fluke	Tropical disorder of the liver and bladder causes blood in urine, diarrhea, weakness, lack of energy, repeated attacks of abdominal pain
Ancylostomiasis	Hookworm	Severe anemia, sometimes symptoms of bronchitis

SPL/Science Source Images

(**Table 4.1**). Although most strains of coliform bacteria found in sewage do not cause disease, testing for these bacteria is a reliable way to indicate the likely presence of **pathogens** in water.

pathogen An agent (usually a microorganism) that causes disease.

The **fecal coliform test** assesses whether *E. coli* is present in water (**Figure 4.4**). A small sample of water is passed through a filter to trap the bacteria, which are then transferred to a petri dish that contains nutrients. After an incubation period, the number of greenish colonies present indicates the number of *E. coli*. Safe drinking water should contain no more than one coliform bacterium per 100 mL of water (about ½ cup); safe swimming water should have no more than 200 per 100 mL of water; and general recreational water (for boating) should have no more than 2000 per

100 mL. In contrast, raw sewage may contain several million coliform bacteria per 100 mL of water. Water pollution and purification are discussed further in Chapter 10.

Environmental Changes and Emerging Diseases

Human health improved significantly in developed countries throughout the 20th century. Over the past two decades, however, diseases associated with poor diet and sedentary lifestyles have increased. Environmental factors remain a significant cause of human disease in many areas of the world. Epidemiologists have established links between human health and human activities that alter the environment. A 2010 World Health Organization report concluded that over 25 percent of disease and

Fecal coliform test • Figure 4.4

This test indicates the likely presence of disease-causing agents in water. A water sample is first passed through a filtering apparatus. **a.** A technician drops a growing medium and a water sample on a filter-lined petri dish. **b.** After incubation, the number of bacterial colonies is counted. Each colony of *Escherichia coli* arose from a single coliform bacterium in the original water sample.

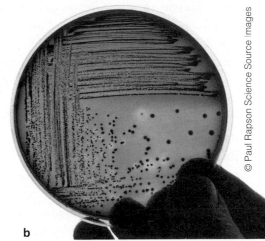

injury worldwide is related to human-caused environmental changes. The environmental component of human health is sometimes direct and obvious, as when people drink unsanitary water and contract dysentery, a waterborne disease that causes diarrhea. Diarrhea, which can be easily treated if resources are available, causes 4 million deaths worldwide each year, mostly in children.

The health effects of many human activities are complex and often indirect. For example, a recent study suggests that cholera, which in 2009 caused more than 4000 deaths in Zimbabwe, is a greater problem as temperatures increase due to climate change. The disruption of natural environments may give disease-causing agents an opportunity to thrive. Cutting down forests, building dams, and expanding agriculture may bring humans into contact with new or previously rare disease-causing agents by increasing the population and distribution of disease-carrying organisms such as mosquitoes (**Figure 4.5**).

Road clearing in the Amazon rain forest • Figure 4.5

The drainage ditches that will be added to each side of the road will hold standing water where mosquito larvae thrive.

Global Locator

NORTH AMERICA

Brazil

SOUTH AMERICA

NATIONAL GEOGRAPHIC

Social factors may also contribute to disease epidemics. Highly concentrated urban populations promote the rapid spread of infectious organisms among large numbers of people (**Figure 4.6**). Global travel also has the potential to contribute to the rapid spread of disease as infected individuals move easily from one place to another.

Consider malaria, a disease that mosquitoes transmit to humans. Each year, between 200 million and 500 million people worldwide contract malaria, resulting in more than 1 million deaths. About 60 different species of the *Anopheles* mosquito transmit the parasites that cause malaria. Each mosquito species thrives in its own unique combination of environmental conditions (such as elevation, amount of precipitation, temperature, relative humidity, and availability of surface water).

Many other diseases are carried by animals, including West Nile virus (mosquitoes), Lyme disease (deer ticks), and the Bubonic plague (rats and fleas). Efforts to control these diseases can have additional environmental impact. For example, since mosquitoes reproduce in stagnant water, they can be controlled by pesticides like DDT or by filling in or altering swamps and ponds. These activities in turn can have effects on other organisms. Also, the animals that carry disease have proved to be highly adaptive—both mosquitoes and malaria, for example, are now resistant to many pesticides and medicines. Another recent concern is pandemic influenza (flu). A **pandemic** disease reaches nearly every part of the world and has the potential to infect almost every person. Avian influenza is a strain of influenza virus that is common in birds. It tends to be difficult for humans to contract because it is usually transferred from bird to human but not from human to human. It is extremely potent once contracted and has a high fatality rate.

In late spring 2009 a strain of the swine flu appeared in Mexico, and by early summer was pandemic, killing thousands of people worldwide. Flu and other diseases generally spread more easily between related species: Humans are more closely related to pigs than to birds. (AIDS is thought to have originated from human contact with diseased monkeys.)

The international response to the 2009 swine flu outbreak continues to be controversial. Some argue that, since far fewer deaths occurred than predicted under worst-case assumptions, we overreacted. Others think that we should have done more, since we did not know that the strain would be relatively mild.

Crowds at an open-air market in Dhaka, Bangladesh • Figure 4.6 _____

The development of cities, and the concentration of people living in them, permit the rapid spread of infectious disease-causing agents.

Justin Guariglia/NG Image Collection

many governments and individuals. The virus often originates in areas that have dense populations of domestic animals, especially chickens and pigs, raised in small cages or close to human households (**Figure 4.7**). In the past decade, large numbers of domestic pigs, cows, and poultry have been killed and burned to prevent or stop disease outbreaks.

Researchers believe that many diseases could become more prevalent as the climate changes, both because temperature increases will extend the range of cold-intolerant diseases and because increased humidity and rainfall will benefit other diseases. For example, mosquito-borne diseases such as malaria and West Nile virus are expected to expand north and south both because winter temperatures will remain above those that kill mosquitoes and because there may be more standing water to harbor mosquito larvae.

People and livestock in close proximity • Figure 4.7

These children in rural Jiangxi, China, live and play near where pigs are raised. Humans and livestock, including chickens and pigs, share enough genetic similarities that some diseases can be transferred from species to species. This includes the swine flu epidemic of 2009, which started in rural Mexico and spread around the world in a matter of months.

Bay Ismoyo/AFP/Getty Images, Inc.

Understanding and controlling an influenza pandemic requires study of the environment that allows the virus to survive and travel, as well as cooperation among

---

CONCEPT CHECK **STOP**

1. **What** is the difference between toxicology and epidemiology?

2. **Why** is the fecal coliform test performed on public drinking water supplies?

3. **How** is the incidence of swine flu related to human activities that alter the environment?

Movement and Fate of Toxicants

LEARNING OBJECTIVES

1. **Distinguish** among persistence, bioaccumulation, and biological magnification of toxicants.

2. **Discuss** the mobility of persistent toxicants in the environment.

3. **Describe** the purpose of the Stockholm Convention on Persistent Organic Pollutants.

Some chemically stable toxicants are particularly dangerous because they resist degradation and readily move around in the environment. These include certain pesticides, radioactive isotopes, heavy metals such as mercury, flame retardants such as PBDEs (polybrominated diphenyl ethers), and industrial chemicals such as PCBs (polychlorinated biphenyls).

Mercury is the only metal that is liquid at room temperature. As such, it moves readily through air, water, and land. Mercury can cause brain and nerve damage, loss of hair, and even death.

The effects of the pesticide **DDT (dichlorodiphenyltrichloroethane)** on many bird species demonstrate the problem. Falcons, pelicans, bald eagles, and many other birds are sensitive to traces of DDT in their tissues.

✓ THE PLANNER

Biological magnification of DDT on a Long Island salt marsh

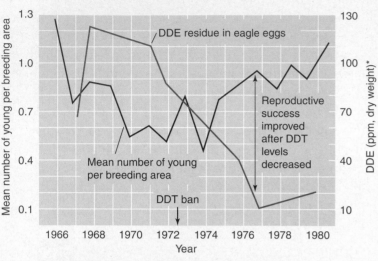

*DDT is converted to DDE in the birds' bodies

Data from Grier, JW "Ban of DDT and subsequent recovery of reproduction in bald eagles." *Science* (1982)

a. A comparison of the number of successful bald eagle offspring with the level of DDT residues in their eggs. (Grier, 1982).

Effects of DDT on birds

b. A bald eagle feeds its chicks.

Klaus Nigge/NG Image Collection

Interpreting Data

In which year was the DDE level in eagle eggs the highest? What was the level in that year?

c. A ring-billed gull forages on a beach.

James Hager/Robert Harding World Imagery/Getty Images

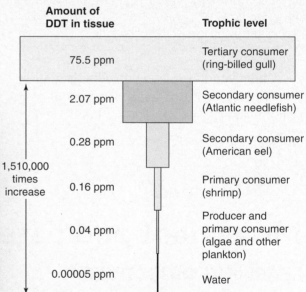

Amount of DDT in tissue	Trophic level
75.5 ppm	Tertiary consumer (ring-billed gull)
2.07 ppm	Secondary consumer (Atlantic needlefish)
0.28 ppm	Secondary consumer (American eel)
0.16 ppm	Primary consumer (shrimp)
0.04 ppm	Producer and primary consumer (algae and other plankton)
0.00005 ppm	Water

1,510,000 times increase

Data from Woodwell, G. M., C. F. Worster Jr. and P. A. Isaacson. "DDT residues in an east coast sanctuary: a case of biological concentration of persistent insecticide," *Science* 156 (May 12, 1967).

d. Note how the level of DDT, expressed as parts per million, increased in the tissues of various organisms as DDT moved through the food chain from producers to consumers (bottom to top of figure). The ring-billed gull at the top of the food chain had approximately 1.5 million times more DDT in its tissues than the water contained (Woodwell et al., 1967).

Substantial evidence indicates that DDT causes these birds to lay eggs with thin, fragile shells that usually break during incubation, causing the chicks' deaths. After 1972, the year DDT was banned in the United States, the reproductive success of many birds began to slowly improve.

The impact of DDT on birds is the result of (1) its persistence, (2) bioaccumulation, and (3) biological magnification. **Persistence** means that the substance is extremely stable and may take many years to break down into a less toxic form. When an organism can't metabolize (break down) or excrete a toxicant, it is simply stored, usually in fatty tissues. Over time, the organism may accumulate high concentrations of the toxicant.

The buildup of a persistent toxicant in an organism is **bioaccumulation** (**Figure 4.8a** and **b**). Organisms at the top of the food chain tend to store greater concentrations of bioaccumulated toxicants in their bodies than those lower on the food chain. As an example of biological magnification,

> **biological magnification** The increase in toxicant concentrations as a toxicant passes through successive levels of the food chain.

consider a food chain studied in a Long Island salt marsh that was sprayed with DDT over several years for mosquito control: algae and plankton shrimp → American eel → Atlantic needlefish → ring-billed gull (**Figure 4.8c** and **d**). All top carnivores, from fishes to humans, are at risk of health problems from biological magnification. Scientists therefore test pesticides to ensure that they do not persist and accumulate in the environment.

Mobility in the Environment

Persistent toxicants tend to move through the soil, water, and air, sometimes long distances. For example, pesticides applied to agricultural lands may be washed into rivers and streams by rain, harming aquatic life (**Figure 4.9**). If the pesticide level in their aquatic ecosystem is high enough, plants and animals may die. At lower pesticide levels, aquatic life may still suffer from symptoms of chronic toxicity such as bone degeneration in fishes.

Mobility of pesticides in the environment • Figure 4.9

The intended pathway of pesticides in the environment is shown in the tan column to the right of the figure, and the actual pathways are shown in the blue column to the left.

WileyPLUS

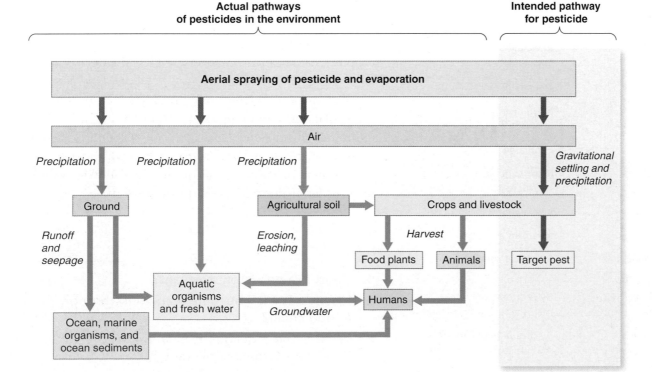

These symptoms may, for example, decrease fishes' competitiveness or increase their chances of being eaten by predators.

Mobility of persistent toxicants is also a risk for humans. In 1994 the Environmental Working Group, a private organization, analyzed five common herbicides (weed-killing chemicals) found in drinking water. It concluded that 3.5 million people in the Midwest face a slightly elevated cancer risk because of exposure to the herbicides. The EPA has since mandated a reduction in use of the five herbicides.

The Global Ban of Persistent Organic Pollutants

persistent organic pollutants (POPs) Persistent toxicants that bioaccumulate in organisms and travel through air and water to contaminate sites far from their source.

The **Stockholm Convention on Persistent Organic Pollutants**, which was adopted in 2001, is an important U.N. treaty that seeks to protect human health and the environment from the 12 most toxic **persistent organic pollutants (POPs)** on Earth (**Table 4.2**). Some

Persistent organic pollutants: The "dirty dozen" • Table 4.2

Persistent organic pollutant	Use
Aldrin	Insecticide
Chlordane	Insecticide
DDT (dichlorodiphenyl-trichloroethane)	Insecticide
Dieldrin	Insecticide
Endrin	Rodenticide and insecticide
Heptachlor	Fungicide
Hexachlorobenzene	Insecticide; fire retardant
Mirex™	Insecticide
Toxaphene™	Insecticide
PCBs (polychlorinated biphenyls)	Industrial chemicals
Dioxins	By-products of certain manufacturing processes
Furans (dibenzofurans)	By-products of certain manufacturing processes

Mosquito net • Figure 4.10

This net was treated with DDT to kill malaria-transmitting mosquitoes. Photographed in Nigeria.

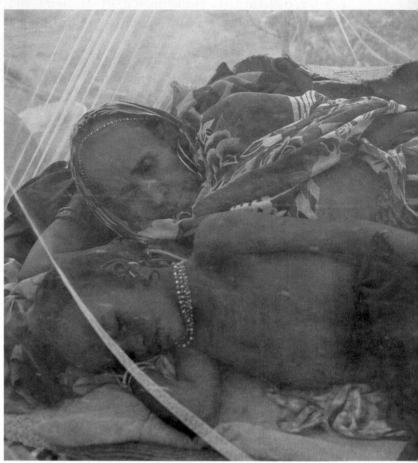

KAREN KASMAUSKI/NG Image Collection

POPs disrupt the endocrine system (discussed later in this chapter), cause cancer, or adversely affect the developmental processes of organisms.

The Stockholm Convention requires countries to develop plans to eliminate the production and use of intentionally produced POPs. A notable exception to this requirement is that DDT is still produced and used to control malaria-carrying mosquitoes in countries where no affordable alternatives exist (**Figure 4.10**).

CONCEPT CHECK STOP

1. **What** is a persistent toxicant?
2. **How** does DDT become magnified through a food chain?
3. **What** is the Stockholm Convention on Persistent Organic Pollutants?

Determining Health Effects of Pollutants

LEARNING OBJECTIVES

1. **Describe** how a dose–response curve is used to determine the health effects of environmental pollutants.

2. **Describe** the most common method of determining whether a chemical causes cancer.

3. **Distinguish** among additive, synergistic, and antagonistic interactions in chemical mixtures.

4. **Explain** why children are particularly susceptible to toxicants.

We assess the toxicity of a pollutant by the dose at which adverse effects are produced. A **dose** of a toxicant is the amount that enters the body of an exposed organism. Doses can be measured in several ways, but all include a quantity (mass or volume). A time component is important as well, since a milligram of a chemical all at once can have very different effects than the same milligram spread over a day or week. Finally, it is useful to know the weight of the person (or other organism) who is exposed: A large adult can tolerate a much larger dose of a toxin than can an infant.

The **response** is the type and amount of damage that exposure to a particular dose causes. A dose may cause death (*lethal dose*) or harm but not death (*sublethal dose*). Lethal doses, which are usually expressed in milligrams of toxicant per kilogram of body weight, vary depending on the organism's age, sex, health, and metabolism, as well as on how the dose was administered (all at once or over a period of time). The lethal doses, for humans, of many toxicants are known through records of homicides and accidental poisonings.

One way to determine acute toxicity is to administer different-sized doses to populations of laboratory animals, measure the responses, and use these data to predict the chemical effects on humans (**Figure 4.11**). The dose that is lethal to 50 percent of a population of test animals is called the **lethal dose–50 percent**, or LD_{50}. It is usually reported in milligrams of chemical toxicant per kilogram of body weight. An inverse relationship exists between the size of the LD_{50} and the acute toxicity of a chemical: The smaller the LD_{50}, the more toxic the chemical, and, conversely, the greater the LD_{50}, the less toxic the chemical (**Table 4.3**). The LD_{50} is determined for new synthetic chemicals—thousands are produced

Laboratory rat • Figure 4.11

The results of chemicals eaten or inhaled by laboratory animals are extrapolated to humans.

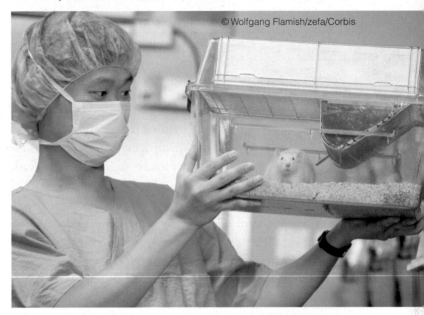

© Wolfgang Flamish/zefa/Corbis

LD_{50} values for selected chemicals (Josten and Wood 1996) • Table 4.3

Chemical	LD_{50} (mg/kg)*
Aspirin	1750
Oxycodone	920
Morphine	600
Caffeine	200
Cocaine	93
Methyl mercury	30
Ethanol	14
Nicotine	9.5
Sodium Cyanide	6.4
Strychnine	0.96

Administered orally to rats.

Interpreting Data
How much caffeine would an 80 kg (175 lb) person have to ingest to reach the LD_{50}.

each year—as a way of estimating their toxic potential. It is generally assumed that a chemical with a low LD_{50} for several species of test animals is also very toxic in humans.

The **effective dose–50 percent**, or ED_{50}, measures a wide range of biological responses, such as stunted development in the offspring of a pregnant animal, reduced enzyme activity, or onset of hair loss. The ED_{50} is the dose that causes 50 percent of a population to exhibit whatever response is under study.

To develop a dose–response curve, scientists first test the effects of high doses and then work their way down to a **threshold level**, the maximum dose that has no measurable effect (or, alternatively, the minimum dose that produces a measurable effect) (**Figure 4.12**). Scientists assume that doses lower than the threshold level are safe.

> **dose–response curve** In toxicology, a graph that shows the effects of different doses on a population of test organisms.

A growing body of evidence, however, suggests that for certain toxicants there is no safe dose. A threshold does not exist for these chemicals, and even the smallest amount causes a measurable response.

Cancer-Causing Substances

Because cancer is so feared, for many years it was the main effect evaluated in chemical risk assessment. Environmental contaminants are linked to many serious health concerns, including other diseases, birth defects, damage to the immune system, reproductive problems, and damage to the nervous system or other body systems. We focus here on risk assessment as it relates to cancer, but noncancer hazards are assessed in similar ways.

The most common method of determining whether a chemical causes cancer is to expose groups of laboratory animals, such as rats, to various large doses and count how many animals develop cancer at the different levels. This method is indirect and uncertain, however. For one thing, although humans and rats are both mammals, they are different organisms and may respond differently to exposure to the same chemical. (Even rats and mice often respond differently to the same toxicant.)

Another problem is that lab rats are exposed to massive doses of the suspected carcinogen relative to their body size, whereas humans are usually exposed to much lower amounts. Researchers must use large doses to cause cancer in a small group of laboratory animals within a reasonable amount of time. Otherwise, such tests would take years, require thousands of test animals, and be prohibitively expensive to produce enough data to have statistically significant results.

> **carcinogen** Any substance (for example, chemical, radiation, virus) that causes cancer.

Dose–response curves • Figure 4.12

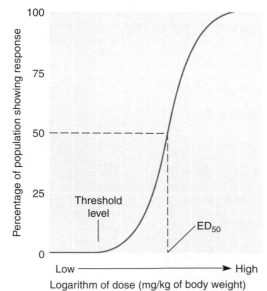

a. This hypothetical dose–response curve demonstrates two assumptions: first, that the biological response increases as the dose is increased; second, that harmful responses occur only above a certain threshold level.

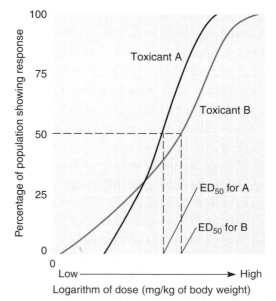

b. Dose–response curves for two hypothetical toxicants, A and B. In this example, A has a lower effective dose–50 percent (ED_{50}) than B. However, at lower doses, B is more toxic than A.

Risk assessment assumes that we can extrapolate (work backward) from the huge doses of chemicals and the high rates of cancer they cause in rats to determine the expected rates of cancer in humans exposed to lower amounts of the same chemicals. However, even if we are reasonably sure that exposure to high doses of a chemical causes the same effects for the same reasons in both rats and humans, we cannot assume that these same mechanisms work at low doses in humans. The body metabolizes small and large doses of a chemical in different ways. For example, enzymes in the liver may break down carcinogens in small quantities, but an excessive amount of carcinogen might overwhelm the liver enzymes.

In short, extrapolating from one species to another and from high doses to low doses is uncertain and may overestimate or underestimate a toxicant's danger. However, animal carcinogen studies provide valuable information: A toxicant that does not cause cancer in laboratory animals at high doses is not likely to cause cancer in humans at levels found in the environment or in occupational settings.

Scientists do not currently have a reliable way to determine whether exposure to small amounts of a substance causes cancer in humans. However, the EPA is planning to change how toxic chemicals are evaluated and regulated. Toxicologists are developing methods to provide direct evidence of the risk involved in exposure to low doses of cancer-causing chemicals (**Figure 4.13**).

Epidemiological evidence, including studies of human groups accidentally exposed to high levels of suspected carcinogens, is also used to determine whether chemicals are carcinogenic. For example, in 1989 epidemiologists in Germany established a direct link between cancer and a group of persistent organic pollutants called *dioxins* (see Table 4.2). They observed the incidence of cancer in workers exposed to high concentrations of dioxins during an accident at a chemical plant in 1953 and found unexpectedly high levels of cancers in their digestive and respiratory tracts.

Risk Assessment of Chemical Mixtures

Humans are frequently exposed to various combinations of chemical compounds, in the air we breathe, the food we eat, and the water we drink. For example, cigarette smoke contains a mixture of chemicals, as does automobile exhaust. Cigarette smoke is a mixture of air pollutants that includes hydrocarbons, carbon dioxide, carbon monoxide, particulate matter, nicotine, cyanide, and a

Measuring low doses of a toxicant (dioxin) in human blood serum • Figure 4.13

Scientists are developing increasingly sophisticated methods of biomonitoring to analyze human tissues and fluids. Photographed at the Centers for Disease Control and Prevention's Environmental Health Laboratory.

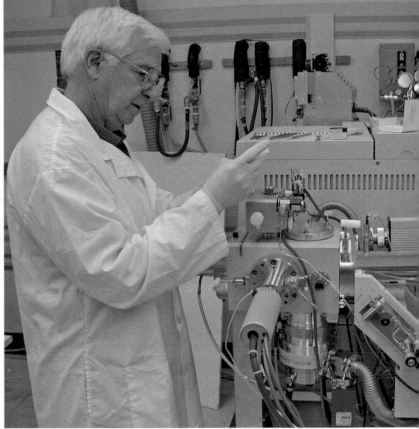

Courtesy Centers for Disease Control

small amount of radioactive materials that come from the fertilizer used to grow the tobacco plants.

The vast majority of toxicology studies have been performed on single chemicals rather than chemical mixtures, and for good reason. Mixtures of chemicals interact in a variety of ways, increasing the level of complexity in risk assessment, a field already complicated by many uncertainties. Moreover, there are simply too many chemical mixtures to evaluate.

Chemical mixtures interact by **additivity**, **synergy**, or **antagonism**. When a chemical mixture is *additive*, the effect is exactly what you would expect, given the individual effects of each component of the mixture. If a chemical with a toxicity level of 1 is mixed with a different chemical, also with a toxicity level of 1, the combined effect of exposure to the mixture is 2. A chemical mixture that is *synergistic* has a greater combined effect than expected; two chemicals, each with a toxicity level of 1, might have a combined toxicity of 3. For example, exposure to either tobacco smoke or inhaled asbestos can cause lung cancer. However smokers who worked in the asbestos industry were found to have much higher rates of lung cancer than expected from a simple additive relationship. An *antagonistic* interaction in a chemical mixture results in a smaller combined effect than expected; for example, two chemicals, each with a toxicity level of 1, might have a combined effect of 1.3.

If toxicological studies of chemical mixtures are lacking, how is risk assessment for chemical mixtures assigned? Toxicologists use *additivity* to assign risk to mixtures—that is, they add the known effects of each compound in the mixture. Such an approach sometimes overestimates or underestimates the actual risk involved, but it is the best method currently available. The alternative—waiting for years or decades until numerous studies have been designed, funded, and completed—is unreasonable.

Children and Chemical Exposure

Children weigh less than adults, tend to interact more with their environments and are undergoing rapid internal changes as they grow. They are also less aware of

EnviroDiscovery
Smoking: A Significant Risk

Tobacco use is the single largest cause of preventable death. Smoking causes serious diseases such as lung cancer, emphysema, and heart disease and is responsible for the premature deaths of nearly half a million people in the United States each year. Cigarette smoking annually causes about 120,000 of the 140,000 deaths from lung cancer in the United States. Smoking also contributes to heart attacks and strokes and to cancers of the bladder, mouth, throat, pancreas, kidney, stomach, voice box, and esophagus.

Passive smoking, which is nonsmokers' chronic breathing of smoke from cigarette smokers, also increases the risk of cancer. Passive smokers suffer more respiratory infections, allergies, and other chronic respiratory diseases than other nonsmokers. Passive smoking is particularly harmful to infants and young children, pregnant women, the elderly, and people with chronic lung disease.

There is good news and bad news about smoking. The good news is that fewer people in highly developed nations such as the United States are smoking. Recent research indicates that bans on smoking in public buildings and restaurants have reduced heart disease–related emergency room visits by as much as 17 percent.

The bad news is that more and more people are taking up the habit in China, Brazil, Pakistan, and other developing nations. Tobacco companies in the United States promote smoking abroad, and a substantial portion of the U.S. tobacco crop is exported. The World Health Organization (WHO) estimates that, worldwide, 5 million people die each year of smoking-related causes. In an attempt to establish a global ban on tobacco advertising, WHO developed the Framework Convention on Tobacco Control. The treaty went into effect in 2005, but as of early 2012, the United States was not among the 167 parties that had joined.

© Petrut Calinescu/Alamy

 A sugarcane worker smokes during a break in his long, hard day

Cigarette use continues to increase in many developing countries, even as consumption of tobacco decreases in the United States. Photographed in the Philippines.

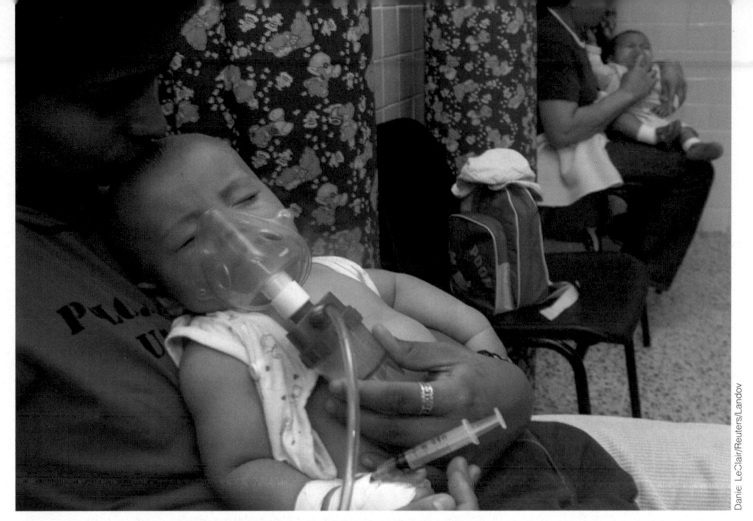

Air pollution and respiratory disease in children • Figure 4.14

A Honduran mother gives oxygen to her baby, who suffers from environmentally linked respiratory disease. Farmers nearby burn land to prepare for the planting season; the resulting smoke triggers breathing problems, mostly in children and the elderly.

potential risks from exposures. Consequently, they are often more susceptible than adults to the effects of chemicals. Consider a toxicant with an LD_{50} of 100 mg/kg. A potentially lethal dose for a child who weighs 11.3 kg (25 lb) is $100 \times 11.3 = 1130$ mg, which is equal to a scant ¼ teaspoon if the chemical is a liquid. In comparison, the potentially lethal dose for an adult who weighs 68 kg (150 lb) is 6800 mg, or about 2 teaspoons. Consequently, policies designed to protect children from chemical exposures must account for these differences.

Children and Pollution Consider the toxicants in air pollution. Air pollution is a greater health threat to children than it is to adults (**Figure 4.14**). Lungs continue to develop throughout childhood, and air pollution restricts lung development. In addition, a child has a higher metabolic rate than an adult and therefore needs more oxygen. To obtain this oxygen, a child breathes more air—about two times as much air per pound of body weight as an adult. This means that a child also breathes more air pollutants into the lungs. A 1990 study in which autopsies were performed on 100 Los Angeles children who died for reasons unrelated to respiratory problems found that more than 80 percent had subclinical lung damage, which is lung disease in its early stages, before clinical symptoms appear. (Los Angeles has some of the worst air quality in the world.)

CONCEPT CHECK STOP

1. **What** is a dose–response curve?
2. **What** is one way that scientists determine whether a chemical causes cancer? What are two problems with this method?
3. **What** are the three ways that chemical mixtures interact?
4. **Why** are children particularly susceptible to toxicants?

The Precautionary Principle

LEARNING OBJECTIVE

1. **Discuss** the precautionary principle as it relates to the introduction of new technologies or products.

You've probably heard the expression "An ounce of prevention is worth a pound of cure." This statement is the heart of the precautionary principle that many politicians and environmental activists advocate. According to the precautionary principle, we should not introduce new technology, practice, or material until it is demonstrated that (a) the risks are small and (b) the benefits outweigh the risks. The precautionary principle puts the burden of proof on the developers of the new technology or substance. The precautionary principle might require that all new industrial chemicals be tested for acute and chronic toxicity before they can be sold.

The precautionary principle is also applied to existing technologies when new evidence suggests that they are more dangerous than originally thought. For example, when observations and experiments suggested that lead added to gasoline as an anti-knock ingredient was contaminating soil, particularly in inner cities near major highways, the precautionary principle led to the phase-out of leaded gasoline (**Figure 4.15**).

To many people the precautionary principle is just common sense, given that science and risk assessment often cannot provide

> **precautionary principle** The idea that new technologies, practices, or materials should not be adopted until there is strong evidence that they will not adversely affect human or environmental health.

Children in the South Bronx of New York City • Figure 4.15

In addition to air pollution, these children are probably exposed to lead in the soil (from leaded gasoline) and in paint from old buildings. Children with even low levels of lead in their blood may suffer from partial hearing loss, hyperactivity, attention deficit, lowered IQ, and learning disabilities.

Andy Levin/Science Source Images

Precautionary principle or economic protectionism? • **Figure 4.16**

Is Europe's ban of U.S. beef the result of its concern over the safety of hormone-treated beef or an excuse to support the European beef industry? Photographed on a range in Wyoming.

definitive answers to policy makers dealing with environmental and public health problems. The developers of a new technology or substance must prove that it is safe instead of society proving that it is harmful after it has already been introduced. However, the precautionary principle does not require that developers provide absolute proof that their product is safe; such proof would be impossible to provide.

Certain laws and decisions in many European Union nations have incorporated the precautionary principle, and some laws in the United States have a precautionary tone. In 2000, **Christine Todd Whitman,** then governor of New Jersey, said in a speech to the National Academy of Sciences:

Policy makers need to take a precautionary approach to environmental protection. . . . We must acknowledge that uncertainty is inherent in managing natural resources, recognize it is usually easier to prevent environmental damage than to repair it later, and shift the burden of proof away from those advocating protection toward those proposing an action that may be harmful.

The precautionary principle has generated much controversy. Some scientists fear that the precautionary principle challenges the role of science and endorses making decisions without the input of science. Some critics contend that its imprecise definition reduces trade and limits technological innovations. For example, several European countries made precautionary decisions to ban beef from the United States and Canada because these countries use growth hormones to make the cattle grow faster. Europeans contend that the growth hormone might harm humans eating the beef, but the ban, in effect since 1989, is widely viewed as protecting the European beef industry (**Figure 4.16**). Another international controversy in which the precautionary principle is involved is the cultivation of genetically modified foods (discussed further in Chapter 14).

Climate change is an area where the precautionary principle is often invoked. Increasing the amounts of greenhouse gases in the atmosphere not only causes global warming, but will lead to a variety of economic,

CASE STUDY

Endocrine Disrupters

Mounting evidence suggests that dozens of widely used industrial and agricultural chemicals are **endocrine disrupters**, which interfere with the normal actions of the endocrine system (the body's hormones) in humans and animals. These chemicals include chlorine-containing industrial compounds known as PCBs and dioxins, the heavy metals lead and mercury, pesticides such as DDT, and certain plastics and plastic additives.

Hormones are chemical messengers that organisms produce to regulate their growth, reproduction, and other important biological functions. Some endocrine disrupters mimic *estrogens*, a class of female sex hormones. Other endocrine disrupters mimic *androgens* (male hormones such as testosterone) or *thyroid hormones*. Like hormones, endocrine disrupters are active at very low concentrations and therefore may cause significant health effects at relatively low doses.

Many endocrine disrupters appear to alter the reproductive development of various animal species. A chemical spill in 1980 contaminated Lake Apopka, Florida's third largest lake, with DDT and other agricultural chemicals that have known estrogenic properties. In the years following the spill, male alligators had low levels of testosterone and elevated levels of estrogen. The mortality rate for eggs in this lake was extremely high, which reduced the alligator population for many years (see photo). Fortunately, a 2006 study indicates that Lake Apopka's alligator population is recovering.

Humans may also be at risk from endocrine disrupters. The number of reproductive disorders, infertility cases, and hormonally related cancers (such as testicular cancer and breast cancer) appears to be increasing. However, we cannot make definite connections between environmental endocrine disrupters and human health problems at this time because of the limited number of human studies. Complicating such assessments is the fact that humans are also exposed to *natural*, hormone-mimicking substances in the plants we eat.

For example, soy-based foods such as bean curd and soymilk contain natural estrogens.

Congress amended the *Food Quality Protection Act* and the *Safe Drinking Water Act* in 1996 to require the Environmental Protection Agency to develop a plan and establish priorities to test thousands of chemicals for their potential to disrupt endocrine systems. Chemicals testing positive are tested further to determine what specific damage, if any, they cause to reproduction and other biological functions. These tests, which may take decades to complete, should reveal the level of human and animal exposure to endocrine disrupters and the effects of this exposure.

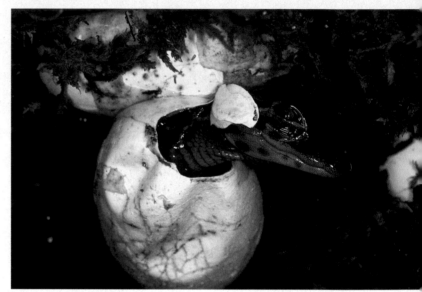

Steve Cooper/Science Source Images

Lake Apopka alligators

A young American alligator hatches from eggs that University of Florida researchers took from Lake Apopka, Florida. Many of the young alligators that hatch have abnormalities in their reproductive systems. This young alligator may not leave any offspring.

environmental, and social changes. Given the certainty that changes are occurring, but high uncertainty about the extent, severity, and timing of the effects, many scientists and policy makers argue that reducing emissions quickly is necessary to avoid future losses and suffering.

CONCEPT CHECK **STOP**

1. **What** is the precautionary principle? What are two criticisms of the precautionary principle?

Summary

1 A Perspective on Risks 74

1. A **risk** is the probability of harm (such as injury, disease, death, or environmental damage) occurring under certain circumstances. **Risk assessment** is the quantitative and qualitative characterization of risks that allows us to compare, contrast, and manage them.

2. Risk assessment characterizes the dose–response relationship between exposures to hazards and the effects of those exposures. These characterizations can be used to help inform decisions about how best to avoid, reduce, or eliminate risks.

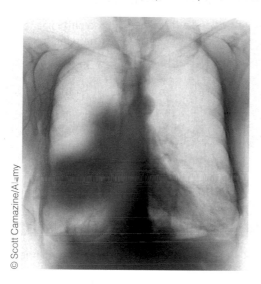

© Scott Camazine/Alamy

2 Environmental Health Hazards 77

1. **Toxicology** is the study of **toxicants**, chemicals that have adverse effects on health. **Epidemiology** is the study of the effects of chemical, biological, and physical agents on the health of human populations.

Bay Ismoyo/AFP/Getty Images, Inc.

2. While most strains of coliform bacteria do not cause disease, the **fecal coliform test** is a reliable way to indicate the likely presence of **pathogens**, or disease-causing agents, in water.

3. Over 25 percent of disease and injury worldwide is related to human-caused environmental changes. The environmental component of human health is sometimes direct, as when people drink unsanitary water and contract a waterborne disease. The health effects of other human activities are complex and indirect, as when climate change allows disease-causing agents to prosper.

3 Movement and Fate of Toxicants 81

1. Some toxicants exhibit **persistence**—they are extremely stable in the environment and may take many years to break down into less toxic forms. **Bioaccumulation** is the buildup of a persistent toxicant in an organism's body. **Biological magnification** is the increase in toxicant concentration as a toxicant passes through successive levels of the food chain.

2. Persistent toxicants do not stay where they are applied but tend to move through the soil, water, and air, sometimes long distances. For example, pesticides applied to agricultural lands may wash into rivers and streams, harming fishes.

3. The Stockholm Convention on Persistent Organic Pollutants requires countries to eliminate the production and use of the 12 worst **persistent organic pollutants (POPs)**. POPs are a group of persistent toxicants that bioaccumulate in organisms and travel thousands of kilometers through air and water, contaminating sites far removed from their source.

Klaus Nigge/NG Image Collection

The Flow of Energy Through Ecosystems

LEARNING OBJECTIVES

1. **Define** *energy*, and state the first and second laws of thermodynamics.

2. **Distinguish** among producers, consumers, and decomposers.

3. **Summarize** how energy flows through a food web and transfers between trophic levels.

4. **Distinguish** between gross primary productivity and net primary productivity.

Energy is the capacity or ability to do work. Organisms require energy to grow, move, reproduce, and maintain and repair damaged tissues. Energy exists as stored energy—called **potential energy**—and as **kinetic energy**, the energy of motion. We can think of potential energy as an arrow on a drawn bow (**Figure 5.2**). When the string is released, this potential energy is converted to kinetic energy as the motion of the bow propels the arrow. Similarly, the grass a bison eats has chemical potential energy, some of which is converted to kinetic energy and heat as the bison runs across the prairie. Thus, energy changes from one form to another.

first law of thermodynamics
A physical law which states that energy cannot be created or destroyed, although it can change from one form to another.

photosynthesis
The biological process that captures light energy and transforms it into the chemical energy of organic molecules, which are manufactured from carbon dioxide and water.

The First and Second Laws of Thermodynamics

Thermodynamics is the study of energy and its transformations. Two laws about energy apply to all things in the universe: the first and second laws of thermodynamics. According to the **first law of thermodynamics**, an organism may utilize energy by converting it from one form to another, but the total energy content of the organism and its surroundings is always the same. An organism can't create the energy it requires to live. Instead, it must capture energy from the environment to use for biological work, a process that involves the transformation of energy from one form to another. In **photosynthesis**, for example,

---

Potential and kinetic energy • Figure 5.2

Potential energy is stored in the drawn bow (**a**) and is converted to kinetic energy (**b**) as the arrow speeds toward its target. Photographed in Athens, Greece, during the 2004 Summer Olympics.

Capturing energy from the environment • Figure 5.3

The sun powers photosynthesis, producing chemical energy stored in the leaves and seeds of this umbrella tree. Photographed in Hanging Rock State Park, North Carolina.

plants absorb the radiant energy of the sun and convert it into the chemical energy contained in the bonds of sugar molecules (**Figure 5.3**). Later, through **cellular respiration**, an animal that eats the plant may transform some of the chemical energy stored in the plant into the mechanical energy of muscle contraction, enabling the animal to walk, run, slither, fly, or swim.

As each energy transformation occurs, some of the energy is changed to heat that is released into the cooler surroundings. No organism can ever use this energy again for biological work; it is "lost" from the biological point of view. However, it isn't gone from a thermodynamic point of view because it still exists in the surrounding physical environment. The use of food to enable you to walk or run doesn't destroy the chemical energy that was once present in the food molecules. After you have performed the task of walking or running, the energy still exists in your surroundings as heat.

According to the **second law of thermodynamics**, the amount of usable energy available to do

second law of thermodynamics A physical law which states that when energy is converted from one form to another, some of it is degraded into heat, a less usable form that disperses into the environment.

work in the universe decreases over time. The second law of thermodynamics is consistent with the first law—that is, the total amount of energy in the universe isn't decreasing with time. However, the total amount of energy in the universe available to do biological work is decreasing over time.

Less usable energy is more diffuse, or disorganized, than more usable energy. *Entropy* is a measure of this disorder or randomness. Organized, usable energy has low entropy, whereas disorganized energy such as low-temperature heat has high entropy. Another way to explain the second law of thermodynamics is that entropy, or disorder, in a system tends to increase over time. As a result of the second law of thermodynamics, no process that requires an energy conversion is ever 100 percent efficient because much of the energy is dispersed as heat, resulting in an increase in entropy. For example, an automobile engine, which converts the chemical energy of gasoline to mechanical energy, is between 20 and 30 percent efficient: Only 20 to 30 percent of the original energy stored in the chemical bonds of the gasoline molecules is actually transformed into mechanical energy, or work.

Organisms are highly organized and at first glance appear to refute the second law of thermodynamics. However, organisms maintain their degree of order over time only with the constant input of energy. That is why plants must photosynthesize and why animals must eat food.

Producers, Consumers, and Decomposers

The organisms of an ecosystem are divided into three categories, based on how they obtain nourishment: producers, consumers, and decomposers. Virtually all ecosystems contain representatives of all three groups, which interact extensively with one another, both directly and indirectly.

Plants and other photosynthetic organisms are **producers** and manufacture large organic molecules from simple inorganic substances, generally carbon dioxide and water, usually using the energy of sunlight. Producers are potential food resources for other organisms because they incorporate the chemicals they manufacture into their own bodies. Plants are the most significant producers on land, and algae and certain types of bacteria are important producers in aquatic environments.

Animals are **consumers**—they consume other organisms as a source of food energy and bodybuilding materials. Consumers that eat producers are *primary*

Food web at the edge of an eastern U.S. deciduous forest • Figure 5.6

THE PLANNER ✓

This food web is greatly simplified compared to what actually happens in nature. Many species aren't included, and numerous links in the web aren't shown.

Key

1. Pitch pine
2. White oak
3. Barred owl
4. Gray squirrel
5. Eastern chipmunk
6. Eastern cottontail
7. Red fox
8. White-tailed deer
9. Red-tailed hawk
10. Eastern bluebird
11. Red-winged blackbird
12. Blackberry
13. American robin
14. Red-bellied woodpecker
15. Red clover
16. Bacteria
17. Worms and ants
18. Moths
19. Deer mouse
20. Spiders
21. Insect larvae
22. Insects
23. Fungi

Think Critically Compare and explain the very different locations of decomposers and producers.

Pyramid of numbers • Figure 5.7

This pyramid is for a hypothetical area of temperate grassland; in this example, 10,000 grass plants support 10 mice, which support one bird of prey. (Note that decomposers are not shown.)

NUMBER OF INDIVIDUALS · TROPHIC LEVEL

1 — Secondary consumer (bird of prey)

10 — Primary consumer (field mouse)

10,000 — Producers (grass)

Think Critically Use what you know about the movement of energy in a food web to explain why there are so many more organisms at the bottom of this pyramid than at the top.

Once an organism uses energy, it is lost as heat (recall the second law of thermodynamics) and is unavailable for any other organism in the ecosystem.

Organisms at each step of a food chain use a large amount of the potential energy available to them. Because this energy is ultimately lost into the environment as heat, the number of steps in any food chain is limited. The longer the food chain, the less energy is available for organisms at the higher trophic levels.

An important feature of energy flow is that most of the energy going from one trophic level to the next in a food chain or food web dissipates into the environment as a result of the second law of thermodynamics. **Ecological pyramids** often graphically represent the relative energy values of each trophic level. A *pyramid of numbers* shows the number of organisms at each trophic level in a given ecosystem, with greater numbers illustrated by a larger area for that section of the pyramid (**Figure 5.7**). In most pyramids of numbers, the organisms at the base of the food chain are the most abundant, and fewer organisms occupy each successive

trophic level. Another type of ecological pyramid, *pyramids of energy*, are shaped similarly, illustrating how energy dissipates into the environment as it moves from one trophic level to another.

Ecosystem Productivity

The original source of energy in all ecosystems is the sun. The **gross primary productivity (GPP)** of an ecosystem is the rate at which energy is captured during photosynthesis. (Gross and net primary productivities are referred to as *primary* because plants occupy the first trophic level in food webs.) Of course, plants respire to provide energy for their own use, and this acts as a drain on productivity. Energy in plant tissues after cellular respiration has occurred is **net primary productivity (NPP)**. That is, NPP is the amount of biomass found in excess of that broken down by a plant's cellular respiration. Only the energy represented by NPP is available as food for an ecosystem's consumers.

Ecosystems differ strikingly in their productivities (**Figure 5.8**). On land, tropical rain forests have the

Estimated annual net primary productivities (NPP) for selected ecosystems • Figure 5.8

NPP is expressed as grams of dry matter produced per square meter per year.

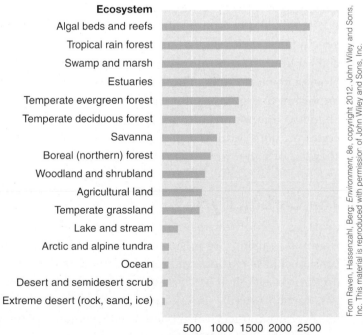

Ecosystem
- Algal beds and reefs
- Tropical rain forest
- Swamp and marsh
- Estuaries
- Temperate evergreen forest
- Temperate deciduous forest
- Savanna
- Boreal (northern) forest
- Woodland and shrubland
- Agricultural land
- Temperate grassland
- Lake and stream
- Arctic and alpine tundra
- Ocean
- Desert and semidesert scrub
- Extreme desert (rock, sand, ice)

500 1000 1500 2000 2500
Estimated net primary productivity (grams/square meter/year)

highest NPP, because of their abundant rainfall, warm temperatures, and intense sunlight. Tundra, with its harsh, cold winters, and deserts, with their lack of precipitation, are the least productive terrestrial ecosystems. The most productive aquatic ecosystems are algal beds and coral reefs. The lack of available nutrient minerals in some regions of the open ocean makes them extremely unproductive, equivalent to aquatic deserts. (Earth's major aquatic and terrestrial ecosystems are discussed in Chapter 6.)

| CONCEPT CHECK | STOP |

1. **What** is the first law of thermodynamics? the second?
2. **Why** is a balanced ecosystem unlikely to contain only producers and consumers? only consumers and decomposers?
3. **How** does energy move through a food web?
4. **How** does gross primary productivity differ from net primary productivity?

The Cycling of Matter in Ecosystems

LEARNING OBJECTIVE

1. **Diagram and explain** the carbon, hydrologic, nitrogen, sulfur, and phosphorus cycles.

In contrast to energy flow, matter, the material of which organisms are composed, moves in numerous cycles from one part of an ecosystem to another—from one organism to another and from living organisms to the abiotic environment and back again. We call these cycles of matter **biogeochemical cycles** because they involve biological, geological, and chemical interactions. Five different biogeochemical cycles of matter—carbon, hydrologic, nitrogen, sulfur, and phosphorus—are representative of all biogeochemical cycles. These five cycles are particularly important to organisms, for these materials make up the chemical compounds of cells. Humans affect all of these cycles on both local and global scales; we conclude the chapter with an example of this human influence.

The Carbon Cycle

Proteins, carbohydrates, and other molecules that are essential to living organisms contain carbon, so organisms must have carbon available to them. Carbon makes up approximately 0.04 percent of the atmosphere as a gas, carbon dioxide (CO_2). It is present in the ocean in several chemical forms, such as carbonate (CO_3^{2-}) and bicarbonate (HCO_3^-), and in sedimentary rocks such as limestone, which consists primarily of calcium carbonate ($CaCO_3$). The global movement of carbon between organisms and the abiotic environment—including the atmosphere, ocean, and sedimentary rock—is known as the *carbon cycle* (**Figure 5.9**).

During photosynthesis, plants, algae, and certain bacteria remove carbon (as CO_2) from the air and fix (incorporate) it into chemical compounds such as sugar. Plants use sugar to make other compounds. Thus, photosynthesis incorporates carbon from the abiotic environment into the biological compounds of producers. Those compounds are usually used as fuel for cellular respiration by the producer that made them, by a consumer that eats the producer, or by a decomposer that breaks down the remains of the producer or consumer. During respiration, sugar is broken down to carbon dioxide that is returned to the atmosphere. A similar carbon cycle occurs in aquatic ecosystems, involving carbon dioxide dissolved in the water.

Sometimes the carbon in biological molecules isn't recycled back to the abiotic environment for quite a while. For example, a large amount of carbon is stored in the wood of trees, where it may stay for several hundred years or even longer. Coal, oil, and natural gas, called *fossil fuels* because they formed from the remains of ancient organisms, are vast deposits of carbon compounds—the end products of photosynthesis that occurred millions of years ago. In *combustion*, organic molecules in wood, coal, oil, and natural gas are burned, with accompanying releases of heat, light, and carbon dioxide. (The

The carbon cycle • Figure 5.9

The movement of carbon between the abiotic environment (the atmosphere and ocean) and living organisms is known as the carbon cycle. Because proteins, carbohydrates, and other living molecules contain carbon, the process is essential to life. Sedimentary rocks and fossil fuels hold almost all of Earth's estimated 10^{23} g of carbon. The values shown for some of the active pools in the global carbon budget are expressed as 10^{15} g of carbon. For example, the soil contains an estimated 1500 x 10^{15} g of carbon.

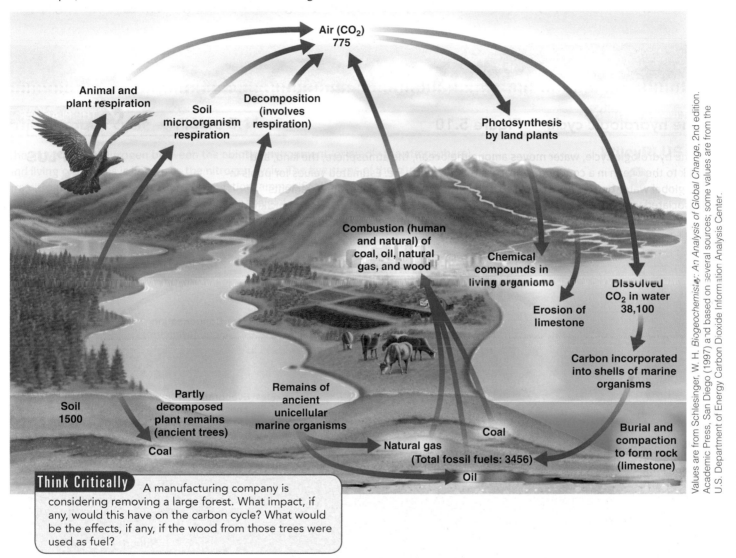

Values are from Schlesinger, W. H. *Biogeochemistry: An Analysis of Global Change*, 2nd edition. Academic Press, San Diego (1997) and based on several sources; some values are from the U.S. Department of Energy Carbon Dioxide Information Analysis Center.

Think Critically A manufacturing company is considering removing a large forest. What impact, if any, would this have on the carbon cycle? What would be the effects, if any, if the wood from those trees were used as fuel?

case study at the end of this chapter explores how global climate change—particularly as driven by the release of carbon dioxide from burning fossil fuels—affects the carbon cycle.)

The thick deposits of shells of marine organisms contain carbon. These shells settle to the ocean floor and are eventually cemented together to form the sedimentary rock limestone. The crust is dynamically active, and over millions of years, sedimentary rock on the bottom of the seafloor

may lift to form land surfaces. The summit of Mount Everest, for example, is composed of sedimentary rock.

The Hydrologic Cycle

In the *hydrologic cycle* water continuously circulates from the ocean to the atmosphere to the land and back to the ocean. It provides a renewable supply of purified water for terrestrial organisms. This cycle results in water

Bacteria are the only organisms involved in each of these steps except assimilation.

Nitrogen-fixing bacteria carry out *nitrogen fixation* in soil and aquatic environments. The process gets its name from the fact that nitrogen is fixed into a form that organisms can use, ammonia (NH_3). Volcanic activity, lightning, and human activities—combustion and industrial processes—also fix considerable nitrogen because all supply enough energy to break apart atmospheric nitrogen.

Nitrogen-fixing bacteria split atmospheric nitrogen and combine the resulting nitrogen atoms with hydrogen. Some nitrogen-fixing bacteria, *Rhizobium*, live inside swellings, or nodules, on the roots of legumes such as beans or peas and some woody plants (**Figure 5.12a**). In moist environments, photosynthetic bacteria called *cyanobacteria* perform most of the nitrogen fixation (**Figure 5.12b**).

During *nitrification*, soil bacteria convert ammonia to nitrate (NO_3^-). The process of nitrification furnishes these bacteria, called nitrifying bacteria, with energy. In *assimilation*, plants absorb ammonia or nitrate through their roots and convert the nitrogen into plant compounds such as proteins. Animals assimilate nitrogen when they consume plants or other animals and convert the proteins into animal proteins.

Ammonification occurs when organisms produce nitrogen-containing waste products such as urine. These substances, plus the nitrogen compounds that occur in dead organisms, are decomposed, releasing the nitrogen into the abiotic environment as ammonia. The bacteria that perform this process are called ammonifying bacteria. Other bacteria perform *denitrification*, in which nitrate is converted back to nitrogen gas. Denitrifying bacteria typically live and grow where there is little or no free oxygen. For example, they are found deep in the soil near the water table, an environment that is nearly oxygen free.

Human activities have disturbed the balance of the global nitrogen cycle. Nitrogen in fertilizers washes into rivers, lakes, and coastal areas, where it stimulates the growth of algae. As these algae die, their decomposition by bacteria robs the water of dissolved oxygen, which in turn causes many fishes and other aquatic organisms to die of suffocation. These no-oxygen conditions have formed large *dead zones* in about 150 coastal areas around the world (see the Chapter 11 case study). Nitrogen compounds are also released into the atmosphere as air pollutants when fossil fuels are burned, contributing to *photochemical smog* (see Chapter 8) and *acid deposition* (see Chapter 9).

The Sulfur Cycle

Scientists are still piecing together how the global sulfur cycle works. Most sulfur is underground in sedimentary rocks and minerals, which over time erode to release sulfur-containing compounds into the ocean (**Figure 5.13**). Sulfur gases enter the atmosphere from natural sources in both the ocean and land. Sea spray delivers sulfates (SO_4^{2-}) into the air, as do forest fires and dust storms. Volcanoes release both hydrogen sulfide (H_2S), a poisonous gas that smells like rotten eggs, and sulfur oxides (SO_x). Hydrogen sulfide reacts with oxygen to form sulfur oxides, and sulfur oxides react with water to form sulfuric acid (H_2SO_4). Although sulfur gases make up a minor part of the atmosphere, the total movement of sulfur to and from the atmosphere is substantial.

A tiny fraction of global sulfur is present in living organisms, where it is an essential component of proteins. Plant roots absorb sulfate and incorporate the sulfur into plant proteins. Animals assimilate sulfur when they

Nitrogen fixation • Figure 5.12

a. Bacteria carry out nitrogen fixation in the nodules of a pea plant's roots.

b. *Nostoc*, a cyanobacterium that fixes nitrogen, grows here on a mossy bank. This particular species forms colonies that range in size from a pinhead to a potato.

STEPHEN SHARNOFF/NG
Image Collection

Dr Jeremy Burgess/Science Source Images

The sulfur cycle • Figure 5.13

The largest sources of sulfur on Earth are sedimentary rock and the ocean. In the sulfur cycle, sulfur compounds are incorporated into organisms and move among them, the atmosphere, the ocean, and land. The values shown in the figure for the global sulfur budget are expressed in units of 10^{12} g of sulfur per year. For example, human-produced gases (air pollution) emit an estimated 90×10^{12} g of sulfur per year into the atmosphere.

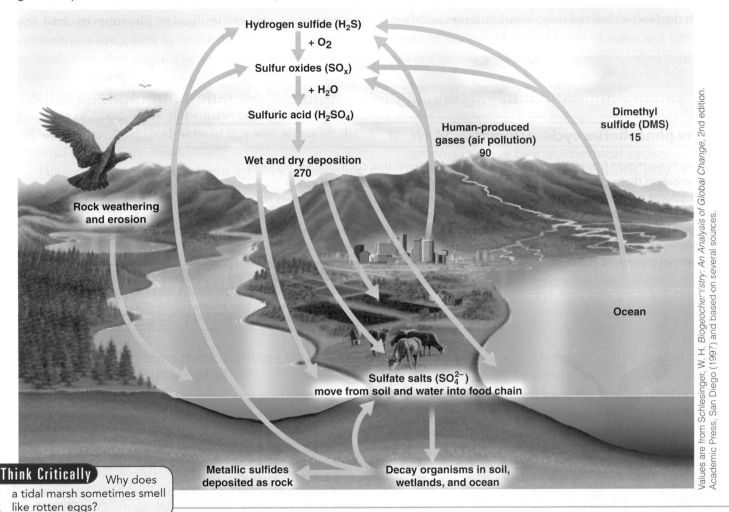

Hydrogen sulfide (H_2S)

+ O_2

Sulfur oxides (SO_x)

+ H_2O

Sulfuric acid (H_2SO_4)

Wet and dry deposition
270

Rock weathering and erosion

Human-produced gases (air pollution)
90

Dimethyl sulfide (DMS)
15

Ocean

Sulfate salts (SO_4^{2-}) move from soil and water into food chain

Metallic sulfides deposited as rock

Decay organisms in soil, wetlands, and ocean

Values are from Schlesinger, W. H. *Biogeochemistry: An Analysis of Global Change*. 2nd edition. Academic Press, San Diego (1997) and based on several sources.

Think Critically Why does a tidal marsh sometimes smell like rotten eggs?

consume plant proteins and convert them to animal proteins. In the ocean, certain marine algae release a compound that bacteria convert to dimethyl sulfide (DMS). DMS is released into the atmosphere, where it helps condense water into droplets in clouds and may affect weather and climate. Atmospheric DMS is converted to sulfate, most of which is deposited into the ocean.

As in the nitrogen cycle, bacteria drive the sulfur cycle. In freshwater wetlands, tidal flats, and flooded soils, which are oxygen deficient, certain bacteria convert sulfates to hydrogen sulfide gas, which is released into the atmosphere, or to metallic sulfides, which are deposited as rock. In the absence of oxygen, other bacteria perform

a type of photosynthesis that uses hydrogen sulfide instead of water. Where oxygen is present, different bacteria oxidize sulfur compounds to sulfates.

Coal, and to a lesser extent oil, contain sulfur. Sulfur dioxide, a major cause of acid deposition, is released into the atmosphere when these fuels are burned and during the smelting of sulfur-containing ores of such metals as copper, lead, and zinc.

The Phosphorus Cycle

Unlike the biogeochemical cycles just discussed, the phosphorus cycle doesn't have an atmospheric component. Phosphorus cycles from the land into living organisms, then

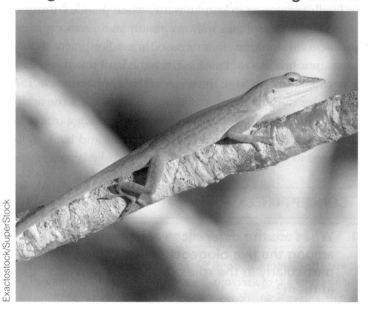

a. The green anole is native to Florida.

b. The brown anole was introduced in Florida.

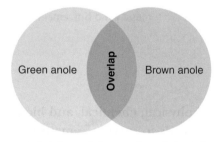

c. The fundamental niches of the two lizards initially overlapped.

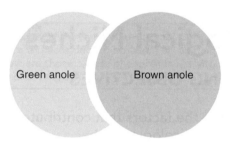

d. The brown anole out-competed the green anole, restricting its realized niche.

An example helps illustrate the difference between fundamental and realized niches. The green anole, a lizard native to Florida and other southeastern states, perches on trees, shrubs, walls, or fences during the day and waits for insect and spider prey (**Figure 5.15a**). In the past, these little lizards were widespread in Florida.

Several years ago, a related species, the brown anole, was introduced from Cuba into southern Florida and quickly became common (**Figure 5.15b**). Suddenly the green anoles became rare, apparently driven out of their habitat by competition from the slightly larger brown anoles. Careful investigation revealed that green anoles were still present but were now confined largely to the wetland vegetation and to the leafy crowns of trees, where they were less obvious.

The habitat portion of the green anole's fundamental niche includes all of the places where it originally lived in Florida: trunks and crowns of trees, exterior house walls,

and many other locations. Where they became established, brown anoles drove green anoles out from all but wetlands and tree crowns, so the green anoles' realized niche—the areas where it could survive—became smaller (**Figure 5.15c** and **d**). Natural communities consist of numerous species, and the interactions among species produce the realized niche of each.

When two species are similar—as are green and brown anoles—their ecological niches may appear to overlap. However, many ecologists think no two species indefinitely occupy the same niche in the same community. Resource partitioning is one way some species avoid or at least reduce niche overlap. **Resource partitioning** is the reduction in competition for environmental resources such as food among coexisting species as a result of the niche of each species differing from the niches of others in one or more ways. Evidence of

WHAT A SCIENTIST SEES

Yellow-rumped warbler   Bay-breasted warbler   Cape May warbler   Black-throated green warbler   Blackburnian warbler

Adapted from MacArthur, R. H. "Population ecology of some warblers of northeastern coniferous forests." *Ecology*, Vol. 39 (1958).

Resource Partitioning

Robert MacArthur's study of five American warbler species is a classic example of resource partitioning. Although it initially appeared that the niches of the species were nearly identical, MacArthur determined that individuals of each species spend most of their feeding time in different portions of spruces and other conifer trees. They also move in different directions through the canopy, consume different combinations of insects, and nest at slightly different times. The photo shows a male Blackburnian warbler.

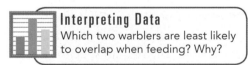

Interpreting Data
Which two warblers are least likely to overlap when feeding? Why?

© Johann Schumacher/Alamy

resource partitioning in animals is well documented and includes studies in tropical forests of Central and South America that demonstrate little overlap in the diets of fruit-eating birds, primates, and bats that coexist in the same habitat. Although fruits are the primary food for several hundred bird, primate, and bat species, the wide variety of fruits available has allowed fruit eaters to specialize, thereby reducing competition. Resource partitioning may also include timing of feeding, location of feeding, nest sites, and other aspects of an organism's ecological niche (see *What a Scientist Sees*).

CONCEPT CHECK	STOP

1. **What** are three aspects of an organism's ecological niche?

2. **What** is resource partitioning?

Interactions Among Organisms

LEARNING OBJECTIVES

1. **Distinguish** among mutualism, commensalism, and parasitism.

2. **Define** *predation* and describe predator–prey relationships.

3. **Define** *competition* and distinguish between intraspecific and interspecific competition.

4. **Discuss** an example of a keystone species.

No organism exists independently of other organisms. The producers, consumers, and decomposers of an ecosystem interact with one another in a variety of ways, and each forms associations with other organisms. Three main types of interactions occur among species in an ecosystem: symbiosis, predation, and competition.

Symbiosis

In **symbiosis**, one species usually lives in or on another species. The partners in a symbiotic relationship may benefit, be unaffected, or be harmed by the relationship. Symbiosis is the result of **coevolution**, the interdependent evolution of two interacting species. Flowering plants and their animal pollinators are an excellent example of coevolution (see Chapter 6 for more on evolution). Bees, beetles, birds, bats, and other animals transport pollen from one plant to another. During the millions of years over which these associations developed, flowering plants evolved several ways to attract animal pollinators. One of the rewards for the pollinator is food—nectar (a sugary solution) and pollen. Plants have a variety of ways to get the pollinator's attention, most involving showy petals and scents.

As plants acquire specialized features to attract pollinators, animals coevolve specialized body parts and behaviors to aid pollination and obtain nectar and pollen as a reward. Coevolution is responsible for the hairy bodies of bumblebees, which catch and hold sticky pollen for transport from one flower to another.

> **symbiosis** An intimate relationship or association between members of two or more species; includes mutualism, commensalism, and parasitism.

Coevolution is also responsible for the long, curved beaks of certain Hawaiian birds that insert their beaks into tubular flowers to obtain nectar (**Figure 5.16**).

The thousands, or even millions, of symbiotic associations that result from coevolution fall into three categories: mutualism, commensalism, and parasitism (summarized in **Figure 5.17a**). One example of **mutualism**, an association in which both organisms benefit, is the interaction between acacia ants and the bull's horn acacia plant (**Figure 5.17b**). The ants make hollow nests out of thorns at the base of the plant's leaves and gain special nutrients from the leaf tips. In return, the ants effectively protect the plant from invertebrate and vertebrate herbivores and clear away competing plants. Both ant and acacia depend on this association for survival.

Commensalism is a symbiotic relationship in which one species benefits and the other is neither harmed nor helped. One example of commensalism is the relationship between a tropical tree and its *epiphytes*, smaller plants such as mosses, orchids, and ferns that live attached to the bark of the tree's branches (**Figure 5.17c**). An epiphyte anchors itself to a tree but typically doesn't obtain nutrients or water directly from the tree. Its location enables it to obtain adequate light, water (as rain dripping down the branches), and required nutrient minerals (which rain washes out of the tree's leaves). The epiphyte benefits from the association, whereas the tree is apparently unaffected.

Parasitism is a symbiotic relationship in which one species (the *parasite*) benefits at the expense of the other (the *host*). Parasitism is a successful lifestyle; more than 100 parasites live in or on the human species (**Figure 5.17d**).

Coevolution • Figure 5.16

This Hawaiian honeycreeper uses its gracefully curved bill to sip nectar from the long, tubular flowers of the lobelia.

WileyPLUS

	Organism 1	Organism 2	Characteristic of relationship
Mutualism	Benefits	Benefits	Each organism depends on the other
Commensalism	Benefits	Not affected	Only one organism depends on the other
Parasitism	Benefits	Harmed	Host harmed, rarely killed; host usually much larger than parasite

a. Categories of Symbiosis.

Interpreting Data
A bee pollinates a plant species while gathering material to make nectar. What type of symbiosis exists between plant and bee?

b. Mutualism. Most common in Central America, the acacia ant gains shelter and nutrients from the acacia plant, in turn protecting the plant from predators. Photographed in Costa Rica.

© WILDLIFE GmbH/Alamy

d. Parasitism. Close-up of body lice feeding on a human arm. Each louse is about 3 mm (0.12 in) long.

Darlyne A. Murawski/NG Image Collection

c. Commensalism Epiphytes are small plants that attach to the branches and trunks of larger trees. Photographed in Tanzania.

© blickwinkel/Alamy

EnviroDiscovery
Bee Colonies Under Threat

Since late 2006, many U.S. beekeepers have experienced major losses in their bee colonies, 30 to 90 percent of total individuals. Similar bee disappearances have occurred in other countries, resulting in losses of millions of bees worldwide. These sudden declines, known in the United States as colony collapse disorder (CCD), are thought to be triggered by a complex mix of factors. Researchers are investigating three major potential causes of CCD:

- The negative effects of pesticides including—according to recent research—some of the most widely used insecticides in the world

- Damage caused by pathogens or parasites, such as *Varroa* and tracheal mites, and a parasitic fly recently reported to be taking over the bodies of Northern California honeybees

- Deaths resulting from viruses, including the Israeli acute-paralysis virus (IAPV), which can be spread by mites

Bees are necessary for the pollination of a variety of important crops, many of which—nearly 100—are potentially threatened by these bee declines. The demand on bees to pollinate crop species has increased even as the number of bee colonies maintained by keepers has dropped. Some agricultural researchers believe that the added stress placed on bee colonies as they are transported to carry out pollination has increased the susceptibility of these colonies to health threats by compromising bees' immune systems. Recent research indicates that improving hive hygiene can help prevent CCD.

Two fruit growers hand pollinate pear trees in Yongchuan, Chongqing, China, a region where local bee populations have vanished.

A parasite, usually much smaller than its host, obtains nourishment from its host, but although a parasite may weaken its host, it rarely kills it quickly. (A parasite would have a difficult life if it kept killing off its hosts!) Some parasites, such as ticks, live outside the host's body; other parasites, such as tapeworms, live within the host.

Predation

Predators kill and feed on other organisms. **Predation** includes both animals eating other animals (for example, herbivore–carnivore interactions) and animals eating plants (producer–herbivore interactions). Predation has resulted in an "arms race," with the coevolution of predator strategies—more efficient ways to catch prey—and prey strategies—better ways to escape the predator. An efficient predator exerts a strong selective force on its prey, and over time the prey species may evolve some sort of countermeasure that reduces the probability of

> **predation** The consumption of one species (the prey) by another (the predator).

its being captured. The countermeasure that the prey acquires in turn may act as a strong selective force on the predator.

Adaptations related to predator–prey interactions include predator strategies (pursuit and ambush) and prey strategies (animal defenses and plant defenses). Keep in mind that such strategies are not "chosen" by the respective predators or prey. New traits arise randomly in a population as a result of mutation and natural selection. (See Chapter 6 to learn more about how traits evolve.)

The cheetah is the world's fastest animal and can sprint at 110 km (68 mi) per hour for short distances (**Figure 5.18a**). Orcas (commonly known as killer whales) hunt in packs and often herd salmon or tuna into a cove so that they are easier to catch. Any trait that increases hunting efficiency, such as the speed of a cheetah or the intelligence of orcas, favors predators that pursue their prey.

Ambush is another effective way to catch prey. The goldenrod spider is the same color as the white or yellow flowers in which it hides (**Figure 5.18b**). This camouflage prevents unwary insects that visit the flower for nectar from noticing the spider until it is too late.

Predation • Figure 5.18

a. The cheetah sprints at high speed to catch prey. Photographed in the Okavango Delta, Botswana, Africa.

b. The goldenrod spider employs camouflage to ambush its prey.

Chris Johns/NG Image Collection

Rich Reid/NG Image Collection

5 Interactions Among Organisms 116

1. **Symbiosis**, an intimate relationship or association between members of two or more species, is the result of **coevolution**, the interdependent evolution of two interacting species. **Mutualism** is a symbiotic relationship in which both species benefit. **Commensalism** is a symbiotic relationship in which one species benefits and the other species is neither harmed nor helped. **Parasitism** is a symbiotic relationship in which one species (the parasite) benefits at the expense of the other (the host).

2. **Predation** is the consumption of one species (the prey) by another (the predator). With coevolution between predator and prey, the predator evolves more efficient ways to catch prey (such as pursuit and ambush), and the prey evolves better ways to escape the predator (such as flight, association in groups, and camouflage).

3. **Competition** is the interaction among organisms that vie for the same resources in an ecosystem (such as food or living space). Competition occurs among individuals within a population (intraspecific competition) and between species (interspecific competition).

4. A **keystone species** is crucial in determining the nature and structure of the entire ecosystem in which it lives. Though present in relatively small numbers, keystone species have disproportionate effects on ecosystems.

Key Terms

- biosphere 99
- community 98
- competition 120
- ecological niche 113
- ecology 98

- ecosystem 98
- energy flow 103
- first law of thermodynamics 100
- landscape 98
- photosynthesis 100

- population 98
- predation 119
- second law of thermodynamics 101
- symbiosis 116

What is happening in this picture?

This dwarf frog in Brazil has an intriguing color pattern.

- Note the two large spots on the frog's rump. What do they resemble? Why would this animal have such conspicuous spots?

- If a hungry bird saw this frog, do you think it would have second thoughts about eating it? Why or why not?

- What other strategies might this frog species use to catch food or to avoid becoming food?

© Photoshot Holdings Ltd/Alamy

Critical and Creative Thinking Questions

1. To function, ecosystems require inputs of energy. Where does this energy come from?

2. After an organism uses energy, what happens to the energy? Is all the energy captured through gross primary productivity available to organisms higher in a food chain? Explain.

3. What is a biogeochemical cycle? Why is the cycling of matter essential to the continuance of life? Why, specifically, is the cycling of nitrogen important to humans?

4. What types of resources might two organisms compete over if those resources are scarce? How might interspecific competition affect two species' ecological niches?

5. Are food chains important in biogeochemical cycles? Explain why or why not.

6. What components of a typical phosphorus cycle were affected by pollution in Lake Washington? What cooperative efforts were involved in correcting this pollution?

7. In both parasitism and predation, one organism benefits at the expense of another. What is the difference between the two relationships?

8. Some biologists think protecting keystone species would help preserve biological diversity in an ecosystem. Do you agree? Explain your answer.

Sustainable Citizen Question

9. How does the role of humans in the carbon cycle influence global climate change? How might your role in the carbon cycle compare to that of a young person on a remote South American farm that uses animal labor rather than machines? Name three changes you could make in your life to reduce your input to the carbon cycle.

10. Describe how the close-up image below of an alpine meadow represents a community.

Geroge F. Mobley/NG Image Collection

11. Ecologists investigating interactions of two species at a study site first counted individuals of Species A and then removed all Species B individuals. Six months later, the ecologists again counted individuals of Species A. Viewing their results as graphed below, what is the likely ecological interaction between Species A and Species B? Explain your answer.

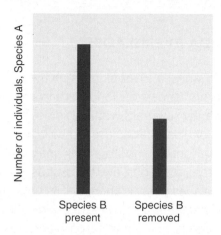

The figure below shows the components of a simple food chain. Use it to answer questions 12–14.

12. Identify the producers, consumers, and decomposers in the food chain. How many trophic levels are represented?

13. Describe or indicate the flow of food and energy within this system.

14. Which forms of energy are present within this chain?

Ecosystems and Evolution

THE FLORIDA EVERGLADES

The Everglades, a "river of grass" in the southernmost part of Florida, is a vast expanse of predominantly sawgrass wetlands dotted with small islands of trees. It is a haven for wildlife, including alligators (see photograph), snakes, panthers, otters, raccoons, and thousands of birds.

The Everglades today is about half its original size of 1.6 million hectares (4 million acres) and suffers from many serious environmental problems. Wading bird populations have dropped 93 percent since 1930 (see graph), and the area is now home to 50 endangered or threatened species. Invasive predator species, including Burmese pythons—former pets—decimate some prey populations.

More than 70 years of engineering projects aimed at protecting the human population from storm-related flooding have reduced the quantity of water flowing into the Everglades, restricting the natural recharging process there. Flood-control measures created dry spaces that were then converted to agricultural or residential use, fragmenting wildlife habitat and polluting the water that does enter.

The Everglades will never return completely to its original condition because there are now too many cities and sugar plantations in the region. However, state and federal governments are working together on the massive Comprehensive Everglades Restoration Plan to eventually restore a more natural water flow to the area, repel invasions of foreign species, and reestablish native species.

graphingactivity

Decline in wading birds in the Everglades, 1930s–1988.

Total number of nesting attempts made by wading birds

	1930s	1940s	1975	1988
160,000				
140,000				
120,000				
100,000				
80,000				
60,000				
40,000				
20,000				
0				

Year or decade

Based on data from *Everglades: The Ecosystem and Its Restoration,* Steven M. Davis and John C. Ogden, eds. Table 23.1, p. 574.

CHAPTER OUTLINE

Earth's Major Biomes   128
- ■ Environmental InSight: How Climate Shapes Terrestrial Biomes
- • Tundra
- • Boreal Forest
- • Temperate Rain Forest
- • Temperate Deciduous Forest
- • Temperate Grassland
- • Chaparral
- ■ EnviroDiscovery: Using Goats to Fight Fires
- • Desert
- • Savanna
- • Tropical Rain Forest

Aquatic Ecosystems   142
- • Freshwater Ecosystems
- ■ What a Scientist Sees: Zonation in a Large Lake
- • Brackish Ecosystems: Estuaries

Population Responses to Changing Conditions over Time: Evolution   147
- • Natural Selection
- ■ Environmental InSight: Evidence for Evolution

Community Responses to Changing Conditions over Time: Succession   151
- • Primary Succession
- • Secondary Succession
- ■ Case Study: Wildfires

CHAPTER PLANNER

- ❏ Study the picture and read the opening story.
- ❏ Scan the Learning Objectives in each section:
  p. 128 ❏   p. 142 ❏   p. 147 ❏   p. 151 ❏
- ❏ Read the text and study all figures and visuals. Answer any questions.

Analyze key features

- ❏ Environmental InSight, p. 129 ❏   pp. 150–151 ❏
- ❏ National Geographic Map, pp. 130–131 ❏
- ❏ EnviroDiscovery, p. 138
- ❏ What a Scientist Sees, p. 143
- ❏ Process Diagram, p. 149 ❏   p. 152 ❏   p. 153 ❏
- ❏ Case Study, p. 154
- ❏ Stop: Answer the Concept Checks before you go on:
  p. 142 ❏   p. 147 ❏   p. 150 ❏   p. 153 ❏

End of Chapter

- ❏ Review the Summary and Key Terms.
- ❏ Answer What is happening in this picture?
- ❏ Answer the Critical and Creative Thinking Questions.

Similar vegetation types can occur at many different locations.

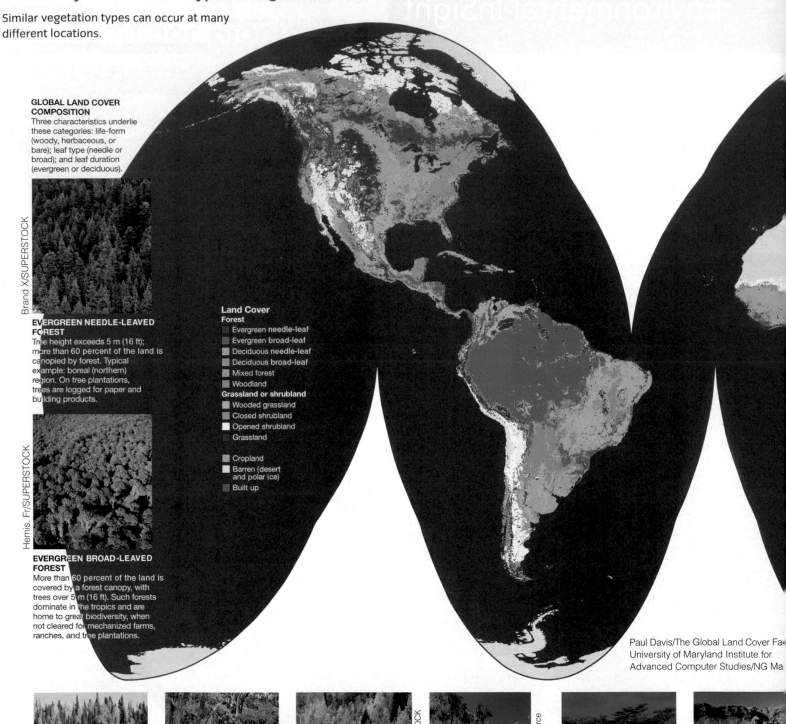

GLOBAL LAND COVER COMPOSITION
Three characteristics underlie these categories: life-form (woody, herbaceous, or bare); leaf type (needle or broad); and leaf duration (evergreen or deciduous).

Brand X/SUPERSTOCK

EVERGREEN NEEDLE-LEAVED FOREST
Tree height exceeds 5 m (16 ft); more than 60 percent of the land is canopied by forest. Typical example: boreal (northern) region. On tree plantations, trees are logged for paper and building products.

Hemis. Fr/SUPERSTOCK

EVERGREEN BROAD-LEAVED FOREST
More than 60 percent of the land is covered by a forest canopy, with trees over 5 m (16 ft). Such forests dominate in the tropics and are home to great biodiversity, when not cleared for mechanized farms, ranches, and tree plantations.

Land Cover
Forest
- Evergreen needle-leaf
- Evergreen broad-leaf
- Deciduous needle-leaf
- Deciduous broad-leaf
- Mixed forest
- Woodland

Grassland or shrubland
- Wooded grassland
- Closed shrubland
- Opened shrubland
- Grassland

- Cropland
- Barren (desert and polar ice)
- Built up

Paul Davis/The Global Land Cover Fac
University of Maryland Institute for
Advanced Computer Studies/NG Ma

Photodisc/SUPERSTOCK

DECIDUOUS NEEDLE-LEAVED FOREST
A forest canopy covers more than 60 percent of the land; tree height exceeds 5 m (16 ft). This class is dominant only in Siberia, taking the form of larch forests.

age fotostock/SUPERSTOCK

DECIDUOUS BROAD-LEAVED FOREST
More than 60 percent of the land is covered by a forest canopy; tree height exceeds 5 m (16 ft). In temperate regions, much of this forest has been converted to cropland.

Polka Dot Images/SUPERSTOCK

MIXED FOREST
Both needle and deciduous types of trees appear. Mixed forest is largely found between temperate deciduous and boreal evergreen forests.

Michael McCoy/Science Source Images

WOODLAND
Land has herbaceous or woody understory; trees exceed 5 m (16 ft) and may be deciduous or evergreen. Highly degraded in long-settled human environments, such as West Africa.

Corbis/SUPERSTOCK

WOODED GRASSLAND
Woody or herbaceous understories are punctuated by trees. Examples are African savanna as well as open boreal border land between trees and tundra.

CLOSED SHRUBLAND
Found where prolonged or dry seasons limit plan growth. This cover is dominated by bushes or shrubs not exceeding 5 (16 ft). Tree canopy is les than 10 percent.

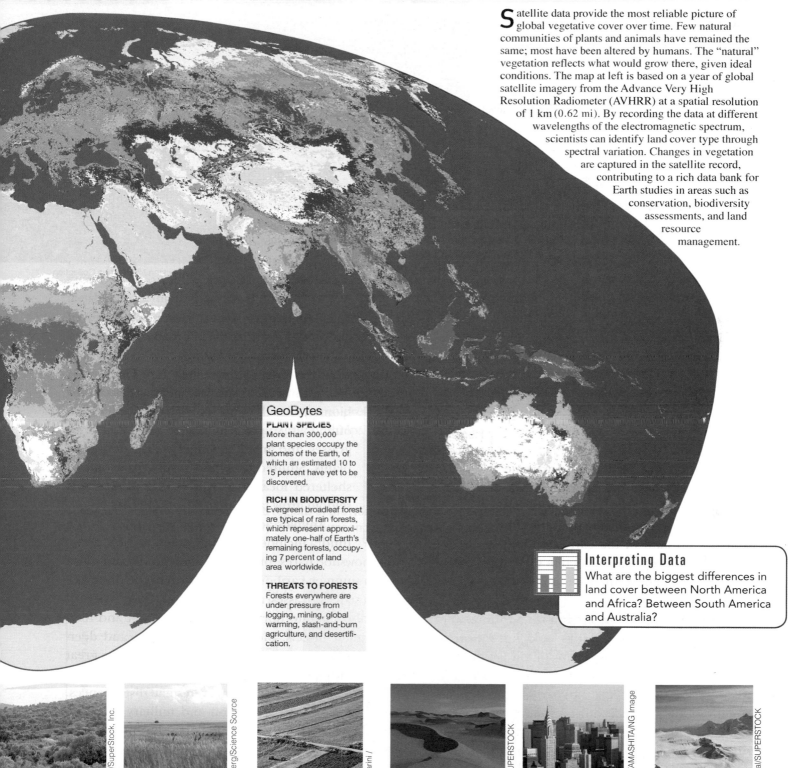

Satellite data provide the most reliable picture of global vegetative cover over time. Few natural communities of plants and animals have remained the same; most have been altered by humans. The "natural" vegetation reflects what would grow there, given ideal conditions. The map at left is based on a year of global satellite imagery from the Advance Very High Resolution Radiometer (AVHRR) at a spatial resolution of 1 km (0.62 mi). By recording the data at different wavelengths of the electromagnetic spectrum, scientists can identify land cover type through spectral variation. Changes in vegetation are captured in the satellite record, contributing to a rich data bank for Earth studies in areas such as conservation, biodiversity assessments, and land resource management.

GeoBytes

PLANT SPECIES
More than 300,000 plant species occupy the biomes of the Earth, of which an estimated 10 to 15 percent have yet to be discovered.

RICH IN BIODIVERSITY
Evergreen broadleaf forest are typical of rain forests, which represent approximately one-half of Earth's remaining forests, occupying 7 percent of land area worldwide.

THREATS TO FORESTS
Forests everywhere are under pressure from logging, mining, global warming, slash-and-burn agriculture, and desertification.

Interpreting Data
What are the biggest differences in land cover between North America and Africa? Between South America and Australia?

PEN SHRUBLAND
rubs are dominant not eeding 2 m (6.5 ft) in ght. They can be ergreen or deciduous. s type occures in niarid or severely cold as.

GRASSLAND
Occurring in a wide range of habitats, this landscape has continuous herbaceous cover. The American Plains and central Russia are the premier examples.

CROPLAND
Crop-producing fields constitute over 80 percent of the land. Temperate regions are home to large areas of mechanized farming; in the developing world, plots are small.

BARREN (DESERT)
The land never has more than 10 percent vegetated cover. True deserts, such as the Sahara, as well as areas succumbing to desertification are examples.

BUILT UP
This land cover type was mapped using digital population data from around the world. It represents the most densely inhabited areas.

BARREN (POLAR ICE)
Permanent snow cover characterizes this class, the greatest examples of which are in the polar regions, as well as on high elevations in Alaska and the Himalayas.

Human population growth and industrial expansion in tropical countries may spell the end of tropical rain forests during the 21st century. Biologists know that many rainforest species will become extinct before they are even identified and scientifically described. (See Chapter 13 for more discussion of the ecological impacts of rainforest destruction.)

CONCEPT CHECK STOP

1. **What** is a biome?

2. **How** do you distinguish between temperate rain forest and tropical rain forest? between savanna and desert?

Aquatic Ecosystems

LEARNING OBJECTIVES

1. **Summarize** the important environmental factors that affect aquatic ecosystems.

2. **Describe** the various aquatic ecosystems, giving attention to the environmental characteristics of each.

The most fundamental division in aquatic ecology is probably between freshwater and saltwater environments. Salinity, which is the concentration of dissolved salts (such as sodium chloride) in a body of water, affects the kinds of organisms present in aquatic ecosystems, as does the amount of dissolved oxygen. Water greatly interferes with the penetration of light, so floating aquatic organisms that photosynthesize must remain near the water's surface, and vegetation anchored to lake floors or streambeds will grow only in relatively shallow water. In addition, low levels of essential nutrient minerals limit the number and distribution of organisms in certain aquatic environments. In this section, we discuss freshwater ecosystems only; because the immense marine environment is so critical to the environmental well-being of Earth, we devote an entire chapter to it (see Chapter 11).

Aquatic ecosystems contain three main ecological categories of organisms: free-floating plankton, strongly swimming nekton, and bottom-dwelling benthos. **Plankton** are usually small or microscopic organisms. They tend to drift or swim feebly, so, for the most part, they are carried about at the mercy of currents and waves. Plankton include *phytoplankton*, photosynthetic algae and cyanobacteria that form the base of most aquatic food webs, and *zooplankton*, animal-like organisms that feed on algae and cyanobacteria and are in turn consumed by newly hatched fish and other small aquatic organisms. **Nekton** are larger, more strongly swimming organisms such as fishes, turtles, and whales. **Benthos** are bottom-dwelling organisms that fix themselves to one spot (sponges and oysters), burrow into the sand (worms and clams), or simply walk about on the bottom (crawfish and aquatic insect larvae).

Freshwater Ecosystems

Freshwater ecosystems include lakes and ponds (standing-water ecosystems), rivers and streams (flowing-water ecosystems), and marshes and swamps (freshwater wetlands). Specific abiotic conditions and characteristic organisms distinguish each freshwater ecosystem. Although freshwater ecosystems occupy only about 2 percent of Earth's surface, they play an important role in the hydrologic cycle: They help recycle precipitation that flows into the ocean as surface runoff. (See Chapter 5 for a detailed explanation of the hydrologic cycle.) Large bodies of fresh water help moderate daily and seasonal temperature fluctuations on nearby land regions, and freshwater habitats provide homes for many species.

Zonation is characteristic of **standing-water ecosystems**. A large lake has three zones: the littoral, limnetic, and profundal zones (see *What a Scientist Sees*). The *littoral zone* is a productive, shallow-water area along the shore of a lake or pond. Emergent vegetation, such as cattails and bur reeds, as well as several deeper-dwelling aquatic plants and algae, live in the littoral zone. Animals here include frogs, turtles, worms, crayfish and other crustaceans, insect larvae, and many fishes. The *limnetic zone* is the open water beyond the littoral zone—that is, away from

standing-water ecosystem A body of fresh water surrounded by land and whose water does not flow; a lake or a pond.

WHAT A SCIENTIST SEES

Zonation in a Large Lake

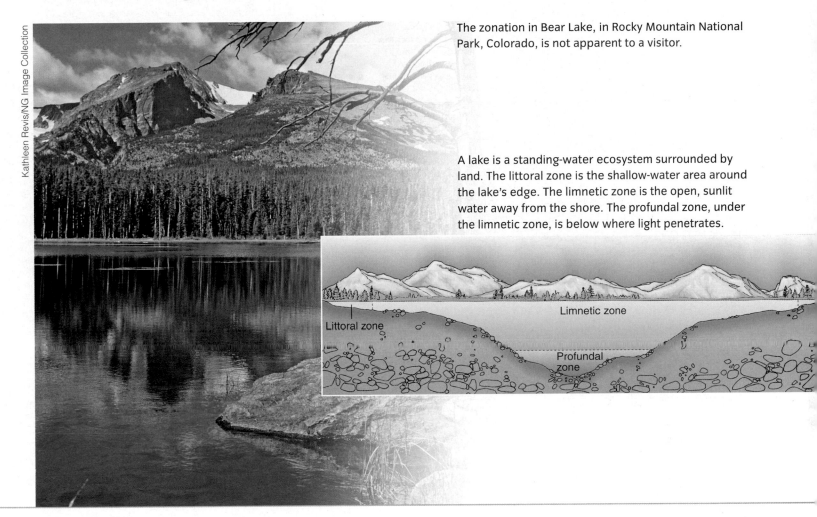

The zonation in Bear Lake, in Rocky Mountain National Park, Colorado, is not apparent to a visitor.

A lake is a standing-water ecosystem surrounded by land. The littoral zone is the shallow-water area around the lake's edge. The limnetic zone is the open, sunlit water away from the shore. The profundal zone, under the limnetic zone, is below where light penetrates.

Littoral zone

Limnetic zone

Profundal zone

the shore. The limnetic zone extends down as far as sunlight penetrates to permit photosynthesis. The main organisms of the limnetic zone are microscopic plankton. Larger fishes also spend most of their time in the limnetic zone, although they may visit the littoral zone to feed and reproduce. The deepest zone, the *profundal zone*, is beneath the limnetic zone of a large lake; smaller lakes and ponds typically lack a profundal zone. Because light does not penetrate effectively to this depth, plants and algae do not live there. Detritus drifts into the profundal zone from the littoral and limnetic zones; bacteria decompose this detritus. This marked zonation is accentuated by **thermal stratification**, in which the temperature changes sharply with depth.

Temperate lakes undergo fall and spring *turnovers*, when changing surface temperatures break down the thermal stratification and the water layers mix. In fall, as surface water cools, its density increases, and eventually it displaces the less dense, warmer, mineral-rich water beneath. The warmer water then rises to the surface where it, in turn, cools and sinks. This process of cooling and sinking continues until the lake reaches a uniform temperature throughout. In the spring, surface ice melts and surface water again sinks to the bottom, resulting in a mixing of the layers. In summer, thermal stratification occurs once again. The mixing of deeper, nutrient-rich water with surface, nutrient-poor water during the fall and spring turnovers brings essential nutrient minerals to the surface and oxygenated water to the bottom.

Human effects on lakes and ponds include *eutrophication*, which is nutrient enrichment of a body of water with inorganic plant and algal nutrients like

Features of a typical river • Figure 6.13

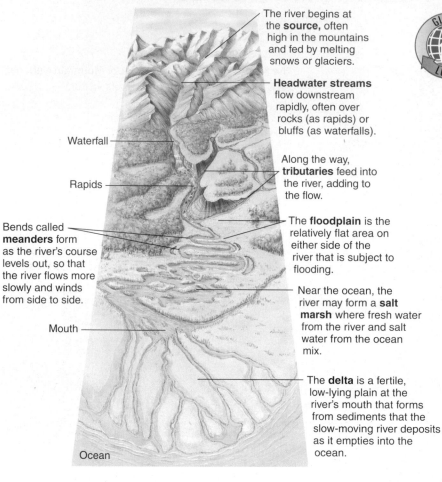

The river begins at the **source,** often high in the mountains and fed by melting snows or glaciers.

Headwater streams flow downstream rapidly, often over rocks (as rapids) or bluffs (as waterfalls).

Along the way, **tributaries** feed into the river, adding to the flow.

Waterfall

Rapids

The **floodplain** is the relatively flat area on either side of the river that is subject to flooding.

Bends called **meanders** form as the river's course levels out, so that the river flows more slowly and winds from side to side.

Near the ocean, the river may form a **salt marsh** where fresh water from the river and salt water from the ocean mix.

Mouth

The **delta** is a fertile, low-lying plain at the river's mouth that forms from sediments that the slow-moving river deposits as it empties into the ocean.

Ocean

a. A river flows from its source to the ocean.

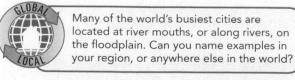

Many of the world's busiest cities are located at river mouths, or along rivers, on the floodplain. Can you name examples in your region, or anywhere else in the world?

Frans Lanting/Corbis

Meanders

Floodplain

b. Aerial view of meanders in the Tambopata River, Peru.

---

nitrates and phosphates. Although eutrophication is a natural process, human activities often accelerate it, such as the runoff of agricultural fertilizers and discharge of treated or untreated sewage. Eutrophication of lakes is discussed in detail in Chapter 10.

Flowing-water ecosystems are highly variable. The surrounding environment changes greatly between a river's source and its mouth (**Figure 6.13**). Certain parts of the stream's course are shaded by forest, while other parts are exposed to direct sunlight. Groundwater may well up through sediments on the bottom in one particular area, making the water temperature cooler in summer or

> **flowing-water ecosystem** A freshwater ecosystem such as a river or stream in which water flows in a current.

warmer in winter than in adjacent parts of the stream or river. The kinds of organisms found in flowing water vary greatly from one stream to another, depending primarily on the strength of the current. In streams with fast currents, some inhabitants have adaptations such as suckers, with which they attach themselves to rocks to prevent being swept away. Some stream inhabitants have flattened bodies to slip under or between rocks. Other inhabitants such as fish are streamlined and muscular enough to swim in the current.

Human activities such as pollution and dam construction have adverse impacts on rivers and streams. These activities damage wildlife habitat and threaten water supplies and fisheries. (See Chapter 10 for more discussion of the environmental effects of dams.)

Freshwater wetlands include **marshes**, dominated by grass-like plants, and **swamps**, dominated by woody trees or shrubs (**Figure 6.14**). Wetland soils are waterlogged for variable periods and are therefore anaerobic (without oxygen). They are rich in accumulated organic materials, partly because anaerobic conditions discourage decomposition.

With their productive plant communities, wetlands provide excellent wildlife habitat for migratory waterfowl and other bird species, as well as for beaver, otters, muskrats, and game fish. In addition to providing unique wildlife habitat, wetlands serve other important environmental functions, known as **ecosystem services**. When rivers flood their banks, wetlands are capable of holding or even absorbing the excess water, thereby helping to control flooding. The floodwater then drains slowly back into the rivers, providing a steady flow of water throughout the year. Wetlands also serve as groundwater recharging areas. One of their most important roles is to trap and hold pollutants in the flooded soil, thereby cleansing and purifying the water.

Although wetlands are afforded some legal protection, they are still threatened by pollution, development, agriculture, and dam construction. (See Chapter 10 for more on threats to freshwater ecosystems.)

Freshwater swamp • Figure 6.14

Freshwater swamps are inland areas covered by water and dominated by trees, such as baldcypress. Photographed in Lake Martin, at the edge of the Atchafalaya Basin, Louisiana.

© Dan Leeth/Alamy Limited

Brackish Ecosystems: Estuaries

> **estuary** A coastal body of water, partly surrounded by land, with access to the open ocean and a large supply of fresh water from a river.

Where the ocean meets the land, there may be one of several kinds of ecosystems: a rocky shore, a sandy beach, an intertidal mud flat, or a tidal **estuary**. Water levels in an estuary rise and fall with the tides; salinity fluctuates with tidal cycles, the time of year, and precipitation. Salinity also changes gradually within the estuary, from fresh water at the river entrance, to *brackish* (somewhat salty) water, to salty ocean water at the mouth of the estuary. Because estuaries undergo significant daily, seasonal, and annual variations in physical factors such as temperature, salinity, and depth of light penetration, estuarine organisms must have a high tolerance for changing conditions.

Estuaries are among the most productive ecosystems in the world. Their high productivity is brought about by nutrient transport from land, tidal action that rapidly circulates nutrients and helps remove waste products, a high level of light that penetrates the shallow water, and the many plants that form the base of a detritus food web.

Temperate estuaries usually feature **salt marshes**, shallow wetlands in which salt-tolerant grasses grow (**Figure 6.15a**). Salt marshes perform many ecosystem services, including providing biological habitats, trapping sediment and pollution, supplying groundwater, and buffering storms by absorbing their energy, which prevents flood damage elsewhere.

Mangrove forests, the tropical equivalent of salt marshes, cover perhaps 70 percent of tropical coastlines (**Figure 6.15b**). Like salt marshes, mangrove forests provide valuable ecosystem services. Their interlacing roots are breeding grounds and nurseries for several commercially important fishes and shellfish, such as mullet, spotted sea trout, crabs, and shrimp. Mangrove branches are nesting sites for many species of birds, such as pelicans, herons, egrets, and roseate spoonbills. Mangrove roots stabilize the submerged soil, thereby preventing coastal erosion and providing a barrier against the ocean during storms.

Estuaries • Figure 6.15

© William A. Bake/CORBIS

a. A salt marsh along Okracoke Harbor, Outer Banks, North Carolina.

© Photoshot/Alamy

b. A mangrove forest in Risong Bay, Palau, Micronesia, with an underwater view of the prop root system. Mangrove roots grow into deeper water as well as into mudflats that are exposed at low tide. Many animals live among the mangroves' complex root system.

Both salt marsh and mangrove forest ecosystems have experienced significant losses due to coastal development. Salt marshes have been polluted—by countless ongoing sources as well as oil spills—and turned into dumping grounds; mangrove forests have been logged unsustainably and used as aquaculture sites. Some countries, such as the Philippines, Bangladesh, and Guinea-Bissau, have lost 70 percent or more of their mangrove forests.

CONCEPT CHECK STOP

1. **Which** environmental factors shape flowing-water ecosystems? standing-water ecosystems?

2. **How** do the characteristics of a freshwater wetland differ from those of an estuary? How does a mangrove swamp differ from a salt marsh?

Population Responses to Changing Conditions over Time: Evolution

LEARNING OBJECTIVES

1. **Define** *evolution.*
2. **Explain** the four conditions necessary for evolution by natural selection to occur.
3. **Describe** various types of evidence that supports evolution.

Scientists think all of Earth's remarkable variety of organisms descended from earlier species by a process known as evolution. The concept of evolution dates back to the time of Aristotle (384–322 B.C.E.), but **Charles Darwin** (1809–1882), a 19th-century naturalist, proposed the mechanism of evolution that today's scientific community still accepts (**Figure 6.16**). As you will see, the environment plays a crucial role in Darwin's theory of evolution.

> **evolution**
> The cumulative genetic changes in populations that occur during successive generations.

It occurred to Darwin that in a population, inherited traits favorable to survival in a given environment tended to be preserved over successive generations, whereas unfavorable traits were eliminated. The result is *adaptation,* an evolutionary modification that improves the chance of survival and reproductive success of a species in a given environment. Eventually the accumulation of many adaptive modifications might result in a new species.

Darwin proposed the theory of evolution by natural selection in his monumental book *The Origin of Species by Means of Natural Selection,* which was published in 1859. Since that time, scientists have accumulated an enormous body of observations and experiments that support Darwin's theory. Although biologists still do not agree completely on some aspects of the evolutionary process, the concept that evolution by natural selection has taken place and is still occurring is now well documented.

Portrait of a young Charles Darwin
• Figure 6.16

JAMES L. STANFIELD/NG Image Collection

Natural Selection

Evolution occurs through the process of **natural selection**. As favorable traits increase in frequency in successive generations, and as unfavorable traits decrease or disappear, the collection of characteristics of a given population changes. Natural selection is the process by which successful traits are passed on to the next generation and unsuccessful ones are weeded out. It consists of four phenomena that occur in the natural world , which can be considered conditions necessary for natural selection to take place:

> **natural selection**
> The tendency of better-adapted individuals—those with a combination of genetic traits best suited to environmental conditions—to survive and reproduce, increasing their proportion in the population.

1. *High reproductive capacity.* Each species produces more offspring than will survive to maturity. Natural populations have the reproductive potential to increase their numbers continuously over time (**Figure 6.17**).

2. *Limits on population growth, or a struggle for existence.* Only so much food, water, light, growing space, and so on are available to a population, and organisms compete with one another for the limited resources available to them. Because there are more individuals than the environment can support, not all of an organism's offspring will survive to reproductive age, including many of the fish yet to hatch in Figure 6.17. Other limits on population growth include predators and diseases.

3. *Heritable variation.* The individuals in a population exhibit variation. Each individual has a unique combination of traits, such as size, color, and ability to tolerate harsh environments. Some traits improve the chances of an individual's survival and reproductive success, whereas others do not. It is important to remember that the variation necessary for evolution by natural selection must be inherited so that it can be passed to offspring.

4. *Differential reproductive success.* Individuals that possess the most favorable combination of characteristics (those that make individuals better adapted to their environment) are more likely than others to survive, reproduce, and pass their traits to the next generation. Sexual reproduction is the key to natural

© Images & Stories/Alamy

High reproductive capacity and limits to population growth • Figure 6.17

A jawfish incubates eggs in his mouth. If all offspring of a jawfish pair survived and in turn reproduced, reefs would be choked with jawfish. Yet this fish species has not overrun the ocean, because individuals must avoid predation and compete for limited resources. Photographed at Dimakya Island, Philippines.

selection: The best-adapted individuals are those that reproduce most successfully, whereas less-fit individuals die prematurely or produce fewer or inferior offspring. Over time, enough changes may accumulate in geographically separated populations (often with slightly different environments) to produce new species (**Figure 6.18**).

One premise on which Darwin based his theory of evolution by natural selection is that individuals transmit traits to the next generation. However, Darwin could not explain *how* this occurs or *why* individuals within a population vary. Beginning in the 1930s and 1940s, biologists combined the principles of genetics with Darwin's theory of natural selection. The resulting unified explanation of evolution is known as the **modern synthesis** (where *synthesis* refers to a combination of parts of previous theories).

The modern synthesis explains Darwin's observation of variation among offspring in terms of **mutation**,

Darwin's finches • Figure 6.18

Charles Darwin was a ship's naturalist on a 5-year voyage around the world. During an extended stay in the Galápagos Islands off the coast of Ecuador, he studied the plants and animals of each island, including 14 species of finches.

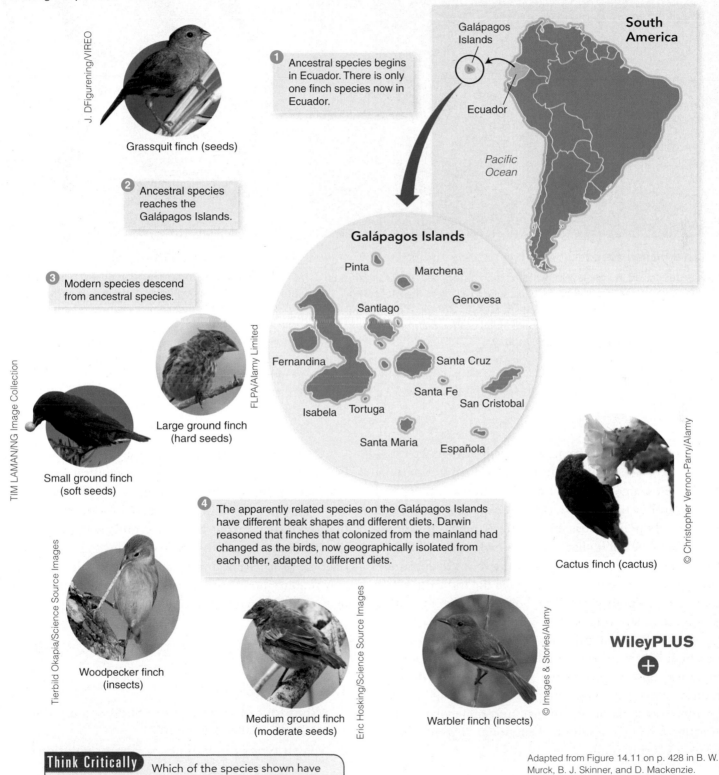

J. DFigurening/VIREO

Grassquit finch (seeds)

1 Ancestral species begins in Ecuador. There is only one finch species now in Ecuador.

Galápagos Islands

South America

Ecuador

Pacific Ocean

2 Ancestral species reaches the Galápagos Islands.

3 Modern species descend from ancestral species.

FLPA/Alamy Limited

Galápagos Islands

Pinta
Marchena
Genovesa
Santiago
Fernandina
Santa Cruz
Santa Fe
San Cristobal
Isabela   Tortuga
Santa Maria
Española

Large ground finch (hard seeds)

TIM LAMAN/NG Image Collection

Small ground finch (soft seeds)

4 The apparently related species on the Galápagos Islands have different beak shapes and different diets. Darwin reasoned that finches that colonized from the mainland had changed as the birds, now geographically isolated from each other, adapted to different diets.

Cactus finch (cactus)

© Christopher Vernon-Parry/Alamy

Tierbild Okapia/Science Source Images

Woodpecker finch (insects)

Eric Hosking/Science Source Images

Medium ground finch (moderate seeds)

Warbler finch (insects)

© Images & Stories/Alamy

WileyPLUS
⊕

Think Critically Which of the species shown have similar beaks? Is this reflected in their diet similarities?

Adapted from Figure 14.11 on p. 428 in B. W. Murck, B. J. Skinner, and D. Mackenzie. *Visualizing Geology*, Hoboken, NJ: John Wiley and Sons, Inc. (2008)

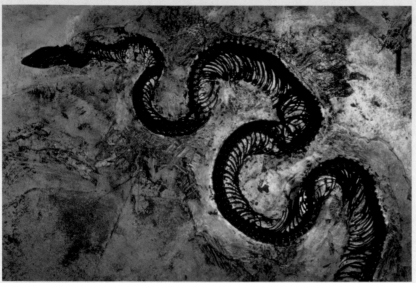

© Jonathan Blair/Corbis

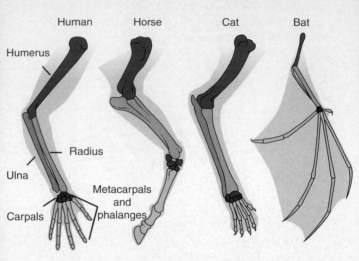

Adapted from Figure 15.13 on p. 244, in S. A. Alters and B. Alters *Biology: Understanding Life*, Hoboken, NJ: John Wiley and Sons, Inc. (2009).

a. The Fossil Record

Fossils deposited in rock layers, which can be dated, show how organisms evolved over time. This well-preserved snake fossil from the Messel Pit, a significant fossil site near the village of Messel, Germany, dates from 47 million years ago. The ancient snake bears both similarities to and differences from snake species living today.

b. Comparative Anatomy

Similarities among organisms demonstrate how they are related. These similarities among four vertebrate limbs illustrate that, while proportions of bones have changed in relation to each organism's way of life, the forelimbs have the same basic bone structure.

or changes in DNA. Mutations provide the genetic variability on which natural selection acts during evolution. Some new traits may be beneficial, whereas others may be harmful or have no effect at all. As a result of natural selection, beneficial strategies, or traits, persist in a population because such characteristics make the individuals that possess them well suited to thrive and reproduce. In contrast, characteristics that make the individuals that possess them poorly suited to their environment tend to disappear in a population.

A vast body of evidence supports evolution, most of which is beyond the scope of this text. This evidence includes observations from the fossil record, comparative anatomy, biogeography (the study of the geographic locations of organisms), and molecular biology (**Figure 6.19**). In addition, evolutionary hypotheses are tested experimentally.

On the basis of these kinds of evidence, virtually all biologists accept the principles of evolution by natural selection, although they don't agree on all the details. They try to better understand certain aspects of evolution, such as the role of chance and how quickly new species evolve. As discussed in Chapter 1, science is an ongoing process, and information obtained in the future may require modifications to certain parts of the theory of evolution by natural selection.

CONCEPT CHECK

1. **What** is evolution?

2. **What** four phenomena or conditions are the basis of natural selection?

3. **Which** types of evidence support evolution?

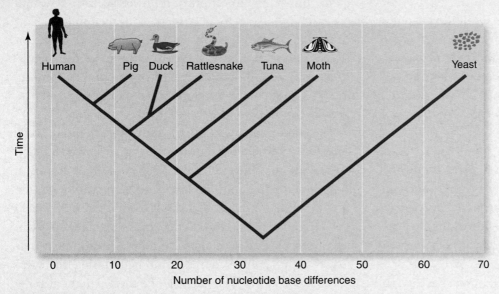

Time

Number of nucleotide base differences

Human · Pig · Duck · Rattlesnake · Tuna · Moth · Yeast

0 10 20 30 40 50 60 70

c. Molecular Biology

The organisms pictured here all share a particular enzyme, but in the course of evolution, mutations have resulted in changes in the gene that codes for that enzyme. This diagram shows the nucleotide base differences in this gene among humans and other organisms. Note that organisms thought to be more closely related to humans have fewer differences than organisms that are more distantly related to humans.

Adapted from Figure 15.19 on p.247, in S. A. Alters and B. Alters *Biology: Understanding Life*, Hoboken, NJ: John Wiley and Sons, Inc. (2009).

 Interpreting Data
Based on the diagram, which organism shown is most closely related to the duck?

Community Responses to Changing Conditions over Time: Succession

LEARNING OBJECTIVES

1. **Define** *ecological succession*.
2. **Distinguish** between primary and secondary succession.

A community of organisms does not spring into existence full blown. By means of **ecological succession**, a given community develops gradually through a sequence of species. Certain organisms colonize an area; over time, others replace them, and eventually the replacements are themselves replaced by still other species. Ecologists first studied succession in three diverse ecosystems: an abandoned field, a northern freshwater bog, and sand dunes.

The actual mechanisms that underlie succession are not clear. In some cases, it may be

ecological succession The process of community development over time, which involves species in one stage being replaced by different species.

that a resident species modified the environment in some way, thereby making it more suitable for a later species to colonize. It is also possible that prior residents lived there in the first place because there was little competition from other species. Later, as more invasive species arrived, the original species were displaced.

Ecologists initially thought that succession inevitably led to a stable and persistent community, known as a *climax community*, such as a forest. But more recently, this traditional view has fallen out of favor. The apparent stability of a "climax" forest is probably the result of how long trees live relative to the human life span. It is now recognized that mature climax communities are not in a state of stable equilibrium but rather in a state of continual disturbance. Over time, a mature community changes in species composition and in the relative abundance of each species, despite the fact that it retains an overall uniform appearance.

Succession is usually described in terms of the changes in the plant species growing in a given area, although each stage of the succession may also have its own kinds of animals and other organisms. Ecological succession is measured on the scale of tens, hundreds, or thousands of years, not the millions of years involved in the evolutionary timescale.

Primary Succession

Primary succession is the change in species composition over time in a previously uninhabited environment (**Figure 6.20**). No soil exists when primary succession begins. Bare rock surfaces, such as recently formed volcanic lava and rock scraped clean by glaciers, are examples of sites where primary succession may take place.

Details vary from one site to another, but on bare rock, lichens are often the most important element in the *pioneer community*, which is the initial community that develops during primary succession. Lichens secrete acids that help break apart the rock, beginning the process of soil formation. Over time, mosses and drought-resistant ferns may replace the lichen community, followed in turn by tough grasses and herbs. Once enough soil accumulates, low shrubs may replace the grasses and herbs; over time, forest trees in several distinct stages would replace the shrubs. Primary succession on bare rock from a pioneer community to a forest community often occurs in this sequence: lichens → mosses → grasses → shrubs → trees.

The concept of succession was developed in the 1880s by Henry Cowles, who studied the process as it occurred on sand dunes along the shores of Lake Michigan, which has been gradually shrinking since the last ice age. The shrinking lake exposed new sand dunes that displayed a series of stages in the colonization of the land. As in many other lake and ocean shore areas, the Lake Michigan sand dune environment is severe, with temperatures ranging from high during the day to low at night. Few plants could tolerate these stresses and the low nutrient content of the sand making up the dunes.

As Cowles observed, grasses are common pioneer plants on Great Lakes dunes. As the grasses extend over

PROCESS DIAGRAM

Primary succession on glacial moraine • Figure 6.20

✓ THE PLANNER

During the past 200 years, glaciers have retreated in Glacier Bay, Alaska. Although these photos were not taken in the same area, they show some of the stages of primary succession on glacial moraine (rocks, gravel, and sand that a glacier deposits).

① After a glacier's retreat, lichens initially colonize the barren landscape, followed by mosses and small shrubs.

Martin Shields/Science Source Images

② At a later date, dwarf trees and shrubs colonize the area.

Charles D. Winters/Science Source Images

③ Still later, spruces dominate the community.

© Mira/Alamy

Think Critically Is it possible for spruce trees to grow directly on the rocks deposited by glaciers? Why or why not?

Secondary succession on an abandoned field in North Carolina • Figure 6.21

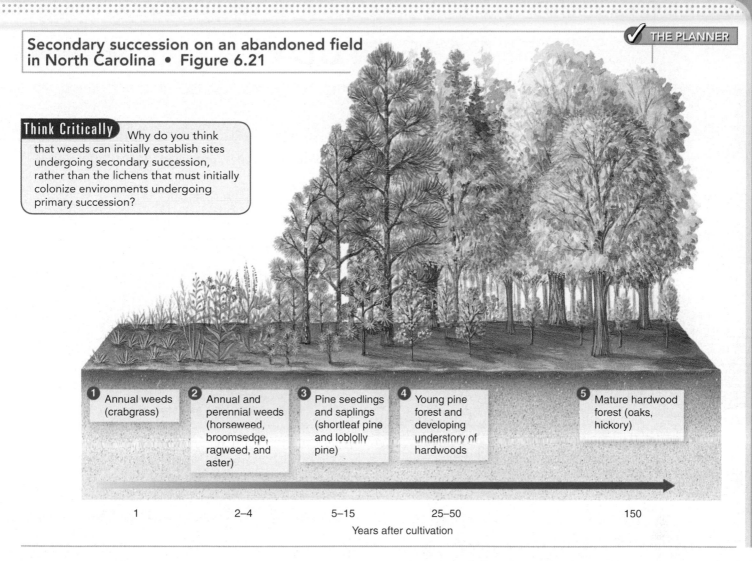

Think Critically Why do you think that weeds can initially establish sites undergoing secondary succession, rather than the lichens that must initially colonize environments undergoing primary succession?

1 Annual weeds (crabgrass)

2 Annual and perennial weeds (horseweed, broomsedge, ragweed, and aster)

3 Pine seedlings and saplings (shortleaf pine and loblolly pine)

4 Young pine forest and developing understory of hardwoods

5 Mature hardwood forest (oaks, hickory)

| 1 | 2–4 | 5–15 | 25–50 | 150 |

Years after cultivation

---

the surface of a dune, their roots hold it in place, helping to stabilize the dune surface. Mat-forming shrubs then invade to further stabilize the dune, followed by a succession of tree species over the course of many years. Primary succession on sand dunes around the Great Lakes might proceed in this sequence: grasses → shrubs → poplars (cottonwoods) → pine trees → oak trees.

Secondary Succession

Secondary succession is the change in species composition that takes place after some disturbance destroys the existing vegetation; soil is already present. A clear-cut forest, open areas caused by a forest fire, and abandoned farmland are common examples of sites where secondary succession occurs. During the summer of 1988, wildfires burned approximately one-third of Yellowstone National Park, a disaster that provided a chance for biologists to study secondary succession in areas that were once forests. Secondary succession in Yellowstone has occurred rapidly, moving from ash-covered forest floor and charred

trees, to lilies and other herbs, and—by ten years after the fires—to a young forest of lodgepole pines, with some Douglas fir seedlings.

Biologists have studied secondary succession on abandoned farmland extensively (**Figure 6.21**). Although it takes more than 100 years for secondary succession to occur at a single site, a single researcher can study old-field succession in its entirety by observing different sites undergoing succession in the same general area. The biologist may examine county tax records to determine when each field was abandoned. Secondary succession on abandoned farmland in the southeastern United States proceeds in this sequence: crabgrass → horseweed, broomsedge, and other weeds → pine trees → hardwood trees.

CONCEPT CHECK   STOP

1. **What** is ecological succession?

2. **How** does primary succession differ from secondary succession?

Wildfires

A wildfire is any unexpected—and unwanted—fire that burns in grass, shrub, and forest areas. Whether started by lightning or by humans, wildfires are an important environmental force in many geographic areas, especially places with wet seasons followed by dry seasons, such as chaparral. Vegetation that grows during the wet season dries to tinder during the dry season. After fire ignites the dry organic material, wind spreads the fire through the area.

At the peak of the wildfire season in the American West and Southwest, an area prone to wildfires, hundreds of new wildfires can break out each day. In 2011, dry conditions triggered U.S wildfires that consumed 3.5 million hectares (8.7 million acres), an annual total of destruction that ranked third in the past 12 years, whereas the number of fires reported in 2011 was about average for that period (see graph). Arizona, New Mexico, and Texas in particular experienced record wildfires (see photos).

Fires have several effects on the environment. First, combustion frees minerals locked in dry organic matter. The ashes left by fire are rich in potassium, phosphorus, calcium, and other nutrient minerals essential for plant growth. Thus, vegetation flourishes after a fire. Second, fire removes plant cover and exposes the soil, which stimulates the germination of seeds that require bare soil and the growth of shade-intolerant plants. Third, fire increases soil erosion because it removes plant cover, leaving soil more vulnerable to wind and water.

Fires were a part of the natural environment long before humans appeared, and many terrestrial ecosystems have adapted to fire. Grasses adapted to wildfire have underground stems and buds. After fire kills the aboveground parts, the untouched underground parts send up new sprouts. Fire-adapted trees such as bur oak and ponderosa pine have thick, fire-resistant bark; others, such as jack pine, depend on fire for successful reproduction because the fire's heat opens the cones and releases the seeds.

Human interference also affects the frequency and intensity of wildfires, even when the goal is fire prevention. When fire is excluded from a fire-adapted ecosystem, organic litter accumulates. As a result, when a fire does occur, it burns hotter and is much more destructive than ecologically helpful. Decades of fire suppression in the West are partly responsible for the massively destructive fires that have occurred there in recent years. *Prescribed burning* is an ecological management tool that allows for controlled burning to reduce organic litter and suppress fire-sensitive trees in fire-adapted areas.

U.S. acres burned in 2011, compared to 2001–2010 average.

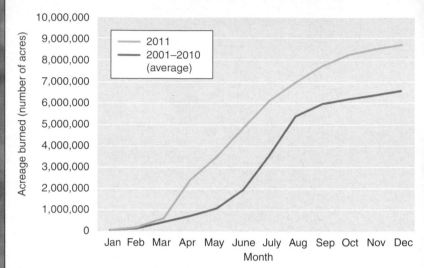

Courtesy of National Oceanic and Atmospheric Administration (NOAA 2011)

A time exposure image highlights the scale of New Mexico's Las Conchas fire, June 27, 2011.

© Albuquerque Journal/ZUMAPress/Corbis

NASA satellite image shows two major wildfires burning in Arizona, June 8, 2011.

NASA Images

Summary

1 Earth's Major Biomes 128

1. A **biome** is a large, relatively distinct terrestrial region with characteristic climate, soil, plants, and animals, regardless of where it occurs; a biome encompasses many interacting ecosystems. Near the poles, temperature is generally the overriding climate factor in determining biome distribution, whereas in temperate and tropical regions, precipitation is more significant.

2. **Tundra** is the treeless biome in the far north that consists of boggy plains covered by lichens and small plants such as mosses; it has harsh, very cold winters and extremely short summers. **Boreal forest** is a region of coniferous forest in the Northern Hemisphere, located just south of the tundra. **Temperate rain forest** is a coniferous biome with cool weather, dense fog, and high precipitation. **Temperate deciduous forest** is a forest biome that occurs in temperate areas where annual precipitation ranges from about 75 cm to 126 cm (30–50 in). **Temperate grassland** is grassland with hot summers, cold winters, and less rainfall than is found in the temperate deciduous forest biome. **Chaparral** is a biome with mild, moist winters and hot, dry summers; vegetation is typically small-leafed evergreen shrubs and small trees. **Desert** is a biome in which the lack of precipitation limits plant growth; deserts are found in both temperate and tropical regions. **Savanna** is tropical grassland with widely scattered trees or clumps of trees. **Tropical rain forest** is a lush, species-rich forest biome that occurs where the climate is warm and moist throughout the year.

2 Aquatic Ecosystems 142

1. In aquatic ecosystems, important environmental factors include salinity, amount of dissolved oxygen, and availability of light for photosynthesis.

2. Freshwater ecosystems include standing-water, flowing-water, and freshwater wetlands. A **standing-water ecosystem** is a body of fresh water surrounded by land and whose water does not flow, such as a lake or pond. A **flowing-water ecosystem** is a freshwater ecosystem such as a river or stream in which the water flows in a current. **Freshwater wetlands** are marshes and swamps—lands that are covered by shallow fresh water at least part of the year; wetlands have a characteristic soil and water-tolerant vegetation. An **estuary** is a coastal body of water, partly surrounded by land, with access to the open ocean and a large supply of fresh water from a river. Water in an estuary is brackish rather than truly fresh. Temperate estuaries usually contain **salt marshes**, whereas tropical estuaries are lined with **mangrove forests**.

3 Population Responses to Changing Conditions over Time: Evolution 147

1. **Evolution** is the cumulative genetic changes in populations that occur during successive generations.

2. **Natural selection** is the tendency of better-adapted individuals—those with a combination of genetic traits best suited to environmental conditions—to survive and reproduce, increasing their proportion in the population. Natural selection is based on four observations established by Charles Darwin: (1) Each species produces more offspring than will survive to maturity. (2) Organisms compete with one another for the resources needed to survive. (3) The individuals in a population exhibit inheritable variation in their traits. (4) Individuals with the most favorable combination of traits are most likely to survive and reproduce, passing their genetic traits to the next generation.

© Danita Delimont/Alamy

Average monthly precipitation in cm

Average monthly temperature in °C

J F M A M J J A S O N D
Months

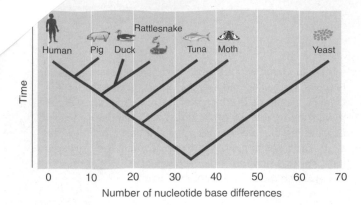

Number of nucleotide base differences

3. Scientific evidence supporting evolution comes from the fossil record, comparative anatomy, biogeography, and molecular biology.

4 Community Responses to Changing Conditions over Time: Succession 151

1. **Ecological succession** is the process of community development over time, which involves species in one stage being replaced by different species.

2. **Primary succession** is the change in species composition over time in an environment that was not previously inhabited by organisms; examples include bare rock surfaces, such as recently formed volcanic lava and rock scraped clean by glaciers. **Secondary succession** is the change in species composition that takes place after some disturbance destroys the existing vegetation; soil is already present. Examples include abandoned farmland and open areas caused by forest fires.

Key Terms

- biome 128
- boreal forest 133
- chaparral 137
- desert 138
- ecological succession 151
- ecosystem services 145
- estuary 146
- evolution 147
- flowing-water ecosystem 144
- freshwater wetlands 145
- natural selection 148
- savanna 140
- standing-water ecosystem 142
- temperate deciduous forest 135
- temperate grassland 136
- temperate rain forest 134
- tropical rain forest 140
- tundra 132

What is happening in this picture?

- This 1994 image from Yellowstone National Park shows young lodgepole pines growing among trees burned in the massive 1988 wildfires. What community process is taking place?

- What type of biome is pictured here? What other biomes are susceptible to fires? How do humans increase the fire risk in these biomes?

François Gohier/Science Source Images

Critical and Creative Thinking Questions

1. What two climate factors are most important in determining an area's characteristic biome?

Sustainable Citizen Question

2. In which biome do you live? Where would you place your biome in the figure below? What human-caused threats are faced by your biome, and how might you help reduce them? Which other biomes might be affected by your lifestyle, such as the foods you eat or other resources you consume?

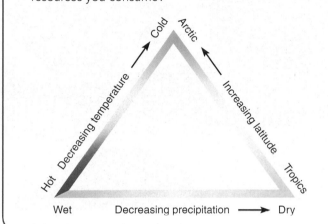

3. What environmental factors are most important in determining the kinds of organisms found in aquatic environments?

4. Distinguish between freshwater wetlands and estuaries and between flowing-water and standing-water ecosystems.

5. Name and compare temperate and tropical estuaries. What types of plants are characteristic of each?

6. During the mating season, male giraffes slam their necks together in fighting bouts to determine which male is stronger and can therefore mate with females. Explain how the long necks of giraffes may have evolved, using Darwin's theory of evolution by natural selection.

7. Explain why evolution, by definition, cannot take place within one individual and during that individual's life span.

8. Describe the process and stages of ecological succession.

9. Which type of ecological succession do you think occurred in the region surrounding Mount St. Helens after the volcano erupted in 1980? Explain your choice by comparing primary and secondary succession.

10. Although most salamanders have four legs, the aquatic salamander shown below resembles an eel. It lacks hind limbs and has very tiny forelimbs. Propose a hypothesis to explain how these salamanders evolved according to Darwin's theory of natural selection.

Joseph T. Collins/Science Source Images

11. How could you test the hypothesis you proposed in question 10? What type of evidence might you produce?

12. Which biome discussed in this chapter is depicted by the information in the graph below? Explain your answer.

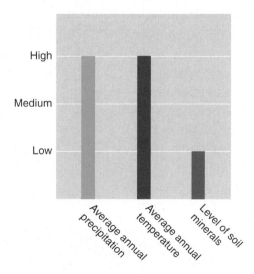

THE PLANNER

7 Human Population Change and the Environment

INDIA'S POPULATION PRESSURES

India is the world's second most populous nation, with a mid-2011 population of 1.24 billion. In the 1950s, it became the first country to establish government-sponsored family planning. India did not experience immediate results from its efforts to control population growth, in part because of the diverse cultures, religions, and customs in different regions of the country. Indians speak 15 main languages and more than 700 dialects, which makes communicating a program of family planning education difficult.

In recent years, India has attempted to integrate economic development and family planning projects. Adult literacy and population education programs have been combined. Multimedia advertisements and education promote voluntary birth control, and contraceptives are more available. India has emphasized that improving health services lowers infant and child mortality rates. These efforts have had an effect: The average number of children born per Indian woman declined from 4.7 in 1980 to 2.6 in 2011 (see graph).

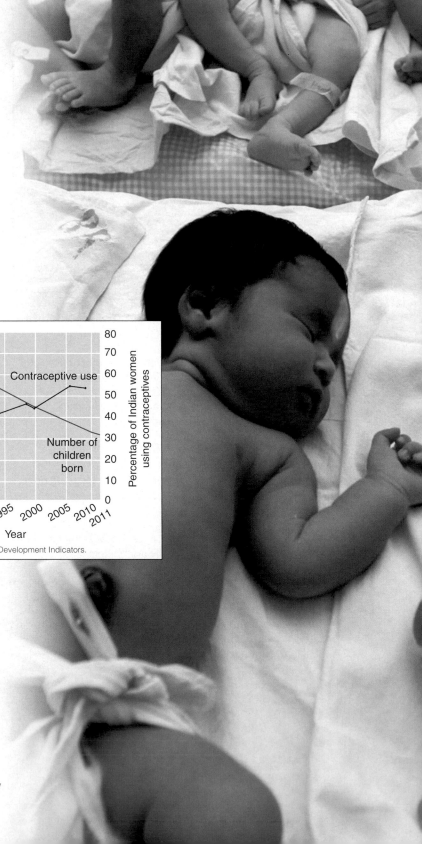

Based on data from World Bank World Development Indicators.

Despite these gains, population pressure has contributed to the deterioration of India's environment in the past few decades, and 76 percent of Indians live below the official poverty level (less than US $2 a day). India's large population exacerbates its poverty, environmental degradation, and economic underdevelopment.

Interpreting Data Question
Between 1985 and 2008, what was the change in contraceptive use? the change in the birth rate?

graphingactivity

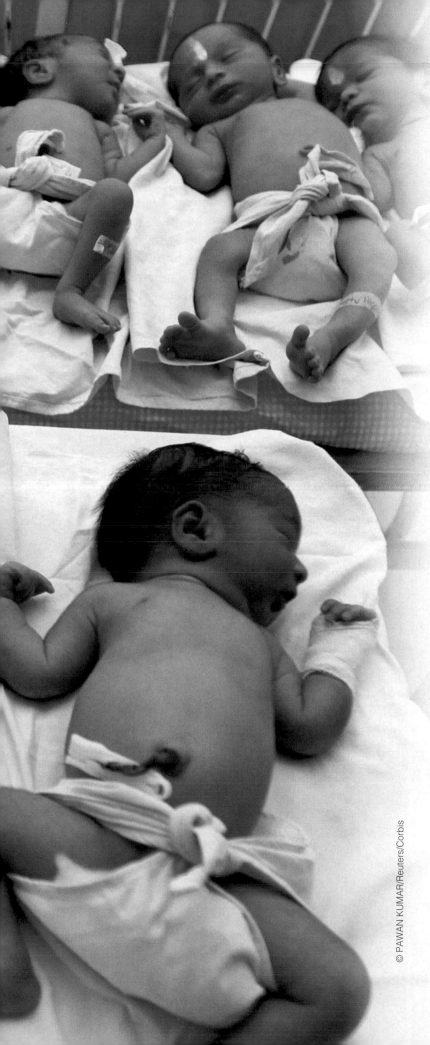

CHAPTER OUTLINE

Population Ecology   160
- How Do Populations Change in Size?
- Maximum Population Growth
- Environmental Resistance and Carrying Capacity

Human Population Patterns   165
- Projecting Future Population Numbers

Demographics of Countries   170
- The Demographic Transition
- ■ Environmental InSight: Demographics of Countries
- Age Structure of Countries

Stabilizing World Population   174
- Culture and Fertility
- The Social and Economic Status of Women
- Family Planning Services
- ■ What a Scientist Sees: Education and Fertility
- ■ EnviroDiscovery: Microcredit Programs
- Government Policies and Fertility

Population and Urbanization   181
- Environmental Problems of Urban Areas
- Environmental Benefits of Urbanization
- Urbanization Trends
- ■ Case Study: Urban Planning in Curitiba, Brazil

CHAPTER PLANNER ✓

- ❑ Study the picture and read the opening story.
- ❑ Scan the Learning Objectives in each section:
  p. 160 ❑   p. 165 ❑   p. 170 ❑   p. 174 ❑   p. 181 ❑
- ❑ Read the text and study all figures and visuals. Answer any questions.

Analyze key features

- ❑ National Geographic Map, pp. 168–169
- ❑ Environmental InSight, p. 171
- ❑ What a Scientist Sees, p. 179
- ❑ EnviroDiscovery, p. 180
- ❑ Case Study, p. 186
- ❑ Stop: Answer the Concept Checks before you go on:
  p. 164 ❑   p. 167 ❑   p. 174 ❑   p. 180 ❑   p. 185 ❑

End of Chapter

- ❑ Review the Summary and Key Terms.
- ❑ Answer What is happening in this picture?
- ❑ Answer the Critical and Creative Thinking Questions.

© PAWAN KUMAR/Reuters/Corbis

Population Ecology

LEARNING OBJECTIVES

1. **Define** *population ecology*.
2. **Explain** the four factors that produce changes in population size.
3. **Define** *biotic potential* and *carrying capacity*.

I ndividuals of a given species are part of a larger organization called a *population*. Populations exhibit characteristics that are distinct from those of the individuals in them. Some of the features characteristic of populations but not of individuals are birth and death rates, growth rates, and age structure. Studying populations of nonhuman species provides insight into some of the processes that affect

population ecology The branch of biology that deals with the number of individuals of a particular species found in an area and how and why those numbers increase or decrease over time.

the growth of human populations. Understanding human population change is important because the size of the human population is central to most of Earth's environmental problems and their solutions.

Scientists who study **population ecology** try to determine the processes common to all populations (**Figure 7.1**). Population ecologists study how a population responds to its environment—such as how individuals in a given population compete for food or other resources, and how predation, disease, and other environmental pressures affect that population. Environmental pressures such as these prevent populations—whether of bacteria or maple trees or giraffes—from increasing indefinitely.

What we learn about one population helps us make predictions about other populations • Figure 7.1

At first glance, the two populations shown here appear to have little in common, but they share many characteristics.

© Organics image library/Alamy Limited

Tom & Pat Leeson/Science Source Library

a. A population of poppies grows in a meadow. Populations of other flowers grow among the red poppies.

b. Walruses congregate on a beach.

How Do Populations Change in Size?

Populations of organisms, whether sunflowers, eagles, or humans, change over time. On a global scale, this change is due to two factors: the rate at which individual organisms produce offspring (the birth rate) and the rate at which individual organisms die (the death rate) (**Figure 7.2a**). In humans, the birth rate (*b*) is usually expressed as the number of births per 1000 people per year and the death rate (*d*) as the number of deaths per 1000 people per year. The **growth rate** (*r*) of a population is the birth rate (*b*) minus the death rate (*d*):

growth rate (r)
The rate of change (increase or decrease) of a population's size, expressed in percentage per year.

$$r = b - d$$

Growth rate is also referred to as *natural increase* in human populations.

If more individuals in a population are born than die, the growth rate is more than zero, and population size increases. If more individuals in a population die than are born, the growth rate is less than zero, and population size decreases. If the growth rate is equal to zero, births and deaths match, and population size is stationary, despite continued reproduction and death.

In addition to birth and death rates, **dispersal**—movement from one region or country to another—affects local populations. There are two types of dispersal: **immigration** (*i*), in which individuals enter a population and increase its size, and **emigration** (*e*), in which individuals leave a population and decrease its size.

The growth rate (*r*) of a local population must take into account birth rate (*b*), death rate (*d*), immigration (*i*), and emigration (*e*) (**Figure 7.2b**). The growth rate equals (birth rate minus death rate) plus (immigration minus emigration):

$$r = (b - d) + (i - e)$$

Maximum Population Growth

Different species have different **biotic potentials** (also called *intrinsic rates of increase*). Several factors influence the biotic potential of a species: the age at which reproduction begins, the fraction of the life span during which an individual can reproduce, the number of reproductive periods per lifetime, and the number of offspring produced during each period of reproduction. These factors, called *life history characteristics*, determine whether a particular species has a large or a small biotic potential.

biotic potential
The maximum rate at which a population could increase under ideal conditions.

Generally, larger organisms, such as blue whales and elephants, have the smallest biotic potentials, whereas microorganisms have the greatest biotic potentials. Under ideal conditions (that is, in an environment with unlimited resources), certain bacteria reproduce by dividing in half every 30 minutes. At this rate of growth, a single bacterium increases to a population of more than 1 million in just 10 hours and exceeds 1 billion in 15 hours. If you plot bacterial population numbers versus time, the graph takes on the characteristic J shape of

---

Factors that interact to change population size • Figure 7.2

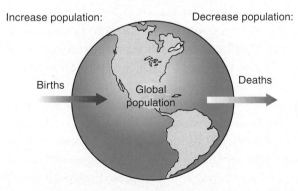

Increase population: Births — Decrease population: Deaths

Global population

a. On a global scale, the change in a population is due to the number of births and deaths.

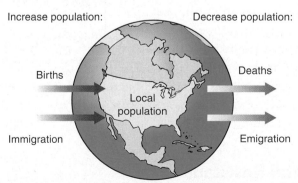

Increase population: Births, Immigration — Decrease population: Deaths, Emigration

Local population

b. In local populations, such as the population of the United States, the number of births, deaths, immigrants, and emigrants affects population size.

Exponential population growth • Figure 7.3

a. *Streptococcus* bacterium in the process of dividing.

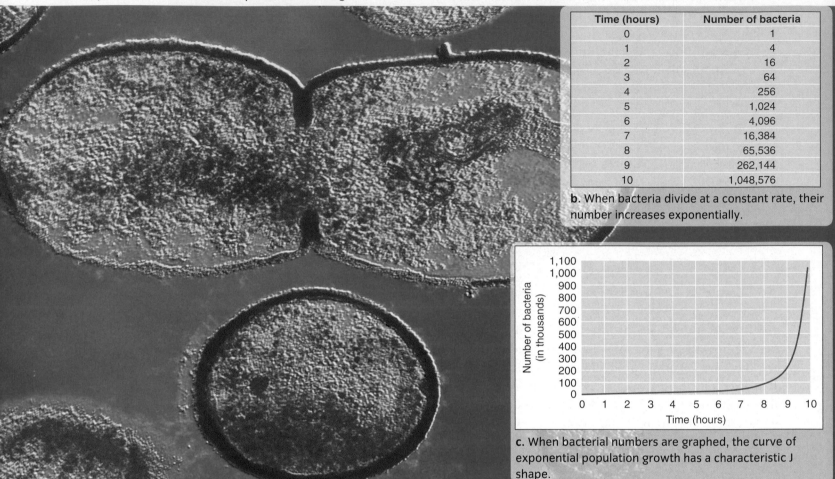

Time (hours)	Number of bacteria
0	1
1	4
2	16
3	64
4	256
5	1,024
6	4,096
7	16,384
8	65,536
9	262,144
10	1,048,576

b. When bacteria divide at a constant rate, their number increases exponentially.

c. When bacterial numbers are graphed, the curve of exponential population growth has a characteristic J shape.

CNRI/Science Photo Library/Science Source

exponential population growth (**Figure 7.3**). When a population grows exponentially, the larger the population gets, the faster it grows. Regardless of species, whenever a population grows at its biotic potential, population size plotted versus time gives the same J-shaped curve. The only variable is time. It may take longer for a dolphin population than for a bacterial population to reach a certain size (because dolphins do not reproduce as rapidly as bacteria), but both populations will always increase exponentially as long as their growth rates remain constant.

> **exponential population growth**
> The accelerating population growth that occurs when optimal conditions allow a constant reproductive rate.

Environmental Resistance and Carrying Capacity

Certain populations—particularly those of bacteria, protists, and certain insects—may exhibit exponential

population growth for a short period. However, organisms don't reproduce indefinitely at their biotic potentials because the environment sets limits, which are collectively called **environmental resistance**. Examples of environmental resistance include such unfavorable environmental conditions as limited food, water, shelter, and other essential resources (resulting in increased competition), as well as increased disease and predation.

Using the earlier example, we find that bacteria never reproduce unchecked for an indefinite period because they run out of food and living space, and poisonous body wastes accumulate in their vicinity. With crowding, bacteria become more susceptible to parasites (high population densities facilitate the spread of infectious organisms such as viruses among individuals) and predators (high population densities increase the likelihood of a predator catching an individual). As the environment deteriorates, the bacteria's birth rate declines and their death rate increases. The environmental conditions might worsen to a point where the death rate exceeds the birth rate, and as a result, the population decreases. Thus, the environment controls population size: As the population increases, so does environmental resistance, which limits population growth.

Over longer periods, environmental resistance may eventually reduce the rate of population growth to nearly zero. This leveling out occurs at or near the environment's **carrying capacity (K)**. In nature, carrying capacity is dynamic and changes in response to environmental changes. An extended drought, for example, might decrease the amount of vegetation growing in an area, and this change, in turn, would lower the carrying capacity for deer and other herbivores in that environment.

carrying capacity (K) The largest population a particular environment can support sustainably (long term), if there are no changes in that environment.

G. F. Gause, a Russian ecologist who conducted experiments in the 1930s, grew a population of *Paramecium* in a test tube (**Figure 7.4a**). He supplied a limited amount of food daily and replenished the media to eliminate the buildup of wastes. Under these conditions, the population increased exponentially at first, but then its growth rate declined to zero, and the population size leveled off.

When a population influenced by environmental resistance is graphed over a long period, the curve has an S shape (**Figure 7.4b**). The curve shows the population's initial exponential increase (note the curve's J shape at the start, when environmental resistance is low). Then the population size levels out as it approaches the carrying capacity of the environment. The rate of population growth is proportional to the amount of existing resources, and competition leads to limited population growth. Although the S curve is an oversimplification of how most populations change over time, it fits some populations studied in the laboratory, as well as a few studied in nature.

A population rarely stabilizes at *K* (carrying capacity), as shown in Figure 7.4, but its size may temporarily rise higher than *K*. It will then drop back to, or below, the

Population growth as carrying capacity is approached • Figure 7.4

a. *Paramecium* is a unicellular microorganism.

Michael Abbey/Science Source

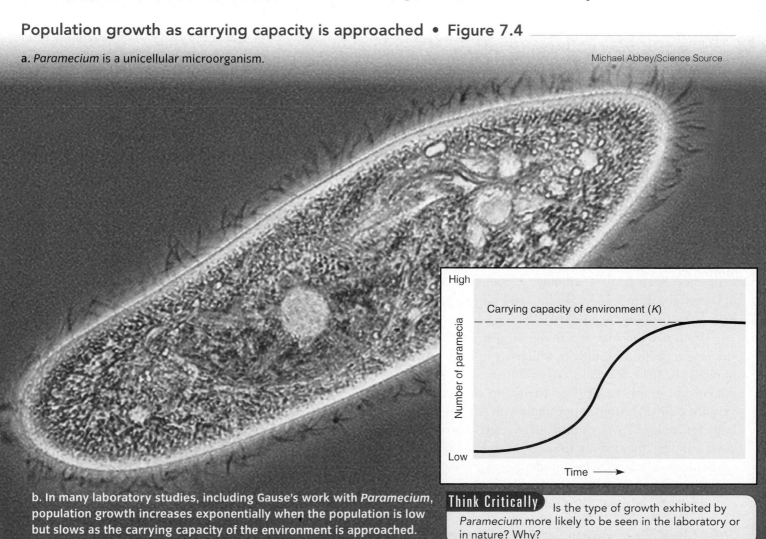

b. In many laboratory studies, including Gause's work with *Paramecium*, population growth increases exponentially when the population is low but slows as the carrying capacity of the environment is approached. This produces a curve with a characteristic S shape.

Think Critically Is the type of growth exhibited by *Paramecium* more likely to be seen in the laboratory or in nature? Why?

A population crash • Figure 7.5

a. A herd of reindeer on one of the Pribilof Islands off the coast of Alaska.

James P. Wright, PE

V.C. Sheffer. "The Rise and Fall of a Reindeer Herd."
1951 *Sci. Month.*, Vol 73.

b. Graph of the reindeer population originally introduced to one of the Pribilof Islands in 1911. Note the population crash, which followed the peak population attained in 1935.

carrying capacity. Sometimes a population that overshoots *K* will experience a *population crash*, an abrupt decline from high to low population density when resources are exhausted. Such an abrupt change is commonly observed in bacterial cultures, zooplankton, and other populations whose resources are exhausted.

The availability of winter forage largely determines the carrying capacity for reindeer, which live in cold northern habitats. In 1911, a small herd of 26 reindeer was introduced on one of the Pribilof Islands in the Bering Sea (**Figure 7.5a**). The herd's population increased exponentially for about 25 years, until there were approximately 2000 reindeer, many more than the island could support, particularly in winter. The reindeer overgrazed the vegetation until the plant life was almost wiped out. Then, in slightly over a decade, as reindeer died from starvation,

the number of reindeer plunged to 8, about one-third the size of the original introduced population and less than 1 percent of the population at its peak (**Figure 7.5b**). Recovery of arctic and subarctic vegetation after overgrazing by reindeer takes 15 to 20 years. During that period, the carrying capacity for reindeer is greatly reduced.

CONCEPT CHECK STOP

1. **What** is population ecology?

2. **How** do each of the following affect population size: birth rate, death rate, immigration, and emigration?

3. **How** do biotic potential and/or carrying capacity produce the J-shaped and S-shaped population growth curves?

Human Population Patterns

LEARNING OBJECTIVES

1. **Summarize** the history of human population growth.

2. **Identify** Thomas Malthus, relate his ideas on human population growth, and explain why he may or may not have been wrong.

3. **Explain** why it is impossible to precisely determine how many people Earth can support—that is, Earth's carrying capacity for humans.

Now that you have examined some of the basic concepts of population ecology, let's apply those concepts to the human population. **Figure 7.6** shows the increase in human population. Reexamine Figure 7.3 and compare the two curves. The characteristic J curve of exponential population growth shown in Figure 7.6 reflects the decreasing amount of time it has taken to add each additional billion people to our numbers. It took tens of thousands of years for the human population to reach 1 billion, a milestone that took place around 1800. It took 130 years to reach 2 billion (in 1930), 30 years to reach 3 billion (in 1960), 15 years to reach 4 billion (in 1975), 12 years to reach 5 billion (in 1987), 12 years to reach 6 billion (in 1999), and 12 years to reach 7 billion (in 2011). Population experts predict that the population will level out during the 21st century, possibly forming an S curve as observed in some other species.

One of the first people to recognize that the human population can't increase indefinitely was **Thomas Malthus** (1766–1834), a British economist. He pointed out that human population growth is not always desirable—a view contrary to the beliefs of his day and to those of many people even today. Noting that human population can increase faster than its food supply, he warned that the inevitable consequences of population growth would be famine, disease, and war. Since Malthus's time, the human population has increased from about 1 billion to 7 billion.

On the surface, it seems that Malthus was wrong. Our population has grown dramatically because geographic expansion and scientific advances have allowed food production to keep pace with population growth. Malthus's ideas may ultimately be proved correct, however, because we don't know whether this increased food production is sustainable. Have we achieved this increase in food production at the environmental cost of reducing the planet's ability to meet the needs of future populations? Many economists suggest that market forces and future technologies will help us prevent resource depletion such as soil degradation and overfishing in the ocean. But the truth is that we still do not know if Malthus was wrong or right.

Our world population was 7 billion in late 2011, an increase of about 95 million from 2010. This increase was not due to a rise in the birth rate (b), although high birth rates are a serious problem in many countries. In fact, the world birth rate has declined slightly during the past 200 years. The population growth is due instead to a dramatic *decrease* in the death rate (d), which has occurred primarily because greater food production, better medical care, and improvements in water quality and sanitation practices have

Human population growth • Figure 7.6

Compare this figure to Figure 7.5b. Do you think what happened to the reindeer on the Pribilof Islands could happen to Earth's human population? Why or why not? (Black Death refers to a devastating disease, probably bubonic plague, that decimated Europe and Asia in the 14th century.)

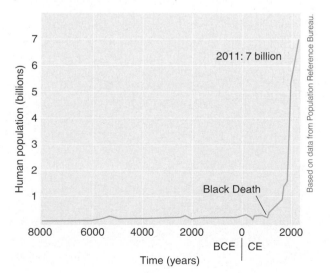

Based on data from Population Reference Bureau.

Advances in global health • Figure 7.7

A child in Bangladesh receives a dose of oral polio vaccine. At one time, polio killed or crippled millions of children each year. Polio is still endemic (constantly present) in Nigeria, India, Afghanistan, and Pakistan, and it sometimes spreads from those countries to other countries.

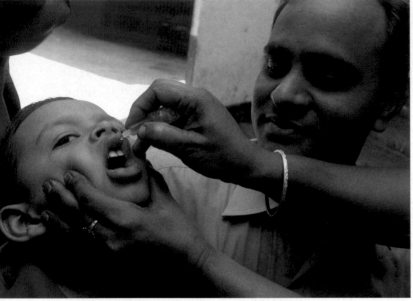

FARJANA KHANGODHULY/Stringer/AFP/Getty Images

increased life expectancy for a great majority of the global population (**Figure 7.7**).

Projecting Future Population Numbers

The human population has reached a turning point. Although our numbers continue to increase, the world growth rate (*r*) has declined slightly over the past sev-

| zero population growth The state in which the population remains the same size because the birth rate equals the death rate. |

eral years, from a peak of 2.2 percent per year in the mid-1960s to the current growth rate of 1.2 percent per year. Population experts at the United Nations and the World Bank project that the growth rate will continue to decrease slowly until **zero population growth** is attained toward the end of the 21st century. Exponential growth of the human population will end, and the S curve may replace the J curve.

The United Nations periodically publishes population projections for the 21st century. The latest (2010) U.N. figures forecast that the human population will reach 9.3 billion in the year 2050 (their "medium" projection), and could range between 8.1 billion (their "low" projection) and 10.6 billion (their "high" projection)

(**Figure 7.8**). The estimates vary depending on fertility changes, particularly in less developed countries, because that is where almost all of the growth will take place.

Population projections must be interpreted with care because they vary depending on what assumptions are made. In projecting that the world population will be 8.1 billion (their low projection) in the year 2050, U.N. population experts assume that the average number of children born to each woman in all countries will have declined to 1.7 by 2045–2050. The average number of children born to each woman on Earth is currently 2.5. If the decline to 1.7 doesn't occur, our population could be significantly higher. If the average number of children born to each woman declines to 2.17 in 2045–2050 instead of 1.5, the 2050 population will be 9.3 billion (the U.N. medium projection).

World population projections to 2050 • Figure 7.8

In 2010 the United Nations made three projections, each based on different fertility rates.

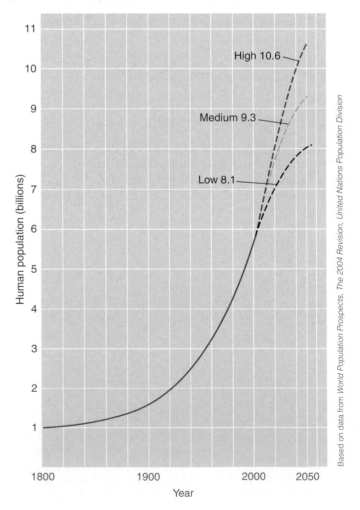

Based on data from *World Population Prospects, The 2004 Revision, United Nations Population Division*

Environmental degradation on a cattle ranch in Brazil • Figure 7.9

Part of the rain forest in the background was cleared for a cattle pasture. After a few years, the pasture became unproductive, and erosion degraded the land further. Photographed in Amazonas State in the Amazon River Basin.

Small differences in fertility, then, produce large differences in population forecasts.

The main unknown factor in any population growth scenario is Earth's *carrying capacity*. Most published estimates of how many people Earth can support range from 4 billion to 16 billion. For example, in 2004, environmental economists in the Netherlands performed a detailed analysis of 69 recent studies of Earth's carrying capacity for humans. Based on current technology, they estimated that 7.7 billion is the upper limit of human population that the world can support. Even the low U.N. projection for 2050 exceeds this value (see Figure 7.8).

These estimates vary widely depending on what assumptions are made about standard of living, resource consumption, technological innovations, and waste generation. If we want all people to have a high level of material well-being equivalent to the lifestyles in highly developed countries, then Earth will support far fewer humans than if everyone lives just above the subsistence level. Unlike with other organisms, environmental constraints aren't the exclusive determinant of Earth's carrying capacity for humans. Human choices and values must be factored into the assessment.

What will happen to the human population when it approaches Earth's carrying capacity? Optimists suggest that a decrease in the birth rate will stabilize the human population. Some experts take a more pessimistic view and predict that our ever-expanding numbers will cause widespread environmental degradation and make Earth uninhabitable for humans as well as other species (**Figure 7.9**). These population researchers contend that a massive wave of human suffering and death will occur. This view doesn't mean we will go extinct as a species, but it projects severe hardship for many people. Some experts think the human population has already exceeded the carrying capacity of the environment, a potentially dangerous situation that threatens our long-term survival as a species.

Global human population trends are summarized in **Figure 7.10** on pages 168 to 169. Note especially regional differences in population growth and density.

CONCEPT CHECK

1. **How** would you describe human population growth for the past 200 years?

2. **Who** was Thomas Malthus, and what were his views on human population growth?

3. **When** determining Earth's carrying capacity for humans, why is it not enough to just consider human numbers?

The human population
• Figure 7.10

Geographers approach the study of human populations from a spatial perspective, asking why density, distribution, resources, births, deaths, and migrations vary from place to place. Earth's population, now above 7 billion, grows by 105 million per year, or 1.2 percent annually. The bulk of the increase occurs in developing countries in Asia, Africa, and Latin America. Physiologic density— the number of people per unit of agricultural land—shows concentrations in Asia, in particular in China and India; in Europe, from Britain into Russia; along the eastern seaboard of the United States; and in West Africa in Nigeria and along the NIle Valley.

Find your location on the world map showing population density. What is the approximate population density per square km in this area? Are there other areas of the world with greater population densities? Where?

NATIONAL GEOGRAPHIC

MAPPING DENSITY

People are not settled evenly on the planet. Some places such as Monaco have a crowded 34,000 people per square kilometer, while others such as Mongolia have a mere 2 people per square kilometer. Concentrations are seen in the cities and around natural resources.

Population density, 2007
People per sq km (sq mi)

- More than 193 (500)
- 58–193 (150–500)
- 10–57 (25–149)
- 1–9 (1–24)
- Less than 1 (1)

○ Megacities with populations of ten million or greater shown, except for Seoul, based on UN urban agglomeration data. Seoul's ranking based on greater metropolitan data.

NORTH AMERICA

New York

Los Angeles

Mexico City

SOUTH AMERICA

Rio de Janeiro
São Paulo

Buenos Aires

A.D. 1 | 50 | 100 | 150 | 200 | 250 | 300 | 350 | 400 | 450 | 500 | 550 | 600 | 650 | 700 | 750 | 800 | 850 | 900 | 950 | 1000 | 1050 | 1100 | 1150 | 1200 | 1250

Year

MEASURING NATIONAL DENSITY

National population density is a measure of the number of people per square kilometer or mile at the country level. Using this type of measurement, some large countries. such as the United States, appear to have uniformly high population densities, even though vast areas of land may be sparsely populated. Other countries, such as Russia, have a low national density because of large rural areas, even though they have large cities, such as Moscow.

Population density, mid-2008
People per sq km (sq mi)

- More than 386 (1,000)
- 97–386 (250–999)
- 29–96 (75–249)
- 10–28 (25–74)
- Less than 10 (25)
- No data available

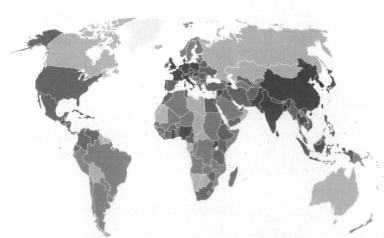

ACCOUNTING FOR ARABLE LAND

Another method of measuring population density is to calculate the number of people per square kilometer or mile of arable, or farmable, land. Using this type of measurement, known as physiologic density, a pattern differing form national population density emerges.

Countries such as Iceland and Egypt have high densities per arable land since their amount of farmable land is low. Other countries, such as Kazakhstan, appear to have low densities because of large areas of arable land.

Arable land population density, 2007
People per sq km of arable land (sq mi)

- More than 1,544 (4,000)
- 502–1,544 (1,300–4,000)
- 251–501 (650–1,299)
- 97–250 (250–649)
- Less than 97 (250)
- No data available

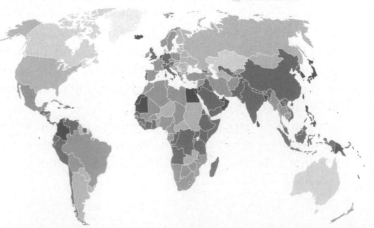

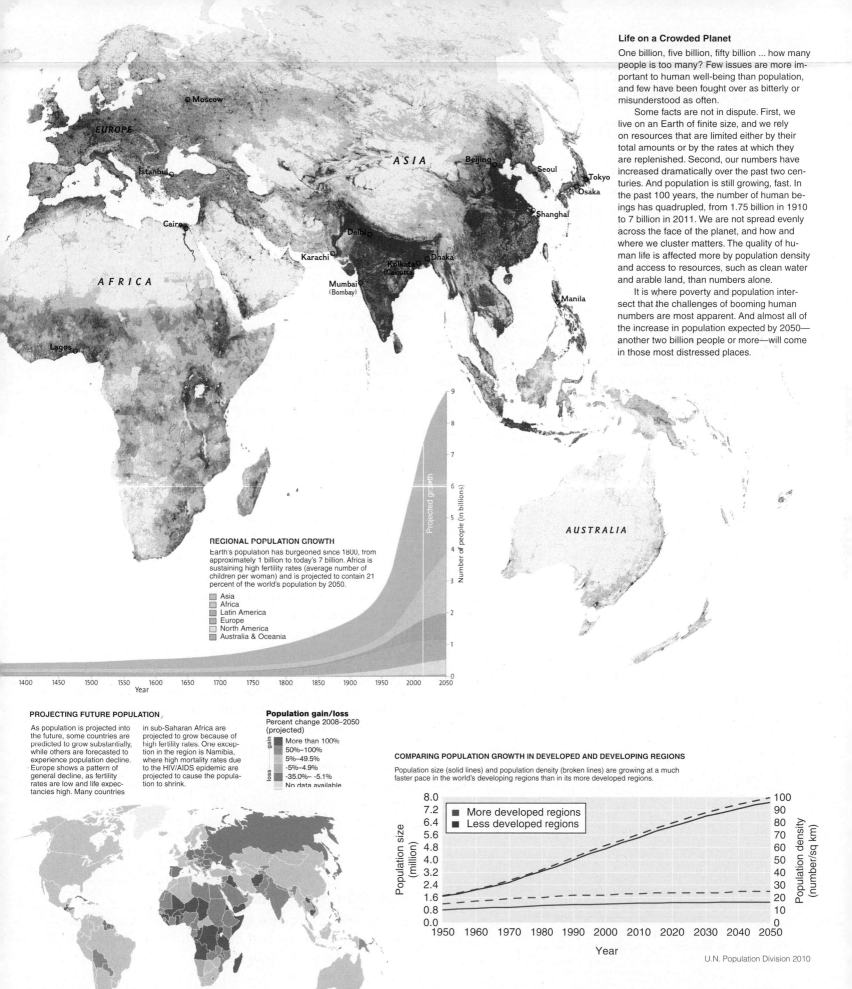

Life on a Crowded Planet

One billion, five billion, fifty billion ... how many people is too many? Few issues are more important to human well-being than population, and few have been fought over as bitterly or misunderstood as often.

Some facts are not in dispute. First, we live on an Earth of finite size, and we rely on resources that are limited either by their total amounts or by the rates at which they are replenished. Second, our numbers have increased dramatically over the past two centuries. And population is still growing, fast. In the past 100 years, the number of human beings has quadrupled, from 1.75 billion in 1910 to 7 billion in 2011. We are not spread evenly across the face of the planet, and how and where we cluster matters. The quality of human life is affected more by population density and access to resources, such as clean water and arable land, than numbers alone.

It is where poverty and population intersect that the challenges of booming human numbers are most apparent. And almost all of the increase in population expected by 2050—another two billion people or more—will come in those most distressed places.

REGIONAL POPULATION GROWTH

Earth's population has burgeoned since 1800, from approximately 1 billion to today's 7 billion. Africa is sustaining high fertility rates (average number of children per woman) and is projected to contain 21 percent of the world's population by 2050.

- Asia
- Africa
- Latin America
- Europe
- North America
- Australia & Oceania

PROJECTING FUTURE POPULATION

As population is projected into the future, some countries are predicted to grow substantially, while others are forecasted to experience population decline. Europe shows a pattern of general decline, as fertility rates are low and life expectancies high. Many countries in sub-Saharan Africa are projected to grow because of high fertility rates. One exception in the region is Namibia, where high mortality rates due to the HIV/AIDS epidemic are projected to cause the population to shrink.

Population gain/loss
Percent change 2008–2050 (projected)

gain
- More than 100%
- 50%–100%
- 5%–49.5%
- -5%–4.9%
loss
- -35.0%– -5.1%
- No data available

COMPARING POPULATION GROWTH IN DEVELOPED AND DEVELOPING REGIONS

Population size (solid lines) and population density (broken lines) are growing at a much faster pace in the world's developing regions than in its more developed regions.

- More developed regions
- Less developed regions

U.N. Population Division 2010

Human Population Patterns   169

Demographics of Countries

LEARNING OBJECTIVES

1. **Define** *demographics* and describe the demographic transition.

2. **Explain** how highly developed and developing countries differ in population characteristics such as infant mortality rate, total fertility rate, replacement-level fertility, and age structure.

W orld population figures illustrate overall trends but don't describe other important aspects of the human population story, such as population differences from country to country (**Table 7.1**). **Demographics** provides information on the populations of various countries. Recall from Chapter 1 that countries are classified into two main groups—highly developed and developing—based on population growth rates, degree of industrialization, and relative prosperity.

> **demographics**
> The applied branch of sociology that deals with population statistics.

Highly developed countries such as the United States, Canada, France, Germany, Sweden,

The world's 10 most populous countries • Table 7.1

Country	2011 Population (in millions)	Population density (per square kilometer)
China	1346	141
India	1241	378
United States	312	32
Indonesia	238	125
Brazil	197	23
Pakistan	177	222
Nigeria	162	176
Bangladesh	151	1046
Russia	143	8
Japan	128	339

Population Reference Bureau

Interpreting Data
Which of these countries is the most crowded? Which is the least crowded?

Australia, and Japan have the lowest birth rates in the world. Indeed, some countries, such as Germany, have birth rates just below those needed to sustain their populations and are declining slightly in numbers.

> **infant mortality rate** The number of deaths of infants under age 1 per 1000 live births.

Highly developed countries also have low **infant mortality rates** (**Figure 7.11a**). The infant mortality rate of the United States was 6.1 in 2011, compared with a world rate of 44. Highly developed countries have longer life expectancies (78 years in the United States versus 70 years worldwide).

Per person GNI PPP is a country's gross national income (GNI) in purchasing power parity (PPP) divided by its population. It indicates the amount of goods and services an average citizen of that particular country could buy in the United States. There is a high average per person GNI PPP in the United States—$45,640—as compared to the worldwide figure of $10,240.

In *moderately developed countries,* such as Mexico, Turkey, Thailand, and most South American nations, birth rates and infant mortality rates are higher than those of highly developed countries, but they are declining. Moderately developed countries have a medium level of industrialization, and their average per person GNI PPPs are lower than those of highly developed countries. *Less developed countries,* such as Bangladesh, Niger, Ethiopia, Laos, and Cambodia, have the shortest life expectancies, the lowest average per person GNI PPPs, the highest birth rates, and the highest infant mortality rates in the world (**Figure 7.11b**).

Replacement-level fertility is usually given as 2.1 children. The number is greater than 2.0 because some infants and children die before they reach reproductive age. Worldwide, the **total fertility rate (TFR)** is currently 2.5, well above the replacement level.

> **replacement-level fertility** The number of children a couple must produce to "replace" themselves.
>
> **total fertility rate (TFR)** The average number of children born to each woman.

The Demographic Transition

Demographers recognize four demographic stages based on their observations of Europe as it became industrialized and urbanized (**Figure 7.11c**). During these stages,

Annie Griffiths Belt/
NG Image Collection

Alberto Ceoloni/ZUMAPRESS/Newscom

a. Infant Mortality Rates in Highly Developed Countries. Nurses care for newborn infants in Israel, a highly developed country with an infant mortality rate of 3.6.

b. Infant Mortality Rates in Developing Countries. This premature Afghan baby was born in a refugee camp. The infant mortality rate in Afghanistan, a less developed country, is 131.

Stage 1 **Preindustrial**	**Stage 2** **Transitional**	**Stage 3** **Industrial**	**Stage 4** **Postindustrial**
Women have many children, but infant mortality rate is high, so population grows very slowly.	Lowered death rate from improved health care and more reliable food and water supplies. Birth rate is still high, and population grows rapidly.	Decline in birth rate slows population growth despite relatively low death rate.	People are better educated and more affluent. They tend to take steps to limit family size. Population grows very slowly or not at all.

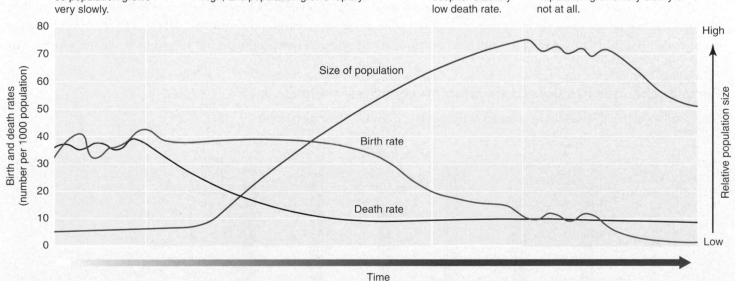

c. The Demographic Transition. Demographers have identified four stages through which a population progresses as its society becomes industrialized.

Interpreting Data
Is population increasing or decreasing in Stage 3 of the demographic transition? Why?

Europe moved from relatively high birth and death rates to relatively low birth and death rates, as a result of industrialization. All highly developed and moderately developed countries with more advanced economies have gone through this **demographic transition**, and demographers assume that the same progression will occur in less developed countries as they industrialize.

Why has the population stabilized in more than 30 highly developed countries in the fourth (postindustrial) demographic stage? The reasons are complex. Declining birth rate is associated with an improvement in living standards. It is difficult to say whether improved socioeconomic conditions have resulted in a decrease in birth rate or whether a decrease in birth rate has resulted in improved socioeconomic conditions. Perhaps both are true. Another reason for the decline in birth rate in highly developed countries is the increased availability of family planning services. Other socioeconomic factors that influence birth rate are increased education, particularly of women, and *urbanization* of society (discussed later in this chapter).

Once a country reaches the fourth demographic stage, is it correct to assume that the country will continue to have a low birth rate indefinitely? We don't know. Low birth rates may be a permanent response to the socioeconomic factors of an industrialized, urbanized society. On the other hand, low birth rates may be a response to socioeconomic factors, such as the changing roles of women in highly developed countries. Unforeseen

> **demographic transition** The process whereby a country moves from relatively high birth and death rates to relatively low birth and death rates.

changes in the socioeconomic status of women and men in the future may again change birth rates. No one knows for sure.

The population in many developing countries is beginning to approach stabilization (**Figure 7.12**). For example, the TFR in Brazil in 1960 was 6.7 children per woman. Today it is 1.9. Worldwide, the TFR in developing countries has decreased from an average of 6.1 children per woman in 1970 to 2.5 today.

Although fertility rates in these countries have declined, many still exceed replacement-level fertility. Consequently, populations in these countries are still increasing. Even when fertility rates equal replacement-level fertility, population growth will still continue for some time. To understand why this is so, let's examine the age structure of various countries.

Age Structure of Countries

A population's **age structure** helps predict future population growth. The number of males and the number of females at each age, from birth to death, are represented in an *age structure diagram*. Each diagram is divided vertically in half, the left side representing the males in a population and the right side the females. The bottom third of each diagram represents prereproductive humans (between 0 and 14 years of age); the middle third, reproductive humans (15 to 44 years); and the top third, postreproductive humans (45 years and older). The

> **age structure** The number and proportion of people at each age in a population.

Fertility changes in selected developing countries • Figure 7.12

Since the 1960s, fertility levels have dropped dramatically in many developing countries.

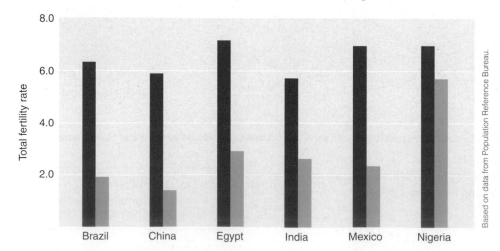

Age structure diagrams • Figure 7.13

Shown are countries with **a** rapid (Ethiopia), **b** slow (United States), and **c** no growth (Italy) or declining population growth.

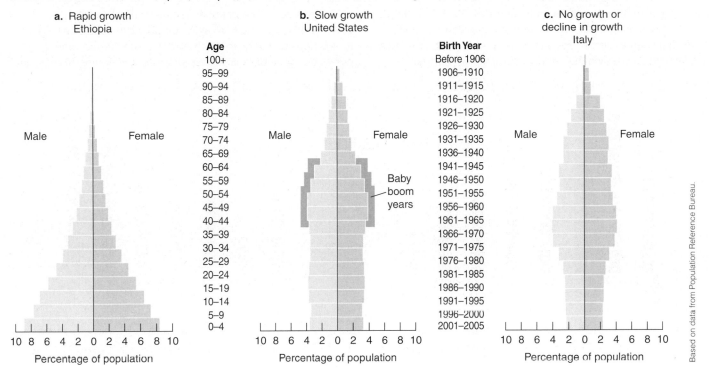

a. Rapid growth
Ethiopia

b. Slow growth
United States

c. No growth or decline in growth
Italy

Percentage of population

widths of these segments are proportional to the population sizes: A broader width implies a larger population. The overall shape of an age structure diagram indicates whether the population is increasing, stable, or shrinking.

The age structure diagram of a country with a high growth rate, based on a high fertility rate—for example, Ethiopia or Guatemala—is shaped like a pyramid (**Figure 7.13a**). The largest percentage of the population is in the prereproductive age group (0 to 14 years of age), so the probability of future population growth is great. A positive **population growth momentum** exists because when all these children mature, they will become the parents of the next generation, and this group of parents will be larger than the previous group. Even if the fertility rate of such a country has declined to replacement level (that is, if couples are having smaller families than their parents did), the population will continue to grow for some time. Population growth momentum, which can be positive or negative, explains how a population's present age distribution affects its future growth.

In contrast, the more tapered bases of the age structure diagrams of countries with slowly growing, stable, or declining populations indicate that a smaller proportion of the population will become the parents of the next generation (**Figure 7.13b** and **c**). The age structure diagram

of a stable population (neither growing nor shrinking) demonstrates that the numbers of people at prereproductive and reproductive ages are approximately the same. A larger percentage of the population is older—that is, postreproductive—than in a rapidly increasing population. Many countries in Europe have stable populations.

In a shrinking population, the prereproductive age group is smaller than either the reproductive or postreproductive age group. Russia, Ukraine, and Germany are examples of countries with slowly shrinking populations.

Worldwide, 27 percent of the human population is under age 15. When these people enter their reproductive years, they have the potential to cause a large increase in the growth rate. Even if the birth rate doesn't increase, the growth rate will increase simply because there are more people reproducing.

Most of the world population increase since 1950 has taken place in developing countries, as a result of the younger age structure and the higher-than-replacement-level fertility rates of their populations. In 1950, 67 percent of the world's population was in developing countries in Africa, Asia (minus Japan), and Latin America. After 1950, the world's population more than doubled in size, but most of that growth occurred in developing countries. As a reflection of this trend, in 2011 the number

Percentages of prereproductive and elderly populations for various regions of the world • Figure 7.14

a. Percentages of the population under age 15 in 2011. The higher this percentage, the greater the potential for population growth. Note the high percentage of young people in Africa, Latin America, and Asia, home to many of the world's developing countries.

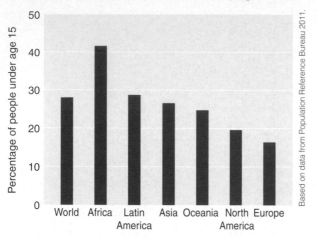

b. Percentages of the population older than 65 in 2011. Lower fertility rates lead to aging populations. Note the larger proportions of elderly in North America and Europe, where population growth rates are typically slow, stagnant, or declining.

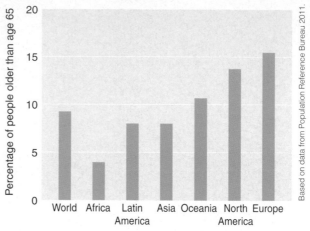

of people in developing countries (including China) increased to 82 percent of the world population. Most of the population increase during the 21st century will take place in developing countries, largely as a result of their younger age structures (**Figure 7.14a**). These countries, with their already limited access to resources, will have economic difficulty supporting such growth.

Declining fertility rates have profound social and economic implications because as fertility rates drop, the percentage of the population that is elderly increases (**Figure 7.14b**). An aging population has a higher percentage of people who are chronically ill or disabled, and these people require more health care and other social services. Because the elderly produce less wealth (most are retired), an aging population reduces a country's productive workforce, increases its tax burden, and

strains its social security, health, and pension systems. To reduce such costs, governments with growing elderly populations may offer incentives to the elderly to work longer before retiring.

Not all characteristics of an elderly population are negative, however. Sociologists have observed that in an aging population the rate of violent crime may decline, as young adults—those most likely to commit crimes—represent an increasingly smaller proportion of the population.

CONCEPT CHECK

1. **What** is the demographic transition?

2. **What** is infant mortality rate? How does it vary in highly developed and developing countries?

Stabilizing World Population

LEARNING OBJECTIVES

1. **Relate** total fertility rates to each of the following: cultural values, social and economic status of women, availability of family planning services, and government policies.

2. **Explain** the link between education and total fertility rates.

ispersal—moving from one place to another—used to be a solution for unsustainable population growth, but not today. As a species, we humans have expanded our range throughout Earth, and few habitable areas remain that have the resources to adequately support a major increase in human population. It is unlikely that death rates will increase substantially in the foreseeable future. Consequently, global human population will not stabilize unless birth rates drop. Cultural traditions, women's social and economic status, family planning, and government policies all influence total fertility rate (TFR).

Culture and Fertility

The values and norms of a society—what is considered right and important and what is expected of a person—are all a part of that society's **culture**. A society's culture, which includes its language, beliefs, and spirituality, exerts a powerful influence over individuals by controlling behavior. Gender—that is, varying roles men and women are expected to fill—is an important part of culture. Different societies have different gender expectations (**Figure 7.15**). With respect to fertility and culture, a couple is expected to have the number of children traditional in their society.

High TFRs are traditional in many cultures. The motivations for having many babies vary from culture to culture, but a major reason for high TFRs is that infant and child mortality rates are high. For a society to endure, it must produce enough children who can survive to reproductive age. If infant and child mortality rates are high, TFRs must be high to compensate. Although world infant and child mortality rates are decreasing, it will take longer for culturally embedded fertility levels to decline. Parents must have confidence that the children they already have will survive before they stop having additional babies. Another reason for the lag in fertility

Varying roles of men and women • Figure 7.15

a. In parts of Latin America, men do the agricultural work. This Argentinian man is harvesting grapes.

b. In sub-Saharan Africa, women do most of the agricultural work in addition to caring for their children. Photographed in South Africa.

Pablo Corral Vega/NG Image Collection

James P. Blair/NG Image Collection

decline is cultural: Changing anything traditional, including large family size, usually takes a long time.

Higher TFRs in some developing countries are also due to the important economic and societal roles of children. In some societies, children usually work in family enterprises such as farming or commerce, contributing to the family's livelihood. The International Labour Organization estimates that, worldwide, about 176 million children between the ages of 5 and 14 worked full time in 2008 (household chores are not counted as labor). This estimate represents 14.5 percent of all children. Almost all of these children live in developing countries (**Figure 7.16**; also see Figure 2.10). About 53 million child laborers do hazardous work such as mining and construction. These child laborers often suffer from chronic health problems caused by the dangerous, unhealthy conditions to which they are exposed. Children who work full time do not have childhoods, nor do they receive education.

When child laborers become adults, they tend to provide support for their aging parents. In contrast, children in highly developed countries have less value as a source of labor because they attend school and because less human labor is required in an industrialized society. Furthermore, highly developed countries provide many social services for the elderly, so the burden of their care doesn't fall entirely on offspring.

Many cultures place a higher value on male children than on female children. In these societies, a woman who bears many sons achieves a high status; thus, the social pressure to have male children keeps the TFR high.

Religious values are another aspect of culture that affects TFRs. Several studies done in the United States point to differences in TFRs among Catholics, Protestants, and Jews. In general, Catholic women have a higher TFR than either Protestant or Jewish women, and women who don't follow any religion have the lowest TFRs of all.

Working child • Figure 7.16

A Pakistani child weaves carpet in a border town near Afghanistan. The carpet industry in Pakistan commonly utilizes child labor.

Think Critically How might a high incidence of child labor in a country be related to that nation's TFR?

AP Photo/Shah Khalid

The observed differences in TFRs may not be the result of religious differences alone. Other variables, such as ethnicity (certain religions are associated with particular ethnic groups) and residence (certain religions are associated with urban or with rural living), complicate any generalizations that might be made.

The Social and Economic Status of Women

Gender inequality exists to varying degrees in most societies: Women don't have the same rights, opportunities, or privileges as men. Gender disparities include the lower political, social, economic, and health status of women compared to men. For example, more women than men live in poverty, particularly in developing countries. In most countries, women are not guaranteed equality in legal rights, education, employment and earnings, or political participation.

Because sons are more highly valued than daughters, girls are often kept at home to work rather than being sent to school (**Figure 7.17a**). In most developing countries, a higher percentage of women are illiterate than men (**Figure 7.17b**). However, definite progress has been made in recent years in increasing literacy in both women and men and in narrowing the gender gap. Fewer young women and men are illiterate than older women and men within a given country.

Worldwide, some 90 million girls aren't given the opportunity to receive a primary (elementary school) education. Laws, customs, and lack of education often limit women to low-skilled, low-paying jobs. In such societies, marriage is usually the only way for a woman to achieve social influence and economic security.

Evidence suggests that the single most important factor affecting high TFRs may be the low status of women in many societies. An effective strategy for reducing population growth, then, is to improve the social and economic status of women.

Let's examine how marriage age and educational opportunities, especially for women, affect fertility. The average age at which women marry affects the TFR; in turn, the laws and customs of a given society affect marriage age. Women who marry are more apt to bear children than women who don't marry, and the earlier a woman marries, the more children she is likely to have.

Gender discrimination • Figure 7.17

a. Nigerian students. Note the number of boys versus girls. Why do you think the school has many more boys than girls? Where are all the girls?

b. Illiteracy percentage of men and women in selected developing countries. A higher percentage of women than men are illiterate.

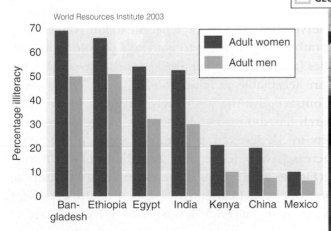

World Resources Institute 2003

Legend:
- Adult women
- Adult men

(Bar chart: Percentage illiteracy vs. country — Bangladesh, Ethiopia, Egypt, India, Kenya, China, Mexico)

Source: Adapted from Data Table 12 in *World Resources 2002–2004: Decision for the Earth: Balance, Voice, and Power.* Washington, D.C.: World Resource Institute (2003)

© INTERFOTO/Alamy

Global Locator

NG Maps

NIGERIA

NATIONAL GEOGRAPHIC

EnviroDiscovery
Microcredit Programs

High population growth rates exacerbate the poverty experienced by many in developing countries. **Microcredit** programs extend small loans ($50 to $500) to very poor people to help them establish businesses that generate income. The poor use these loans for a variety of projects. Some have purchased used sewing machines to make clothing faster than sewing by hand. Others have opened small grocery stores after purchasing used refrigerators to store food so that it does not spoil.

The **Foundation for International Community Assistance (FINCA)** is a not-for-profit agency that administers a global network of microcredit banks. FINCA uses *village banking*, in which a group of very poor neighbors guarantees one another's loans, administers group lending and saving activities, and provides mutual support. These village banks give autonomy to local people.

FINCA primarily targets women because an estimated 70 percent of the world's poorest people are women. FINCA believes that the best way to alleviate the effects of poverty and hunger on children is to provide their mothers with a means of self-employment. A woman's status in the community is raised as she begins earning income from her business (see photo).

RAFIQUR RAHMAN/Reuters/Landov LLC

Microcredit.
This Bangladeshi woman feeds chickens at her poultry farm. She received her first microcredit loan to buy a few chickens and has built the farm into a thriving business.

contraceptives. Fertility declines are occurring in developing countries where contraceptives are readily available. Beginning in the 1970s, use of contraceptives in East Asia and many areas of Latin America increased significantly, and these regions experienced corresponding declines in birth rates.

Family planning centers provide information and services primarily to women. As a result, in the male-dominated societies of many developing countries, such services may not be as effective as they could otherwise be. Polls of women in developing countries reveal that many who say they don't want additional children still don't practice any form of birth control. When asked why they don't use birth control, these women frequently respond that their husbands or in-laws want additional children.

Government Policies and Fertility

The involvement of governments in childbearing and child rearing is well established. Laws determine the minimum age at which people may marry and the amount of compulsory education they receive. Governments may allot portions of their budgets to family planning services, education, health care, old-age security, or incentives for smaller or larger family size. The tax structure, including additional charges or allowances based on family size, also influences fertility.

In recent years, the governments of at least 78 developing countries in Africa, Asia, Latin America, and the Caribbean have taken measures to limit population growth. Most countries sponsor family planning projects, which are integrated with health care, education, economic development, and efforts to improve women's status.

CONCEPT CHECK

1. **What** is family planning? What effect does family planning have on fertility rates?
2. **What** is the relationship between fertility rates and educational opportunities for women?

Population and Urbanization

LEARNING OBJECTIVES

1. **Define** *urbanization* and describe trends in the distribution of people in rural and urban areas.

2. **Describe** some of the problems associated with rapid growth rates in large urban areas.

3. **Explain** how compact development makes a city more livable.

The geographic distribution of people in rural areas, towns, and cities significantly influences the social, environmental, and economic aspects of population growth. During recent history, the human population has become increasingly urbanized. **Urbanization** involves the movement of people from rural to urban areas as well as the transformation of rural areas into urban areas. When Europeans first settled in North America, the majority of the population consisted of farmers in rural areas. As of 2011, approximately 79 percent of the U.S. population lived in cities.

urbanization A process whereby people move from rural areas to densely populated cities.

How many people does it take to make an urban area or city? The answer varies from country to country. According to the U.S. Bureau of the Census, a location with 2500 or more people qualifies as an urban area. One important distinction between rural and urban areas isn't how many people live there but how people make a living. Most people residing in rural areas have occupations that involve harvesting natural resources—such as fishing, logging, and farming. In urban areas, most people have jobs that are not connected directly with natural resources.

Cities have grown at the expense of rural populations for several reasons. With advances in agriculture, fewer farmers support an increased number of people. Also, in many developing countries, a few wealthy people own most of the land, and poor farmers are denied access to it. Consequently, people in rural settings have fewer employment opportunities. Cities have traditionally provided more jobs than rural areas because cities are sites of industry, economic development, educational and cultural opportunities, and technological advancements—all of which generate income.

Cities are urban ecosystems, and scientists study the effects of humans on the urban environment. This research focuses on the ecological effects of human settlement rather than the interactions among humans themselves. Study of urban ecosystems is complicated because the flow of energy, water, and other resources into and out of the city is linked to the flow of money and the human population (**Figure 7.20**). Often political power is connected to better environmental quality of specific (wealthy) neighborhoods.

Every city is unique in terms of size, climate, culture, and economic development. Although there is no such thing as a typical city, certain traits are common to city populations in general. One basic characteristic of city populations is their far greater heterogeneity with respect

The city as a dynamic ecosystem • Figure 7.20

The human population in an urban environment requires inputs from the surrounding countryside and produces outputs that flow into surrounding areas. Not shown in this figure is the internal cycling of materials and energy within the urban system.

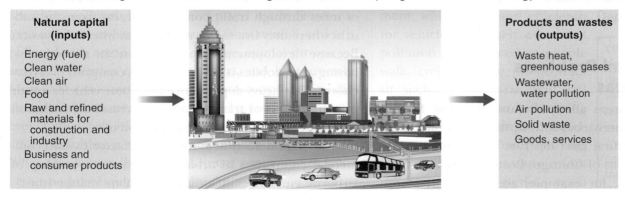

Natural capital (inputs)		Products and wastes (outputs)
Energy (fuel) Clean water Clean air Food Raw and refined materials for construction and industry Business and consumer products		Waste heat, greenhouse gases Wastewater, water pollution Air pollution Solid waste Goods, services

CASE STUDY

Urban Planning in Curitiba, Brazil

Livable cities aren't restricted to highly developed countries. Curitiba, a Brazilian city of 3.1 million people, provides a good example of *compact development* in a moderately developed country. Curitiba's city officials and planners have had notable successes in public transportation, traffic management, land-use planning, waste reduction and recycling, and community livability.

The city developed an inexpensive, efficient mass transit system that uses clean, modern buses that run in high-speed bus lanes. High-density development was largely restricted to areas along the bus lines, encouraging population growth where public transportation was already available. About 2 million people use Curitiba's mass transportation system each day.

Since the 1970s, Curitiba's population has more than tripled, yet traffic has declined by 30 percent. Curitiba doesn't rely on automobiles as much as comparably sized cities do, so it has less traffic congestion and significantly cleaner air, both of which are major goals of compact development. Instead of streets crowded with vehicular traffic, the center of Curitiba is a *calcadao*, or "big sidewalk," that consists of 49 downtown blocks of pedestrian walkways connected to bus stations, parks, and bicycle paths.

Curitiba was the first city in Brazil to use a special low-polluting fuel that contains a mixture of diesel fuel, alcohol, and soybean extract. In addition to burning cleanly, this fuel provides economic benefits for people in rural areas who grow the soybeans and grain used to make the alcohol.

Over several decades, Curitiba purchased and converted flood-prone properties along rivers in the city to a series of interconnected parks crisscrossed with bicycle paths. This move reduced flood damage and increased the per person amount of "green space" from 0.5 m² (5.4 ft²) in 1950 to 50 m² (540 ft²) today, a significant accomplishment considering Curitiba's rapid population growth during the same period.

Another example of Curitiba's creativity is its labor-intensive garbage purchase program, in which poor people exchange filled garbage bags for bus tokens, surplus food (eggs, butter, rice, and beans), or school notebooks. This program encourages garbage pickup from the unplanned shantytowns (which garbage trucks can't access) that surround the city. Curitiba supplies more services to these unplanned settlements than most cities do. It tries to provide water, sewer, and bus service for them.

These changes didn't happen overnight. Urban planners can carefully reshape most cities over several decades to make better use of space and to reduce dependence on motor vehicles. City planners and local and regional governments are increasingly adopting measures to provide the benefits of compact development in the future.

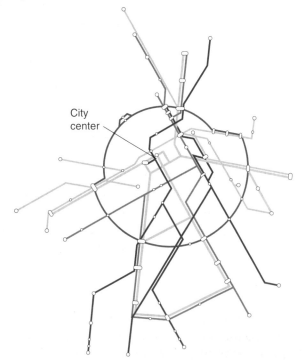

City center

a. Curitiba's bus network, arranged like the spokes of a wheel, has concentrated development along the bus lines, saving much of the surrounding countryside from development.

© Pete M. Wilson/Alamy

b. The downtown area of Curitiba has open terraces lined with shops and restaurants.

Summary

1 Population Ecology 160

1. **Population ecology** is the branch of biology that deals with the number of individuals of a particular species found in an area and how and why those numbers change over time.

2. The **growth rate (r)** is the rate of change (increase or decrease) of a population's size, expressed in percentage per year. On a global scale, growth rate is due to the **birth rate (b)** and the **death rate (d)**: $r = b - d$. **Emigration (e)**, the number of individuals leaving an area, and **immigration (i)**, the number of individuals entering an area, also affect a local population's growth rate.

3. **Biotic potential** is the maximum rate a population could increase under ideal conditions. **Exponential population growth** is the accelerating population growth that occurs when optimal conditions allow a constant reproductive rate for limited periods. Eventually, the growth rate decreases to around zero or becomes negative because of **environmental resistance**, unfavorable environmental conditions that prevent organisms from reproducing indefinitely at their biotic potential. The **carrying capacity (K)** is the largest population a particular environment can support sustainably (long term) if there are no changes in that environment.

2 Human Population Patterns 165

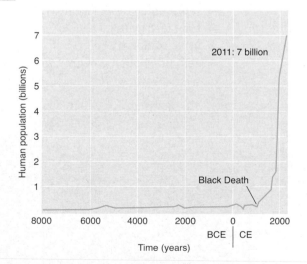

1. It took thousands of years for the human population to reach 1 billion (around 1800). Since then, the population has grown exponentially, reaching 7 billion in late 2011. Although our numbers continue to increase, the growth rate (r) has declined slightly over the past several years. The population should reach **zero population growth**, in which it remains the same size because the birth rate equals the death rate, toward the end of the 21st century.

2. **Thomas Malthus** was a British economist who said that the human population increases faster than its food supply, resulting in famine, disease, and war. Malthus's ideas appear to be erroneous because the human population has grown from about 1 billion in his time to 7 billion today, and food production has generally kept pace with population. But Malthus may ultimately be proved correct because we don't know whether our increase in food production is sustainable.

3. Estimates of Earth's carrying capacity for humans vary widely depending on what assumptions are made about standard of living, resource consumption, technological innovations, and waste generation. In addition to natural environmental constraints, human choices and values determine Earth's carrying capacity for humans.

3 Demographics of Countries 170

1. **Demographics** is the applied branch of sociology that deals with population statistics. As a country becomes industrialized, it goes through a demographic transition as it moves from relatively high birth and death rates to relatively low birth and death rates.

2. The **infant mortality rate** is the number of deaths of infants under age 1 per 1000 live births. The **total fertility rate (TFR)** is the average number of children born to each woman. **Replacement-level fertility** is the number of children a couple must produce to "replace" themselves. **Age structure** is the number and proportion of people at each age in a population. A country can have replacement-level fertility and

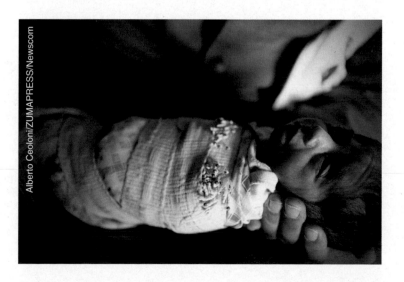

Alberto Ceoloni/ZUMAPRESS/Newscom

still experience population growth if the largest percentage of the population is in the prereproductive years. In contrast to developing countries, highly developed countries have low infant mortality rates, low total fertility rates, and an age structure in which the largest percentage of the population isn't in the prereproductive years.

4 Stabilizing World Population 174

1. Four factors are most responsible for high total fertility rates: high infant and child mortality rates, the important economic and societal roles of children in some cultures, the low status of women in many societies, and a lack of health and family planning services. The single most important factor affecting high TFRs is the low status of women. The governments of many developing countries are trying to limit population growth.

2. Education of women decreases the total fertility rate, in part by delaying the first childbirth. Education increases the likelihood that women will know how to control their fertility. Education also increases women's career options, which provide ways of achieving status besides having babies.

5 Population and Urbanization 181

1. **Urbanization** is the process whereby people move from rural areas to densely populated cities. In developing nations, most people live in rural settings, but their rates of urbanization are rapidly increasing.

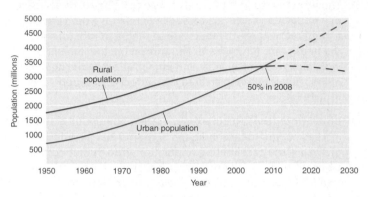

2. Rapid urbanization makes it difficult to provide city dwellers with basic services such as housing, water, sewage, and transportation systems.

3. **Compact development** is the design of cities so that tall, multiple-unit residential buildings are close to shopping and jobs, and all are connected by public transportation.

Key Terms

- age structure 172
- biotic potential 161
- carrying capacity (K) 163
- compact development 183
- demographic transition 172
- demographics 170
- exponential population growth 162
- growth rate (r) 161
- infant mortality rate 170
- population ecology 160
- replacement-level fertility 170
- total fertility rate (TFR) 170
- urbanization 181
- zero population growth 166

© Frances Roberts/Alamy

What is happening in this picture?

- Pedestrians stroll along lower Manhattan's High Line park, constructed along an abandoned elevated rail line. What advantages does such a space provide urban residents?

- What problems are associated with abandoned spaces in cities?

- How might a space like the High Line benefit the natural environment?

Critical and Creative Thinking Questions

1. How does the study of population ecology help us understand why some populations grow, some remain stable, and others decline?

2. The growth rates of various populations are usually expressed in percentages. Why are percentages advantageous in comparing growth rates?

3. The human population has grown as we have increased our global carrying capacity. In your opinion, can the global carrying capacity continue to increase? Explain your answer.

4. Why has human population growth, which increased exponentially for centuries, started to decline in the past few decades?

5. Malthus originally suggested that the population of England would collapse because it could not continue to increase its production of food. Why did this not happen?

6. What is carrying capacity? Do you think carrying capacity applies to people as well as to other organisms? Why or why not?

7. What can the governments of developing countries do to help their countries experience the demographic transition?

8. If you were to draw an age structure diagram for Poland, with a total fertility rate of 1.3, which of the following overall shapes would the diagram have? Explain why a country like Poland faces a population decline even if its fertility rate were to start increasing today.

9. Explain the rationale behind this statement: It is better for highly developed countries to spend millions of dollars on family planning in developing countries now than to have to spend billions of dollars on relief efforts later.

10. Which factor do you think would have a larger effect on total fertility rate: the increased education of men or of women? Explain your answer.

11. What are two serious problems associated with the rapid growth of large urban areas? Explain why they are serious.

12. In cities utilizing compact development, motor vehicle use is reduced. What are some alternatives to motor vehicles?

13. Should the rapid increase in world population be of concern to the average citizen in the United States? Why or why not?

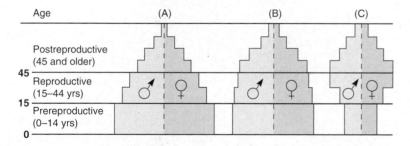

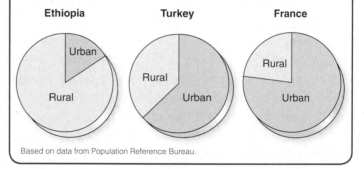

Sustainable Citizen Question

14. Urbanization varies from one country to another (see figure). Local and national government agencies in the three countries represented below strive to provide services to their populations. How might each of their efforts differ, and why? How do you think the United States compares to these countries? What do you believe to be the biggest problem faced by the United States, as related to population growth or urbanization, and how would you propose to address it?

Based on data from Population Reference Bureau.

Air and Air Pollution

LONG-DISTANCE TRANSPORT OF AIR POLLUTION

Persistent toxic compounds are found in the Yukon (in northwestern Canada) and in other pristine arctic regions, far from where they were originally produced. This occurs through the *global distillation effect*, in which chemicals enter the atmosphere in warm regions and move to areas at higher, cooler latitudes, where they are deposited on the surface. The chemicals are then available to be absorbed, inhaled, or ingested by organisms at these distant locations.

Chemicals concentrate in the body fat of animals at the top of food chains, including humans (see Chapter 4). When an Inuit woman consumes a single bite of raw whale skin, she ingests more toxic PCBs than scientists think should be consumed in a week (*see* photograph). Five times as much PCB is found in the breast milk of Inuit women than in the milk of women who live in southern Canada. Persistent organic pesticides also concentrate in the milk of polar bear mothers (see graph).

Atmospheric conditions also cause pollutants from Asia to move east across the Pacific Ocean. In 1998 a Chinese dust storm produced a visible cloud of particulate matter that transferred toxic metals from ore smelters in Manchuria to the United States.

Around the world, the air we breathe can be contaminated with a variety of pollutants. Because air pollution causes many health and environmental problems, most highly developed nations and many developing nations have policies and regulations limiting emissions from transportation, industry, and even households.

graphingactivity

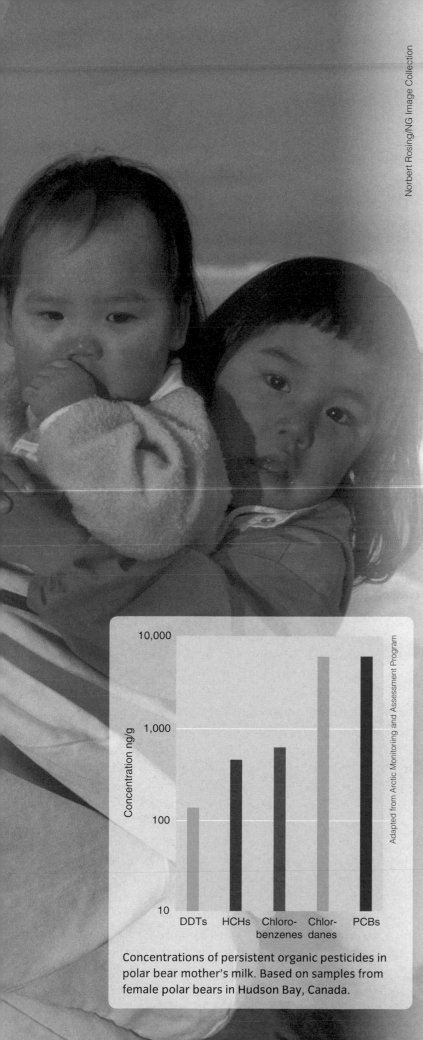

Norbert Rosing/NG Image Collection

Concentrations of persistent organic pesticides in polar bear mother's milk. Based on samples from female polar bears in Hudson Bay, Canada.

Adapted from Arctic Monitoring and Assessment Program

CHAPTER OUTLINE

The Atmosphere   192
- ■ Environmental InSight: The Atmosphere
- • Atmospheric Circulation

Types and Sources of Air Pollution   196
- • Major Classes of Air Pollutants
- • Sources of Outdoor Air Pollution
- ■ What a Scientist Sees: Air Pollution from Volcanoes

Effects of Air Pollution   201
- • Air Pollution and Human Health
- • Urban Air Pollution
- • How Weather and Topography Affect Air Pollution
- ■ EnviroDiscovery: Air Pollution May Affect Precipitation
- • Urban Heat Islands and Dust Domes

Controlling Air Pollutants   206
- • The Clean Air Act
- • Air Pollution in Developing Countries

Indoor Air Pollution   209
- • Radon
- ■ Case Study: Curbing Air Pollution in Chattanooga

CHAPTER PLANNER ✓

- ❑ Study the picture and read the opening story.
- ❑ Scan the Learning Objectives in each section:
  p. 192 ❑   p. 196 ❑   p. 201 ❑   p. 206 ❑   p. 209 ❑
- ❑ Read the text and study all figures and visuals. Answer any questions.

Analyze key features

- ❑ Environmental InSight, p. 193
- ❑ Process Diagram, p. 195
- ❑ What a Scientist Sees, p. 199
- ❑ EnviroDiscovery, p. 203
- ❑ Case Study, p. 212
- ❑ Stop: Answer the Concept Checks before you go on:
  p. 195 ❑   p. 200 ❑   p. 205 ❑   p. 209 ❑   p. 211 ❑

End of Chapter

- ❑ Review the Summary and Key Terms.
- ❑ Answer What is happening in this picture?
- ❑ Answer the Critical and Creative Thinking Questions.

The Atmosphere

LEARNING OBJECTIVES

1. **Define** *atmosphere* and list the major gases comprising the atmosphere.

2. **Briefly describe** the four major concentric layers of the atmosphere.

3. **Explain** the causes of wind, including the Coriolis effect.

Oxygen and nitrogen are the predominant gases in the **atmosphere**, accounting for about 99 percent of dry air (**Figure 8.1**). Other gases make up the remaining 1 percent. In addition, water vapor (the most variable gas in the atmosphere) and trace amounts of air pollutants are present in the air. The atmosphere becomes less dense as it extends outward into space.

> **atmosphere** The gaseous envelope surrounding Earth.

Ulf Merbold, a German space shuttle astronaut, felt differently about the atmosphere after viewing it in space (**Figure 8.2**): "For the first time in my life, I saw the horizon as a curved line. It was accentuated by a thin seam of dark blue light—our atmosphere. Obviously, this wasn't the 'ocean' of air

Composition of the atmosphere • Figure 8.1

Nitrogen and oxygen form most of the atmosphere. Air also contains water vapor and various pollutants (methane, ozone, dust particles, microorganisms, and chlorofluorocarbons [CFCs]).

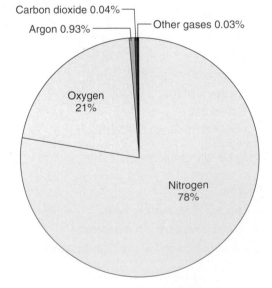

Carbon dioxide 0.04%
Argon 0.93%
Other gases 0.03%
Oxygen 21%
Nitrogen 78%

Roger Harris/Photo Researchers, Inc.

The atmosphere • Figure 8.2

The "ocean of air" is a thin blue layer that separates the planet from the blackness of space.

I had been told it was so many times in my life. I was terrified by its fragile appearance." The atmosphere is composed of four major concentric layers—the troposphere, stratosphere, mesosphere, and thermosphere (**Figure 8.3**). These layers vary in altitude and temperature, depending on the latitude and season.

The atmosphere performs several valuable **ecosystem services**. First, it protects Earth's surface from most of the sun's ultraviolet (UV) radiation and x-rays, and from lethal exposures to cosmic rays from space. Life as we know it would cease to exist without this shielding. Second, atmospheric greenhouse gases absorb some of the heat reradiated from Earth's surface, which keeps the lower atmosphere within the range of temperatures that support life.

Organisms depend on the atmosphere for existence, but they also maintain and, in certain instances, modify its composition. Atmospheric oxygen is thought to have increased to its present level as a result of billions of years of photosynthesis. Over the course of a year, oxygen-producing photosynthesis and oxygen-using cellular respiration roughly balance, although carbon dioxide levels have increased each year over the past century (see Chapter 9).

b. A Thunderstorm in New Mexico. During a lightning flash, a negative charge moves from the bottom of the cloud to the ground, followed by an upward-moving charge along the same channel. The expansion of air around the lightning strike produces sound waves, or thunder.

WileyPLUS

Kenneth Garrett/NG Image Collection

a. Layers of Atmosphere.

Thermosphere
Extends to 480 km (300 mi)
Gases in extremely thin air absorb x-rays and short-wave radiation, raising the temperature to 1000°C (1800°F) or more. The thermosphere is important in long-distance communication because it reflects outgoing radio waves back to Earth without the use of satellites. Auroras occur here.

Mesosphere
Extends to 80 km (50 mi)
Directly above the stratosphere, temperatures drop to the lowest in the atmosphere—as low as −138°C (−216°F). Meteors often burn up from friction with air molecules in the mesosphere.

Stratosphere
Extends to 50 km (30 mi)
Steady wind occurs but no turbulence; commercial jets fly here. Contains a layer of ozone that absorbs much of the sun's damaging ultraviolet (UV) radiation. Temperature increases with increasing altitude because absorption of UV radiation by ozone layer heats the air.

c. An Aurora in the Northern Hemisphere. Electrically charged particles from the sun collide with the gas molecules in the thermosphere, releasing energy visible as light of different colors.

Antony Spencer/E+/Getty Images

Troposphere
Average thickness: 12 km (7.5 mi)
16 km (10 mi) thick at equator
8 km (5 mi) thick at poles
Layer of atmosphere closest to Earth's surface. Temperature decreases with increasing altitude. Weather, including turbulent wind, storms, and most clouds, occurs here.

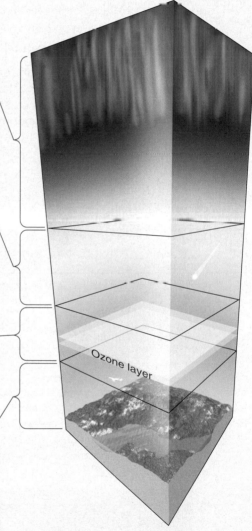

Ozone layer

Atmospheric Circulation

The amount of solar energy that reaches different areas on Earth varies over the course of each year and from place to place around the globe. This variation creates differences in temperature, which then drive the circulation of the atmosphere. The very warm regions near the equator heat the air, which expands and rises (**Figure 8.4**). As this warm air rises, it cools, spreads, and then sinks again. Much of it recirculates almost immediately to the same areas it has left. The remainder of the heated air splits and flows toward the poles. The air chills enough to sink to the surface at about 30 degrees north and south latitudes. This descending air splits and flows over the surface.

Similar upward movements of warm air and its subsequent flow toward the poles also occur at higher latitudes farther from the equator. At the poles, the air cools, sinks, and flows back toward the equator, generally beneath the currents of warm air that simultaneously flow toward the poles. These constantly moving currents transfer heat from the equator toward the poles and cool the land over which they pass on their return. This continuous circulation moderates temperatures over Earth's surface.

Atmospheric circulation and heat exchange • Figure 8.4

a. In atmospheric convection, heating of the ground surface heats the air, producing an updraft of less dense, warm air. The convection process ultimately causes air currents that mix warmer and cooler parts of the atmosphere.

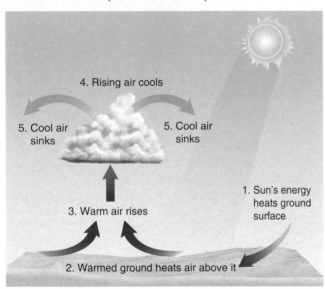

4. Rising air cools
5. Cool air sinks
5. Cool air sinks
1. Sun's energy heats ground surface
3. Warm air rises
2. Warmed ground heats air above it

c. Spring sandstorm in China. This NASA satellite image shows a massive sandstorm in China. The sand passed over Japan a few days later.

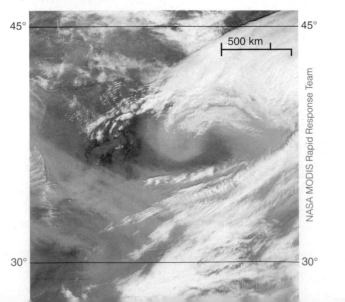

45°
500 km
45°
30°
30°

NASA MODIS Rapid Response Team

b. Atmospheric circulation transports heat from the equator to the poles (left side of figure). The greatest solar energy input occurs at the equator, heating air most strongly in that area. The air rises, travels toward the poles, and cools in the process so that much of it descends again at around 30 degrees latitude in both hemispheres. At higher latitudes, the patterns of air circulation are more complex.

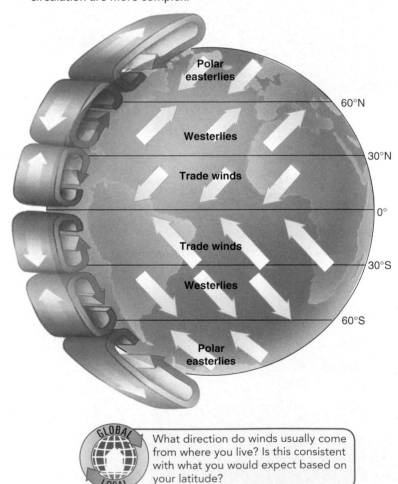

Polar easterlies
60°N
Westerlies
30°N
Trade winds
0°
Trade winds
30°S
Westerlies
60°S
Polar easterlies

GLOBAL LOCAL

What direction do winds usually come from where you live? Is this consistent with what you would expect based on your latitude?

The Coriolis effect • Figure 8.5

Viewed from the North Pole, the Coriolis effect appears to deflect ocean currents and winds to the right. From the South Pole, the deflection appears to be to the left.

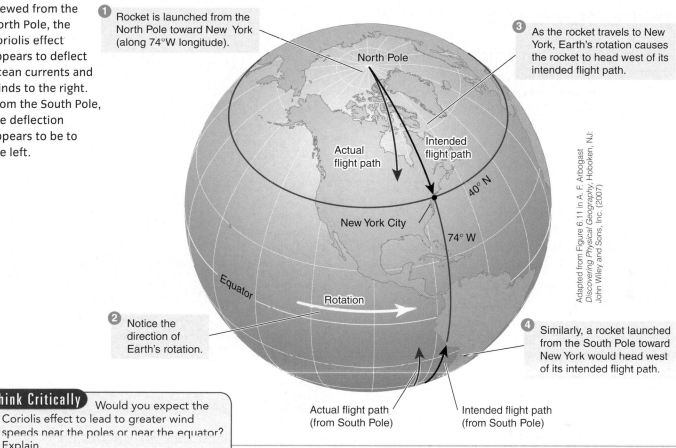

1 Rocket is launched from the North Pole toward New York (along 74°W longitude).

3 As the rocket travels to New York, Earth's rotation causes the rocket to head west of its intended flight path.

North Pole

Actual flight path

Intended flight path

40° N

New York City

74° W

Equator

Rotation

2 Notice the direction of Earth's rotation.

4 Similarly, a rocket launched from the South Pole toward New York would head west of its intended flight path.

Actual flight path (from South Pole)

Intended flight path (from South Pole)

Adapted from Figure 6.11 in A. F. Arbogast *Discovering Physical Geography*, Hoboken, NJ: John Wiley and Sons, Inc. (2007)

Think Critically Would you expect the Coriolis effect to lead to greater wind speeds near the poles or near the equator? Explain.

In addition to these global circulation patterns, the atmosphere features smaller-scale horizontal movements, or **winds**. The motion of wind, with its eddies, lulls, and turbulent gusts, is difficult to predict. It results partly from fluctuations in atmospheric pressure and partly from the planet's rotation.

The gases that constitute the atmosphere have weight and exert a pressure—about 1013 millibars (14.7 lb per in²) at sea level. Air pressure is variable, depending on altitude, temperature, and humidity. Winds tend to blow from areas of high atmospheric pressure to areas of low pressure, and the greater the difference between the high- and low-pressure areas, the stronger the wind.

As a result of the **Coriolis effect**, Earth's rotation from west to east also influences the direction of wind. To visualize the Coriolis effect, imagine that a rocket is launched from the North Pole toward New York (**Figure 8.5**).

Coriolis effect The tendency of moving air or water to be deflected from its path and swerve to the right in the Northern Hemisphere and to the left in the Southern Hemisphere.

The atmosphere has three **prevailing winds**—major surface winds that blow more or less continually (see Figure 8.4). Prevailing winds from the northeast near the North Pole, or from the southeast near the South Pole, are called *polar easterlies*. Winds that blow in the middle latitudes from the southwest in the Northern Hemisphere or from the northwest in the Southern Hemisphere are called *westerlies*. Tropical winds from the northeast in the Northern Hemisphere or from the southeast in the Southern Hemisphere are called *trade winds*.

CONCEPT CHECK 🛑 STOP

1. **What** gases make up the atmosphere?

2. **What** two layers of the atmosphere are closest to Earth's surface? How do they differ from one another?

3. **What** factors cause wind, and how do they relate to the Coriolis effect?

The Atmosphere   195

Types and Sources of Air Pollution

LEARNING OBJECTIVES

1. **Define** *air pollution* and distinguish between primary and secondary air pollutants.

2. **List** the seven major classes of air pollutants and describe their characteristics and sources.

Air pollution can come from natural sources, such as smoke from a forest fire ignited by lightning or gases from an erupting volcano. However, human activities release many kinds of substances into the atmosphere and contribute greatly to global air pollution. Some of these substances are harmful when they are inhaled or settle on land and surface waters, and some substances are harmful because they alter the chemistry of the atmosphere.

Although many different air pollutants exist, we focus on the seven most important classes from a regulatory perspective: particulate matter, nitrogen oxides, sulfur oxides, carbon oxides, hydrocarbons, ozone, and air toxics.

Air pollutants are often divided into two categories, primary and secondary (**Figure 8.6**). **Primary air pollutants** are released directly from a source into the atmosphere. They include carbon oxides, nitrogen oxides, sulfur dioxide, particulate matter, and hydrocarbons.

Ozone, sulfur trioxide, and several acids are called **secondary air pollutants** because they are formed from chemical reactions that take place in the atmosphere.

> **air pollution** Various chemicals (gases, liquids, or solids) present in the atmosphere in high enough levels to harm humans, other organisms, or materials.

> **primary air pollutants** Harmful chemicals that enter directly into the atmosphere due to either human activities or natural processes.

> **secondary air pollutants** Harmful chemicals that form in the atmosphere when primary air pollutants react chemically with one another or with natural components of the atmosphere.

Major Classes of Air Pollutants

Particulate matter consists of dusts and mists—thousands of different types of solid and liquid particles suspended in the atmosphere. Particulate matter includes soil particles, soot, lead, asbestos, sea salt, and sulfuric acid droplets. Some particulate matter has toxic or carcinogenic effects.

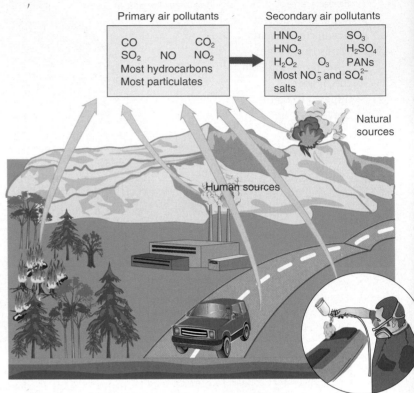

Primary and secondary air pollutants
• **Figure 8.6**

Primary air pollutants are emitted, unchanged, from a source directly into the atmosphere, whereas secondary air pollutants are produced from chemical reactions involving primary air pollutants.

Particulate matter can scatter and absorb sunlight, reducing visibility. Urban areas receive less sunlight than rural areas, partly as a result of greater quantities of particulate matter in the air. Particulate matter can corrode metals, erode buildings and sculptures when the air is humid, and soil clothing and draperies. Lead particles, which are heavy, tend to travel relatively short distances before settling on the ground or a water surface.

All particulate matter eventually settles out of the atmosphere, but microscopic particles can remain suspended in the atmosphere for weeks or even years. Trace amounts of hundreds of different chemicals bind to these microscopic particles; inhaling the particles introduces the chemicals, some of which are toxic, into the human body. Microscopic particles are considered more dangerous than larger particles because they are inhaled more deeply into the lungs.

The Environmental Protection Agency (EPA) samples microscopic particulate matter at 1,000 locations around the United States because its composition varies with location and season. Particulate matter includes:

Pollutant	Category	Characteristics
Dust particles	Primary	Solid particles
Lead (Pb)	Primary	Solid particles
Sulfuric acid (H_2SO_4)	Secondary	Liquid droplets

Nitrogen oxides are gases produced by chemical interactions between nitrogen and oxygen when a source of energy, such as fuel combustion, produces high temperatures. Collectively known as NO_x, nitrogen oxides consist mainly of nitric oxide (NO), nitrogen dioxide (NO_2), and nitrous oxide (N_2O). Nitrogen oxides inhibit plant growth and, when inhaled, aggravate health problems such as asthma. They are involved in the production of *photochemical smog* (discussed later in the chapter) and acid deposition (see Chapter 9). Nitrous oxide is associated with global warming, and it depletes ozone in the stratosphere (again, see Chapter 9). Nitrogen oxides cause metals to corrode and textiles to fade and deteriorate. Nitrogen oxides include:

Pollutant	Category	Characteristics
Nitrogen dioxide (NO_2)	Primary	Reddish-brown gas
Nitric oxide (NO)	Primary	Colorless gas

Sulfur oxide gases result from chemical interactions between sulfur and oxygen. Sulfur dioxide (SO_2), a colorless, nonflammable gas with a strong, irritating odor, is emitted as a primary air pollutant. Sulfur trioxide (SO_3) is a secondary air pollutant that forms when sulfur dioxide reacts with oxygen in the air. Sulfur trioxide, in turn, reacts with water to form another secondary air pollutant, sulfuric acid. Sulfur oxides play a major role in acid deposition, and they corrode metals and damage stone and other materials. Sulfuric acid and other sulfur oxides damage plants and irritate the respiratory tracts of humans and other animals. Sulfur oxides include:

Pollutant	Category	Characteristics
Sulfur dioxide (SO_2)	Primary	Colorless gas with strong odor
Sulfur trioxide (SO_3)	Secondary	Reactive colorless gas

Carbon oxides are the gases carbon monoxide (CO) and carbon dioxide (CO_2). Carbon monoxide is a colorless, odorless, and tasteless gas produced in larger quantities than any other atmospheric pollutant except carbon dioxide. Carbon monoxide is poisonous. Because hemoglobin in blood has a stronger affinity for carbon monoxide than for oxygen, carbon monoxide reduces the blood's ability to transport oxygen. Carbon dioxide, also colorless, odorless, and tasteless, is associated with global climate change. Carbon oxides are:

Pollutant	Category	Characteristics
Carbon monoxide (CO)	Primary	Colorless, odorless gas
Carbon dioxide (CO_2)	Primary	Colorless, odorless gas

Hydrocarbons are a diverse group of organic compounds that contain only the elements hydrogen and carbon. Small hydrocarbon molecules, such as methane (CH_4), are gaseous at room temperature. Methane is colorless and odorless and is the principal component of natural gas. (The odor of natural gas comes from sulfur compounds deliberately added so that humans can detect the gas's presence.) Medium-sized hydrocarbons, such as benzene (C_6H_6), are liquids at room temperature, although many are volatile and may evaporate easily. The largest hydrocarbons, such as the waxy fuel paraffin, are solids at room temperature. The many different hydrocarbons have a variety of effects on human and animal health. Some cause no adverse effects, some injure the respiratory tract, and others cause cancer. All except methane contribute to the production of photochemical smog. Methane is linked to global warming. Hydrocarbons include:

Pollutant	Category	Characteristics
Methane (CH_4)	Primary	Colorless, odorless gas
Benzene (C_6H_6)	Primary	Liquid with sweet smell

Ozone (O_3) is a form of oxygen considered a pollutant in one part of the atmosphere but an essential component of another. In the stratosphere, oxygen reacts with solar UV radiation to form ozone. Stratospheric ozone protects Earth's surface from receiving harmful levels of solar UV radiation. Unfortunately, certain human-made pollutants, such as chlorofluorocarbons (CFCs), react with stratospheric ozone, breaking it down into molecular oxygen (O_2). As a result of this breakdown, more solar UV reaches Earth's surface.

Unlike stratospheric ozone, ozone in the troposphere—the layer of atmosphere closest to Earth's surface—is a human-made air pollutant. (Ground-level, or tropospheric, ozone does not replenish the ozone depleted from the stratosphere because tropospheric ozone breaks down to form oxygen long before it drifts up to the stratosphere.) Ozone in the troposphere is a secondary air pollutant formed when sunlight triggers reactions between nitrogen oxides and volatile hydrocarbons. The most harmful component of photochemical smog, ozone reduces air visibility and causes health problems. Ozone also reduces plant vigor, and chronic ozone exposure (of long duration) lowers crop yields (**Figure 8.7**). Chronic exposure to ozone is one possible contributor to forest decline, and ground-level ozone is associated with global warming. As discussed:

Pollutant	Category	Characteristics
Ozone (O_3)	Secondary	Pale blue gas with irritating odor

Hazardous air pollutants (HAPs), or **air toxics**, include hundreds of other air pollutants—such as chlorine (Cl_2), lead, hydrochloric acid, formaldehyde, radioactive substances, and fluorides. HAPs are present in very low concentrations, although it is possible to have high local concentrations of specific pollutants. They are potentially harmful and may pose long-term health risks to people

Ozone damage • Figure 8.7

A scientist measures the effects of ozone on the growth and productivity of several plants. Plants exposed to ozone generally exhibit damaged leaves, reduced root growth, and reduced productivity. Photographed in Birmensdorf, Switzerland.

who live and work around chemical factories, incinerators, or other facilities that produce or use them. To limit the release of more than 180 HAPs, the Clean Air Act Amendments of 1990 (discussed later in this chapter) regulate the pollutant emissions of both large and small businesses. Hazardous air pollutants include:

Pollutant	Category	Characteristics
Chlorine (Cl_2)	Primary	Yellow-green gas
Formaldehyde	Primary	Colorless gas with pungent odor

Sources of Outdoor Air Pollution

Not all air pollution is human generated. Throughout Earth's history, volcanoes have released particulate matter and sulfur oxides (see *What a Scientist Sees*). Plants can also contribute to air pollution, producing a variety of hydrocarbons in response to heat. The hydrocarbon isoprene, for example, may protect leaves from high temperatures. However, isoprene and other hydrocarbons are volatile and evaporate into the air, where they interact with other substances to affect atmospheric chemistry.

WHAT A SCIENTIST SEES

Air Pollution from Volcanoes

Volcanoes occur where the hot magma inside Earth reaches the surface. Active volcanoes can release large quantities of pollutants.

a. Mount Pinatubo. When the Philippine volcano Mount Pinatubo erupted in 1991, it released huge amounts of particulate matter.

USGS

Interpreting Data
What was the average temperature from 1992 to 1993? from 1996 to 1998? from 1987 to 1998?

Mount Pinatubo erupts, June 1991

Courtesy of National Aeronautics and Space Administration (NASA)

AFP Photo/Halldor Kolbeins/NewsCom

b. Global Average Temperature, 1987 to 1998. Climate scientists observed that the years following Mount Pinatubo's eruption were cooler than previous and subsequent years. This brief cooling period temporarily interrupted a longer-term warming trend (Chapter 9).

c. The Eyjafjallajokull Volcano. In 2010, the Eyjafjallajokull volcano in Iceland erupted. The resulting ash cloud disrupted air traffic for days.

of mountains (the side toward which the wind blows) are prime candidates for temperature inversions. The Los Angeles Basin, for example, lies between the Pacific Ocean on the west and mountains to the north and east. During the summer the sunny climate produces a layer of warm dry air at upper elevations. A region of upwelling occurs just off the Pacific coast, bringing cold ocean water to the surface and cooling the ocean air. As this cool air blows inland over the basin, the mountains block its movement further. Thus, a layer of warm, dry air overlies cool air at the surface, producing a temperature inversion.

Air Pollution in Los Angeles

Los Angeles, California, has some of the worst smog in the world. Its location, in combination with its sunny climate, is conducive to the formation of stable temperature inversions that trap photochemical smog near the ground, sometimes for long periods (**Figure 8.10**). Passenger vehicles, heavy-duty trucks, and buses are the source of more than half of the smog-producing emissions in Los Angeles.

In 1969 California became the first state to enforce emissions standards on motor vehicles, largely because of the air pollution problems in Los Angeles. Today Los Angeles has stringent smog controls that regulate everything from low-emission alternative fuels (such as compressed natural gas) for buses to lawn mower emissions to paint vapors. Using the cleanest emission-reduction equipment available significantly reduces emissions from large industrial and manufacturing sources, including oil refineries and power plants. California has no coal-fired power plants; most of its power plants burn natural gas. Future pollution reductions

Peak ozone concentration in southern California, 1960–2010 • Figure 8.11

Peak ozone is the highest level of ozone recorded on any single day during the year. Average daily ozone, number of days above federal and state standards, and other measures show similar patterns. Air quality has improved steadily over the past half century but still presents a health threat.

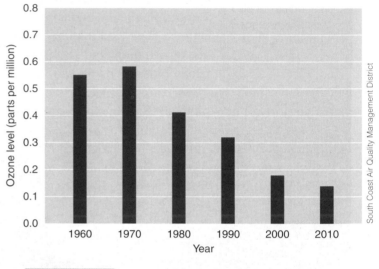

South Coast Air Quality Management District

Think Critically How do these ozone levels compare to where you live? (Search the EPA website for data.)

will come in part from requiring auto manufacturers to sell ultra-low-emission cars.

After several decades devoted to improving its air quality, Los Angeles now has the cleanest skies it has had since the 1950s (**Figure 8.11**). Despite the impressive progress, Los Angeles still exceeds federal air quality standards on more days than almost any other metropolitan region in the United States. Los Angeles experienced 120 days above the federal ozone standard in 2008, down from 203 days in 1977.

Urban Heat Islands and Dust Domes

Streets, rooftops, and parking lots in areas of high population density absorb solar radiation during the day and radiate heat into the atmosphere at night. Heat from human activities such as fuel combustion is also highly concentrated in cities. The air in urban areas therefore forms urban heat islands in the surrounding suburban and rural areas (**Figure 8.12**).

urban heat island Local heat buildup in an area of high population.

Photochemical smog • Figure 8.10

Ozone is a major constituent of the photochemical smog depicted here. Photographed in Los Angeles, California, on a day when air pollution exceeded federal air quality standards.

Justin Lambert/The Image Bank/Getty Image

Urban heat island • Figure 8.12

This figure shows how temperatures might vary on a summer afternoon. The city stands out as a heat island against the surrounding rural areas.

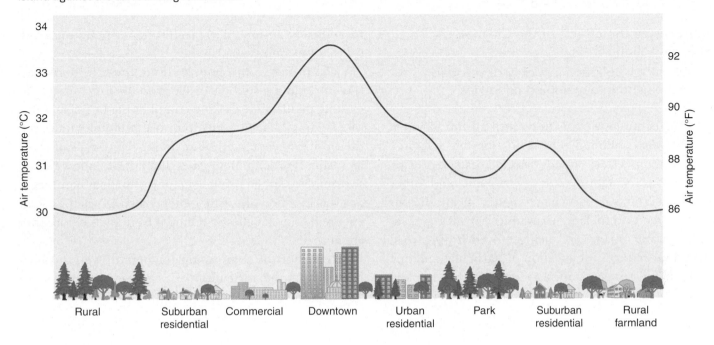

| Rural | Suburban residential | Commercial | Downtown | Urban residential | Park | Suburban residential | Rural farmland |

Urban heat islands also contribute to the buildup of pollutants, especially particulate matter, in the form of **dust domes** over cities (**Figure 8.13a**). Pollutants concentrate in a dust dome because convection

> **dust dome** A dome of heated air that surrounds an urban area and contains a lot of air pollution.

(the vertical motion of warmer air) lifts pollutants into the air, where they remain because of the somewhat stable air masses the urban heat island produces. If wind speeds increase, the dust dome moves downwind from the city, and the polluted air spreads over rural areas (**Figure 8.13b**).

Urban heat islands affect local air currents and weather conditions. Cities located in valleys or upwind of mountain ranges are particularly susceptible to buildup of pollutants on low wind days. For example, urban heat islands may increase the number of thunderstorms over the city during summer months. The uplift of warm air over the city produces a low-pressure cell that draws in cooler air from the surroundings. As the heated air rises, it cools, causing water vapor to condense into clouds and produce thunderstorms.

Dust dome • Figure 8.13

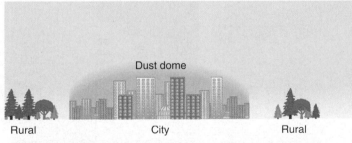

a. A dust dome of pollutants forms over a city when the air is somewhat calm and stable.

b. When wind speeds increase, the pollutants move downwind from the city.

CONCEPT CHECK

1. **What** are some of the health effects of exposure to air pollution?

2. **What** are urban heat islands? What are dust domes?

Controlling Air Pollutants

LEARNING OBJECTIVES

1. **Summarize** the effects of the Clean Air Act on U.S. air pollution.
2. **Contrast** air pollution in highly developed countries and in developing countries.

Technology exists to control all the forms of air pollution discussed in this chapter. Smokestacks fitted with electrostatic precipitators, fabric filters, scrubbers, and other technologies remove particulate matter from the air (**Figure 8.14**). Careful land-excavating activities, such as sprinkling water on dry soil being moved during road construction, reduces the creation of particulate matter. Many of the measures that increase energy efficiency and conservation also reduce air pollution. Smaller, more fuel-efficient automobiles produce fewer polluting emissions, for example.

The primary objections to using these technologies is cost. Adding scrubbers or electrostatic precipitators to existing plants increases the cost of doing business. Likewise, adding catalytic converters to automobiles increases their price. Several methods exist for removing sulfur oxides from flue (chimney) gases, but it is often less expensive simply to switch to a low-sulfur fuel such as natural gas or even to a non–fossil fuel energy source, such as solar energy. Sulfur can also be removed from fuels before they are burned.

Reduction of combustion temperatures in automobiles lessens the formation of nitrogen oxides. Use of mass transit reduces automobile use, thereby decreasing

Electrostatic precipitator • Figure 8.14

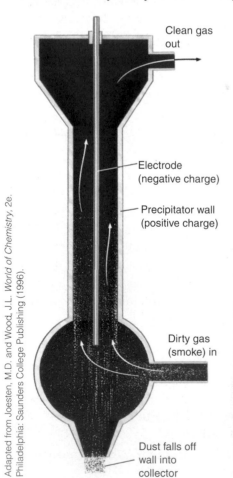

Clean gas out

Electrode (negative charge)

Precipitator wall (positive charge)

Dirty gas (smoke) in

Dust falls off wall into collector

Adapted from Joesten, M.D. and Wood, J.L. *World of Chemistry*, 2e. Philadelphia: Saunders College Publishing (1996).

a. In an electrostatic precipitator, the electrode imparts a negative charge to particulates in the dirty gas. These particles are attracted to the positively charged precipitator wall and then fall off into the collector.

Image Source / Getty Images

b. Industrial stacks without emission control devices can release substantial amounts of particulate matter and other pollutants.

Kodda/ Shutterst

c. Effective emission control devices can reduce particulate matter and other pollutants.

nitrogen oxide emissions. Nitrogen oxides produced during high-temperature combustion processes in industry can be removed from smokestack exhausts.

Modification of furnaces and engines to provide more complete combustion helps control the production of both carbon monoxide and hydrocarbons. Catalytic afterburners, used immediately following combustion, oxidize most unburned gases. The use of catalytic converters to treat auto exhaust reduces carbon monoxide and volatile hydrocarbon emissions about 85 percent over the life of the car. Careful handling of petroleum and hydrocarbons, such as benzene, reduces air pollution from spills and evaporation.

The Clean Air Act

There is good news and bad news about air pollution in the United States. The bad news is that many locations throughout the country still have unacceptably high levels of one or more air pollutants. Moreover, health experts estimate that air pollution causes the premature deaths of thousands of people in the United States each year. The good news is that overall air quality has improved since 1970.

This improvement is largely due to the U.S. **Clean Air Act (CAA)**, first passed in 1970 and updated and amended in 1977 and 1990. The CAA authorizes the Environmental Protection Agency (EPA) to apply and enforce the CAA by establishing limits on the amount of specific air pollutants permitted everywhere in the United States. Individual states must meet deadlines to reduce air pollution to acceptable levels. States may pass more stringent pollution controls than the EPA authorizes, but they can't mandate weaker limits than those stipulated in the CAA.

The EPA has focused on six air pollutants—lead, particulate matter, sulfur dioxide, carbon monoxide, nitrogen oxides, and ozone—and established maximum acceptable concentrations for each. The most dramatic improvement so far has been in the amount of lead in the atmosphere, which showed a 98 percent decrease between 1970 and 2000, primarily because of the switch from leaded to unleaded gasoline.

Atmospheric levels of the other pollutants, with the exception of particulate matter, have also declined (**Figure 8.15**). For example, between 1980 and 2010, sulfur dioxide emissions declined 83 percent. During this same time, U.S. gross domestic product increased more than 120 percent, energy consumption increased 23 percent, and vehicle miles increased about 95 percent. As will be discussed in Chapter 9, emissions of carbon dioxide have increased substantially since 1970.

Emissions in the United States, 1970 and 2011
• Figure 8.15

Carbon monoxide, sulfur dioxide, volatile organic compounds (many of which are hydrocarbons), and nitrogen oxides showed decreases; only particulate matter did not decline. "PM = 10" applies to particles less than or equal to 10 μm (10 micrometers). Since 1990 the EPA has also monitored PM = 2.5, which are very small particles less than or equal to 2.5 μm.

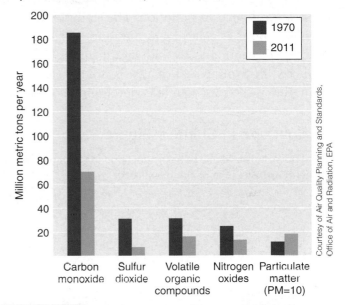

Courtesy of Air Quality Planning and Standards, Office of Air and Radiation, EPA

Think Critically Why have emissions of particulate matter increased since 1970, when all other emissions have gone down?

The Clean Air Act of 1970 and its amendments in 1977 and 1990 required progressively stricter controls of motor vehicle emissions. The provisions of the Clean Air Act Amendments of 1990 include the development of "superclean" cars, which emit lower amounts of nitrogen oxides and hydrocarbons, and the use of cleaner-burning gasoline in the most polluted cities in the United States. More recent automobile models do not produce as many pollutants as older models. Yet despite the increasing percentage of newer automobile models on the road, air quality has not improved in some areas of the United States because of the large increase in the number of cars being driven.

The Clean Air Act Amendments of 1990 focus on industrial airborne toxic chemicals in addition to motor vehicle emissions. The Clean Air Act Amendments of 1990 required a 90 percent reduction in the atmospheric emissions of 189 toxic chemicals. To comply with these requirements, both small businesses (such as dry cleaners) and large manufacturers (such as chemical companies) installed pollution-control equipment if they had not already done so.

Federal courts also play a role in interpreting the CAA. In 2008 the U.S. Supreme Court ruled that the EPA must regulate carbon dioxide under the CAA, a ruling the EPA is working to implement.

Air Pollution in Developing Countries

As developing nations become more industrialized, they also produce more air pollution. The leaders of most developing countries believe they must industrialize rapidly to compete economically with more highly developed countries. Environmental quality is usually a low priority in the race for economic development. Outdated technologies are adopted because they are less expensive, and air pollution laws, where they exist, are not enforced. Thus, air quality is deteriorating rapidly in many developing nations.

Many cities and towns in China burn so much low-quality coal for heating and industry that residents see the sun only a few weeks of the year (**Figure 8.16**). The rest of the time residents are choked in a haze of orange-colored coal dust. In other developing countries, such as India and Nepal, wood or animal dung is burned indoors, often in poorly designed stoves with little or no outside ventilation, thereby exposing residents to serious indoor air pollution (discussed in the next section).

Air pollution in China • Figure 8.16

A coal-powered steel mill releases pollution in Liaoning Province, China. All forms of pollution are increasing as China becomes industrialized.

Global Locator

CHINA

NATIONAL GEOGRAPHIC

NG Maps

George Steinmetz/NG Image Collection

GLOBAL LOCAL

What are the major sources of air pollution where you live?

The growing number of automobiles in developing countries is also contributing to air pollution, particularly in urban areas. Many vehicles in these countries are 10 or more years old and have no pollution-control devices. Motor vehicles produce about 60 to 70 percent of the air pollutants in urban areas of Central America, and they produce 50 to 60 percent in urban areas of India. The most rapid proliferation of motor vehicles worldwide is currently occurring in Latin America, Asia, and eastern Europe.

Lead pollution from heavily leaded gasoline is an especially serious problem in developing nations. The gasoline refineries in these countries are generally not equipped to remove lead from gasoline. (The United States was in the same situation until federal law mandated that U.S. refineries upgrade their equipment by 1986.) In Cairo, Egypt, for example, many children have blood lead levels more than two times higher than the level considered at-risk in the United States. Lead can retard children's growth and cause brain damage.

According to the World Health Organization, the five cities with the worst air pollution are Beijing, China; New Delhi, India; Santiago, Chile; Ulaanbator, Mongolia; and Cairo, Egypt. Respiratory disease is now the leading cause of death for children worldwide. More than 80 percent of these deaths occur in children under age 5 who live in cities in developing countries.

CONCEPT CHECK

1. **What** is the U.S. Clean Air Act, and how has it reduced outdoor air pollution?
2. **Where** is air pollution worse: in highly developed nations or in developing countries? Why?

Indoor Air Pollution

LEARNING OBJECTIVES

1. **Summarize** at least four sources of indoor air pollution and explain their role in sick building syndrome.
2. **Describe** the effects of indoor air pollution in developing countries.
3. **Explain** why radon gas is an indoor health hazard.

People around the world spend much of their time indoors, and contaminated indoor air can lead to substantial health problems. In rural areas, and particularly in developing countries, cooking with solid fuels (wood, coal, peat, and dung) can have serious health impacts. This has led the World Health Organization to determine that burning solid fuels is among the 10 greatest threats to human health.

The most common contaminants of indoor air in highly developed countries are radon, cigarette smoke, carbon monoxide, nitrogen dioxide (from gas stoves), formaldehyde (from carpet, fabrics, and furniture), household pesticides, cleaning solvents, ozone (from photocopiers), and asbestos. In addition, viruses, bacteria, fungi (yeasts, molds, and mildews), dust mites, pollen, and other organisms are often found in heating, air-conditioning, and ventilation ducts. Because illnesses from indoor air pollution usually resemble common ailments such as colds, influenza, or upset stomachs, they are often not recognized.

Health officials are paying increasing attention to **sick building syndrome**. The Labor Department estimates that more than 20 million employees are exposed to health risks from indoor air pollution. The EPA estimates that annual medical costs for treating the health effects of indoor air pollution in the United States exceed $1 billion. When lost work time and diminished productivity are added to health care costs, the

sick building syndrome Eye irritations, nausea, headaches, respiratory infections, depression, and fatigue caused by indoor air pollution.

Indoor air pollution • Figure 8.17

Cooking indoors with open fires or traditional cooking stoves results in dangerous levels of indoor air pollution.

Priti Bhatt/Flickr / Getty Images

total annual cost to the economy may be as much as $50 billion. Fortunately, most building problems are relatively inexpensive to alleviate.

Indoor air pollution is a particularly serious health hazard in developing countries, where many people burn fuels such as firewood or animal dung indoors to cook and heat water (**Figure 8.17**). Smoke from indoor cooking contains carbon monoxide, particulates, hydrocarbons, and other hazardous air pollutants such as formaldehyde and benzene. Women and children are harmed the most by indoor cooking, which can cause acute lower respiratory infections, pneumonia, eye infections, and lung cancer. The World Health Organization estimates that smoke from indoor cooking kills 1.6 million people each year.

Radon

Radon is a colorless, odorless, tasteless radioactive gas produced naturally as a result of the radioactive decay of uranium in Earth's crust. In the United States, radon has become an increasingly important indoor air contaminant, especially as laws and changes in habits have reduced secondhand cigarette smoke exposure. Radon

seeps through the ground and enters buildings, where it sometimes accumulates to dangerous levels (**Figure 8.18**). Radon emitted into the atmosphere gets diluted and dispersed and is of little consequence outdoors.

Radiation associated with radon decay does not penetrate deeply into body tissue. Consequently, only ingested or inhaled radon harms the body. The National Research Council of the National Academy of Sciences estimates that residential exposure to radon causes 12 percent of all lung cancers—between 15,000 and 22,000 lung cancers annually. Cigarette smoking exacerbates the risk from radon exposure; about 90 percent of radon-related cancers occur among current or former smokers.

According to the EPA, about 6 percent of U.S. homes have high enough levels of radon to warrant corrective action—a radon level above 4 picocuries per liter of air. The curie is a standard measure of radiation dose; a picocurie is one billionth of a curie. As a standard of reference, outdoor radon concentrations range from 0.1 to 0.15 picocuries per liter of air worldwide. The highest radon levels in the United States are found in homes across southeastern Pennsylvania into northern New Jersey and New York.

How radon infiltrates a house • Figure 8.18

Cracks in basement walls or floors, openings around pipes, and pores in concrete blocks provide some of the entries for radon.

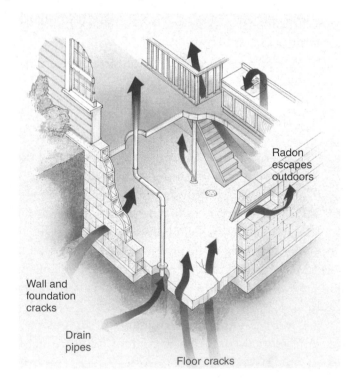

Radon escapes outdoors

Wall and foundation cracks

Drain pipes

Floor cracks

Homes may contain higher levels of air pollutants than outside air, even near polluted industrial sites.

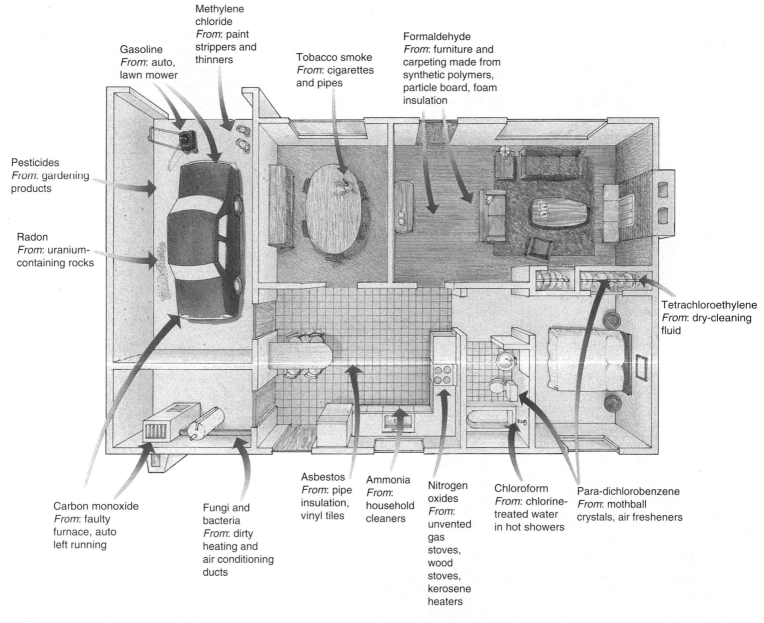

Methylene chloride
From: paint strippers and thinners

Gasoline
From: auto, lawn mower

Tobacco smoke
From: cigarettes and pipes

Formaldehyde
From: furniture and carpeting made from synthetic polymers, particle board, foam insulation

Pesticides
From: gardening products

Radon
From: uranium-containing rocks

Tetrachloroethylene
From: dry-cleaning fluid

Carbon monoxide
From: faulty furnace, auto left running

Fungi and bacteria
From: dirty heating and air conditioning ducts

Asbestos
From: pipe insulation, vinyl tiles

Ammonia
From: household cleaners

Nitrogen oxides
From: unvented gas stoves, wood stoves, kerosene heaters

Chloroform
From: chlorine-treated water in hot showers

Para-dichlorobenzene
From: mothball crystals, air fresheners

Ironically, efforts to make our homes more energy efficient have increased the hazard of indoor air pollutants, including radon. Drafty homes waste energy but allow radon to escape outdoors so it does not build up inside. Every home should be tested for radon because levels vary widely from home to home, even in the same neighborhood. Generally, testing and corrective actions are reasonably priced. However, some corrective actions can be expensive, costing thousands of dollars.

Figure 8.19 summarizes many possible sources of air pollution in homes.

CONCEPT CHECK STOP

1. **What** are some common indoor air contaminants?

2. **Why** is indoor air pollution such a serious health hazard in developing countries?

3. **How** does radon gas enter buildings?

CASE STUDY

Curbing Air Pollution in Chattanooga

During the 1960s, the federal government gave Chattanooga, Tennessee, the dubious distinction of having the worst air pollution in the United States. The air was so dirty in this manufacturing city that sometimes people driving downtown had to turn on their headlights in the middle of the day. The orange air soiled their white shirts so quickly that many businesspeople brought extra ones to work. To compound the problem, the mountains surrounding the city kept the pollutants produced by its inhabitants from dispersing.

Today the air in this scenic midsized city of 200,000 people is clean, and Chattanooga ranks high among U.S. cities in terms of air quality (*see* photograph). City and business leaders are credited with transforming Chattanooga's air. Soon after the passage of the federal Clean Air Act of 1970, the city established an air pollution control board to enforce regulations controlling air pollution. New local regulations allowed open burning by permit only, placed limits on industrial odors and particulate matter, outlawed visible automotive emissions, and set a cap on sulfur content in fuel, which controlled the production of sulfur oxides. Businesses installed expensive air pollution-control devices. The city started an emissions-free electric bus system. Chattanooga also decided to recycle its solid waste rather than build an emissions-producing incinerator.

In 1984 the EPA declared Chattanooga in attainment for particulate matter; this designation meant particulate levels had been below the federal health limit for one year. The city reached attainment status for ozone in 1989. Since then, the city's levels for all seven EPA-regulated air pollutants have been lower than federal standards require.

In the early 2000s, Chattanoogans continued to move their city toward *sustainability*. The city plans to convert a run-down business district into a community in which people live near their places of work. Businesses located in this district will form an **industrial ecosystem** in which the wastes of one business are raw materials for another business.

Chattanooga, Tennessee

Chattanooga's air quality has improved dramatically during the past several decades.

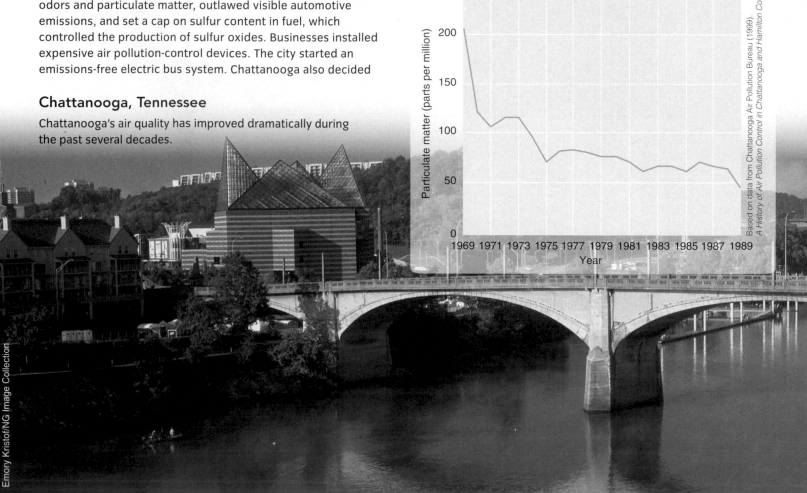

Particulate matter in Chattanooga, Tennessee, 1969 to 1989. Over a twenty year period, effective air pollution control measures reduced the concentration of particulate matter in Chattanooga's air by more than a factor of four.

Based on data from Chattanooga Air Pollution Bureau (1999). A History of Air Pollution Control in Chattanooga and Hamilton County.

Emory Kristof/NG Image Collection

Summary

1 The Atmosphere 192

1. Oxygen (21 percent) and nitrogen (78 percent) are the main gases in the **atmosphere**, the gaseous envelope surrounding Earth. Argon, carbon dioxide, other gases, water vapor, and trace amounts of various air pollutants are also present.

2. The **troposphere**, the layer of atmosphere closest to Earth's surface, extends to a height of approximately 12 km (7.5 mi). Temperature decreases with increasing altitude, and weather occurs in the troposphere. In the **stratosphere**, there is a steady wind but no turbulence. The stratosphere contains an ozone layer that absorbs much of the sun's UV radiation. The **mesosphere**, directly above the stratosphere, has the lowest temperatures in the atmosphere. The **thermosphere** has steadily rising temperatures and gases that absorb x-rays and short-wave UV radiation. The thermosphere reflects outgoing radio waves back toward Earth without the aid of satellites.

3. The **Coriolis effect** is the tendency of moving air or water to be deflected from its path and swerve to the right in the Northern Hemisphere and to the left in the Southern Hemisphere.

air pollutant in the lower atmosphere (troposphere) but an essential part of the stratosphere. Tropospheric ozone reduces visibility, causes health problems, stresses plants, and is associated with global warming. Some air pollutants are called **air toxics**, or **hazardous air pollutants**, because they are potentially harmful and may pose long-term health risks to people who are exposed to them; chlorine, lead, hydrochloric acid, formaldehyde, radioactive substances, and fluorides are examples. Air pollutants come from transportation, fuel combustion, industrial processes, and other sources.

2 Types and Sources of Air Pollution 196

1. **Air pollution** consists of various chemicals (gases, liquids, or solids) present in the atmosphere in high enough levels to harm humans, other organisms, or materials. **Primary air pollutants** are harmful chemicals that enter the atmosphere directly due to either human activities or natural processes; examples include carbon oxides, nitrogen oxides, sulfur dioxide, particulate matter, and hydrocarbons. **Secondary air pollutants** are harmful chemicals that form in the atmosphere when primary air pollutants react chemically with each other or with natural components of the atmosphere; ozone and sulfur trioxide are examples.

2. **Particulate matter**—solid particles and liquid droplets suspended in the atmosphere—corrodes metals, erodes buildings, soils fabrics, and can damage the lungs. **Nitrogen oxides** are gases associated with photochemical smog, acid deposition, global warming, and stratospheric ozone depletion; they also corrode metals and fade textiles. **Sulfur oxides** are gases associated with acid deposition; they corrode metals and damage stone and other materials. **Carbon oxides** include the gases carbon monoxide, which is poisonous, and carbon dioxide, which is linked to global warming. **Hydrocarbons** are solids, liquids, or gases associated with photochemical smog and global warming; some are dangerous to human health. **Ozone** is a secondary

3 Effects of Air Pollution 201

1. Exposure to low levels of air pollutants irritates the eyes and causes inflammation of the respiratory tract. Many air pollutants suppress the immune system, increasing susceptibility to infection. Exposure to air pollution during respiratory illnesses may result in the development of chronic respiratory diseases, such as emphysema and chronic bronchitis.

2. **Industrial smog** refers to smoke pollution. **Photochemical smog** is a brownish-orange haze formed by chemical reactions involving sunlight, nitrogen oxides, and hydrocarbons. Ozone is a principal component of photochemical smog. A **temperature inversion** is a layer of cold air temporarily trapped near the ground by a warmer upper layer; during a temperature inversion, polluting gases and particulate matter remain trapped in high concentrations close to the ground. An **urban heat island** is local heat buildup in an area of high population. Urban heat islands affect local air currents and weather conditions and contribute to the buildup of pollutants, especially particulate matter, in the form of a **dust dome**, a dome of heated air that surrounds an urban area and contains a lot of air pollution.

4 Controlling Air Pollutants 206

1. Improvements in U.S. air quality since 1970 are largely due to the **Clean Air Act**, which authorizes the EPA to set limits on specific air pollutants. Individual states must meet deadlines to reduce air pollution to acceptable levels and can't mandate weaker limits than those stipulated in the Clean Air Act.

2. Air quality in the United States has slowly improved since passage of the Clean Air Act. The most dramatic improvement is the decline of lead in the air, although levels of sulfur oxides, ozone, carbon monoxide, volatile compounds, and nitrogen oxides have also declined. Air quality is deteriorating in developing nations as a result of rapid industrialization, growing numbers of automobiles, and a lack of emissions standards.

5 Indoor Air Pollution 209

1. Indoor air pollution includes radon, cigarette smoke, nitrogen dioxide (from gas stoves), and formaldehyde (from carpet, fabrics, and furniture). These contribute to a variety of symptoms referred to as **sick building syndrome**.

2. Burning solid fuels indoors in developing countries leads to diseases including respiratory and eye infections, particularly among women and children.

3. Radon, a colorless, odorless, tasteless radioactive gas enters buildings from the ground. In some locations, indoor radon can pose a significant health threat.

Key Terms

- air pollution 196
- atmosphere 192
- Coriolis effect 195
- dust dome 205

- photochemical smog 201
- primary air pollutants 196
- secondary air pollutants 196
- sick building syndrome 209

- temperature inversion 202
- urban heat island 204

What is happening in this picture?

- This Nepalese woman is preparing a meal inside a poorly ventilated room. Cooking meals can take up many hours each day. In this picture, where is the smoke most visible? What does this imply for the health of women, who do much of the cooking in developing countries?

- Young children in developing countries tend to spend much of their time with their mothers; in fact, an infant may be strapped to the mother while she cooks. Explain what sorts of health effects you might expect these children to suffer as a result.

Global Locator

NEPAL

NG Maps

NATIONAL GEOGRAPHIC

Sean White/Design Pics/Perspectives/Getty Images

Critical and Creative Thinking Questions

1. What two gases comprise most of the atmosphere? Which two gasses have the greatest effect on life on Earth?

2. The atmosphere of Earth has been compared to the peel covering an apple. Explain the comparison.

3. What basic forces determine the circulation of the atmosphere? Describe the general directions of atmospheric circulation.

4. Distinguish between primary and secondary air pollutants. Give examples of each.

5. Distinguish between mobile and stationary sources of air pollution.

6. In what ways does agriculture impact air quality?

7. The graphs below represent air pollutant measurements taken at two different locations. Which location is indoors, and which is outdoors? Explain your answer.

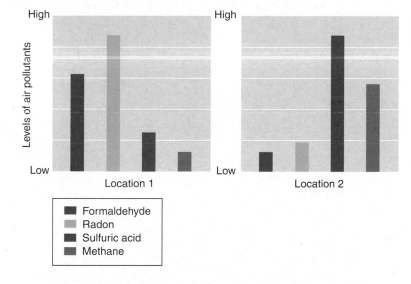

8. One of the most effective ways to reduce the threat of radon-induced lung cancer is to quit smoking. Explain.

9. What air pollutants do the 1990 amendments to the Clean Air Act target?

10. During a formal debate on the hazards of air pollution, one team argues that ozone is helpful to the atmosphere, and the other team argues that it is destructive. Under which conditions is each team's arguments correct?

11. This figure shows phase I vapor recovery from an underground gasoline storage tank. Before phase I vapor recovery was developed, gasoline vapors were vented directly into the air. Now the vapor is vented through Hose B into the truck and returned to the gasoline depot, where it is condensed or burned. Which of the following does gasoline vapor recovery control: photochemical smog, urban heat islands, or dust domes? Explain.

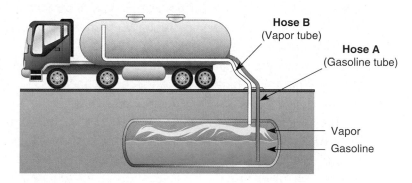

12. The graph below shows air pollutant levels in a city in the Northern Hemisphere, measured throughout a year. Is this city likely to be found in a developing country or in a developed country? Why?

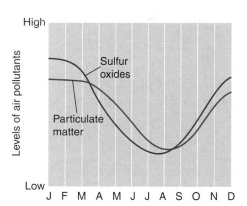

Sustainable Citizen Question

13. Identify the major sources of particulate matter near where you live. What actions could you take to reduce particulate matter? What laws or regulations limit particulate matter in your region?

THE PLANNER

Global Atmospheric Changes

MELTING ICE AND RISING SEA LEVELS

Powerful evidence that Earth is warming comes from the melting of continental and polar ice. Globally, the mean thickness of glacial ice diminished by 14 m (46 ft) from 1955 to 2005 (see graph). In 2002, an iceberg roughly twice the size of Rhode Island broke off from the Antarctic Peninsula. The Antarctic ice pack has retreated and thinned, losing 40 percent of its volume since 1980.

The Muir Glacier in Alaska was once enormous, with a huge vertical front from which icebergs calved into Glacier Bay. Today, the Muir Glacier has shrunk to a fraction of its former size (see photograph, taken in 2004; the inset shows approximately the same location in 1903).

Human-caused climate change is an established phenomenon. Within the scientific community, the question is no longer whether climate change will occur. Rather, we are concerned about how and whether we can reduce the rate of changes that have already begun, and prepare for those changes we cannot avoid. The biggest culprit in climate change is an increase in atmospheric carbon dioxide (CO_2), which is generated primarily through the burning of fossil fuels.

In this chapter we examine the challenges of global atmospheric changes: climate change, ozone depletion, and acid deposition. Changes in economics, politics, energy use, agriculture, and human behavior will be necessary to address these issues.

graphingactivity

C. L. Andrews/National Geographic Stock

Molnia, Bruce F. 2004 Muir Glacier: From the Glacier Photograph Collection. Boulder, Colorado USA: National Snow and Ice Data Center/World Data Center for Glaciology. Digital Media.

CHAPTER OUTLINE

The Atmosphere and Climate 218

- Solar Radiation and Climate
- Precipitation
- ▨ What a Scientist Sees: Rain Shadow

Global Climate Change 222

- Causes of Global Climate Change
- Effects of Global Climate Change
- ▨ Environmental InSight: The Effects of Global Climate Change
- Dealing with Global Climate Change: Mitigation and Adaptation

Ozone Depletion in the Stratosphere 231

- Causes of Ozone Depletion
- Effects of Ozone Depletion
- ▨ Environmental InSight: The Ozone Layer
- ▨ EnviroDiscovery: Links Between Climate and Atmospheric Change
- Reversing Ozone Layer Thinning

Acid Deposition 234

- How Acid Deposition Develops
- Effects of Acid Deposition
- The Politics of Acid Deposition
- Facilitating Recovery from Acid Deposition
- ▨ Environmental InSight: The Effects of Acid Deposition
- ▨ Case Study: International Implications of Global Climate Change

CHAPTER PLANNER ✓

❏ Study the picture and read the opening story.

❏ Scan the Learning Objectives in each section:
p. 218 ❏   p. 222 ❏   p. 231 ❏   p. 234 ❏

❏ Read the text and study all figures and visuals. Answer any questions.

Analyze key features:

❏ Process Diagram, p. 219 ❏   p. 224 ❏

❏ What a Scientist Sees, p. 221

❏ Environmental InSight, p. 228 ❏   p. 232 ❏   p. 236 ❏

❏ EnviroDiscovery, p. 233

❏ Case Study, p. 238

❏ Stop: Answer the Concept Checks before you go on:
p. 221 ❏   p. 230 ❏   p. 233 ❏   p. 238 ❏

End of Chapter:

❏ Review the Summary and Key Terms.

❏ Answer What is happening in this picture?

❏ Answer the Critical and Creative Thinking Questions.

(Dyurgerov, Mark B. and Mark F. Meier (2005). "Glaciers and the Changing Earth System: A 2004 Snapshot". Institute of Arctic and Alpine Research, Occasional Paper 58

Interpreting Data

Does this reduction appear to be constant, accelerating, or decelerating? Explain.

The Atmosphere and Climate

LEARNING OBJECTIVES

1. **Distinguish** between weather and climate, and explain what determines Earth's climate.

2. **Summarize** the effects of solar energy on Earth's temperature.

3. **Provide** several reasons for regional precipitation differences.

Weather refers to the conditions in the atmosphere at a given place and time; it includes temperature, atmospheric pressure, precipitation, cloudiness, humidity, and wind. Weather changes from one hour to the next and from one day to the next.

Earth's overall **climate** is determined by several factors: the sun's intensity, Earth's distance from the sun, tilt of the Earth relative to its rotational axis, distribution of water and landmasses across Earth's surface, and composition of gases in Earth's atmosphere. Most of these factors change very slowly—for example, millions of years ago, Earth's surface had only one landmass, while now it has several. A few billion years ago, the sun was only about 25 percent as intense as it is now.

> **climate** The typical weather patterns that occur in a place over a period of years.

However, only two of these factors change over years or decades: solar intensity and atmospheric composition. Solar intensity changes in cycles, and over the past decade, the sun has been slightly less intense than over the previous century. We would expect this to lead to a slight cooling of the atmosphere.

As we explore later in this chapter, however, changes in the atmosphere's composition have caused Earth's temperature to increase—along with other local and global changes—over the past century. Across Earth, the two most important factors that define an area's overall climate are *temperature*—both average temperature and temperature variability—and both average and seasonal *precipitation*. Latitude, elevation, topography, quantity and types of vegetation, distance from the ocean, and geographic location all influence climate. Other climate factors include weather conditions such as wind, humidity, fog, cloud cover, and, in some areas, lightning. Unlike weather, which changes rapidly, climate generally changes slowly, over hundreds or thousands of years.

Because each region's climate is relatively constant for many years, organisms have adapted to them. The

Tropical climate • Figure 9.1

Tropical climates occur in a region that spans the equator, from 15 to 25° latitude north to 15 to 25° latitude south. Photographed at the Na Pali Coast, Kauai, Hawaiian Islands.

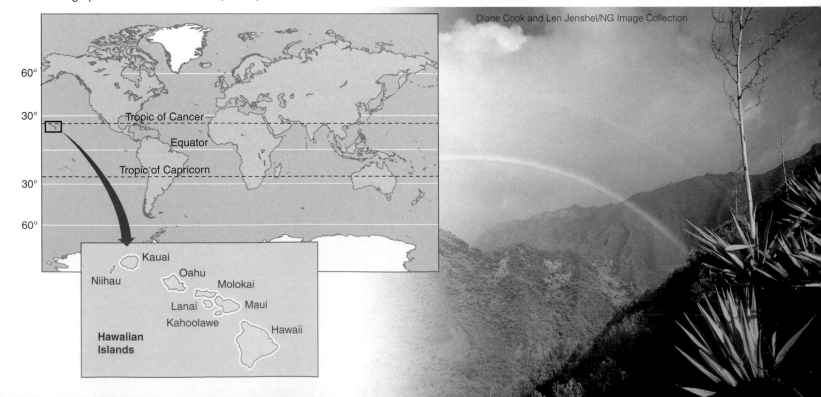

Diane Cook and Len Jenshel/NG Image Collection

many kinds of organisms on Earth are here in part because of the large number of different climates—from cold, snow-covered polar climates to tropical climates that are hot and have rain almost every day (**Figure 9.1**).

Solar Radiation and Climate

The sun makes life on Earth possible. Sunshine, or **insolation**, warms the planet, including the atmosphere, to habitable temperatures. Without the sun's energy, all water on Earth would be frozen, including the ocean.

Photosynthetic organisms capture the sun's energy and use it to make the food molecules almost all forms of life require. Most of our fuels, such as wood and **fossil fuels** (oil, coal, and natural gas), originated as solar energy captured by photosynthetic organisms. The amount of insolation that reaches Earth is a key factor in the overall climate. The uneven distribution of the total insulation around the globe is a major determinant of regional climates.

Clouds, and to a lesser extent snow, ice, and the ocean, reflect away about 31 percent of the solar radiation that falls on Earth (**Figure 9.2**). The remaining 69 percent is

---

PROCESS DIAGRAM

Fate of solar radiation that reaches Earth • Figure 9.2

 **THE PLANNER**

Most of the energy that the sun produces never reaches Earth. The solar energy that reaches Earth warms the planet's surface, drives the hydrologic cycle and other biogeochemical cycles, produces our climate, and powers almost all life through photosynthesis. On average, about 1366 watts per square meter (W/m²) enter Earth's atmosphere, resulting in about 1000 W/m² at Earth's surface.

WileyPLUS
⊕

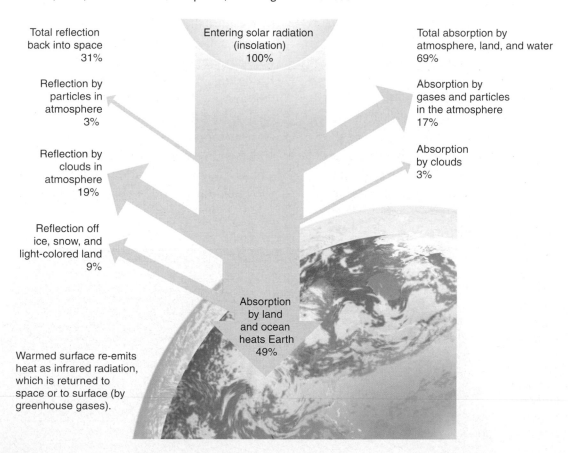

Total reflection back into space
31%

Reflection by particles in atmosphere
3%

Reflection by clouds in atmosphere
19%

Reflection off ice, snow, and light-colored land
9%

Entering solar radiation (insolation)
100%

Total absorption by atmosphere, land, and water
69%

Absorption by gases and particles in the atmosphere
17%

Absorption by clouds
3%

Absorption by land and ocean heats Earth
49%

Warmed surface re-emits heat as infrared radiation, which is returned to space or to surface (by greenhouse gases).

Think Critically Ice is highly reflective compared to water and land. How will this diagram change as ice caps and glaciers melt?

Solar intensity and latitude • Figure 9.3

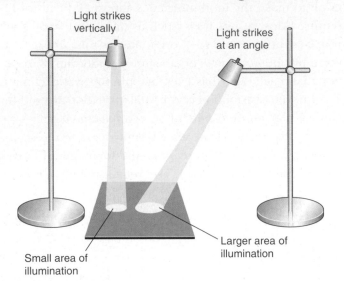

Light strikes vertically

Light strikes at an angle

Small area of illumination

Larger area of illumination

infrared radiation Electromagnetic radiation with wavelengths longer than those of visible light but shorter than microwaves; perceived as invisible waves of heat.

absorbed and runs the hydrologic cycle, carbon cycle, and other biogeochemical cycles; drives winds and ocean currents; powers photosynthesis; and warms the planet. Ultimately, all this energy returns to space as long-wave infrared radiation (heat).

Temperature Changes with Latitude and Season

Earth's roughly spherical shape and the tilt of its axis produce a great deal of variation in the exposure of the surface to solar energy (**Figure 9.3**). Sunlight that shines vertically near the equator (represented by the desk lamp on the left) is concentrated on Earth's surface. As one moves toward the poles, the light hits the surface more and more obliquely (represented by the lamp on the right), spreading the same amount of insolation over larger and larger areas. Because the sun's energy does not reach all places uniformly, temperature varies locally.

Earth's inclination on its axis (23.5 degrees from a line drawn perpendicular to the orbital plane) determines the seasons. During half the year (March 21 or 22 to September 22 or 23) the Northern Hemisphere tilts toward the sun, and during the other half (September 22 or 23 to March 21 or 22) it tilts away from the sun (**Figure 9.4**). The Southern Hemisphere tilts the opposite way, so that summer in the Northern Hemisphere corresponds to winter in the Southern Hemisphere.

Precipitation

Precipitation refers to any form of water, such as rain, snow, sleet, or hail, that falls from the atmosphere. Differences in precipitation depend on three factors:

1. **The amount of water vapor in the atmosphere.** *Equatorial uplift* of warm, moisture-laden air produces heavy rainfall in some areas of the tropics. High surface water temperatures cause vast quantities of water to evaporate from tropical parts of the ocean. Prevailing winds blow the resulting moist air over landmasses. The moist air continues to rise as it is heated by the sun-warmed land surface. As the air rises, it cools, which decreases its moisture-holding ability. When the air reaches its saturation point— when it can't hold any additional water vapor—clouds form and water is released as precipitation.

Progression of seasons • Figure 9.4

Earth's inclination on its axis remains the same as it travels around the sun. Thus, the sun's rays hit the Northern Hemisphere obliquely during its winter months and more directly during its summer. In the Southern Hemisphere, the sun's rays are oblique during its winter, which corresponds to the Northern Hemisphere's summer. At the equator, the sun's rays are approximately vertical on March 21 or 22 and September 22 or 23.

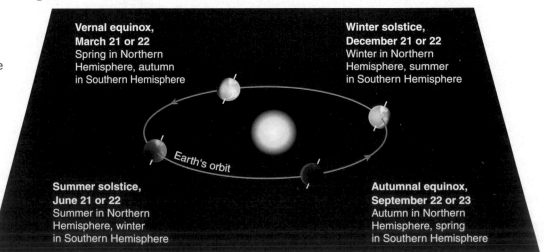

Vernal equinox, March 21 or 22 Spring in Northern Hemisphere, autumn in Southern Hemisphere

Winter solstice, December 21 or 22 Winter in Northern Hemisphere, summer in Southern Hemisphere

Earth's orbit

Summer solstice, June 21 or 22 Summer in Northern Hemisphere, winter in Southern Hemisphere

Autumnal equinox, September 22 or 23 Autumn in Northern Hemisphere, spring in Southern Hemisphere

WHAT A SCIENTIST SEES

Rain Shadow

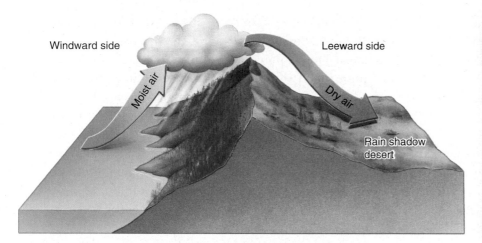

Windward side

Leeward side

Moist air

Dry air

Rain shadow desert

© Michael T. Sedam/CORBIS

a. A rain shadow refers to arid or semiarid land that occurs on the far side (leeward side) of a mountain. Prevailing winds blow warm, moist air from the windward side. Air temperature cools as it rises, releasing precipitation, so dry air descends on the leeward side. Such a rain shadow exists east of the Cascades.

b. Proxy Falls is in the Cascade Range, which divides the states of Washington and Oregon into a moist western region and an arid region east of the mountains.

2. **Geographic location.** The rising air from the equator eventually descends to Earth near the Tropic of Cancer and Tropic of Capricorn (latitudes 23.5 degrees north and 23.5 degrees south, respectively). By then most of its moisture has precipitated, and the dry air returns to the equator. Over land, this dry air produces some of the great tropical deserts, such as the Sahara Desert. Air also dries out as it travels long distances over landmasses. Near the windward coasts of continents (the side from which the wind blows), rainfall may be heavy. However, in the temperate zones (the areas between the tropics and the polar zones), continental interiors are usually dry because they are far from the ocean, which replenishes water in the air passing over it.

3. **Topographic features.** When flowing air encounters mountains, it flows up and over them, cooling as it gains altitude. Because cold air holds less moisture than does warm air, clouds form and precipitation occurs, primarily on the mountains' windward slopes.

The air mass is warmed as it moves down on the other side of the mountain, reducing the chance of precipitation of the remaining moisture. This situation exists on the West Coast of North America, where precipitation falls on the western slopes of mountains close to the coast (see *What a Scientist Sees*). The dry land on the side of the mountains away from the prevailing wind—in this case, east of the mountain range—is called a **rain shadow**. The dry conditions of a rain shadow often occur on a regional scale.

CONCEPT CHECK STOP

1. **How** do you distinguish between weather and climate?

2. **What** effect does solar energy have on Earth's temperature?

3. **What** are some of the environmental factors that produce areas of precipitation extremes, such as rain forests and deserts?

Global Climate Change

LEARNING OBJECTIVES

1. **List** the five main greenhouse gases and describe the enhanced greenhouse effect.

2. **Discuss** some of the potential effects of global climate change.

3. **Give** examples of strategies to mitigate or adapt to global climate change.

Earth's average temperature is based on daily measurements taken at several thousand land-based meteorological stations around the world, as well as data from weather balloons, orbiting satellites, transoceanic ships, and hundred of sea-surface buoys with temperature sensors. These data show that the years 1998 and 2001 through 2010 were the hottest since records began in the 1880s, and 2010 tied with 2005 as the hottest years on record. According to the National Oceanic and Atmospheric Administration (NOAA), global temperatures in those years may have been the highest in the last millennium. (Although widespread thermometer records have been assembled only since the mid-19th century, scientists reconstruct earlier temperatures using indirect climate evidence in tree rings, lake and ocean sediments, small air bubbles in ancient ice, and coral reefs.) The last two decades of the 20th century were its warmest (**Figure 9.5**).

Other evidence also suggests an increase in global temperature. Several studies indicate that spring in the Northern Hemisphere now comes about six days earlier than it did in 1959, and autumn comes five days later. Since 1949, the United States has experienced an increased frequency of heat waves, resulting in increased heat-related deaths among elderly and other vulnerable people. In the past few decades, the sea level has risen, glaciers worldwide have retreated, and while hurricanes may not have become more frequent, higher ocean surface temperatures have made them increasingly severe.

Scientists around the world have been researching global climate change for the past 50 years. As the evidence has accumulated, those most qualified to address the issue have reached a strong consensus that the 21st century will experience significant climate change. Almost all of that change will be attributed to anthropogenic (caused by humans) release of greenhouse gases into the atmosphere.

In response to this growing consensus, governments around the world organized the United Nations Intergovernmental Panel on Climate Change (IPCC). With input from hundreds of climate experts, the IPCC provides a definitive scientific assessment of global climate change. In its most recent (2007) report, the IPCC

Mean annual global temperature, 1960 to present • Figure 9.5

Data are presented as surface temperatures (°C) for each year since 1960. The measurements, which naturally fluctuate, show the warming trend of the past several decades. The dip in global temperatures in the early 1990s was caused by the eruption of Mount Pinatubo in 1991.

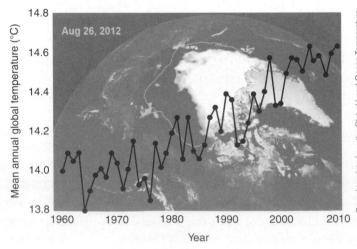

Based on data from the Global Land-Ocean Temperature Index, Goddard Institute of Space Studies (NASA). Photo from Scientific Visualization Studio, NASA Goddard Space Flight Center.

projected a 1.8° to 4.0°C (3.2° to 7.2°F) increase in global temperature by the year 2100. The IPCC predicts that we will observe higher maximum temperatures and more hot days over nearly all land areas, higher minimum temperatures, fewer frost days, fewer cold days, and an increase in the heat index. We may also experience more intense precipitation events over many areas, an increased risk of drought in the continental interiors in the mid-latitudes, and stronger hurricanes in some coastal areas.

Causes of Global Climate Change

Carbon dioxide (CO_2) and certain trace gases, including methane (CH_4), nitrous oxide (N_2O), chlorofluorocarbons (CFCs), and tropospheric ozone (O_3), accumulate in the atmosphere as a result of human activities. The concentration of atmospheric carbon dioxide has increased from about 288 parts per million (ppm) approximately 200 years ago (before the Industrial Revolution began) to 394 ppm in 2011 (**Figure 9.6**). According to the U.N. Food and Agriculture Organization, burning carbon-containing fossil fuels accounts for about 70 to 75 percent of human-generated CO_2 increase. The remaining 25 to 30 percent is released through deforestation, particularly when people fell or burn tropical rain forests. By 2050 the concentration of atmospheric CO_2 may be double what it was in the 1700s.

The combustion of gasoline in your car's engine releases not only CO_2 but also N_2O, which triggers the production of tropospheric ozone. Various industrial processes, land-use conversion, and the use of fertilizers also produce nitrous oxide. CFCs (discussed later in the chapter as they relate to depletion of the stratospheric ozone layer) are chemicals released into the atmosphere from old, leaking refrigerators and air conditioners. Methane is produced by the decomposition of carbon-containing organic material by anaerobic bacteria in moist places as varied as rice paddies, sanitary landfills, and the intestinal tracts of large animals. Tundra that is thawing due to warmer global temperatures also contributes methane—an example of an unfortunate feedback.

Global climate change occurs because these gases absorb infrared radiation—that is, heat—given off by Earth's surface. This absorption slows the natural flux of heat into space, warming the lower atmosphere.

Carbon dioxide (CO_2) in the atmosphere, 1958 to present • Figure 9.6

Note the steady increase in the concentration of atmospheric CO_2 since 1958, when measurements began at the Mauna Loa Observatory in Hawaii. This location was selected because it is far from urban areas where factories, power plants, and motor vehicles emit CO_2. The seasonal fluctuations correspond to winter (a high level of CO_2), when plants are not actively growing and absorbing CO_2, and summer (a low level of CO_2), when plants are growing and absorbing CO_2.

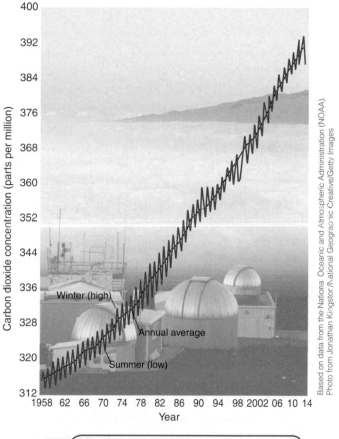

Based on data from the National Oceanic and Atmospheric Administration (NOAA). Photo from Jonathan Kingston/National Geographic Creative/Getty Images

Interpreting Data
If this rate of CO_2 increase continues, in what year will the concentration exceed 450 ppm? Explain your answer.

Because CO_2 and other gases trap the sun's infrared radiation somewhat like glass does in a greenhouse, they are called **greenhouse gases**. Greenhouse gases accumulating in the atmosphere as a result of human activities

greenhouse gases
Gases—including water vapor, carbon dioxide, methane, and certain other gases—that absorb infrared radiation.

PROCESS DIAGRAM

Enhanced greenhouse effect • Figure 9.7

The buildup of carbon dioxide (CO_2) and other greenhouse gases warms the atmosphere by absorbing some of the outgoing infrared (heat) radiation. Some of the heat in the warmed atmosphere is transferred back to Earth's surface, warming the land and ocean.

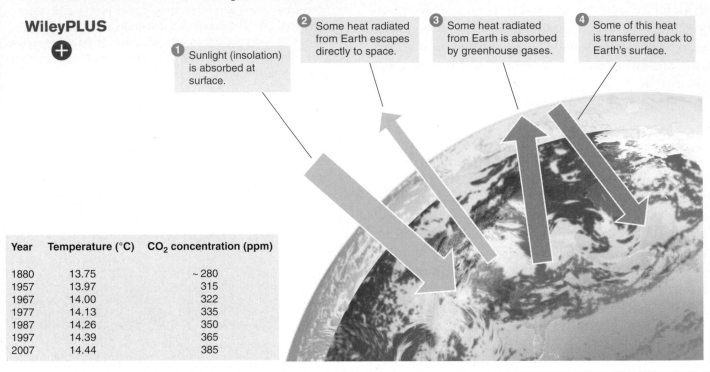

WileyPLUS

1 Sunlight (insolation) is absorbed at surface.

2 Some heat radiated from Earth escapes directly to space.

3 Some heat radiated from Earth is absorbed by greenhouse gases.

4 Some of this heat is transferred back to Earth's surface.

Year	Temperature (°C)	CO_2 concentration (ppm)
1880	13.75	~ 280
1957	13.97	315
1967	14.00	322
1977	14.13	335
1987	14.26	350
1997	14.39	365
2007	14.44	385

are causing an **enhanced greenhouse effect** (**Figure 9.7**).

Radiative forcing is the term used to describe how gases affect heat in the atmosphere. When the atmosphere contains more heat, its temperature is greater. While radiative forcing by some gases can be negative (causing temperature to decrease), the greenhouse gases released as we burn fossil fuels cause positive radiative forcing, and atmospheric temperature rises (**Figure 9.8**). For example, atmospheric aerosols tend to cool the atmosphere. **Aerosols**, which come from both natural and human sources, are particles so small they remain suspended in the atmosphere for days, weeks, or even months. Sulfur haze is an aerosol that reflects sunlight back into space, reducing the amount of solar energy reaching Earth's surface, and thereby cooling the atmosphere. Sulfur haze significantly moderates

enhanced greenhouse effect
Additional atmospheric warming produced as human activities increase atmospheric concentrations of greenhouse gases.

radiative forcing
For greenhouse gases, the capacity to retain heat in Earth's atmosphere.

warming in industrialized parts of the world. Sulfur emissions come from the same smokestacks that emit CO_2. Volcanic eruptions also eject sulfur particles into the atmosphere (**Figure 9.9**).

This cooling effect, however, is much weaker than the enhanced greenhouse effect. Human-produced sulfur emissions typically remain in the atmosphere for less than a year and rarely disperse globally. Greenhouse gases can remain in the atmosphere for hundreds of years. And carbon dioxide and other greenhouse gases help warm the planet 24 hours a day, whereas sulfur haze cools the planet only during the daytime. In addition, sulfur emissions are a respiratory irritant and cause acid deposition (discussed later in this chapter). Most nations are trying to reduce their sulfur emissions, not maintain or increase them.

Radiative forcing of different gases
• Figure 9.8

Radiative forcing is measured in watts per square meter (W/m²). Stratospheric ozone and aerosols cause the atmosphere to retain less energy; the other gases cause it to retain more. Overall, humans are causing a net increase of atmospheric radiative forcing.

Interpreting Data
Which factors contribute most to climate change? What is the total radiative forcing of all factors on this graph combined?

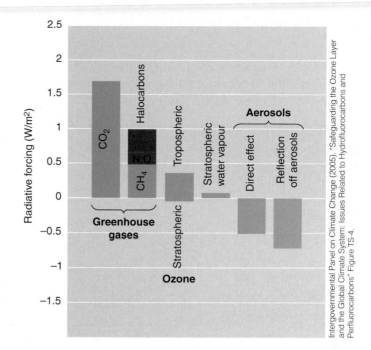

Intergovernmental Panel on Climate Change (2005). "Safeguarding the Ozone Layer and the Global Climate System: Issues Related to Hydrofluorocarbons and Perfluorocarbons" Figure TS-4.

Global Locator

PHILIPPINES

NG Maps

NATIONAL GEOGRAPHIC

Volcanic eruption • Figure 9.9

The eruption of Mount Pinatubo in the Philippines in 1991 injected massive amounts of sulfur into the atmosphere. Because sulfur haze reduces the amount of sunlight reaching the surface, this eruption caused Earth to cool temporarily. Compared to temperatures during the rest of the 1990s, global temperatures in 1992 and 1993 were relatively cool.

Arctic Sea Ice Volume, 1979–2011
• Figure 9.10 _____

This figure depicts seasonal (winter high and summer low) extent of polar ice cap volume. Some models predict that the North Pole will be clear of ice during summers as early as 2020.

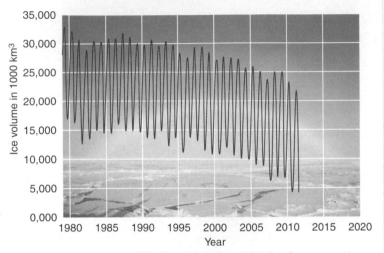

Based on data from United States National Snow and Ice Data Center, reported in 2011. Photo from Sue Flood/Stone/Getty Images

Interpreting Data
What is the overall trend of ice cap volume from 1980 to 2011? What causes the variability over the course of each year?

Effects of Global Climate Change

One clearly observed effect of climate change to date has been melting land and ocean ice. The ice cap at the North Pole, for example, has decreased consistently over the past three decades (**Figure 9.10**). Other observed and potential effects of global climate change include sea-level rise, changes in precipitation patterns, and impacts on agriculture, human health, and other organisms.

Two factors contribute to sea-level rise. First, as the Antarctic ice cap and continental glaciers melt, the amount of water in the ocean increases. Second, as water heats up, it expands. During the 20th century, the sea level rose 10 to 20 cm (4 to 8 in). Climate scientists estimate an additional 20 to 50 cm (7.9 to 20 in) rise by 2100. As sea level rises, small island nations such as the Maldives, a low-lying chain of islands in the Indian Ocean, will be increasingly vulnerable to saltwater intrusion and storm surges (**Figure 9.11**).

Predicted sea-level rise, as with predicted temperature and precipitation changes, is based on computer models of future climate conditions, or General Circulation Models (GCMs). As with all models, GCMs are based on a range of variables, none of which can be known with absolute certainty. Consequently, modelers typically present ranges of expected outcomes, based on what they know and what they consider most likely to be true. Computer models of weather changes caused by global climate change indicate that precipitation patterns will change, causing some areas to have more frequent droughts. At the same time, heavier snow and rainstorms may cause more frequent flooding in other areas. These changes could lower the availability and quality of fresh water in many locations, particularly in areas that are currently arid or semiarid, such as the Sahel region just south of the Sahara Desert.

Global climate change will have mixed effects on agriculture. The rise in sea level will inundate some river deltas, which are fertile agricultural lands. Certain agricultural pests and disease-causing organisms will probably proliferate and reduce crop yields. Increased frequency and duration of droughts will be a particularly serious problem, and lack of water for drinking and agriculture may force millions of people to relocate. However, agricultural productivity may increase in some areas.

Low-lying island • Figure 9.11 _____

Residents of this low-lying island, which is part of the Maldives (in the Indian Ocean), may need to relocate as sea level continues to rise.

James L. Stanfield/National Geographic Stock

In a climate-warmed world, the mosquito that spreads malaria could expand into temperate areas. Photographed in Ariquemes, Brazil, where most of the town's inhabitants suffer from malaria.

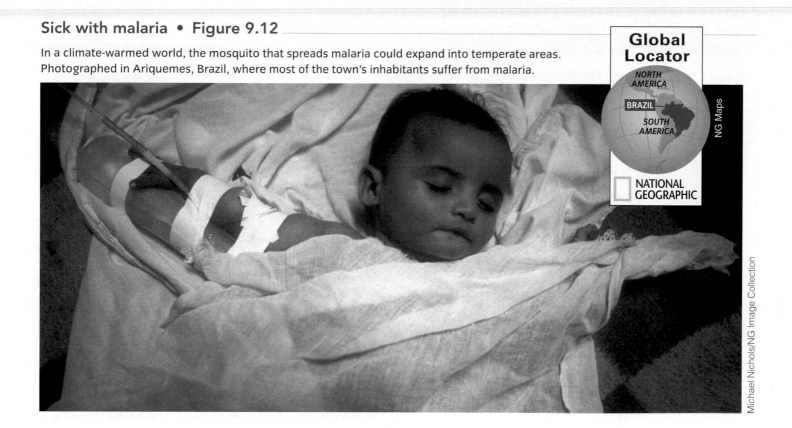

Currently, most evidence linking climate warming to disease outbreaks is circumstantial. Nonetheless, data linking climate warming and human health problems are accumulating. More frequent and more severe heat waves during summer have increased the number of heat-related illnesses and deaths. Mosquitoes and other disease carriers are expected to expand their range into the newly warm areas and spread dengue fever, schistosomiasis, yellow fever, and malaria (**Figure 9.12**). According to the World Health Organization, during 1998, the fourth warmest year on record, the incidence of malaria, Rift Valley fever, and cholera surged in developing countries.

An increasing number of studies report measurable changes in the biology of plant and animal species as a result of climate warming. Climate change also affects populations, communities, and ecosystems. In *Environmental InSight: The Effects of Global Climate Change*, we report on the results of several of the hundreds of studies conducted thus far.

Rising temperatures in the waters around Antarctica have led to a decline in the populations of shrimp-like krill and Antarctic silverfish, major food sources for Adélie penguins, reducing Adélie penguin populations (**Figure 9.13a**). Warmer temperatures also cause higher rates of reproductive failure in these penguins by producing puddles of melted snow that kill developing chick embryos at egg-laying sites.

Worldwide, many frog populations have plummeted, these include Puerto Rico's national symbol, the tiny tree frog known as coqui (**Figure 9.13b**). Warmer temperatures and more frequent dry periods have stressed the coqui, making them more vulnerable to infection by a lethal fungus.

Ecosystems considered at greatest risk of climate-change loss are polar seas, coral reefs, mountain ecosystems, coastal wetlands, and tundra. Water temperature increases of 1° to 2°C (1.8° to 3.6°F) cause coral bleaching, which contributes to the destruction of coral reefs (**Figure 9.13c**). In 1998, when tropical waters were some of the warmest ever recorded, about 10 percent of the world's corals died.

As atmospheric CO_2 increases, some of it dissolves in the ocean, producing carbonic acid (**Figure 9.13d**). The consequent acidification could be disastrous for shelled sea animals, particularly zooplankton at the base of the marine food web; the acid would attack and dissolve away their shells. Increased acidity also exacerbates coral bleaching. It could take centuries for the atmosphere–ocean CO_2 balance to stabilize, so even if emissions cease, the ocean will continue to acidify.

a. Warmer temperatures in Antarctica threaten the Adélie penguin's food supply and reduce its reproductive success.

© FLPA/Alamy

b. A coqui tree frog in Puerto Rico. These once-ubiquitous little frogs have become rarer, an indirect casualty of climate change.

© Thomas R. Fletcher/Alamy

Peter Scoones /Science Source Images

c. Ocean warming and acidification stress corals, causing them to become bleached. Photographed near the Maldives in the Indian Ocean.

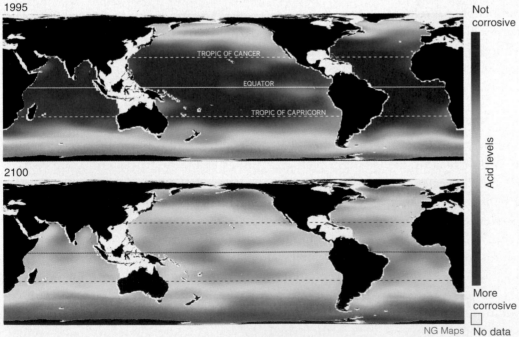

1995

TROPIC OF CANCER
EQUATOR
TROPIC OF CAPRICORN

2100

Not corrosive

Acid levels

More corrosive

 No data

NG Maps

How might plants and animals where you live be affected by a 2°C increase in the lowest winter temperature each year? the same increase in the highest summer temperature?

GLOBAL LOCAL

 Interpreting Data
In which parts of the ocean will organisms that are sensitive to acidification be most affected? least affected?

d. Scientific models project that ocean water will become increasingly acidic if human-produced CO_2 levels continue to rise. Shown are computer models for 1995 and 2100 from the NOAA Pacific Marine Environmental Laboratory.

WileyPLUS
⊕

Dealing with Global Climate Change: Mitigation and Adaptation

There are basically two ways to manage global climate change: mitigation and adaptation. *Mitigation* is the moderation or postponement of global climate change through measures that reduce greenhouse gas emissions. *Adaptation* is a planned response to changes caused by global climate change. Because the impacts of climate change have already begun, some combination of mitigation and adaptation is necessary to avoid severe or disruptive effects. The earlier we begin implementing strategies for mitigation and adaptation, the more effective they are likely to be.

We will have to deal with all greenhouse gases as we develop strategies to address global climate change, but we focus on CO_2 because it is produced in the greatest quantity and has the largest total effect. Carbon dioxide has an atmospheric lifetime of more than a century, so emissions produced today will still be around in the 22nd century. The extent and severity of global climate change will depend on the amount of additional greenhouse gas emissions we add to the atmosphere.

The rate of increase of atmospheric greenhouse gases from fossil fuels depends on such factors as economic conditions, policy choices, population growth, and technology changes. To accommodate uncertainty about these factors, climate scientists consider several **scenarios**, or possible futures. One such scenario, which assumes no intentional reduction in emissions, predicts that greenhouse gases will be at least triple the preindustrial concentration by 2100. Even in the most optimistic scenario, greenhouse gases in 2100 will probably be at least 550 ppm, or roughly twice the preindustrial concentration. Other uncertainties include how quickly the ocean can absorb heat and carbon dioxide, and how cloud cover will change. For these and other reasons, scientists are uncertain about *how much* change will occur during the next 90 years, but highly certain that temperature, precipitation, and sea level will continue to change significantly, with dramatic effects on life on Earth.

Mitigation of Global Climate Change Because most of the CO_2 that human activities produce comes from burning coal, oil, and natural gas, climate change is essentially an energy issue. Developing alternatives to fossil fuels offers a solution to warming caused by CO_2 emissions. We address alternatives to fossil fuels, including solar, hydroelectric, wind, and nuclear power, in Chapters 17 and 18.

Reducing energy use (for example, by driving less) and increasing efficiency (for example, by switching to hybrid cars) will reduce our output of CO_2, and will help mitigate global climate change. Energy-pricing strategies, such as carbon taxes and the elimination of energy subsidies, are other policies that could mitigate global climate change. Most experts think that using current technologies and developing such policies could significantly reduce greenhouse gas emissions with little cost to society.

Planting and maintaining forests also mitigates global climate change. Like other green plants, trees remove carbon dioxide from the air and incorporate the carbon into organic matter through photosynthesis. Reasonable estimates suggest that trees could remove 10 to 15 percent of the excess CO_2 in the atmosphere, but only through enormous plantings, so such efforts should not be considered a substitute for cutting emissions of greenhouse gases.

Many countries are investigating **carbon management**. Several experimental power plants currently capture CO_2 from their flue gases, but the technology is new (**Figure 9.14**). Technological innovations that more efficiently trap CO_2 from smokestacks would help mitigate global climate change and allow us to continue using fossil fuels (while they last) for energy. The carbon could be sequestered in geologic formations or in depleted oil or natural gas wells on land.

> **carbon management** Ways to separate and capture the CO_2 produced during the combustion of fossil fuels and then sequester (store) it.

Additional strategies to mitigate climate change include the following:

- Planting trees on degraded land
- Increasing efficiency of coal-fired power plants
- Replacing coal-fired power plants with nuclear power, hydropower, wind power, or even natural gas
- Increasing fuel economy of motor vehicles
- Redesigning cities to reduce reliance on single-occupant vehicles
- Insulating buildings to reduce the need for heating in the winter and cooling in the summer
- Improving management of agricultural soils

Adaptation to Global Climate Change Because the overwhelming majority of climate experts think human-induced global climate change will continue, government planners and social scientists are developing

Vattenhall carbon capture and storage pilot project • Figure 9.14

In the Vattenhall carbon capture and storage pilot project, coal is burned in a gasifier **1**. Particulates, sulfur, and other contaminants are removed **2**. After that, CO_2 is absorbed and removed **3**; the hot, CO_2-free gas is then used to generate electricity, first in a gas turbine **4** and then in a heat recovery steam generator **5**. Meanwhile, the CO_2 is piped away for storage underground **6**.

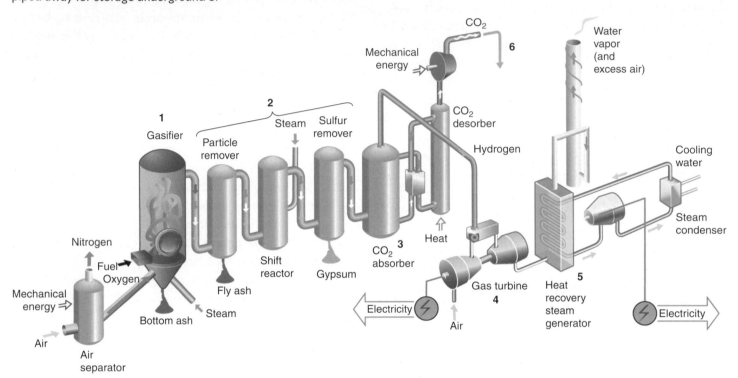

strategies to help various regions and sectors of society adapt to climate warming. One of the most pressing issues is rising sea level. People living in coastal areas could be moved inland, away from the dangers of storm surges, although the societal and economic costs would be great. Another extremely expensive alternative is the construction of massive sea walls to protect coastal land. Rivers and canals that spill into the ocean could be channeled to prevent *saltwater intrusion* into fresh water and agricultural land.

We must also adapt to shifting agricultural zones. Countries with temperate climates are evaluating semi-tropical crops to determine the best substitutes for traditional crops as the climate warms. Large lumber companies are developing heat- and drought-resistant strains of trees that will be harvested when global climate change may be well advanced. Evaluating such problems and finding and implementing solutions now will ease future stresses of climate warming.

Adaptation to global climate change is under study at several locations around the United States. One of the problems identified in a New York City study involves its sewer system. The waterways for storm runoff normally close during high tides. As the sea level rises in response to global climate change, the waterways will have to be shut during many low tides, which will increase the risk of flooding during storms (because excess water will not drain away). City planners will have to rebuild the storm runoff system or find some other way to prevent flooding. Evaluating such problems and implementing solutions now will ease future stresses of climate warming.

CONCEPT CHECK   **STOP**

1. **What** are greenhouse gases?

2. **How** will climate change affect agriculture? wildlife?

3. **What** are two examples of each of the approaches to manage global climate change: mitigation and adaptation?

Ozone Depletion in the Stratosphere

LEARNING OBJECTIVES

1. **Describe** the importance of the stratospheric ozone layer.

2. **Explain** how ozone thinning takes place and relate some of its harmful effects.

3. **Relate** how the international community is working to protect the ozone layer.

Although ozone (O_3) is a human-made pollutant in the troposphere, it is a naturally produced, essential component in the stratosphere, which encircles our planet some 10 to 45 km (6 to 28 mi) above the surface. The ozone layer shields Earth's surface from much of the high-energy **ultraviolet (UV) radiation** coming from the sun (**Figure 9.15a** and **b** in *Environmental Insight: The Ozone Layer*). If ozone disappeared from the stratosphere, Earth would become uninhabitable for most forms of life, including humans.

A slight **ozone thinning** occurs naturally over Antarctica for a few months each year. In 1985, however, the thinning was first observed to be greater than it should have been if natural causes were the only factor inducing it. This increased thinning, which occurs each September, is commonly referred to as the "ozone hole" (**Figure 9.15c**). There, ozone levels decrease as much as 70 percent each year.

During the subsequent two decades the ozone-thinned area continued to grow, and by 2006 it had reached the record size of 29.5 million km^2 (11.4 million mi^2), which is larger than the North American continent. A smaller thinning was also detected in the stratospheric ozone layer over the Arctic. In addition, world levels of stratospheric ozone have been decreasing for several decades (**Figure 9.15d**). According to the National Center for Atmospheric Research, ozone levels over Europe and North America have dropped almost 10 percent since the 1970s.

> **ultraviolet (UV) radiation** Radiation from the part of the electromagnetic spectrum with wavelengths just shorter than visible light; can be lethal to organisms at high levels of exposure.

> **ozone thinning** The removal of ozone from the stratosphere by human-produced chemicals or natural processes.

Causes of Ozone Depletion

The primary chemicals responsible for ozone loss in the stratosphere are a group of industrial and commercial compounds called **chlorofluorocarbons (CFCs)**. Scientists first discovered that CFCs can deplete stratospheric ozone in the mid-1970s. Chlorofluorocarbons such as Freon were used as propellants for aerosol cans and coolants in air conditioners and refrigerators. Other CFCs were used as solvents and as foam-blowing agents for insulation and packaging (Styrofoam, for example).

> **chlorofluorocarbons (CFCs)** Human-made organic compounds that contain chlorine and fluorine; now banned because they attack the stratospheric ozone layer.

Other compounds that destroy ozone include *halons*, used as fire retardants; *methyl bromide*, a pesticide; *methyl chloroform* and *carbon tetrachloride*, industrial solvents; and *nitrous oxide*, released from the burning of fossil fuels (particularly coal) and from the breakdown of nitrogen fertilizers in the soil.

Effects of Ozone Depletion

With depletion of the ozone layer, more UV radiation reaches the Earth's surface. Increased levels of UV radiation may disrupt ecosystems. For example, the productivity of Antarctic phytoplankton, the microscopic drifting algae that are the base of the Antarctic food web, has declined due to increased exposure to UV radiation. (The UV radiation inhibits photosynthesis.) Biologists have documented direct UV damage to natural populations of Antarctic fish. Widespread decline of amphibian populations may be linked to increased UV radiation. Because organisms are interdependent, the negative effect on one species has ramifications throughout the ecosystem.

Excessive exposure to UV radiation is linked to several health problems in humans, including eye cataracts, weakened immunity, and skin cancer. Exposure to any amount of UV radiation increases the risk of skin cancer. Radiation can cause mutations in cells that then reproduce abnormally and rapidly. This creates a particular concern for younger people, since cancers often only appear years after exposure. For this reason, many

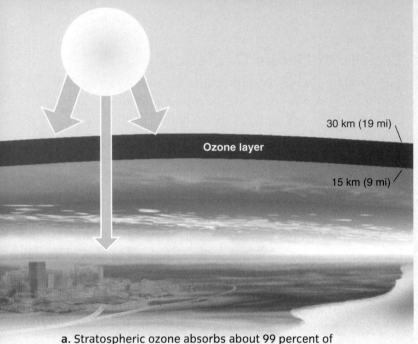

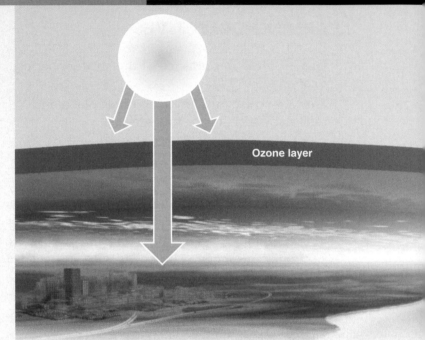

30 km (19 mi)

Ozone layer

15 km (9 mi)

Ozone layer

a. Stratospheric ozone absorbs about 99 percent of incoming solar ultraviolet (UV) radiation, effectively shielding the surface.

b. When stratospheric ozone is present at reduced levels, more high-energy UV radiation penetrates the atmosphere to the surface, where its presence harms organisms.

c. Ozone depletion. A computer-generated image of part of the Southern Hemisphere, taken in September 2012, reveals ozone thinning (the purple area over Antarctica). The ozone-thin area is not stationary but moves about as a result of air currents.

d. Average yearly ozone column over New Zealand and annual melanoma rate in New Zealand, 1970 to 2006. Located in the Southern Hemisphere, New Zealand is particularly vulnerable to increasing UV radiation due to ozone thinning. Melanoma is a common form of skin cancer that can be caused by exposure to UV radiation.

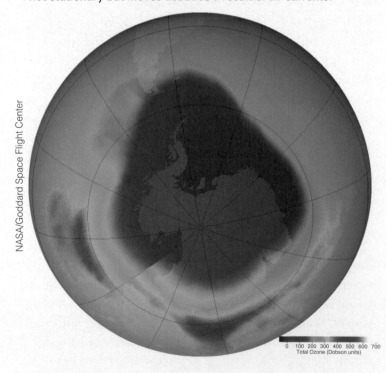

NASA/Goddard Space Flight Center

0 100 200 300 400 500 600 700
Total Ozone (Dobson units)

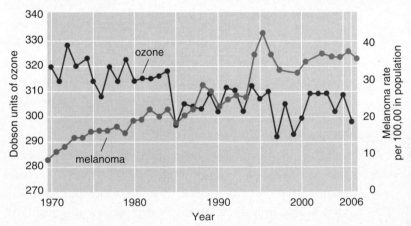

Based on data from New Zealand National Institute of Water and Atmospheric Research

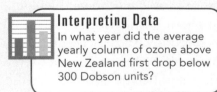

Interpreting Data
In what year did the average yearly column of ozone above New Zealand first drop below 300 Dobson units?

EnviroDiscovery
Links Between Climate and Atmospheric Change

Most environmental studies examine a single issue, such as acid deposition, global climate change, or ozone depletion. In the past few years, however, some researchers have been exploring the interactions of all three problems simultaneously. One study of such interactions found that North American lakes may be more susceptible to damage from UV radiation than the thinning of the ozone hole would indicate. The reason: Organic matter in the lakes, which absorbs some UV radiation and protects the lakes' plant and fish life, is affected by acid deposition and global climate change. Acid deposition reacts with organic matter in lakes, causing it to settle to the lake floor, where it does not absorb as much of the UV radiation as it once did. And a warmer climate increases evaporation, which reduces the amount of organic matter washed into lakes by streams.

Several studies report a link between human-caused climate warming and polar ozone depletion. Greenhouse gases that warm the troposphere also contribute to stratospheric cooling, presumably because heat trapped in the troposphere is not available to warm the stratosphere. The stratospheric temperature has been dropping for the past several years, and these lower temperatures provide better conditions for ozone-depleting chemicals to attack stratospheric ozone. Record ozone holes over Antarctica are attributed to cooler stratospheric temperatures. Some scientists speculate that if the cooling trend in the stratosphere continues, recovery of the ozone layer may be delayed. This means climate warming could prolong ozone depletion in the stratosphere despite the success of the Montreal Protocol.

Scientists now know environmental problems can't be studied as separate issues because they often interact in surprisingly subtle ways. As global climate change, ozone depletion, and acid deposition are studied further, it is likely that other interactions will be discovered.

---

states have banned indoor tanning for children. Malignant melanoma, the most dangerous type of skin cancer, is increasing faster than any other type of cancer.

Reversing Ozone Layer Thinning

In 1978 the United States, the world's largest user of CFCs, banned the use of CFC propellants in products such as antiperspirants and hair sprays. Although this ban was a step in the right direction, it did not solve the problem. Most nations did not follow suit, and propellants represented only a small portion of all CFC use.

In 1987 representatives from many countries met in Montreal to sign the **Montreal Protocol**, an agreement that originally stipulated a 50 percent reduction of CFC production by 1998. Despite this effort, stratospheric ozone continued to thin over the heavily populated mid-latitudes of the Northern Hemisphere, and the Montreal Protocol was modified to include even stricter limits on CFC production.

Industrial companies that manufacture CFCs quickly developed substitutes, such as hydrofluorocarbons (HFCs) and hydrochlorofluorocarbons (HCFCs). HFCs do not attack ozone, although they are potent greenhouse gases. HCFCs attack ozone but are less destructive than the chemicals they are replacing.

CFC, carbon tetrachloride, and methyl chloroform production was almost completely phased out in the United States and other highly developed countries in 1996, except for a relatively small amount exported to developing countries. Developing countries were on a different timetable and phased out CFC use in 2005. Methyl bromide was phased out in highly developed countries, which were responsible for 80 percent of its global use, in 2005. HCFCs will be phased out in 2030.

Unfortunately, CFCs are extremely stable compounds and will probably continue to deplete stratospheric ozone for several decades. Human-exacerbated ozone thinning will reappear over Antarctica each year, although the area and degree of thinning will gradually decline over time, until full recovery takes place sometime after 2050.

CONCEPT CHECK 🛑

1. **What** is the stratospheric ozone layer, and how does it protect life on Earth?
2. **What** is stratospheric ozone thinning? What role do CFCs play in ozone thinning?
3. **How** have governments responded to ozone thinning?

Acid Deposition

LEARNING OBJECTIVES

1. **Define** *acid deposition* and explain how acid deposition develops.
2. **Relate** examples of the effects of acid deposition.

What do fishless lakes in the Adirondack Mountains, recently damaged Mayan ruins in southern Mexico, and dead trees in the Czech Republic have in common? All these problems are the result of acid precipitation or, more properly, **acid deposition** (**Figure 9.16a**). Acid deposition has been around since the Industrial Revolution began. Robert Angus Smith, a British chemist, coined the term *acid rain* in 1872 after he noticed that buildings in areas with heavy industrial activity were being worn away by rain.

> **acid deposition** A type of air pollution that includes sulfuric and nitric acids in precipitation, as well as dry acid particles that settle out of the air.

Acid precipitation, including acid rain, sleet, snow, and fog, poses a serious threat to the environment. Until recently, industrialized countries in the Northern Hemisphere had been hurt the most, especially the Scandinavian countries, central Europe, Russia, and North America. In the United States alone, the annual damage from acid deposition is estimated at $10 billion. Acid deposition is now recognized as a global problem because it also occurs in developing countries as they become industrialized. For example, Chinese scientists have reported that acid deposition affects 40 percent of their country. In the two decades from 1990 to 2010, the amount of sulfur dioxide released from burning high-sulfur coal tripled in China (**Figure 9.16b**).

Sources and effects of acid deposition • Figure 9.16

b. China has abundant supplies of high-sulfur coal.

a. These carved stone slabs from the Mayan Palace at Palenque in Chiapas, Mexico have been damaged by acidic deposition.

How Acid Deposition Develops

The processes that lead to acid deposition begin when sulfur dioxide and nitrogen oxides are released into the atmosphere (**Figure 9.17a**). At the high temperatures experienced in, for example, automobile engines, atmospheric nitrogen (N_2) combines with oxygen (O_2) to form oxides of nitrogen (NO_x). Somewhat more than half of all NO_x comes from mobile sources. The rest comes from stationary facilities, such as coal and natural gas-burning power plants. Coal-burning power plants, large smelters, and industrial boilers are the main sources of sulfur dioxide emissions.

Wind carries sulfur dioxide and nitrogen oxides, released into the air from tall smokestacks, for long distances. Tall smokestacks allow England to "export" its acid deposition problem to the Scandinavian countries and the midwestern United States to "export" its acid emissions to New England and Canada.

In the atmosphere, sulfur dioxide and nitrogen oxides react with water to produce dilute solutions of sulfuric acid (H_2SO_4), nitric acid (HNO_3), and nitrous acid (HNO_2). The acidic water returns to Earth's surface in the form of precipitation or particulates, which together are called acid deposition.

Effects of Acid Deposition

Acid deposition affects both physical and biological components of the atmosphere. The link between acid deposition and declining aquatic animal populations, particularly fish, is well established, but other animals are also adversely affected. Birds living in areas with pronounced acid deposition are at increased risk of laying eggs with thin, fragile shells that break or dry out before the chicks hatch. The inability to produce strong eggshells is attributed to reduced calcium in the birds' diets. Calcium is less available to the food chain because in acidic soils calcium becomes soluble and is washed away, with little left for plant roots to absorb.

Acid deposition also has a serious effect on forest ecosystems. In the Black Forest of Germany, for example, up to 50 percent of trees surveyed are dead or severely damaged. This **forest decline** appears to result from a combination of stressors, including tropospheric ozone, UV radiation (which is more intense at higher altitudes), insect

forest decline A gradual deterioration and eventual death of many trees in a forest.

attack, drought, and acid deposition. When one or more stressors weaken a tree, an additional stressor, such as air pollution, may be decisive in causing the tree's death (**Figure 9.17b** and **c**).

Acid deposition can also damage agriculture, and it corrodes metals, building materials, and statues (**Figure 9.17d**). It eats away at historically important structures, such as the Washington Monument in Washington, DC, and ancient Mayan ruins in southern Mexico.

The Politics of Acid Deposition

Acid deposition is hard to combat because it does not occur only in the locations where acidic gases are emitted. It is entirely possible for sulfur and nitrogen oxides released in one spot to return to Earth's surface hundreds of kilometers from their source.

The United States has wrestled with this issue. Several states in the Midwest and East—Illinois, Indiana, Missouri, Ohio, Pennsylvania, Tennessee, and West Virginia—produce between 50 and 75 percent of the acid deposition that contaminates New England and southeastern Canada. Legislation formulated to deal with acid deposition has led to arguments about who should pay for the installation of expensive devices to reduce emissions of sulfur and nitrogen oxides.

In international disputes, these issues are magnified even more. For example, gases from coal-burning power plants in England move eastward with prevailing winds and return to the surface as acid deposition in Sweden and Norway. Similarly, emissions from mainland China produce acid deposition in Japan, Taiwan, North Korea, and South Korea.

Facilitating Recovery from Acid Deposition

Although the science and the politics surrounding acid deposition are complex, the basic concept of control is straightforward: Reducing emissions of sulfur and nitrogen oxides curbs acid deposition. Simply stated, if sulfur and nitrogen oxides are not released into the atmosphere, they cannot come down as acid deposition. Much of the sulfur released to the atmosphere comes from burning coal. One way to avoid acid rain is to reduce energy use, or switch to cleaner fuels. Alternatively, sulfur can be removed from the coal, either directly, before the coal is burned, or by installing scrubbers in the

a. Acid deposition.
Sulfur dioxide and nitrogen oxide emissions react with water vapor in the atmosphere to form acids that return to the surface as either dry or wet deposition.

Forest Decline.
Acid deposition is one of several stressors that may interact, contributing to the decline and death of trees.

b. Healthy Sitka spruce branch.

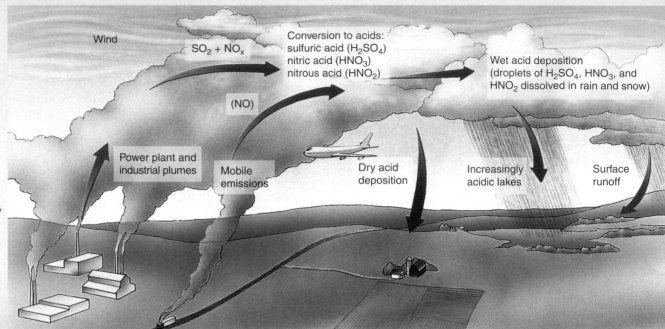

Wind

$SO_2 + NO_x$

Conversion to acids:
sulfuric acid (H_2SO_4)
nitric acid (HNO_3)
nitrous acid (HNO_2)

Wet acid deposition
(droplets of H_2SO_4, HNO_3, and HNO_2 dissolved in rain and snow)

(NO)

Power plant and industrial plumes

Mobile emissions

Dry acid deposition

Increasingly acidic lakes

Surface runoff

Randy Wells/Stone/Getty Images

c. Sitka spruce branch exhibiting the effects of forest decline. Photographed in Black Forest, Germany.

© Superclic/Alamy

WileyPLUS
⊕

Maurice E. Landre/Science Source Images

d. Acid rain damage.

Acidic stream • Figure 9.18

This mountain stream in the Adirondack Mountains of New York may be acidic because years of acid precipitation have altered soil chemistry.

© America/Alamy

smokestacks of coal-fired power plants. The resulting decrease in acid deposition prevents surface waters and soil from becoming more acidic than they already are.

Rainfall in parts of the Midwest, Northeast, and Mid-Atlantic regions is less acidic today than it was two decades ago, as a result of governmental policies that require cleaner-burning power plants and the use of reformulated gasoline. Many power plants in the Ohio Valley switched from high-sulfur to low-sulfur coal. However, solving one environmental problem often creates others. While the move to low-sulfur coal reduced sulfur emissions, it contributed to the problem of global climate change. Because low-sulfur coal has a lower heat value than high-sulfur coal, more of it must be burned—and more CO_2 emitted—to generate a given amount of electricity. Low-sulfur coal also contains higher levels of mercury and other trace metals, so burning it adds more of these hazardous pollutants to the air.

However, a 1991 agreement between the United States and Canada, along with legislation within each country, has reduced acid deposition by as much as 35 percent in some areas.

Despite the fact that the United States, Canada, and many European countries have reduced sulfur emissions, acid precipitation remains a serious problem. Acidified forests and bodies of water have not recovered as quickly as hoped. Trees in the U.S. Forest Service's Hubbard Brook Experimental Forest in New Hampshire, an area damaged by acid deposition, have grown little, even following two decades of declining emissions. Many northeastern streams and lakes, such as those in New York's Adirondack Mountains, remain acidic (**Figure 9.18**). A likely reason for the slow recovery is that the past 30 or more years of acid rain have profoundly altered soil chemistry in many areas. Essential plant minerals such as calcium and magnesium have washed away from forest and lake soils. Because soils take hundreds or even thousands of years to develop, it may take that long for them to recover from the effects of acid rain.

Many scientists are convinced that ecosystems will not recover from acid rain damage until substantial reductions in nitrogen oxide emissions occur. Nitrogen oxide emissions are harder to control than sulfur dioxide emissions because motor vehicles produce a substantial portion of nitrogen oxides. Engine improvements may help reduce nitrogen oxide emissions, but as the human population continues to grow, the increasing number of motor vehicles will probably offset any engineering gains. Dramatic cuts in nitrogen oxide emissions will require a reduction in high-temperature energy generation, especially in gasoline and diesel engines.

CONCEPT CHECK

1. **What** is acid deposition, and what are the main sources of atmospheric acid?
2. **What** are the harmful effects of acid deposition on materials, aquatic organisms, and soils?

CASE **STUDY**

International Implications of Global Climate Change

Various social, economic, and political factors complicate international efforts to deal with global climate change. Although highly developed countries have historically been the major producers of greenhouse gases, many developing countries are rapidly increasing production as they industrialize. But because developing countries have less technical expertise and fewer economic resources, they are often less able to respond to the challenges of global climate change.

The difference between total emissions from a country and the per person emissions from that country creates tensions among nations, especially between highly developed and developing countries. Most developing countries view fossil fuels as their route to industrial development and resist pressure from highly developed nations to decrease fossil fuel consumption.

Developing countries argue that it would be most fair to limit CO_2 on a per person basis, since highly developed countries such as the United States, France, and Japan emit several times as much CO_2 per person than do developing countries such as China, India, and Kenya (see figure). However, as both population and per person energy use increase in developing countries, their total CO_2 emissions are increasing rapidly. The average person in the United States is responsible for more than five times as much CO_2 as the average person in China, but China has surpassed the United States as the largest total emitter.

The international community recognizes that it must stabilize and decrease CO_2 emissions, but progress is slow. At least 174 nations, including the United States, signed the U.N. Framework Convention on Climate Change developed at the 1992 Earth Summit, which established goals for future international policies. In 1997 representatives from 160 countries determined timetables for reductions at a meeting in Kyoto, Japan. By 2005 enough countries had ratified the Kyoto Protocol for it to come into force. Political and economic concerns prevented the United States from joining the Kyoto Protocol, and those countries that have signed on have had limited success in meeting its provisions. Current international negotiations acknowledge that stopping climate change is not an option. Instead, conversation focuses on limiting the amount of change. It appears that the global community may set a maximum global temperature increase of 2°C (3.8°F) between now and 2100 as an achievable target.

Per person carbon dioxide (CO_2) emission estimates for selected countries, 1990 and 2008

Currently, industrialized nations produce a disproportionate share of CO_2 emissions. As developing nations such as China and India industrialize, however, their per person CO_2 emissions increase.

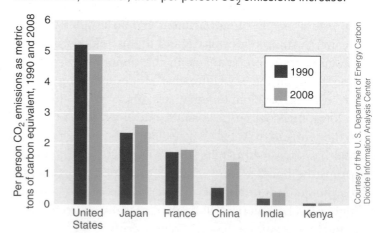

Summary

1 The Atmosphere and Climate 218

1. **Weather** is the condition in the atmosphere at a given place and time; it includes temperature, atmospheric pressure, precipitation, cloudiness, humidity, and wind. Earth's overall **climate** is determined by the sun's intensity, Earth's distance from the sun, tilt of the Earth relative to its rotational axis, distribution of water and landmasses across Earth's surface, and composition of gases in Earth's atmosphere . The typical weather patterns that occur over a period of years determine a region's climate. The two most important factors that define an area's climate are temperature and precipitation.

2. Sunlight, or **insolation**, is the primary (almost sole) source of energy available in the biosphere. The sun's energy runs the hydrologic cycle, drives winds and ocean currents, powers photosynthesis, and warms the planet. Of the solar energy that reaches Earth, 31 percent is immediately reflected away, and the remaining 69 percent is absorbed. Ultimately, all absorbed solar energy is radiated into space as **infrared radiation**, electromagnetic radiation with wavelengths longer than those of visible light but shorter than microwaves.

3. **Precipitation** is greatest where warm air passes over the ocean, absorbing moisture, and is then cooled, such as when mountains force humid air upward. Deserts develop in the rain shadows of mountain ranges or in continental interiors.

2 Global Climate Change 222

1. **Greenhouse gases** are gases that absorb infrared radiation; they include carbon dioxide, methane, nitrous oxide, chlorofluorocarbons, and tropospheric ozone. The **enhanced greenhouse effect** is the additional warming produced as human activities increase the amount of gases that absorb infrared radiation. **Radiative forcing** is the term used to describe the ability of different gases to cause the atmosphere to retain heat.

2. Global climate change will continue to cause sea level to rise, precipitation patterns to alter, extinction of many species, and problems for agriculture. It could result in the displacement of millions of people, thereby increasing international tensions.

3. Mitigation (slowing down the rate of global climate change) and adaptation (making adjustments to live with climate change) are two ways to address climate change. Mitigation includes developing alternatives to fossil fuels; increasing energy efficiency of automobiles and appliances; planting and maintaining forests; and instigating **carbon management**, by finding ways to separate and capture the CO_2 produced during the combustion of fossil fuels and then sequester it. Adaptation includes strategies to help various regions and sectors of society prepare for warmer temperatures, higher sea level, and altered precipitation patterns.

3 Ozone Depletion in the Stratosphere 231

1. Ozone (O_3) Is a human-made pollutant in the troposphere but a naturally produced, essential component in the stratosphere. The stratosphere contains a layer of ozone that shields the surface from much of the Sun's **ultraviolet (UV) radiation**, that part of the electromagnetic spectrum with wavelengths just shorter than those of visible light; UV radiation is a high-energy form of radiation that can cause skin cancer in humans, and be lethal to organisms at high levels of exposure.

2. **Ozone thinning** is the natural and human-caused removal of ozone from the stratosphere. The primary chemicals responsible for ozone thinning in the stratosphere are **chlorofluorocarbons (CFCs)**, human-made organic aerosol compounds that contain chlorine and fluorine. CFCs are now banned because they attack the stratospheric ozone layer. Ozone thinning causes excessive exposure to UV radiation, which can increase cataracts, weaken immunity, and cause skin cancer in humans. Increased levels of UV radiation may also disrupt ecosystems.

3. The **Montreal Protocol** is an international agreement that has phased out much CFC production worldwide, leading to decreased stratospheric ozone thinning.

4 Acid Deposition 234

1. **Acid deposition** is a type of air pollution that includes sulfuric and nitric acids in precipitation as well as dry acid particles that settle out of the air. Acid deposition develops

when sulfur and nitrogen oxides are released into the air, where they react to form acids and then return to surface waters and soil.

2. Acid deposition kills aquatic organisms, changes soil chemistry, and may contribute to **forest decline**, a gradual deterioration and eventual death of many trees in a forest.

Maurice E. Landre/Science Source Images

Key Terms

- acid deposition 234
- carbon management 229
- chlorofluorocarbons (CFCs) 231
- climate 218

- enhanced greenhouse effect 224
- forest decline 235
- greenhouse gases 223
- infrared radiation 220

- ozone thinning 231
- radiative forcing 224
- ultraviolet (UV) radiation 231

What is happening in this picture?

- This scientist is drilling into the Antarctic ice sheet to remove an ice core. Do you think the ice deep within the sheet is old or relatively young? Explain your answer.

- Some of the deeper samples were laid down thousands of years ago, when the climate was much cooler. The ice contains bubbles of air. Based on what you have learned in this chapter, do you think the level of carbon dioxide in the air bubbles in the oldest ice is higher or lower than the level in today's atmosphere? Explain your answer.

- If the scientists compared CFCs in the air bubbles in the oldest ice with today's levels of CFCs, what do you think they would find?

Maria Stenzel/NG Image Collection

Critical and Creative Thinking Questions

1. How does the sun affect temperature at different latitudes?

2. On the basis of what you know about the nature of science, can we say with how much certainty that the increased production of greenhouse gases is causing global climate change? Explain.

3. Biologists who study plants growing high in the Alps found that plants adapted to cold-mountain conditions migrated

up the peaks as fast as 3.7 m (12.1 ft) per decade during the 20th century, apparently in response to climate warming. Assuming that warming continues during the 21st century, what will happen to the plants if they reach the tops of the mountains?

4. Will it be easier for societies to mitigate climate change or to adapt to a changed climate? Explain your answer.

5. Distinguish between the benefits of the ozone layer in the stratosphere and the harmful effects of ozone at ground level.

6. What is the Montreal Protocol, and what environmental problem is it designed to correct?

7. Discuss some of the possible causes of forest decline. How might these factors interact to speed the rate of decline?

8–10. This map shows one model of how warmer global temperatures might alter precipitation in the United States in the next 100 years. Colors indicate the percent change in annual precipitation per century.

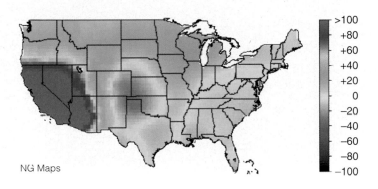

NG Maps

8. Name three states with climates that may become significantly wetter. Name three states with climates that may become significantly drier. Explain your answer.

9. Locate the state where you live. Will it be wetter, drier, or about the same?

10. In what ways might the projected changes in precipitation in the next 100 years impact U.S. agriculture? the U.S. economy?`

11. Consider the figures below, which depict total annual CO_2 production and per person CO_2 production for the United States, China, India, and Brazil over the past 30 years. Some people argue that China and India have the biggest responsibility to cut emissions, because they represent the greatest long–term emissions sources. Others argue that because the United States has such high per person emissions, it bears the greatest responsibility. Which, if either, of these arguments do you find more compelling? Why? Does the historically low total and per person emissions from countries like Brazil affect your decision?

a.

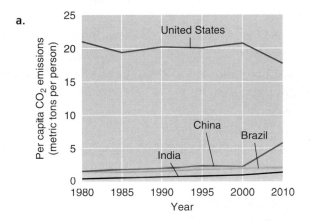

b.

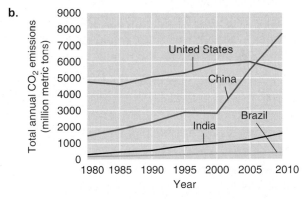

Based on data from the United Nations Statistics Division, Millennium Development Goals Indicators.

THE PLANNER

Freshwater Resources and Water Pollution

WATER: A LIMITED GLOBAL RESOURCE

Although three-fourths of Earth's surface is covered with water, substantially less than 1 percent is available for human use (see graph). Around the world, about 1.1 billion people live without adequate access to water—many have fewer than 10 L (about 2.6 gal) of clean water per day. In contrast, the typical American uses about 340 L (90 gal) each day.

The U.N. Development Program (UNDP) estimates that inhabitants of the slums in Lagos, Nigeria, pay 5 to 10 times as much for water as do those in wealthier neighborhoods (see photograph, where a woman fills a tub from a communal tank). In some places, the poor spend as much as 20 percent of their income on water.

In developed nations, complex systems are in place to make water available: Utility companies purchase, transport, clean, and distribute water. The total cost might be large, but coordination makes the cost per person relatively low for a reliable water supply. Contrast this with the slums of Lagos, where there is little money to develop and maintain infrastructure, so water from pipes is sporadic, and it might be contaminated. Water might be purchased from a vendor, but supplies and prices are unpredictable, and the quality of the water is unknown. Boiling the water to kill off biological contaminants requires energy, which is also in limited supply.

The UNDP asserts that access to enough safe water—at least 20 L (5.2 gal) per day—should be considered a basic human right. The agency has accordingly proposed strategies to make water available at a low cost, develop water infrastructure, and hold water providers accountable for consistency and safety. While enough water exists in the world for all people, problems of distribution and quality assurance, increasing global population, and water supply disruption due to climate change make universal access to water an issue that will face us for decades to come.

graphingactivity

Seawater 97.5%

Fresh water 2.5%

Other water (lakes, rivers, soil moisture, atmosphere) 0.03%

Ice caps and glaciers 1.97%

Groundwater 0.5%

Distribution of Earth's water

Interpreting Data
Which of the categories displayed is diminishing due to rising temperatures associated with global climate change? How will this impact the other categories?

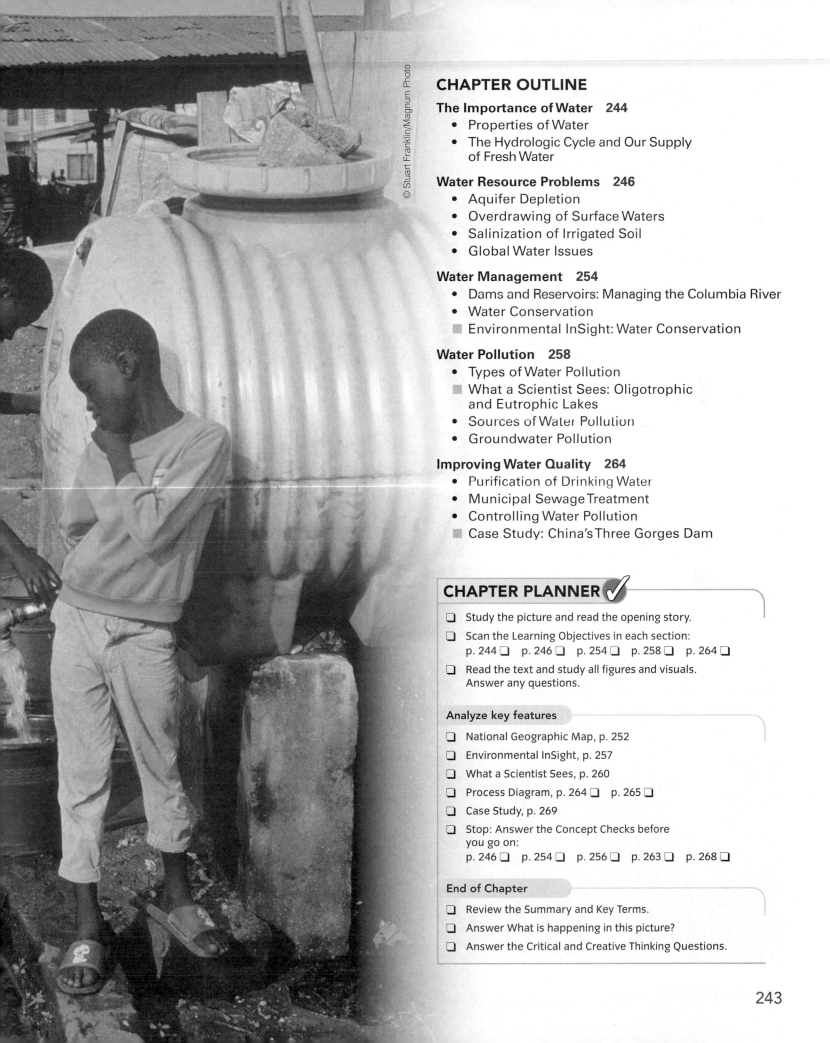

© Stuart Franklin/Magnum Photo

CHAPTER OUTLINE

The Importance of Water 244
- Properties of Water
- The Hydrologic Cycle and Our Supply of Fresh Water

Water Resource Problems 246
- Aquifer Depletion
- Overdrawing of Surface Waters
- Salinization of Irrigated Soil
- Global Water Issues

Water Management 254
- Dams and Reservoirs: Managing the Columbia River
- Water Conservation
- ■ Environmental InSight: Water Conservation

Water Pollution 258
- Types of Water Pollution
- ■ What a Scientist Sees: Oligotrophic and Eutrophic Lakes
- Sources of Water Pollution
- Groundwater Pollution

Improving Water Quality 264
- Purification of Drinking Water
- Municipal Sewage Treatment
- Controlling Water Pollution
- ■ Case Study: China's Three Gorges Dam

CHAPTER PLANNER ✓

❏ Study the picture and read the opening story.

❏ Scan the Learning Objectives in each section:
p. 244 ❏   p. 246 ❏   p. 254 ❏   p. 258 ❏   p. 264 ❏

❏ Read the text and study all figures and visuals. Answer any questions.

Analyze key features

❏ National Geographic Map, p. 252

❏ Environmental InSight, p. 257

❏ What a Scientist Sees, p. 260

❏ Process Diagram, p. 264 ❏   p. 265 ❏

❏ Case Study, p. 269

❏ Stop: Answer the Concept Checks before you go on:
p. 246 ❏   p. 254 ❏   p. 256 ❏   p. 263 ❏   p. 268 ❏

End of Chapter

❏ Review the Summary and Key Terms.

❏ Answer What is happening in this picture?

❏ Answer the Critical and Creative Thinking Questions.

The Importance of Water

LEARNING OBJECTIVES

1. **Describe** the structure of a water molecule and explain how hydrogen bonds form between adjacent water molecules.

2. **List** the unique properties of water.

3. **Explain** how processes of the hydrologic cycle allow water to circulate through the abiotic environment.

L ife on planet Earth would be impossible without water. All life forms, from unicellular bacteria to multicellular plants and animals, contain water. Humans are composed of approximately 70 percent water by body weight. We depend on water for our survival as well as for our convenience: We drink it, cook with it, wash with it (**Figure 10.1**), travel on it, and use an enormous amount of it for agriculture, manufacturing, mining, energy production, and waste disposal.

Although Earth has plenty of water, about 97 percent of it is salty and not consumable by most terrestrial organisms (see graph in the chapter opener). Fresh water is distributed unevenly, resulting in serious regional water supply problems and conflicts. Water experts predict that by 2025, more than one-third of the human population will live in areas where there isn't enough fresh water for drinking and irrigation.

Properties of Water

Water is composed of molecules of H_2O, each consisting of two atoms of hydrogen and one atom of oxygen. Water molecules are **polar**—that is, one end of the molecule has a positive electrical charge, and the other end has a negative charge (**Figure 10.2**). The negative (oxygen) end of one water molecule is attracted to the positive (hydrogen) end of another water molecule, forming a **hydrogen bond** between the two molecules. Hydrogen bonds are the basis

Young brick workers in India bathe with water from an irrigation pipe • Figure 10.1

Jodi Cobb/NG Image Collection

Chemical properties of water • Figure 10.2

a. Each water molecule consists of two hydrogen atoms and one oxygen atom. Water molecules are polar, with positively and negatively charged areas.

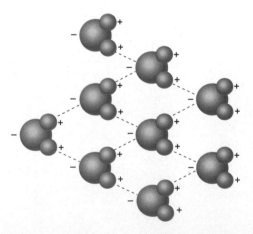

b. The polarity causes hydrogen bonds (represented by dashed lines) to form between the positive areas of one water molecule and the negative areas of others. Each water molecule forms up to four hydrogen bonds with other water molecules.

for many of water's physical properties, including its high melting/freezing point (0°C, 32°F) and high boiling point (100°C, 212°F). Because most of Earth has a temperature between 0°C and 100°C, most water exists in the liquid form organisms need.

Water absorbs a great deal of solar heat without substantially increasing in temperature. This high heat capacity allows the ocean to have a moderating influence on climate, particularly along coastal areas; the ocean does not experience the wide temperature fluctuations common on land.

Water is a *solvent*, meaning that it can dissolve many materials. In nature, water is never completely pure because it contains dissolved gases from the atmosphere and dissolved mineral salts from the land. Water's abilities as a solvent have a major drawback: Many of the substances that dissolve and are transported in water cause water pollution.

The Hydrologic Cycle and Our Supply of Fresh Water

In the **hydrologic cycle**, water continuously circulates through the environment, from the ocean to the atmosphere to the land and back to the ocean (see **Figure 10.3**; also see Figure 5.10 for a more thorough discussion of all components of the hydrologic cycle). The result is a balance of the water resources in the ocean, on the land, and in the atmosphere. The hydrologic cycle provides a continual renewal of the supply of fresh water on land, which is essential to terrestrial organisms.

Surface water is water found in streams, rivers, lakes, ponds, reservoirs, and **wetlands** (areas of land covered with water for at least part of the year). The **runoff** of precipitation from the land replenishes surface waters and is considered a renewable, although finite, resource. A **drainage basin**, or **watershed**, is the area of land drained by a single river or stream. Watersheds range in size from less than 1 km² for a small stream to a huge portion of the continent for a major river system such as the Mississippi River.

> **surface water** Precipitation that remains on the surface of the land and does not seep down through the soil.
>
> **runoff** The movement of fresh water from precipitation and snowmelt to rivers, lakes, wetlands, and the ocean.

Earth contains underground formations that collect and store water. This water originates as rain or melting snow that slowly seeps into the soil. It works its way down through cracks and spaces in sand, gravel, or rock until an impenetrable layer stops it; there it accumulates as **groundwater**. Groundwater flows through permeable sediments or rocks slowly, typically covering distances of several millimeters to a few meters per day. This process of downward movement and accumulation is called *groundwater recharge*. Eventually groundwater is discharged into rivers, wetlands, springs, or the ocean. Thus, surface water

> **groundwater** The supply of fresh water under Earth's surface that is stored in underground aquifers.

Two important components of the hydrologic cycle • Figure 10.3

a. Liquid and solid precipitation continuously falls from the atmosphere to the land and ocean.

b. Evaporation continuously moves water vapor from the land and ocean into the atmosphere.

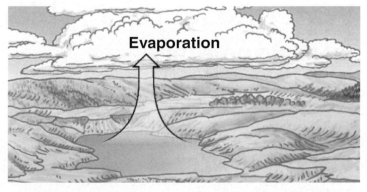

Adapted from Figure 2.8 on p. 28 in Stahler, A. and A. Strahler. *Physical Geography: Science and Systems of the Human Environment.* Hoboken, NJ: John Wiley & Sons, Inc. (2002).

Groundwater • Figure 10.4

Excess surface water seeps downward through soil and porous rock layers until it reaches impermeable rock or clay. An unconfined aquifer holds groundwater recharged by surface water directly above it. A confined aquifer stores groundwater between impermeable layers.

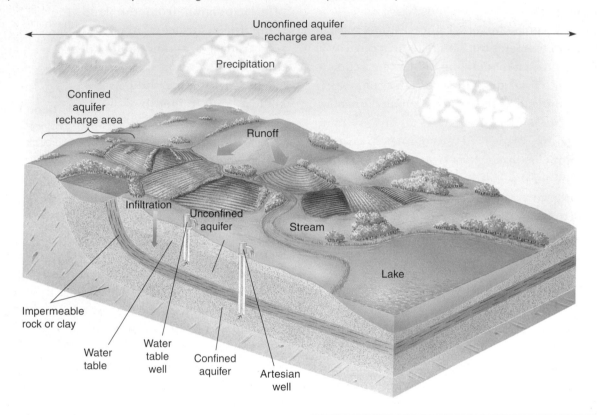

and groundwater are interrelated parts of the hydrologic cycle. **Aquifers** are underground reservoirs in which groundwater is stored (**Figure 10.4**).

Most groundwater is considered a nonrenewable resource because it has taken hundreds or even thousands of years to accumulate, and usually only a small portion of it is replaced each year by seepage of surface water.

CONCEPT CHECK STOP

1. **How** do hydrogen bonds form between adjacent water molecules?

2. **What** are two unique properties of water?

3. **How** do processes in the hydrologic cycle affect the accumulation of groundwater?

Water Resource Problems

LEARNING OBJECTIVES

1. **Relate** some of the problems caused by aquifer depletion, overdrawing of surface waters, and salinization of irrigated soil.

2. **Contrast** the water problems associated with the Ogallala Aquifer and the Colorado River Basin.

3. **Describe** the role of international cooperation in managing shared water resources.

Water resource problems fall into three categories: too much water, too little water, and poor-quality water. Flooding occurs when a river's discharge cannot be contained within its normal channel. Today's floods are more disastrous in terms of property loss than those of the past because humans often remove water-absorbing plant cover from the soil and construct buildings on floodplains. (A **floodplain** is the area bordering a river channel that has the potential to flood.) These activities increase the likelihood of both floods and flood damage.

When a natural area—that is, an area undisturbed by humans—is inundated with heavy precipitation, the plant-protected soil absorbs much of the excess water. What the soil cannot absorb runs off into the river, which may then spill over its banks onto the floodplain. Because rivers meander, the flow is slowed, and the swollen waters rarely cause significant damage to the surrounding area. (See Figure 6.13 for a diagram of a typical river, including its floodplain.)

When an area is developed for human use, construction projects replace much of this protective plant cover. Buildings and paved roads don't absorb water, so runoff, usually in the form of storm sewer runoff, is significantly greater in developed areas (**Figure 10.5**). People who build homes or businesses on the floodplain of a river will most likely experience flooding at some point (**Figure 10.6**).

Arid lands, or deserts, are fragile ecosystems in which plant growth is limited by lack of precipitation. **Semiarid lands** receive more precipitation than deserts but are subject to frequent and prolonged droughts.

Farmers increase the agricultural productivity of arid and semiarid lands with irrigation. Irrigation of these

Flooding in Queensland, Australia • Figure 10.6

A neighborhood in Rockhampton, Queensland, in northeastern Australia, is inundated by the swollen Fitzroy River on January 4, 2011. The region has experienced severe seasonal flooding in recent years, suffering in 2011 at least 35 deaths and costing the nation as much as $30 billion.

© Janie Barrett/Pool/epa/Corbis

How development changes the natural flow of water • Figure 10.5

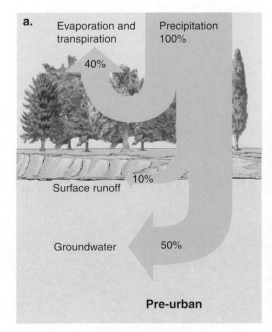

a. Evaporation and transpiration / Precipitation 100% / 40% / Surface runoff 10% / Groundwater 50% / **Pre-urban**

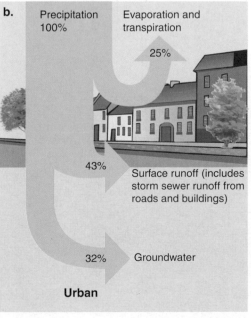

b. Precipitation 100% / Evaporation and transpiration 25% / 43% / Surface runoff (includes storm sewer runoff from roads and buildings) / 32% Groundwater / **Urban**

Control of Water Pollution from Urban Runoff. Paris: Organization for Economic Cooperation and Development (1986).

Shown is the fate of precipitation in Ontario, Canada, before (**a**) and after (**b**) urbanization. After Ontario was developed, surface runoff increased substantially, from 10 percent to 43 percent.

lands has become increasingly important worldwide in efforts to produce enough food for burgeoning populations (**Figure 10.7**). Since 1955, the amount of irrigated land has more than tripled; Asia has more agricultural land under irrigation than do other continents, primarily in China, India, and Pakistan. Water use for irrigation will probably continue to increase in the 21st century, but at a slower rate than in the last half of the 20th century.

Population growth in arid and semiarid regions intensifies water shortage. More people need food, so additional water resources are diverted for irrigation. Also, the immediate need for food prompts people to remove natural plant cover to grow crops on marginal lands subject to frequent drought, which in turn reduces water absorption into soils when rains do come.

aquifer depletion
The removal of groundwater faster than it can be recharged by precipitation or melting snow.

saltwater intrusion The movement of seawater into a freshwater aquifer near the coast.

Aquifer Depletion

Aquifer depletion from excessive removal of groundwater lowers the **water table**, the upper surface of the saturated zone of groundwater (see Figure 10.4). Prolonged aquifer depletion drains an aquifer dry, effectively eliminating it as a water resource. Even areas with high rainfall can experience aquifer depletion if humans remove more groundwater than can be recharged. In addition, aquifer depletion from porous sediments causes **subsidence**, or sinking, of the land above it. **Saltwater intrusion** occurs along coastal areas when groundwater is depleted faster than it recharges. Saltwater intrusion is also occurring in low-lying areas due to sea-level rise associated with global climate change. Well water in such areas eventually becomes too salty for human consumption or other freshwater uses.

Agricultural use of water • Figure 10.7

These fields in Kansas use center-pivot irrigation, which minimizes evaporative water loss and gives fields a distinctive circular shape. Each circle is the result of a long irrigation pipe that extends along the radius from the circle's center to its edge and slowly rotates, spraying the crops. This satellite photo, taken in June, shows wheat fields (bright yellow), corn fields (dark green), and newly emerging sorghum (light green).

Courtesy NASA

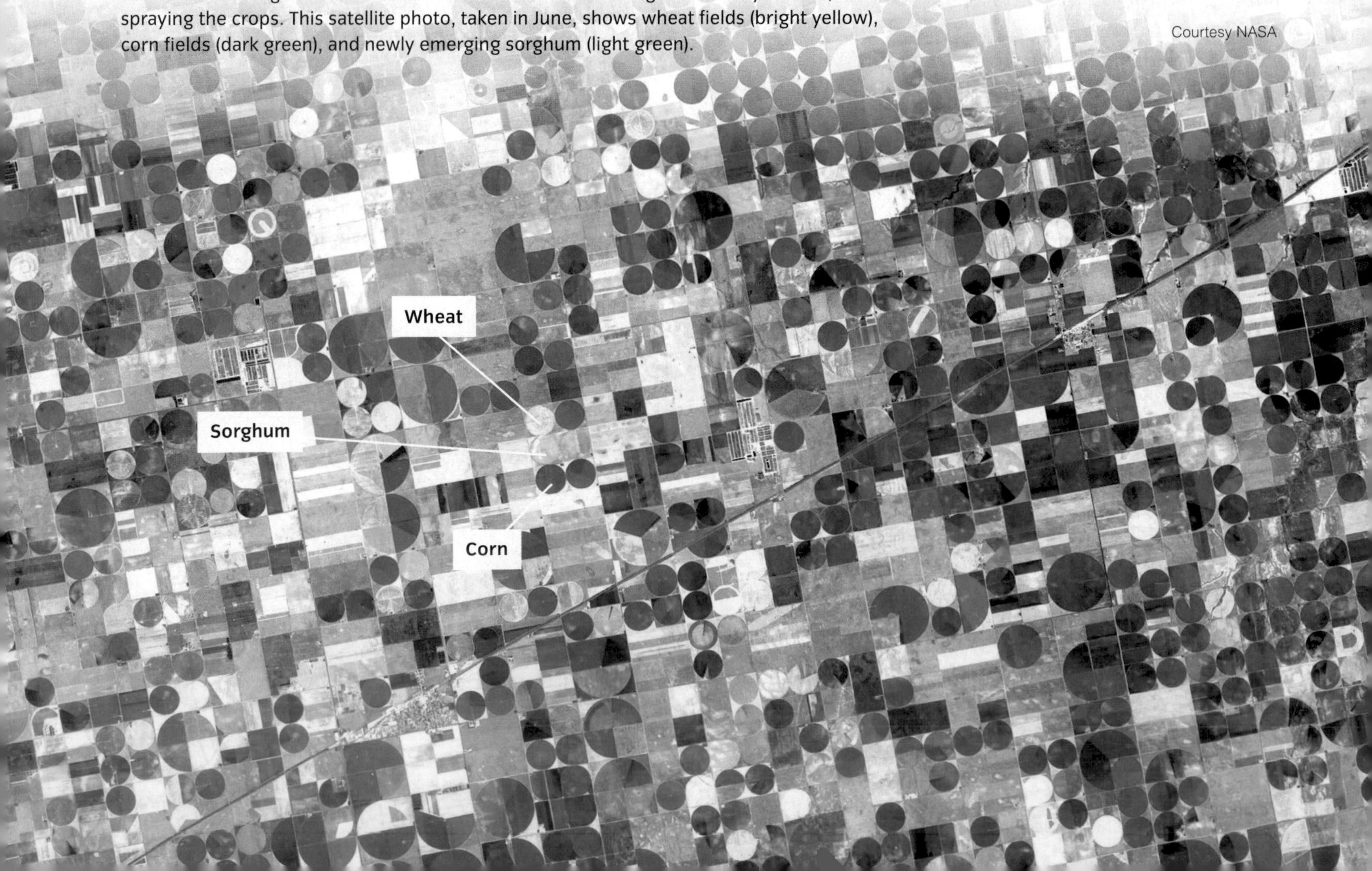

The Ogallala Aquifer The High Plains cover 6 percent of U.S. land but produce more than 15 percent of the nation's wheat, corn, sorghum, and cotton and almost 40 percent of its livestock. This productivity requires approximately 30 percent of the irrigation water used in the United States. Farmers on the High Plains rely on water from the **Ogallala Aquifer**, the largest groundwater deposit in the world (**Figure 10.8**).

In some areas farmers are drawing water from the Ogallala Aquifer as much as 40 times faster than nature replaces it. This rapid depletion has lowered the water table more than 30 m (100 ft) in some places. Most hydrologists (scientists who study water supplies) predict that groundwater will eventually drop in all areas of the Ogallala to a level uneconomical to pump. Their goal is to postpone that day through water conservation, including the use of water-saving irrigation systems.

Ogallala Aquifer • Figure 10.8

This massive deposit of groundwater lies under eight states, with extensive portions in Texas, Kansas, and Nebraska. Water in the Ogallala Aquifer takes hundreds or even thousands of years to renew after it is withdrawn to grow crops and raise cattle.

Courtesy of the U.S. Geological Survey.

Thickness of water layer
- 0.0 – 30.0 m
- 30.1 – 120.0 m
- 120.1 – 350.0 m

Overdrawing of Surface Waters

Removing too much fresh water from a river or lake can have disastrous consequences in local ecosystems. Growing human populations place demands on water sources that are not sustainable. In the arid American Southwest, it is not unusual for 70 percent or more of surface water to be removed.

When surface waters are overdrawn, wetlands dry up. **Estuaries**, where rivers empty into seawater, become saltier when surface waters are overdrawn, which reduces their productivity. Wetlands and estuaries, which serve as breeding grounds for many species of birds and other animals, also play a vital role in the hydrologic cycle. When these resources are depleted, the ensuing water shortages and reduced productivity have economic as well as ecological ramifications.

The increased use of U.S. surface water for agriculture, industry, and personal consumption since the 1960s has caused many water supply and quality problems. Some regions that have grown in population during this period—for example, California, Nevada, Arizona, Georgia (metropolitan Atlanta), and Florida—have placed correspondingly greater burdens on their water supplies. If water consumption in these and other areas continues to increase, regional problems with availability of surface waters will become more serious, even in places that have never experienced water shortages.

Nowhere in the country are water problems as severe as they are in the West and Southwest. Much of this large region is arid or semiarid, receiving less than 50 cm (20 in) of precipitation annually. With the rapid expansion of the population there during the past 25 years, municipal, commercial, and industrial uses now compete heavily with irrigation for available water. Much of the water used in the West and Southwest originates as snow in the Rocky Mountains and the Sierra Nevada; climate change appears to be causing reduced snowfall—and thus making less total water available for a growing population.

The Colorado River Basin One of the most serious water supply problems in the United States is in the Colorado River Basin. The river's headwaters are formed from snowmelt in Colorado, Utah, and Wyoming, and major tributaries—collectively called the upper Colorado—extend throughout these states. The lower Colorado River runs through part of Arizona and then along the border between Arizona and both Nevada and California before crossing into Mexico and emptying into the Gulf of California.

The Colorado River system provides water for more than 30 million people, including those in the cities of Denver, Las Vegas, Albuquerque, Phoenix, Los Angeles, and San Diego, with plans in Utah to divert Colorado River water to Salt Lake City. It irrigates 1.4 million hectares (3.5 million acres) of fruit, vegetable, and field crops worth $1.5 billion per year. The Colorado River has 49 dams, 11 of which produce electricity by hydropower. The river produces $1.25 billion per year in revenues from the recreation industry.

An international agreement with Mexico, along with federal and state laws, severely restricts the use of the Colorado's waters. The most important of all the treaties regulating use of Colorado River water is the 1922 **Colorado River Compact**, which stipulates an annual allotment of 7.5 million acre-feet of water to the lower Colorado (California, Nevada, Arizona, and New Mexico) and the remainder to the upper Colorado (Colorado, Utah, and Wyoming). Each acre-foot equals 326,000 gal (1.2 million liters), enough for about eight people for 1 year. However, the Colorado River Compact overestimated the average annual flow of the Colorado River, and it locked that estimate into the multistate agreement. Mexico also receives a share of the Colorado, as stipulated by a 1944 treaty.

Population growth in the upper Colorado region exacerbates the heavy demand already placed on the river by states through which the lower Colorado flows, particularly California. Consequently, the Colorado River water is often completely consumed before it can reach the Pacific Ocean in Mexico, causing serious problems for the ecosystem and inhabitants of the Colorado River delta (**Figure 10.9**). To compound the problem, as more and more water is used, the lower Colorado becomes increasingly salty—in some places saltier than the ocean—as it flows toward Mexico.

In 2003 California agreed to limit its water withdrawals from the Colorado River to quantities specified in the Colorado River Compact. Also, some California farmers agreed to sell some water they would normally use for irrigation and use the money earned to update their irrigation systems so that they would make more efficient use of their water.

Salinization of Irrigated Soil

Although irrigation improves the agricultural productivity of arid and semiarid lands, it often causes salt to accumulate in the soil, a phenomenon called **salinization**. Irrigation water contains small amounts of dissolved salts. Normally, through precipitation runoff, rivers carry away salt. Irrigation water, however, normally soaks into the soil and does

> **salinization** The gradual accumulation of salt in soil, often as a result of improper irrigation methods.

not run off into rivers. The continued application of such water, season after season, year after year, leads to the gradual accumulation of salt in the soil. Given enough time, the salt concentration can rise to such a high level that plants are poisoned or their roots become dehydrated. Thus, salt hurts soil productivity and, in extreme cases, renders soil unfit for crop production.

Colorado River Delta at the Gulf of California • Figure 10.9

As a result of diversion for irrigation and other uses in the United States, the Colorado River often dries up before reaching the Gulf of California in Mexico.

© Pete Mcbride/National Geographic Society/Corbis

The satellite images show the Aral Sea in 1976 and 2011. As water was diverted for irrigation, the sea level subsided.

1976

NASA/USGS

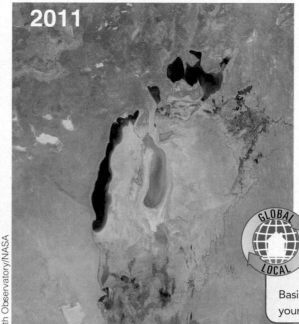

2011

Earth Observatory/NASA

Global Locator

ARAL SEA

ASIA

NG Maps

NATIONAL GEOGRAPHIC

How do water resource conflicts associated with the Aral Sea compare to those involving the Colorado River Basin? to water resource issues in your state or region?

Global Water Issues

As the world's population continues to increase, global water problems are becoming more serious. Earth's people and its water resources are often not concentrated in the same places. In India, where approximately 20 percent of the world's population has access to 4 percent of the world's fresh water, approximately 8000 villages have no local water supply. Water supplies are precarious in much of China, due to population pressures. In many parts of the country, water table levels are dropping; one-third of the wells in Beijing have gone dry. Much of the water in the Yellow River, one of China's main water basins, is diverted for irrigation, depriving downstream areas of water. Mexico is facing an unprecedented drought. The main aquifer supplying Mexico City is dropping rapidly, and the water table is falling fast in Guanajuato, an agricultural state. As of 2012, an estimated 2 million Mexicans lacked access to water.

As the needs of the growing human population deplete freshwater supplies, less water will be available for crops. Local famines often arise from water shortages. In early 2012, a looming drought in West Africa had already cut grain supplies by half in Chad and Mauritania, raising food prices and stimulating efforts to develop long-range approaches to drought relief.

Sharing Water Resources Among Countries In the 1950s, the then Soviet Union began diverting water that feeds into the Aral Sea to irrigate nearby desert areas. Over five decades, the Aral Sea all but disappeared (**Figure 10.10**); its total volume dropped 80 percent, and much of its biological diversity vanished. Millions of people living in the Aral Sea's watershed have developed serious health problems, probably due in part to storms lifting into the air toxic salts from the receding shoreline.

Following the breakup of the Soviet Union in 1991, responsibility for saving the Aral Sea shifted to the five Asian countries that share the Aral basin—Uzbekistan, Kazakstan, Kyrgyzstan, Turkmenistan, and Tajikistan. These nations' cooperative restoration efforts were backed by the World Bank and the U.N. Environment Program. At this time, recovery of the Aral Sea is mixed. Due in part to dam construction, the Northern Aral Sea experienced more than a 30 percent increase in surface area between 2003 and 2010 (though water levels dwindled somewhat in 2011), and salinity levels have been cut in half. The Southern Aral Sea, however, has shown little improvement over the past 20 years; water flow there is restricted by the dam. Like the Aral Sea, many of Earth's other watersheds cross political boundaries and face management issues associated with their shared use; water availability varies greatly worldwide (**Figure 10.11**).

The World's 10 Largest Watersheds

Watershed	Region	Area of watershed (thousand km²)
Amazon	South America	6145
Congo	Africa	3731
Nile	Africa	3255
Mississippi	North America	3202
Ob	Asia	2972
Paraná	South America	2583
Yenisey	Asia	2554
Lena	Asia	2307
Niger	Africa	2262
Yangtze	Asia	1722

Source: Water Resources of the World, World Resources Institute (2010).

PRIMARY WATERSHEDS AND CRITICAL AREAS

Earth's bodies of fresh water can cover enormous areas and cross many political boundaries. Watersheds are Earth's rain barrels. They collect precipitation and filter it as they channel it to streams, rivers, lakes, and aquifers. Many watersheds are stressed by human activities.

Interpreting Data
Which continent appears to have the best access to fresh water? the poorest access?

ACCESS TO CLEAN FRESH WATER

Access to clean fresh water is critical for human health. Yet in many regions, potable water is becoming scarce because of heavy demands and pollution. Especially worrisome is the poisoning of aquifers—a primary source of water for nearly one-third of the world—by sewage, pesticides, and heavy metals.

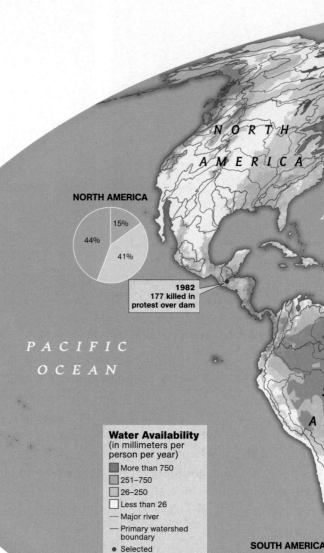

Water Availability
(in millimeters per person per year)
- More than 750
- 251–750
- 26–250
- Less than 26
- Major river
- Primary watershed boundary
- Selected water dispute

1982
177 killed in protest over dam

NORTH AMERICA
44% 15% 41%

SOUTH AMERICA
19% 13% 68%

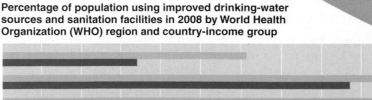

Percentage of population using improved drinking-water sources and sanitation facilities in 2008 by World Health Organization (WHO) region and country-income group

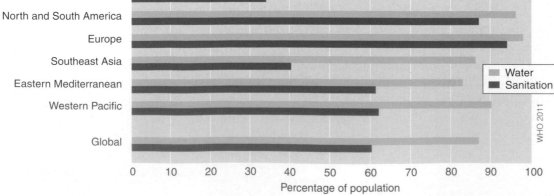

Percentage of population

Water
Sanitation

WHO 2011

WATER

It's as vital to life as air. Yet fresh water is one of the rarest resources on Earth. Only 2.5 percent of Earth's water is fresh, and of that, the usable portion for humans is less that 1 percent of all fresh water, or 0.01 percent of all water on Earth. Water is constantly recycling through Earth's hydrologic cycle. But population growth and pollution are combining to make less and less available per person per year, while global climate change adds new uncertainty. Availability of fresh water varies tremendously around the world, inevitably triggering conflicts over water resources.

ARCTIC OCEAN

Greenland

EUROPE

EUROPE

29%   16%

55%

ASIA

ASIA

9%   9%

82%

PACIFIC

OCEAN

2010
Pakistan irrigation dispute between tribes kills 116

2007
Water conflicts between herders and farmers in Burkina Faso

1978 onward
Egypt threatens Ethiopia over Nile plans

1999
Villagers killed in Yemen water clash

2007
Injuries at protest over allocation of water to industry in India

AFRICA

AFRICA

10%   4%

86%

2004–2006
250 killed in violence over water shortage in Ethiopia

2006
Sri Lankan rebels cut water supplies to villages

2005
90 killed in dispute over water rights in Kenya

INDIAN

OCEAN

AUSTRALIA

2007
Australian man murdered in fight over water restrictions

GLOBAL IRRIGATED AREAS AND WATER WITHDRAWALS
Since 1970, global water withdrawals have correlated with the rise in irrigated area. Some 70 percent of withdrawals are for agriculture, mostly for irrigation that helps produce 40 percent of the world's food.

OCEANIA

17%

10%

73%

ANTARCTICA

Freshwater Withdrawal
(as a percentage of total water utilization)

Agricultural   Domestic

Industrial

GLOBAL IRRIGATED AREAS AND WATER WITHDRAWALS
Since 1970 global water withdrawals have correlated with the rise in irrigated area. Some 70 percent of withdrawals are for agriculture, mostly for irrigation that helps produce 40 percent of the world's food.

Rhine River Basin • Figure 10.12

The Rhine River drains five European countries. (The green area represents the drainage basin.) Water management of such a river requires international cooperation.

Three-fourths of the world's 200 or so major watersheds are shared between at least two nations. International cooperation is required to manage rivers that cross international borders. The heavily populated drainage basin for the Rhine River in Europe spans five countries—Switzerland, Germany, France, Luxembourg, and the Netherlands (**Figure 10.12**). All five nations recognize that international cooperation is essential to conserve and protect the supply and quality of the Rhine River. Together they formed the International Commission for Protection of the Rhine, which in 1987 initiated a 15-year Rhine Action Programme. Their efforts have paid off: The main sources of pollution have been eliminated, and water in the Rhine River today is almost as pure as drinking water; long-absent fishes have returned; and projects are under way to restore riverbanks, control flooding, and clean up remaining pollutants.

CONCEPT CHECK STOP

1. **What** problems are associated with overdrawing surface water? with aquifer depletion?

2. **What** issues surround water problems of the Ogallala Aquifer? the Colorado River Basin?

3. **How** does international cooperation affect shared water resources?

Water Management

LEARNING OBJECTIVES

1. **Define** *sustainable water use.*

2. **Contrast** the benefits and drawbacks of dams and reservoirs.

3. **Give** examples of water conservation in agriculture, industry, and individual homes and buildings.

The main goal of water management is to provide a sustainable supply of high-quality water. **Sustainable water use** means careful human use of water resources so that water is available for future generations and for existing non-human needs.

Water supplies are obtained by building dams, diverting water, or removing salt from seawater or salty groundwater, through a process called *desalinization.* Conservation, which includes reusing water, recycling water, and improving water-use efficiency, augments water supplies and is an important aspect of sustainable water use. Economic policies are also important in managing water sustainably: When water is inexpensive, it tends to be wasted. Raising the price of water to reflect the actual cost generally promotes its more efficient use.

sustainable water use The wise use of water resources, without harming the essential functioning of the hydrologic cycle or the ecosystems on which present and future humans depend.

Dams and Reservoirs: Managing the Columbia River

Dams generate electricity and ensure a year-round supply of water in areas with seasonal precipitation or snowmelt, often for populations that have outgrown other water sources, but many people think their costs outweigh their benefits. In recent years scientists have come to understand how dams alter river ecosystems. Heavy sediment deposition can occur in the reservoir behind a dam, and the water that passes over a dam does not have its normal sediment load. As a result, the river floor downstream of a dam is scoured, producing a deep-cut channel that is a poor habitat for aquatic organisms.

The Columbia River, the fourth-largest river in North America, illustrates the impact of dams on natural fish communities. There are more than 100 dams in the Columbia River system, 19 of which are major generators of hydroelectric power (**Figure 10.13**). The Columbia River system supplies municipal and industrial water to several major urban areas in the northwestern United States and irrigation water for more than 1.2 million hectares (3 million acres) of agricultural land.

As is often the case in natural resource management, one particular use of the Columbia River system may have a negative impact on other uses. The dams that generate electricity and control floods have adversely affected fish populations. The salmon population in the Columbia River system is only a fraction of what it was before the watershed was developed. The many dams that impede salmon migrations are widely considered the most significant factor in salmon decline. Projects to rebuild salmon populations have not proved particularly successful.

To protect remaining natural salmon habitats, several streams in the Columbia River system are off-limits for dam development. Many dams have fish ladders to allow some of the adult salmon to bypass the dams and continue their upstream migration (**Figure 10.14**). Underwater

Grand Coulee Dam on the Columbia River • Figure 10.13

Shown are the dam and part of its reservoir, Lake Roosevelt. Dams provide electricity generation, flood control, and water recreation opportunities, but they disrupt or destroy natural river habitats and are expensive to build.

Courtesy U. S. Dept. of Energy

Fish ladder • Figure 10.14

This ladder is located at the Bonneville Dam on the Oregon side of the Columbia River. Fish ladders help migratory fishes to bypass dams in their migration upstream. Despite the installation of fish ladders, the salmon population remains low.

© Philip James Corwin/Corbis

screens and passages are being installed to steer young salmon (smolts) away from turbine blades, and at some sites the smolts are transported around dams.

Water Conservation

Today there is more competition than ever before among water users with different priorities (see pie charts on map in Figure 10.11), and water conservation measures are necessary to guarantee sufficient water supplies.

Reducing Agricultural Water Waste Irrigation generally makes inefficient use of water. Traditional irrigation methods involve flooding the land or diverting water to fields through open channels. Plants absorb only about 40 percent of the water that flood irrigation supplies to the soil; the rest of the water usually evaporates into the atmosphere, seeps into the ground, or leaves the fields as runoff transporting sediment.

One of the most important innovations in agricultural water conservation is **microirrigation**, also called **drip** or

| **microirrigation** |
| A type of irrigation that conserves water by piping it to crops through sealed systems. |

trickle irrigation, in which pipes with tiny holes bored in them convey water directly to individual plants (**Figure 10.15a**). Microirrigation substantially reduces the water needed to irrigate crops— usually by 40 percent to 60 percent compared to traditional irrigation—and also reduces the amount of salt that irrigation water leaves in the soil.

Other measures that can save irrigation water include using lasers to level fields, and employing computer-controlled technology to place hoses and time water release, all of which allow more even water distribution, and making greater use of recycled wastewater. A drawback of such techniques is their cost, which makes them unaffordable for most farmers in highly developed countries, let alone subsistence farmers in developing nations.

Reducing Water Waste in Industry Electric power generators and many industrial processes require water. In the United States, five major industries— chemical products, paper and pulp, petroleum and coal, primary metals, and food processing—consume almost 90 percent of industrial water.

Stricter pollution-control laws provide some incentive for industries to conserve water. Industries usually recapture, purify, and reuse water to reduce their water use and their water treatment costs.

The potential for industries to conserve water is enormous. In 2010, for example, Jackson Family Wines in California began implementing a water recycling system estimated to eventually save the winery up to 6 million gallons of water annually and to greatly reduce their energy usage. In northeast China, a methanol plant reduces its overall water consumption by trading investments in local irrigation and water conservation projects for agricultural water-use quotas (**Figure 10.15b**).

International companies also have to consider water issues where they locate plants. Ford Motor Company reduced its global water use by 62 percent between 2000 and 2010. During the same period, its plant in Mexico's Sonoran Desert doubled production while cutting water consumption by 40 percent. The company has also installed complex water treatment systems that allow for reuse of 65 percent of its wastewater at plants in India and China—countries facing great water demands.

Reducing Municipal Water Waste Like industries, regions and cities—and the households within them—recycle or reuse water to reduce consumption (**Figure 10.15c**). For example, homes and other buildings can be modified to collect and store gray water. *Gray water* is water that has already been used in sinks, showers, washing machines, and dishwashers. Gray water is recycled to flush toilets, wash cars, or sprinkle lawns. In contrast to water *recycling*, wastewater *reuse* occurs when water is collected and treated before being redistributed. The reclaimed water is generally used for irrigation.

Cities also decrease water consumption by providing consumer education, requiring water-saving household fixtures, developing economic incentives to save water, and repairing leaky water supply systems. Also, increasing the price of water to approach its true cost promotes water conservation.

The average person in the United States uses 295 L (78 gal) of water per day at home on indoor uses. As a water user, you have a responsibility to use water carefully and wisely. The cumulative effect of many people practicing personal water conservation measures has a significant impact on overall water consumption.

CONCEPT CHECK   **STOP**

1. **What** is *sustainable water use*?
2. **What** are the benefits of dams on the Columbia River? the drawbacks?
3. **How** can individuals conserve and manage water resources?

✓ THE PLANNER

Richard Nowitz/NG Image Collection

© Peng Zhaozhi/Xinhua Press/Corbis

a. Microirrigation. Close-up of a drip irrigation pipe system releasing water directly between young plants, eliminating much of the waste associated with traditional methods of irrigation. Photographed on an experimental farm in the Negev Desert, Israel.

b. Industrial Water Conservation. A technician at a methanol plant in Lingwu, China, operates a pump involved in water recycling. The plant, part of Shenhua Ningxia Coal Industry Group, reduces its water consumption in part by trading water conservation technology, such as renovating old irrigation facilities, for agricultural water quotas in the region.

Bathroom
Install water-saving shower and faucets and low-flush toilets. Or use a water-displacement device in the tank of a conventional toilet. Fix leaky fixtures.
Modify personal habits: Avoid leaving the faucet running while shaving or brushing teeth. Take shorter showers.

Kitchen
Use a dishwasher, with a full load. It requires less water than washing dishes by hand. Avoid peak usage times.

Laundry room
Choose a high-efficiency washing machine to use less water and spin more water out of the clothes.

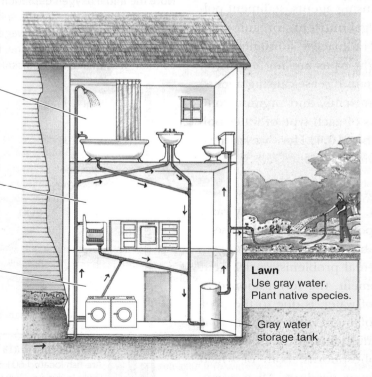

Lawn
Use gray water.
Plant native species.

Gray water storage tank

c. Conserving Water at Home. In your bathroom, kitchen, and laundry room, as well as on your lawn, you can take many steps to limit water use. Individual homes and buildings can be modified to collect and store "gray water"— water already used in sinks, showers, washing machines, and dishwashers—when clean water is not required: in flushing toilets, washing the car, and sprinkling the lawn, or, especially in the Southwest, irrigating golf courses. Permits to install gray water systems vary from state to state. Arizona and other states with severe water shortages are more flexible than other states about allowing gray water systems. Home owners can also conserve water used for landscaping by planting native species adapted to their local climate.

The Ocean and Fisheries

DEPLETING BLUEFIN TUNA STOCKS

Stocks of the giant, or Atlantic, bluefin tuna, highly prized for sushi, are classified as depleted in the Mediterranean Sea by the U.N. Food and Agriculture Organization. Once harvested sustainably through traditional trapping, Mediterranean bluefins are now fished—often illegally—at approximately four times the sustainable rate. Spotter aircraft locate fish stocks and alert huge fishing fleets, whose ships (see inset photograh) cast purse seines, which envelop schools of fish and cinch to close around them. Captured bluefins are fattened in offshore pens (see large photograph) before being butchered for market. The enormous economic value of the huge bluefin places it at great risk. Only recently have Mediterranean nations begun implementing conservation measures to protect the species, including a ban on the purse seine harvest of bluefins in Mediterranean and east Atlantic waters. But although the International Commission for the Conservation of Atlantic Tunas (ICCAT) places yearly catch limits—quotas—on the fishery, illegal harvests result in the total trade exceeding these quotas by approximately 30 percent (see graph). At these catch rates, ICCAT estimates that the Mediterranean bluefin stocks have less than a 24 percent chance of rebuilding by 2022. Conservation efforts have accelerated, however: In late 2011, ICCAT agreed to implement electronic documentation of catches, to help eliminate illegal trade.

Overfishing, the harvesting of fishes faster than they can reproduce, is not limited to the Mediterranean. Worldwide, more than 80 percent of fish species have been overfished, with some U.S. commercial fish stocks depleted by as much as 95 percent, as demand for fish has grown and harvesting methods have become more sophisticated. Ecologists and economists estimate that if overfishing and ocean pollution aren't curbed, populations of virtually all harvested seafood species could collapse by 2048.

Jose Cort/Courtesy NOAA

WileyPLUS

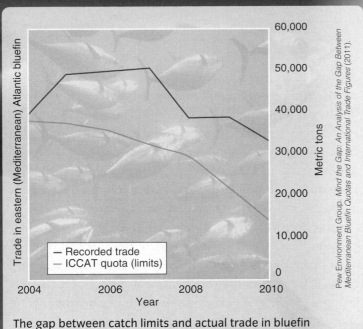

60,000

50,000

40,000

Metric tons

30,000

20,000

10,000

0

Trade in eastern (Mediterranean) Atlantic bluefin

— Recorded trade
— ICCAT quota (limits)

2004   2006   2008   2010
Year

Pew Environment Group. *Mind the Gap: An Analysis of the Gap Between Mediterranean Bluefin Quotas and International Trade Figures* (2011).

The gap between catch limits and actual trade in bluefin tuna in the Mediterranean.

Interpreting Data

In metric tons, what was the ICCAT quota in 2009? How much more bluefin tuna was actually traded that year?

CHAPTER OUTLINE

The Global Ocean 274
- Patterns of Circulation in the Ocean
- Environmental InSight: Ocean Currents
- Ocean–Atmosphere Interaction

Major Ocean Life Zones 278
- The Intertidal Zone: Transition Between Land and Ocean
- The Benthic Environment
- The Neritic Province: From the Shore to 200 Meters
- EnviroDiscovery: Otters in Trouble
- The Oceanic Province: Most of the Ocean

Human Impacts on the Ocean 284
- Marine Pollution and Deteriorating Habitat
- World Fisheries
- Environmental InSight: Human Impacts on the Ocean
- What a Scientist Sees: Modern Commercial Fishing Methods
- Shipping, Ocean Dumping, and Plastic Debris
- Coastal Development
- Human Impacts on Coral Reefs
- Offshore Extraction of Mineral and Energy Resources
- What a Scientist Sees: Ocean Warming and Coral Bleaching
- Climate Change, Sea-Level Rise, and Warmer Ocean Temperatures

Addressing Ocean Problems 291
- Future Actions
- Case Study: The Dead Zone in the Gulf of Mexico

CHAPTER PLANNER ✔

- ❏ Study the picture and read the opening story.
- ❏ Scan the Learning Objectives in each section:
  p. 274 ❏   p. 278 ❏   p. 284 ❏   p. 291 ❏
- ❏ Read the text and study all figures and visuals. Answer any questions.

Analyze key features

- ❏ Environmental InSight, p. 275 ❏   p. 285 ❏
- ❏ Process Diagram, p. 276
- ❏ EnviroDiscovery, p. 282
- ❏ What a Scientist Sees, p. 286 ❏   p. 289 ❏
- ❏ Case Study, p. 293
- ❏ Stop: Answer the Concept Checks before you go on:
  p. 278 ❏   p. 283 ❏   p. 290 ❏   p. 292 ❏

End of Chapter

- ❏ Review the Summary and Key Terms.
- ❏ Answer What is happening in this picture?
- ❏ Answer the Critical and Creative Thinking Questions.

Franco Banfi/Water Frame/Getty Images

The Global Ocean

LEARNING OBJECTIVES

1. **Describe** the global ocean and its significance to life on Earth.

2. **Discuss** the roles of winds and the Coriolis effect in producing global water flow patterns, including gyres.

3. **Define** *El Niño–Southern Oscillation (ENSO)* and *La Niña* and describe some of their effects.

T he ocean is a vast wilderness, much of it unknown. It teems with life—from warm-blooded mammals such as whales to soft-bodied invertebrates such as jellyfish. The ocean is essential to Earth's hydrologic cycle, which provides us with water. It affects cycles of matter on land, influences our climate and weather, and provides foods that enable millions of people to survive. The ocean dominates Earth, and its condition determines the future of life on our planet. If the ocean dies, then we do as well. Yet we lack a full understanding of many oceanic processes—there remains much for us to discover.

The global ocean is a huge body of salt water that surrounds the continents and covers almost three-fourths of Earth's surface. It is a single, continuous body of water, but geographers divide it into four sections separated by the continents: the Pacific, Atlantic, Indian, and Arctic oceans. The Pacific is the largest: It covers one-third of Earth's surface and contains more than half of Earth's water.

Patterns of Circulation in the Ocean

The persistent prevailing winds blowing over the ocean produce currents, mass movements of surface–ocean water (**Figure 11.1a**). The prevailing winds generate **gyres**, circular ocean currents. In the North Atlantic

> **gyres** Large, circular ocean current systems that often encompass an entire ocean basin.

Ocean, the tropical trade winds tend to blow toward the west, whereas the westerlies in the mid-latitudes blow toward the east. This helps establish a clockwise gyre in the North Atlantic. That is, the trade winds produce the westward North Atlantic Equatorial Current in the tropical North Atlantic Ocean. When this current reaches the North American

continent, it is deflected northward, where the westerlies begin to influence it. As a result, the current flows eastward in the mid-latitudes until it reaches the landmass of Europe. Here some water is deflected toward the pole and some toward the equator. The water flowing toward the equator comes under the influence of trade winds again, producing the circular gyre. Although surface–ocean currents and winds tend to move in the same direction, there are many variations to this general rule.

The **Coriolis effect** influences the paths of ocean currents just as it does the winds (see Figure 8.5). Earth's rotation from west to east causes surface ocean currents to swerve to the right in the Northern Hemisphere, helping establish the circular, clockwise pattern of water currents. In the Southern Hemisphere, ocean currents swerve to the left, thereby moving in a circular, counterclockwise pattern.

Vertical Mixing of Ocean Water Variations in the **density** (mass per unit volume) of seawater—caused by wind–driven temperature differences between water layers—affect deep-ocean currents. Cold, salty water is denser than warmer, less salty water. (The density of water increases with decreasing temperature down to 4°C.) Through the **ocean meridional circulation**, colder, salty ocean water sinks and flows under warmer, less salty water, generating currents far below the surface. Deep-ocean currents often travel in different directions and at different speeds than do surface currents, in part because the Coriolis effect is more pronounced at greater depths. **Figure 11.1b** shows the present global circulation of shallow and deep currents—the **ocean conveyor belt**—that transfers heat and salt, moving cold, salty deep-sea water from higher to lower latitudes, where it warms up. Note that the Atlantic Ocean gets its cold deep water from the Arctic Ocean, whereas the Pacific Ocean and Indian Ocean get theirs from the water surrounding Antarctica.

The ocean conveyer belt affects regional and possibly global climate. As the Gulf Stream and North Atlantic Drift push into the North Atlantic, they deliver an immense amount of heat from the tropics to Europe (**Figure 11.1c**). As this shallow current transfers its heat to the atmosphere, the water becomes denser and

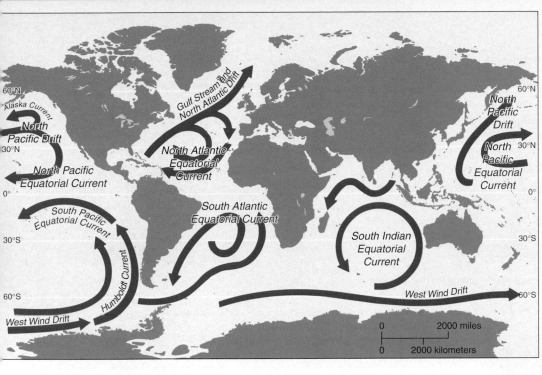

a. Surface–Ocean Currents. Winds largely cause the basic pattern of ocean currents. The main ocean current flow—clockwise in the Northern Hemisphere and counterclockwise in the Southern Hemisphere—results partly from the Coriolis effect.

Adapted from Figure 12.5A on p. 347 in Murck, B.W., B.J. Skinner, and D. Mackenzie. *Visualizing Geology*, Hoboken NJ: John Wiley and Sons, Inc. (2008).

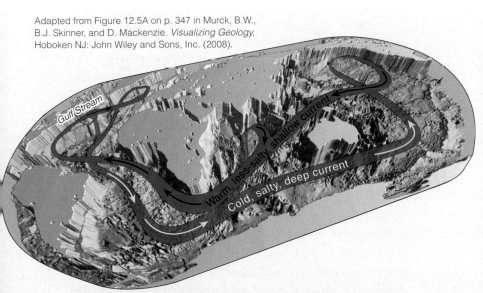

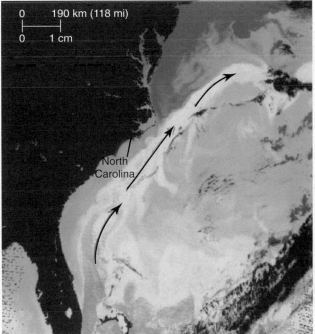

Courtesy NASA

b. The Ocean Conveyor Belt. This loop consists of deep-ocean currents that flow in the opposite direction from surface currents, transferring heat and salt. Vertical motions associated with the meridional overturning circulation drive the conveyor: Cold, salty water near Antarctica and the Arctic Ocean sinks and eventually flows northward into the Pacific Ocean, where it wells up, eventually becoming warmer and fresher. Cold, salty, deep water in the Atlantic Ocean comes from the Arctic Ocean. The ocean conveyor belt affects regional and global climate.

c. The Gulf Stream. The Gulf Stream is a well-known regional link in the ocean conveyor belt. In this satellite image, the colors represent the water's surface temperature: red = warmest, and blue = coolest. The Gulf Stream flows northeast along the North Carolina coast and then out to sea, toward Europe.

Think Critically Which currents appear to have the most effect on the coasts of North America?

PROCESS DIAGRAM

El Niño–Southern Oscillation (ENSO) • Figure 11.2

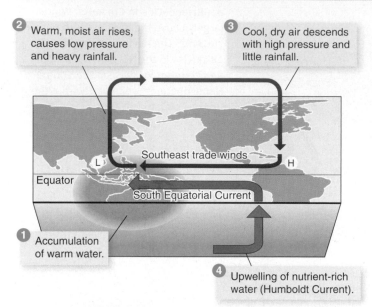

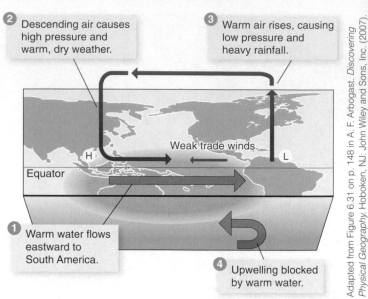

Adapted from Figure 6.31 on p. 148 in A. F. Arbogast. *Discovering Physical Geography.* Hoboken, NJ: John Wiley and Sons, Inc. (2007).

a. Normal climate conditions
ENSO events depend on the relationship of atmospheric circulation to surface water flow in the Pacific. Normal conditions occur when strong easterly flow pushes warm water into the western Pacific.

Think Critically What are the differences in air circulation patterns across the Arctic Ocean and over North America and Asia during normal climate conditions and ENSO conditions?

b. ENSO conditions
An ENSO event occurs when easterly flow weakens, allowing warm water to collect along the South American coast. Note the relationship between precipitation and the location of pressure systems. During an ENSO event, northern areas of the contiguous United States are typically warmer during winter, whereas southern areas are cooler and wetter.

WileyPLUS

sinks. The deep current flowing southward in the North Atlantic is, on average, 8°C (14.4°F) cooler than the shallow current flowing northward.

Scientific evidence indicates that the ocean conveyor belt shifts from one equilibrium state to another. Historically, these shifts are linked to major changes in global climate.

Ocean–Atmosphere Interaction

The ocean and the atmosphere are strongly linked, with wind from the atmosphere affecting the ocean currents and heat from the ocean affecting atmospheric circulation. One of the best examples of the interaction between ocean and atmosphere is the **El Niño–Southern Oscillation (ENSO)** event, which is responsible for much of Earth's interannual (from one year to the next) climate variability. As a result of ENSO, some areas are drier, some wetter,

El Niño–Southern Oscillation (ENSO)
A periodic, large-scale warming of surface waters of the tropical eastern Pacific Ocean that temporarily alters both ocean and atmospheric circulation patterns.

some cooler, and some warmer than usual. Normally, westward-blowing trade winds restrict the warmest waters to the western Pacific near Australia (**Figure 11.2a**). Every three to seven years, however, the trade winds weaken, and the warm mass of water expands eastward to South America, increasing surface temperatures in the usually cooler east Pacific (**Figure 11.2b**). Ocean currents, which normally flow westward in this area, slow down, stop altogether, or even reverse and go eastward.

The name for this phenomenon, El Niño (in Spanish, "the boy child"), refers to the Christ child: The warming usually reaches the fishing grounds off Peru just before Christmas. Most ENSOs last between one and two years.

ENSO can devastate the fisheries off South America. Normally, the colder, nutrient-rich deep water is about 40 m (130 ft) below the surface and **upwells** (comes to the surface) along the coast, partly in response

276 CHAPTER 11 The Ocean and Fisheries

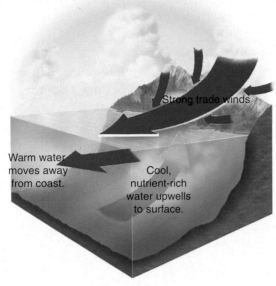

a. Coastal upwelling, where deeper waters come to the surface, occurs in the Pacific Ocean along the South American coast. Upwelling provides nutrients for microscopic algae, which in turn support a complex food web.

b. Coastal upwelling weakens considerably during years with El Niño–Southern Oscillation (ENSO) events, temporarily reducing fish populations.

to strong trade winds (**Figure 11.3a**). During an ENSO event, however, the colder, nutrient-rich deep water is about 150 m (490 ft) below the surface, and the warmer surface temperatures and weak trade winds prevent upwelling (**Figure 11.3b**). The lack of nutrients in the water results in a severe decrease in the populations of anchovies and many other marine fishes. During the 1982–1983 El Niño, one of the most damaging to fish populations, the anchovy population decreased 99 percent. Other species, such as shrimp and scallops, thrive during an ENSO event.

ENSO alters global air currents, directing unusual, sometimes dangerous, weather to areas far from the tropical Pacific where it originates. By one estimate, the 1997–1998 ENSO, the strongest on record, caused more than 20,000 deaths and $33 billion in property damage worldwide. It resulted in heavy snows in parts of the western United States; ice storms in eastern Canada; torrential rains that flooded Peru, Ecuador, California, Arizona, and western Europe; and droughts in Texas, Australia, and Indonesia. An ENSO-caused drought—the worst in 50 years—particularly hurt Indonesia. Fires, many deliberately set to clear land for agriculture, got out of control and burned an area in Indonesia the size of New Jersey.

Climate scientists observe and monitor sea surface temperatures and winds to better understand and predict the timing and severity of ENSO events. The **TAO/TRITON array** consists of 70 moored buoys in the tropical Pacific Ocean. These instruments collect oceanic and weather data during normal conditions and El Niño events. The data are transmitted to scientists onshore by satellite.

Scientists at the National Oceanic and Atmospheric Administration's Climate Prediction Center forecasted the 1997–1998 ENSO six months in advance, using data from TAO/TRITON. Such forecasts give governments time to prepare for the extreme weather changes associated with ENSO.

La Niña El Niño isn't the only periodic ocean temperature event to affect the tropical Pacific Ocean. **La Niña** (in Spanish, "the girl child") occurs when the surface water temperature in the eastern Pacific Ocean becomes unusually cool and westbound trade winds become unusually strong. La Niña often, but not always, occurs after an El Niño event and is considered part of the natural oscillation of ocean temperature.

During the spring of 1998, the surface water of the eastern Pacific cooled 6.7°C (12°F) in just 20 days.

Like ENSO, La Niña affects weather patterns around the world, but its effects are more difficult to predict. In the contiguous United States, La Niña typically causes wetter-than-usual winters in the Pacific Northwest, warmer weather in the Southeast, and drought conditions in the Southwest. Drought and a warmer-than-average winter in much of the United States in 2011–2012 are attributed in part to La Niña's effects. Atlantic hurricanes are stronger and more numerous than usual during a La Niña event.

Scientists have recently determined that flu pandemics are more likely to occur after La Niña events. They are considering a link between the two phenomena, suspecting that weather shifts during La Niña might alter migratory patterns of flu-carrying birds, including their stopovers and interactions with other species.

CONCEPT CHECK

1. **What** is the global ocean, and how does it affect Earth's environment?

2. **How** are the Coriolis effect, prevailing winds, and surface–ocean currents related?

3. **What** is the El Niño–Southern Oscillation (ENSO)? What are some of its global effects?

Major Ocean Life Zones

LEARNING OBJECTIVE

1. **Describe** and distinguish among the four main ocean life zones.

The immense marine environment is subdivided into several zones (**Figure 11.4**):

- The intertidal zone (between low and high tides)

- The benthic (ocean floor) environment

- The two provinces—neritic and oceanic—of the pelagic (ocean water) environment

The *neritic province* is that part of the pelagic environment from the shore to where the water reaches a depth of 200 m (650 ft). It overlies the continental shelf. The *oceanic province* is that part of the pelagic environment where the water depth is greater than 200 m, beyond the continental shelf.

The Intertidal Zone: Transition Between Land and Ocean

Although high levels of light, nutrients, and oxygen make the intertidal zone a biologically productive habitat, it is a stressful one. On sandy intertidal beaches, inhabitants must contend with a constantly shifting environment that threatens to engulf them and gives them little protection against wave action. Consequently, most sand-dwelling organisms are active burrowers. They usually lack adaptations to survive drying out or exposure because they follow the tides up and down the beach.

Rocky shores provide fine anchorage for seaweeds and marine animals, but these organisms are exposed to wave action when submerged during high tides and exposed to temperature changes and drying out when in contact with the air during low tides (**Figure 11.5**).

A rocky-shore inhabitant generally has some way of sealing in moisture, perhaps by closing its shell (if it has one), and a means of anchoring itself to the rocks. For example, mussels have tough, threadlike anchors secreted by a gland in the foot, and barnacles secrete a tightly bonding glue that hardens underwater. Rocky-shore intertidal algae usually have thick, gummy coats, which dry out slowly when exposed to air, and flexible bodies not easily broken by wave action. Some organisms hide in burrows or under rocks or crevices at low tide. Some small crabs run about the splash line, following it up and down the beach.

The Benthic Environment

Most of the benthic environment consists of sediments (mainly sand and mud) where many

> **intertidal zone** The area of shoreline between low and high tides.
>
> **benthic environment** The ocean floor, which extends from the intertidal zone to the deep-ocean trenches.

Zonation in the ocean • Figure 11.4

The intertidal zone, the benthic environment, and the pelagic environment make up the ocean. The pelagic environment consists of the neritic and oceanic provinces. (The slopes of the ocean floor aren't as steep as shown; they are exaggerated here to save space.)

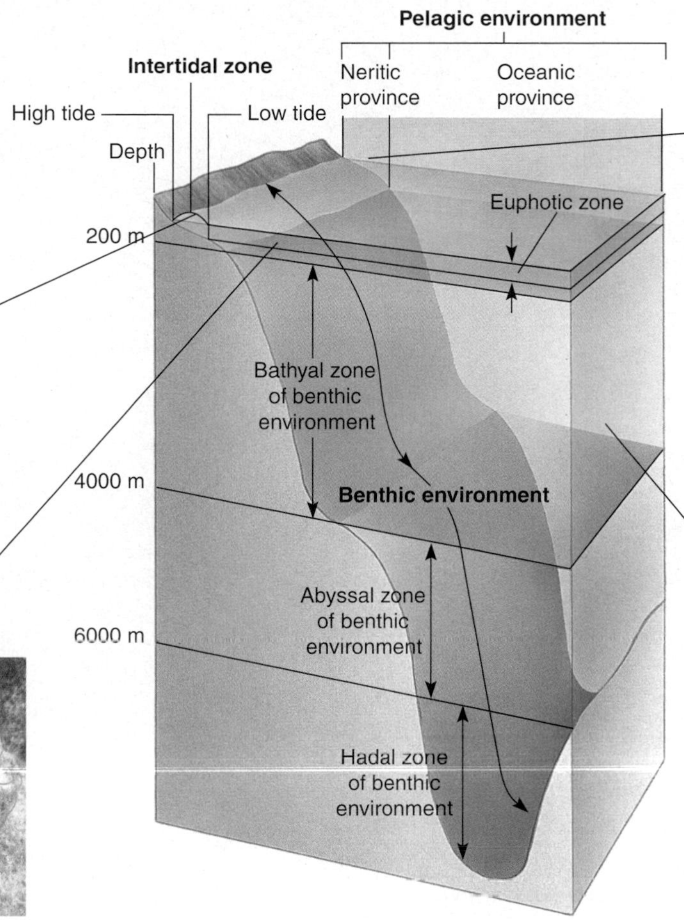

Intertidal zone:

Bill Curtsinger/NG Image Collection

Rockweed (brown algae)

Shallow benthic environment:

Mary Beth Ange o/Science Source

Rough file clam

Pelagic environment

Intertidal zone

Neritic province | Oceanic province

High tide — Low tide

Depth

200 m

Euphotic zone

Bathyal zone of benthic environment

4000 m

Benthic environment

6000 m

Abyssal zone of benthic environment

Hadal zone of benthic environment

Neritic province:

Tom Brakefield/Stockbyte/Getty Images

Bottlenose dolphins

Oceanic province:

Dr Paul A Zahl/Photo Researchers/Getty Images

Saber-toothed viperfish

Zonation along a rocky shore • Figure 11.5

Tide zones	Community zonation pattern of rocky shores

Anemones, tube worms, hermit crabs

Supratidal (splash) zone

Cyanobacteria, sea hair (*Ulothrix*), rough periwinkles

Tide pool

Level of highest tide

Intertidal zone

Acorn barnacles, rock barnacles, mussels, limpets, periwinkles, oysters, brown algae

Level of lowest tide

Subtidal zone

Brown algae, mussels, sea stars, brittle stars, sea urchins, spider crabs

Three zones are shown: the supratidal, or "splash" zone, which is never fully submerged; the intertidal zone, which is fully submerged at high tide; and the subtidal zone (part of the benthic environment), which is always submerged. Representative organisms are listed for each of these zones.

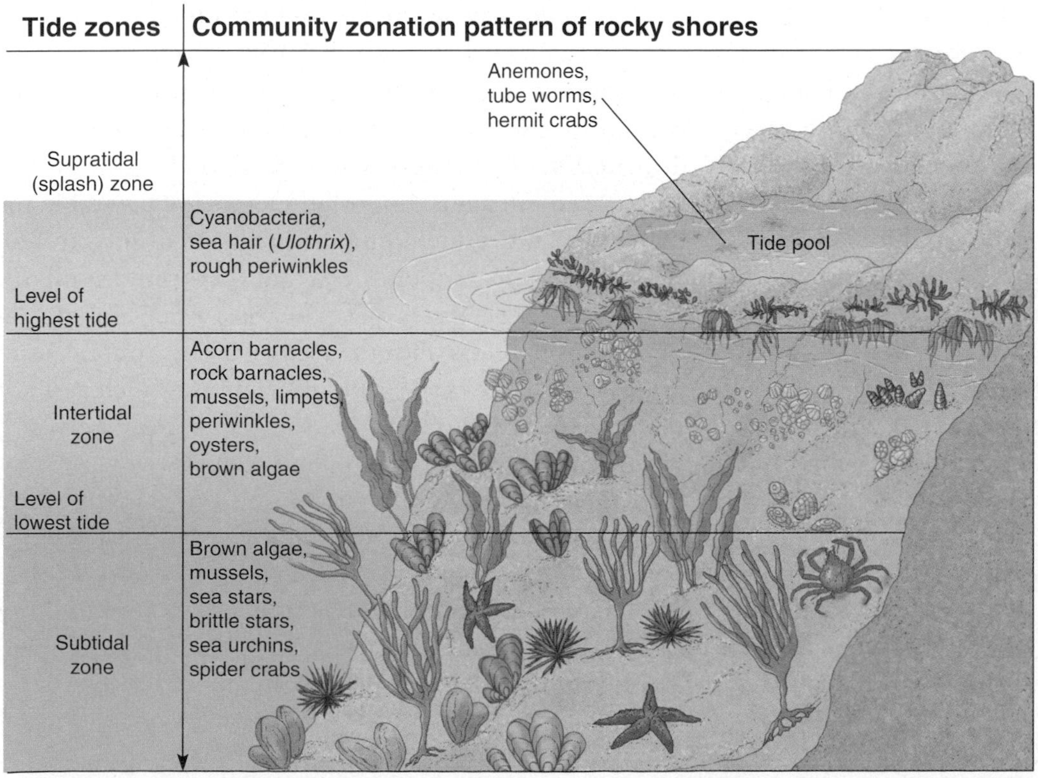

Adapted from Figure 14.1 in Karleskint, G. *Introduction to Marine Biology. Philadelphia*: Harcourt College Publishers (1998).

279

Coral reefs • Figure 11.6

Adapted from Marangos, J. E., M. P. Crosby, and J. W. McManus. "Coral reefs and biodiversity: a critical and threatened relationship." *Oceanography, Vol. 9* (1996).

Tim Laman/NG Image Collection

b. A coral reef in Fiji has a variety of soft corals as well as several fish species.

a. This map shows the distribution of coral reefs around the world. There are more than 6000 of them worldwide.

bottom-dwelling animals, such as worms and clams, burrow. Bacteria are common in marine sediments, found even at depths more than 500 m (1625 ft) below the ocean floor. The deeper parts of the benthic environment are divided into three zones, from shallowest to deepest: the bathyal, abyssal, and hadal zones. The communities in the relatively shallow benthic zone that are particularly productive include coral reefs, sea grass beds, and kelp forests.

Corals are small, soft-bodied animals similar to jellyfish and sea anemones. Corals live in hard cups, or shells, of limestone (calcium carbonate) that they produce using the minerals dissolved in ocean water. When the coral animals die, the tiny cups remain, and a new generation of coral animals grows on top of these. Over thousands of generations, a **coral reef** forms from the accumulated layers of limestone. Most coral reefs consist of colonies of millions of individual corals.

Coral reefs are found in warm (usually greater than 21°C [70°F]), shallow seawater (**Figure 11.6a**). The living portions of coral reefs grow in shallow waters where light penetrates. The tiny coral animals require light

for **zooxanthellae** (symbiotic algae) that live and photosynthesize in their tissues. In addition to obtaining food from the zooxanthellae that live inside them, coral animals capture food at night with stinging tentacles that paralyze **plankton** (small or microscopic organisms carried by currents and waves) and small animals that drift nearby. The waters where coral reefs grow are often poor in nutrients, but other factors are favorable for high productivity, including the presence of zooxanthellae, appropriate temperatures, and year-round sunlight.

Coral reef ecosystems are the most diverse of all marine environments (**Figure 11.6b**). They contain hundreds of species of fishes and invertebrates, such as giant clams, snails, sea urchins, sea stars, sponges, flatworms, brittle stars, sea fans, shrimp, and spiny lobsters. Australia's Great Barrier Reef occupies only 0.1 percent of the ocean's surface, but 8 percent of the world's fish species live there. The multitude of relationships and interactions that occur at coral reefs is comparable only to those of the tropical rain forest. As in the rain forest, competition is intense, particularly for light and space to grow.

Sea grass bed • Figure 11.7

Turtle grasses form underwater meadows that are ecologically important for shelter and food for many organisms. Photographed in the Caribbean Sea, the Cayman Islands.

Raul Touzon /NG Image Collection

Kelp forest • Figure 11.8

Underwater kelp forests are ecologically important because they support many kinds of aquatic organisms. Photographed off the coast of California.

Tim Laman/NG Image Collection

Coral reefs are ecologically important because they both provide habitat for many kinds of marine organisms and protect coastlines from shoreline erosion. They provide humans with seafood, pharmaceuticals, and recreation and tourism dollars.

Sea grasses are flowering plants adapted to complete submersion in salty ocean water. They occur only in shallow water (to depths of 10 m, or 33 ft) where they receive enough light to photosynthesize efficiently. Extensive beds of sea grasses occur in quiet temperate, subtropical, and tropical waters. Eelgrass is the most widely distributed sea grass along the coasts of North America; the world's largest eelgrass bed is in Izembek Lagoon on the Alaska Peninsula. The most common sea grasses in the Caribbean Sea are manatee grass and turtle grass (**Figure 11.7**).

Sea grasses have a high primary productivity and are ecologically important in shallow marine areas. Their roots and rhizomes help stabilize sediments, reducing erosion, and they provide food and habitat for many marine organisms. In temperate waters, ducks and geese eat sea grasses, and in tropical waters, manatees, green turtles, parrot fish, sturgeon fish, and sea urchins eat them. These herbivores consume only about 5 percent of sea grasses. The remaining 95 percent eventually enters the detritus food web and is

decomposed when the sea grasses die. The decomposing bacteria are in turn consumed by animals such as mud shrimp, lugworms, and mullet (a type of fish).

Kelps, known to reach lengths of 60 m (200 ft), are the largest and most complex of all algae commonly called seaweeds (**Figure 11.8**). Kelps, which are brown algae, are common in cooler temperate marine waters of both the Northern and Southern Hemispheres. They are especially abundant in relatively shallow waters (depths of about 25 m, or 82 ft) along rocky coastlines. Kelps are photosynthetic and are the primary food producers for the kelp "forest" ecosystem. Kelp forests provide habitats for many marine animals. Tube worms, sponges, sea cucumbers, clams, crabs, fishes, and sea otters find refuge in the algal fronds. Some animals eat the fronds, but kelps are mainly consumed in the detritus food web. Bacteria that decompose kelp provide food for sponges, tunicates, worms, clams, and snails. The diversity of life supported by kelp beds almost rivals that found in coral reefs.

neritic province
The part of the pelagic environment that overlies the ocean floor from the shoreline to a depth of 200 m (650 ft).

The Neritic Province: From the Shore to 200 Meters

The two main divisions of the pelagic environment are the neritic and oceanic provinces. Organisms that live in the neritic province are

EnviroDiscovery
Otters in Trouble

Sea otters play an important role in their environment. They feed on sea urchins, thereby preventing the urchins from eating kelp, which allows kelp forests to thrive. Now scientists have uncovered an alarming decline in sea otter populations in western Alaska's Aleutian Islands—a stunning 90 percent crash since 1990—that in turn poses wide-ranging threats to the coastal ecosystem there. The population of sea urchins in these areas is exploding, and kelp forests are being devastated. Strong evidence identifies killer whales, or orcas, as the culprits. Orcas generally feed on sea lions, seals, and fishes of all sizes. Sea otters, the smallest marine mammal species, are more like a snack to the orca than a desirable meal. So why are the orcas now choosing sea otters? Biologists suggest that it is because seal and sea lion populations have collapsed across the north Pacific.

In a scenario that is partly documented and partly speculative, the starting point of this disastrous chain of events is a drop in fish stocks, possibly caused by overfishing or climate change. With their food fish in decline, seal and sea lion populations have suffered, and orcas have looked elsewhere for food. Even the terrestrial food chain has been affected, as bald eagles shift away from fish and baby otter prey where otters are scarce. The change in the orcas' feeding behavior has transformed the food chain of kelp forests, with orcas disrupting the otters' role as predators. In 2009, the U.S. Fish and Wildlife Service designated more than 15,000 km^2 (nearly 6000 mi^2) of the Aleutian Islands as critical habitat for the threatened sea otter. The areas are near shore and in kelp beds, where they might offer protection from predators.

Otters in Alaskan waters

---

all floaters or swimmers. The upper level of the pelagic environment is the **euphotic zone**, which extends from the surface to a maximum depth of 150 m (490 ft) in the clearest open ocean water (see Figure 11.4). Sufficient light penetrates the euphotic zone to support photosynthesis.

Large numbers of **phytoplankton** (microscopic algae) produce food by photosynthesis and are the base of food webs. **Zooplankton**, including tiny crustaceans, jellyfish, comb jellies, and the larvae of barnacles, sea urchins, worms, and crabs, feed on phytoplankton in the euphotic zone. Zooplankton are consumed by plankton-eating **nekton** (any marine organism that swims freely), such as herring, sardines, baleen whales, manta rays, and squid (**Figure 11.9**). These in turn become prey for carnivorous nekton such as sharks, tuna, porpoises, and toothed whales. Nekton are mostly confined to the shallower neritic waters (less than 60 m, or 195 ft, deep), near their food.

Neritic province • Figure 11.9

The opalescent squid—here photographed at night—is abundant in the eastern Pacific, particularly off the coast of California, where individuals gather by the thousands to breed. These animals are active predators of planktonic crustaceans and small fish.

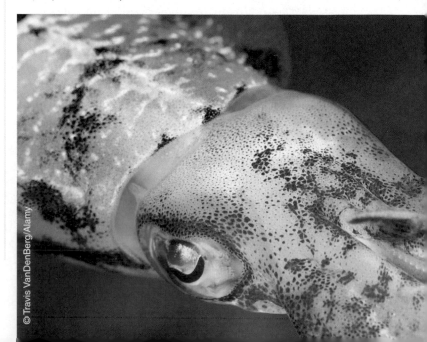

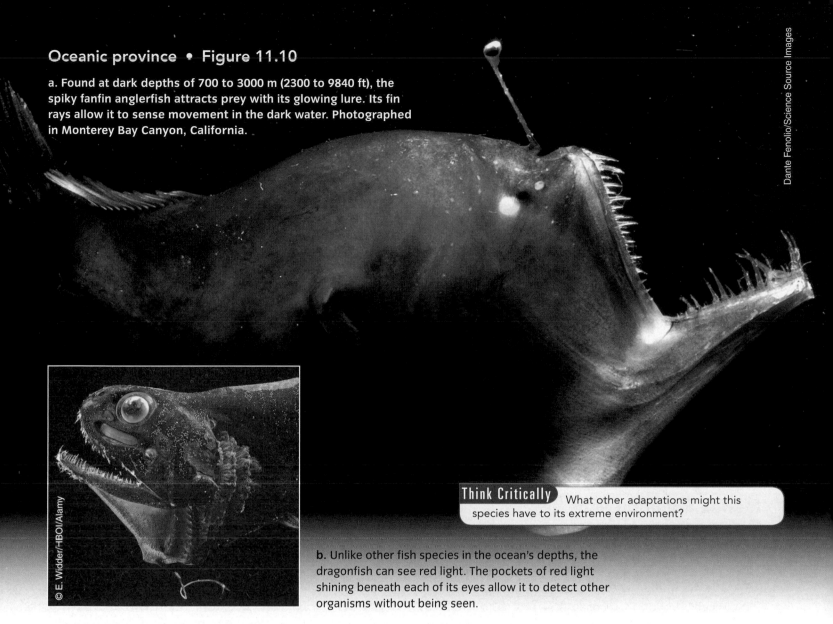

a. Found at dark depths of 700 to 3000 m (2300 to 9840 ft), the spiky fanfin anglerfish attracts prey with its glowing lure. Its fin rays allow it to sense movement in the dark water. Photographed in Monterey Bay Canyon, California.

Dante Fenolio/Science Source Images

© E. Widder/HBOI/Alamy

Think Critically What other adaptations might this species have to its extreme environment?

b. Unlike other fish species in the ocean's depths, the dragonfish can see red light. The pockets of red light shining beneath each of its eyes allow it to detect other organisms without being seen.

The Oceanic Province: Most of the Ocean

The **oceanic province** is the largest marine environment, representing about 75 percent of the ocean's water; it is the open ocean that does not overlie the continental shelf. Most of the oceanic province is loosely described as the "deep sea." (The average depth of the ocean is 4000 m, more than 2 mi.) All but the shallowest waters of the oceanic province have cold temperatures, high pressure, and an absence of sunlight. These environmental conditions are uniform throughout the year.

Most organisms of the deep waters of the oceanic province depend on **marine snow**, organic debris that drifts down into their habitat from the upper, lighted regions of the oceanic province. Organisms of this little-known realm are filter

oceanic province
The part of the pelagic environment that overlies the ocean floor at depths greater than 200 m (650 ft).

feeders, scavengers, and predators. Many are invertebrates, some of which attain great sizes. The giant squid measures up to 18 m (59 ft) in length, including its tentacles.

Fishes of the deep waters of the oceanic province are strikingly adapted to darkness and scarcity of food (**Figure 11.10**). An organism that encounters food infrequently must eat as much as possible when food is available. Adapted to drifting or slow swimming, animals of the oceanic province often have reduced bone and muscle mass. Many of these animals have light-producing organs to locate one another for mating or food capture.

| CONCEPT CHECK | STOP |

1. **What** are the four main life zones in the ocean, and how do they differ from one another?

Human Impacts on the Ocean

LEARNING OBJECTIVES

1. **Contrast** fishing and aquaculture and relate the environmental challenges of each activity.

2. **Identify** the human activities that contribute to marine pollution and describe their effects.

3. **Explain** how global climate change could potentially alter the ocean conveyor belt.

The ocean is so vast, it's hard to imagine that human activities could harm it. Such is the case, however. Fisheries and aquaculture, marine shipping, marine pollution, coastal development, offshore mining, and global climate change all contribute to the degradation of marine environments. Scientists estimate that in 2008, less than 4 percent of the ocean remained unaffected by human activities, and 41 percent had experienced serious harm (**Figure 11.11a**).

Marine Pollution and Deteriorating Habitat

One of the great paradoxes of human civilization is that the same ocean that provides food to a hungry world is used as a dumping ground. Coastal and marine ecosystems receive pollution from land, from rivers emptying into the ocean, and from atmospheric contaminants that enter the ocean via precipitation. Offshore mining and oil drilling pollute the neritic province with oil and other contaminants. Pollution increasingly threatens the world's fisheries. Events such as accidental oil spills—such as the devastating *Deepwater Horizon* oil spill in the Gulf of Mexico in 2010—and the deliberate dumping of litter pollute the water. The World Resources Institute estimates that about 80 percent of global ocean pollution comes from human activities on land. In 2003 the Pew Oceans Commission, composed of scientists, economists, fishermen, and other experts, verified the seriousness of ocean problems in a series of studies. Some of their findings are shown in **Figure 11.11b**.

World Fisheries

The ocean contains valuable food resources. About 90 percent of the world's total marine catch is fishes, with clams, oysters, squid, octopus, and other molluscs representing 6 percent of the total catch. Crustaceans, including lobsters, shrimp, and crabs, make up about 3 percent, and marine algae constitute the remaining 1 percent.

Fleets of deep-sea fishing vessels obtain most of the world's marine harvest. Numerous fishes are also captured in shallow coastal waters and inland waters. According to the U.N. Food and Agriculture Organization (FAO), the world annual fish harvest increased substantially, from 19 million tons in 1950 to a high of nearly 95 million tons in 2000, and 90 million tons in 2011, the latest year for which data are available.

Problems and Challenges for the Fishing Industry No nation lays legal claim to the open ocean. Consequently, resources in the ocean are more susceptible to overuse and degradation than land resources, which individual nations own and for which they feel responsible.

The most serious problem for marine fisheries is that many species have been harvested to the point that their numbers are severely depleted. This generally causes a fishery to become unusable for commercial or sport fishermen, as well as for the other marine species that rely on it as part of the food web. Large predatory fish such as tuna, marlin, and swordfish have declined by 90 percent since the 1950s, according to Canadian researchers who analyzed data from ocean and coastal regions around the world. Scientists have found that dramatically depleted fish populations recover only slowly. Some show no real increase in population size up to 15 years after the fishery has collapsed (see Figure a in *What a Scientist Sees*).

According to the FAO, approximately 87 percent of the world's fish stocks are considered depleted, fully exploited, or overexploited. The three areas with the largest number of depleted fish stocks are the northeastern and northwestern Atlantic Ocean and the Mediterranean Sea (see chapter opener). Fisheries have experienced such pressure for two reasons. First, the growing human population requires protein in its diets, leading to a greater demand for fish. Second, technological advances allow us to fish so efficiently that every single fish is often removed from an area (see Figure b in *What a Scientist Sees*).

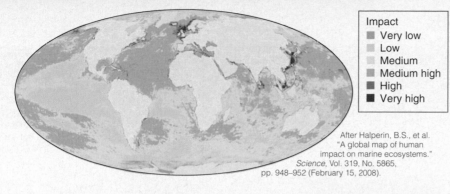

Impact
- Very low
- Low
- Medium
- Medium high
- High
- Very high

After Halperin, B.S., et al. "A global map of human impact on marine ecosystems." *Science*, Vol. 319, No. 5865, pp. 948–952 (February 15, 2008).

a. Mapping Human Impacts. In 2008, an international team of marine scientists mapped effects of 17 human activities on the ocean. Almost no location remains unaffected, and 41 percent of the ocean has been seriously altered by multiple activities.

Interpreting Data
Which regions exhibit the greatest impacts? Which are least affected? Is there a relationship between site location and status? Explain.

b. Major Threats to the Ocean.

Climate change
Example: Coral reefs and polar seas are particularly vulnerable to increasing temperatures.

Aquaculture
Example: Fish farms produce wastes that can pollute ocean water and harm marine organisms.

Invasive species
Example: Organisms are transported and released from ships in ballast water, which contains foreign crabs, mussels, worms, and fishes.

Overfishing
Example: Populations of many commercial fish species are severely depleted.

Nonpoint source pollution (runoff from land)
Example: Agricultural runoff (fertilizers, pesticides, and livestock wastes) pollutes water.

Point source pollution
Example: Passenger cruise ships dump sewage, shower and sink water, and oily bilge water.

Bycatch
Example: Fishermen unintentionally kill dolphins, sea turtles, and sea birds.

Habitat destruction
Example: Trawl nets (fishing equipment pulled along the ocean floor) destroy habitat.

Coastal development
Example: Developers destroy important coastal habitat, such as salt marshes and mangrove swamps.

Think Critically Under which of the major threats would you place the 2010 *Deepwater Horizon* oil spill in the Gulf of Mexico, or floating debris from the 2011 tsunami in Japan?

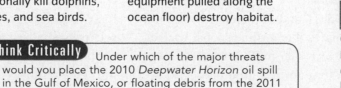

WHAT A SCIENTIST SEES

Modern Commercial Fishing Methods

a. The important cod fishery on Georges Bank, off the coast of Massachusetts, has collapsed due to overfishing. The U.S. Commerce Department closed large portions of Georges Bank in 1994, but cod are only slowly recovering.

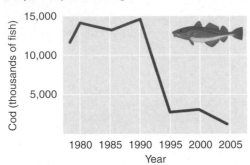

Based on data by O'Brien, L N. Shepherd, and L. Col. *Assessment of the Georges Bank Atlantic Cod Stocks for 2005.* Northwest Fisheries Science Center Ref. Document 06-10 (June 2006).

b. Scientific evidence indicates that modern methods of harvesting fish are so effective that many fish species have become rare. Sea turtles, dolphins, seals, whales, and other aquatic organisms are accidentally caught and killed in addition to the target fish. The depth of longlines is adjusted to catch open-water fishes such as sharks and tuna or bottom fishes such as cod and halibut. Purse seines catch anchovies, herring, mackerel, tuna, and other fishes that swim near the water's surface. Trawls catch cod, flounder, red snapper, scallops, shrimp, and other fishes and shellfish that live on or near the ocean floor. Drift nets catch salmon, tuna, and other fishes that swim in ocean waters.

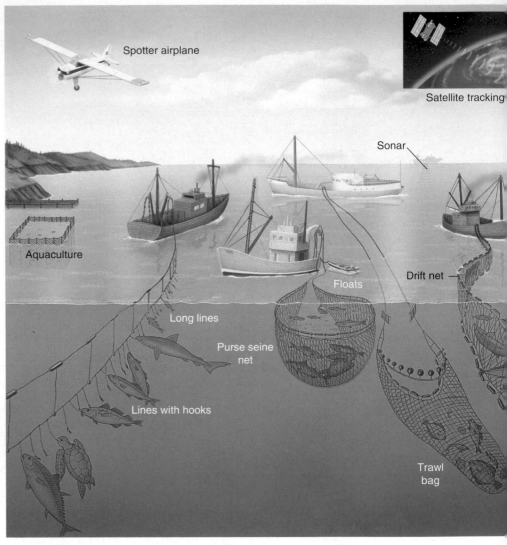

Fishermen tend to concentrate on a few fish species with high commercial value, such as menhaden, salmon, tuna, and flounder, and other species, collectively called **bycatch**, are unintentionally caught and then discarded. Although bycatch is extremely difficult to determine globally—it is defined differently in different places, and statistics are often not available—the FAO estimates that annual global bycatch exceeds 7 million metric tons (7.7 million tons). Most of these unwanted animals that are dumped back into the ocean are dead or soon die because they are crushed by the fishing gear or are out of the water too long. The

> **bycatch** The fishes, marine mammals, sea turtles, seabirds, and other animals caught unintentionally in a commercial fishing catch.

United States and other countries are trying to significantly reduce the amount of bycatch and develop uses for the bycatch that remains.

In response to harvesting, many nations extended their limits of jurisdiction to 320 km (200 mi) offshore. This action removed most fisheries from international use because more than 90 percent of the world's fisheries are harvested in relatively shallow waters close to land. This policy was supposed to prevent overharvesting by allowing nations to regulate the amounts of fishes and other seafood harvested from their waters. However, many countries also have a policy of **open management**, in which all fishing boats of that country are given unrestricted access to fishes in national waters.

Aquaculture: Fish Farming

Aquaculture is more closely related to agriculture on land than it is to the fishing industry. Aquaculture is carried out both in fresh and marine water; the cultivation of marine organisms is sometimes called **mariculture**. According to the FAO, growth in world aquaculture production outpaced that of fishing, increasing substantially, from 544,000 metric tons (600,000 tons) in 1950 to 57.7 million metric tons (63.6 million tons) in 2011 (**Figure 11.12a**).

> **aquaculture**
> The growing of aquatic organisms (fishes, shellfish, and seaweeds) for human consumption.

Aquaculture differs from fishing in several respects. For one thing, although highly developed nations harvest more fishes from the ocean, developing nations produce much more seafood by aquaculture. Developing nations have an abundant supply of cheap labor, which is a requirement of aquaculture because it is labor intensive, like land-based agriculture. Another difference between fishing and aquaculture is that the limit on the size of a catch in fishing is the size of the natural population, whereas the limit on aquacultural production is primarily the size of the area in which organisms can be grown.

In aquacultural "fish farms," fish populations are concentrated in a relatively small area and produce higher than normal concentrations of waste that pollute the adjacent water and harm other organisms. Aquaculture also causes a net loss of wild fish because many of the fishes farmed are carnivorous. Sea bass and salmon, for example, eat up to 5 kg (11 lb) of wild fish to gain 1 kg (2.2 lb) of weight.

Deep-water, offshore aquacultural facilities, sometimes called "ocean ranches," are becoming more common (**Figure 11.12b**). Ocean ranches, which increasingly use cutting-edge technologies such as submersible cages with robotic surveillance, may avoid damaging coastlines but often lack the pollution-restricting oversight associated with other aquaculture operations. Also, caged populations are more genetically homogenous than wild ones; if the two groups interbreed, genetic diversity of wild populations could be diminished. The introduced organisms may also outcompete wild species.

Shipping, Ocean Dumping, and Plastic Debris

Millions of ships dump oily ballast and other wastes overboard in the neritic and oceanic provinces. The U.N. International Maritime Organization's International Convention for the Prevention of Pollution from Ships (MARPOL) bans marine pollution arising from the shipping

Growth in world aquaculture • Figure 11.12

a. In recent years, fish harvest by aquaculture has continued to increase, while fishing (wild catch) has leveled off.

b. Cobia are raised in deep-water cages in this underwater fish farm off Puerto Rico. The open-water circulation reduces the waste problems common in shallow-water aquaculture.

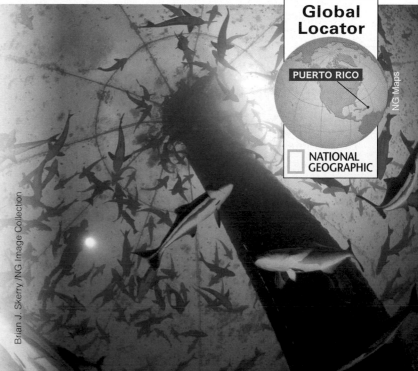

Global Locator

PUERTO RICO

NATIONAL GEOGRAPHIC

NG Maps

Brian J. Skerry /NG Image Collection

[Graph: Million metric tons of seafood (y-axis, 0–200) vs Year (x-axis, 1950–2020), showing "Aquaculture production" and "Wild catch" curves]

FAO *State of World Fisheries and Aquaculture* (2012).

Interpreting Data
During which time periods did aquaculture experience its most dramatic change?

industry. MARPOL regulations specifically address six types of marine pollution caused by shipping: oil, noxious liquids, harmful packaged substances, sewage, garbage, and air pollution released by ships. The most recently adopted MARPOL regulations, which entered into force in early 2013, strive to reduce greenhouse emissions associated with shipping, and to improve energy efficiency of ships. Unfortunately, MARPOL is not well enforced in the open ocean.

In the past, U.S. coastal cities such as New York dumped their sewage sludge into the ocean. Disease-causing viruses and bacteria from human sewage contaminated shellfish and other seafood and posed an increasing threat to public health. The **Ocean Dumping Ban Act** barred ocean dumping of sewage and industrial waste, beginning in 1991.

Huge quantities of trash containing plastics are released into the ocean, sometimes accidentally, from coastal communities or cargo ships. Plastics don't biodegrade; they photodegrade, which means that exposure to light breaks them down into smaller and smaller pieces that exist for an indefinite period. This trash collects in certain areas of the open ocean defined by atmospheric pressure systems. For example, in the north Pacific gyre—halfway between Hawaii and the U.S. mainland—researchers are monitoring a continuous array of floating plastics dubbed the "Eastern Pacific garbage patch." The size of this area is difficult to assess because its boundaries shift, and the debris it contains is mostly made up of tiny, floating plastic pieces not visible by satellite image.

Not only are marine mammals and birds susceptible to being entangled in and strangled by larger pieces of plastic, but the many filter-feeding organisms near the bottom of the ocean food chain constantly ingest the smaller degraded pieces (**Figure 11.13**). These plastic pieces may absorb and transport hazardous chemicals such as PCBs. Scientists have yet to determine whether these substances are incorporated into marine food webs when organisms that ingest the plastic are eaten by other organisms.

Coastal Development

Development of resorts, cities, industries, and agriculture along coasts alters or destroys many coastal ecosystems, including mangrove forests, salt marshes, sea grass beds, and coral reefs. Many coastal areas are overdeveloped, highly polluted, and overfished. Although more than 50 countries have coastal management strategies, their goals are narrow and usually deal only with the economic development of the thin strips of land that directly border the oceans. Coastal management plans generally

Plastic pollution in the ocean • Figure 11.13

a. Larger plastic debris in the ocean can injure or strangle larger marine organisms such as this penguin.

© Photoshot Holdings Ltd/Alamy

Bud Lehnhausen/Science Source

b. Once plastic in oceans degrades into tiny fragments such as these taken from a Costa Rican beach, it is ingested by filter-feeding organisms.

don't integrate the management of both land and water, nor do they take into account the main cause of coastal degradation—sheer human numbers.

Perhaps as many as 60 percent of the world's population live within 150 km (93 mi) of a coastline. Demographers project that three-fourths of all humans will live in that area by 2025. To prevent the world's natural coastal areas from becoming urban sprawl or continuous strips of tourist resorts during the 21st century, coastal management strategies must be developed that take into account projections of human population growth and distribution.

Human Impacts on Coral Reefs

Although coral formations are important ecosystems, they are being degraded and destroyed. Large reef formations are being damaged in 90 of the 109 countries where they are found. According to the latest report of the Global Coral Reef Monitoring Network, coral reefs in

Asia and the Indian Ocean are at the greatest risk, with 54 percent of the reefs lost or critically threatened and an additional 25 percent at moderate risk.

How do we harm coral reefs? In some areas, silt washing downstream from clear-cut inland forests has smothered reefs. Overfishing (particularly the removal of top predators), damage by scuba divers and snorkelers, pollution from ocean dumping and coastal runoff, oil spills, boat groundings, anchor draggings, fishing with dynamite or cyanide, hurricane damage, disease, reclamation, tourism, warming ocean temperatures, and the mining of corals for building material take a heavy toll.

Since the late 1980s, corals in the tropical Atlantic and Pacific have suffered extensive **bleaching** (see *What a Scientist Sees*), in which stressed corals expel their zooxanthellae, becoming pale or white in color. The most likely environmental stressor is warmer seawater temperatures attributed to global climate change (water only about 1 degree above average can lead to bleaching). Although many coral reefs have not recovered from bleaching, some have. Another threat of warmer ocean temperatures is the recent and rapid widespread acidification of ocean water, caused by excessive amounts of climate-warming CO_2 dissolving in the ocean and forming a dilute acid. Acidified seawater might cause the calcium carbonate skeletons of coral animals (and shells of crabs, oysters, clams, and many other marine species) to thin or, in extreme cases, dissolve completely away.

Offshore Extraction of Mineral and Energy Resources

Large deposits of minerals lie on the ocean floor. **Manganese nodules**—small rocks the size of potatoes that contain manganese and other minerals, such as copper, cobalt, and nickel—are widespread on the ocean floor,

WHAT A SCIENTIST SEES

WileyPLUS

Ocean Warming and Coral Bleaching

a. Bleached coral off the Maldive Islands, in the Indian Ocean. Scientists have linked coral bleaching to ocean warming. Warmer than usual temperatures stress the coral animals, causing them to lose their zooxanthellae. Without their algae, the corals can't get enough food, and they die.

Peter Scoones / Science Source

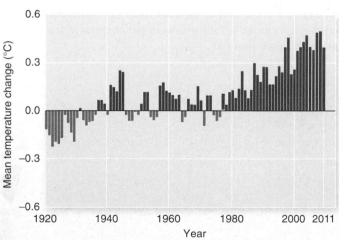

Adapted from National Assessment Synthesis Team, *Climate Change Impacts on the United States: The Potential Consequences of Climate Variability and Change* (Report for the U.S. Global Change Research Program). Cambridge, UK: Cambridge University Press (2001) and based on data from the National Oceanic and Atmospheric Administration.

b. This 1920–2011 time series of annual global mean temperature changes of the ocean surface indicates that the ocean has warmed, particularly during the past 25 years. Most warming has occurred in shallow waters where corals live. Mean temperature anomaly data relative to 1901–2000.

Interpreting Data
If this warming trend continues, what will the mean temperature change be by 2050?

most abundantly in the Pacific. Dredging manganese nodules from the ocean floor would adversely affect sea life, and the current market value for these minerals wouldn't cover the expense of obtaining them using existing technology. Furthermore, it isn't clear which countries have legal rights to minerals in international waters. Despite these concerns, many experts think that deep-sea mining will be technologically feasible in a few decades, and several industrialized nations such as the United States have staked claims in a region of the Pacific known for its large number of nodules. To date, none have been mined.

Such potential exploitations of the ocean floor are controversial. Many people think it is inevitable that minerals will be mined from the floor of the deep sea, but others think the seabed should be declared off-limits because of the potential ecological havoc that mining could cause on the diverse life forms inhabiting the ocean floor.

Offshore reserves of oil have long been tapped as a major source of energy. However, obtaining oil and gas resources from the seafloor generally poses a threat to fishing (**Figure 11.14**). Fishermen and conservationists worry that Congress may allow oil and gas wells to threaten fisheries such as the Georges Bank fishery,

Threats of energy exploration to marine life • Figure 11.14

A slick of crude oil and dispersants clotting the oil spreads across a stretch of the Gulf of Mexico in May 2010, following the *Deepwater Horizon* spill. Such disasters can potentially disrupt or destroy affected fisheries.

© CHRISTOPHER BERKEY/epa/Corbis

What natural resources are extracted in your region? How are issues related to these resources similar to or different from those involved in offshore energy and mineral extraction?

which is already suffering due to decades of overfishing (see *What a Scientist Sees*). The environmental concerns associated with extracting offshore energy resources are discussed in Chapter 17.

Climate Change, Sea-Level Rise, and Warmer Ocean Temperatures

Scientists recognize that the ocean is warming and changing along with global climate, but the unprecedented nature of these events hampers accurate predictions of future consequences. Unanticipated effects from a globally warmed world will undoubtedly occur, however. For example, there could be a disruption of the ocean conveyor belt, which transports heat around the globe (see Figure 11.1b). Evidence from seafloor sediments and Greenland ice indicates that the ocean conveyor belt shifts from one equilibrium state to another in a relatively short period (a few years to a few decades). Scientists are concerned that human activities may affect this equilibrium and cause an abrupt climate shift. Models suggest that climate warming, with its associated freshwater melting off the Greenland ice sheet, could weaken or even— as a worst case—shut down the ocean conveyor belt in as short a period as a decade. Such changes in the ocean conveyor belt could cause major cooling in Europe, while greater climate warming could occur elsewhere.

Sea-level rise, caused by melting ice sheets and glaciers as well as seawater expanding as it warms, threatens coastal areas and the large populations that live there. Coastal erosion, wetlands loss, flooding risks, and saltwater intrusion are all likely to increase.

Until recently, climate scientists couldn't predict whether human-induced global climate change would affect El Niño and La Niña events in the tropical Pacific Ocean. Recent computer models indicate greater extremes of drying and heavy rainfall during El Niño events. Scientists are still uncertain whether El Niño events will occur more frequently with global climate change.

CONCEPT CHECK	STOP

1. **What** are some of the harmful environmental effects associated with the fishing industry? with aquaculture?

2. **How** does the widespread use of plastics contribute to ocean pollution?

3. **How** might the effect of global climate change on the ocean alter the ocean conveyor belt?

Addressing Ocean Problems

LEARNING OBJECTIVES

1. **Describe** international initiatives that address problems in the global ocean.
2. **Explain** strategies proposed to correct ocean problems in the future.

The many different threats to the world's oceans are attributed to a range of local, regional, national, and global sources. Problems in the ocean are complex and therefore require complex solutions.

Industrialized countries' interest in removing manganese nodules from the ocean floor, first expressed in the 1960s, triggered the formation of an international treaty, the **U.N. Convention on the Law of the Sea (UNCLOS)**. UNCLOS, which became effective in 1994, is generally considered a "constitution for the ocean," and its focus is the protection of ocean resources. As of late 2012, 164 countries had joined the treaty and are bound to its requirements. (The United States had not yet ratified UNCLOS but voluntarily observes its provisions.) The provisions of UNCLOS are binding only for international waters, not for territorial waters, so seabed mining is not prohibited in territorial waters. For example, hydrothermal vent systems in deep territorial waters off Papua New Guinea contain gold, zinc, copper, and silver, and exploration efforts are currently under way to determine methods for extracting these resources.

In 1995 the United Nations approved the **U.N. Fish Stocks Agreement**, the first international treaty to regulate marine fishing. The treaty went into effect in 2001. Because the overfishing problem continues to escalate, the United Nations has sponsored other fishery protection pacts.

The Magnuson Fishery Conservation Act, which went into effect in 1977, regulates marine fisheries in the United States. This law established eight regional fishery councils, each of which developed a management plan for its region. Until 1996, the act was not particularly successful because managers were often pressured to set quotas too high and the National Marine Fisheries Service estimated that more than one-third of U.S. fish stocks were being fished at higher levels than could be sustained. In 1996 the act was amended and reauthorized as the **Magnuson-Stevens Fishery Conservation and Management Act** or Sustainable Fisheries Act. It requires the regional councils and the National Marine Fisheries Service to protect essential fish habitat for more than 600 fish species, reduce overfishing, rebuild the populations of overfished species, and minimize bycatch. Fishing quotas, restrictions of certain types of fishing gear, limits on the number of fishing boats, and closure of fisheries during spawning periods are some of the management tools used to reduce overfishing. The 2007 reauthorization of the Magnuson-Stevens Act strengthened controls on illegal and unreported fishing in U.S. waters.

Future Actions

A 2004 report by the U.S. Commission on Ocean Policy, the first comprehensive review of federal ocean policy in 35 years, recommended three primary strategies for improving the ocean and coasts:

- **Create a new ocean policy to improve decision making**. Currently, a number of agencies and committees manage U.S. waters, and their respective goals often conflict. The commission recommends strengthening and reorganizing the National Oceanic and Atmospheric Administration (NOAA) and consolidating other federal ocean programs under it.

- **Strengthen science and generate information for decision makers**. There is a critical need for high-quality research on how marine ecosystems function and how human activities affect them.

- **Enhance ocean education to instill in citizens a stewardship ethic**. Environmental education should be part of the curriculum at all levels and should include a strong marine component.

Ensuring the recovery of depleted fisheries may require the establishment of networks of "no-take" reserves and a substantial reduction of fishing fleets. Governments will also have to reduce or remove subsidies that help support the fishing industry. (A **subsidy** is a form of government support given to a business or an institution to promote the activity performed by that business or institution.) Government subsidies encourage modernization and expansion of fishing fleets.

Many scientists think the best way to halt and reverse destruction of the ocean is to adopt an *ecosystem-based approach* to manage ocean environments. This means

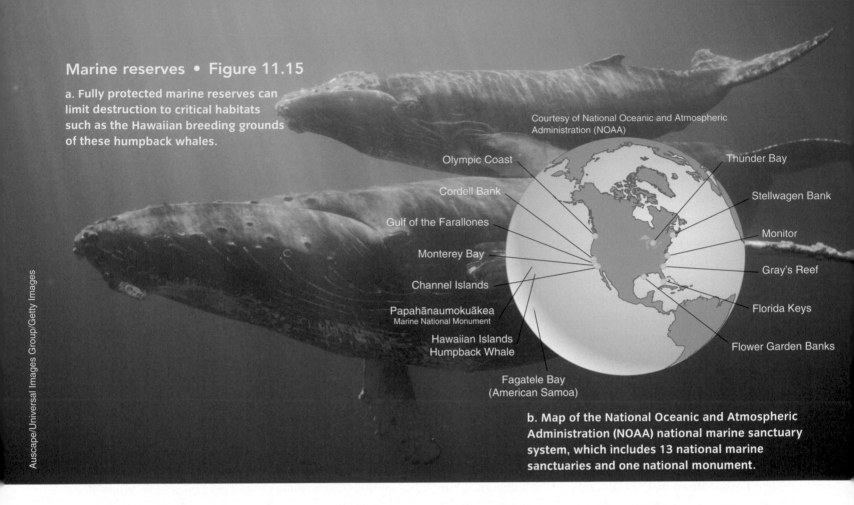

Marine reserves • Figure 11.15

a. Fully protected marine reserves can limit destruction to critical habitats such as the Hawaiian breeding grounds of these humpback whales.

Courtesy of National Oceanic and Atmospheric Administration (NOAA)

Olympic Coast

Cordell Bank

Gulf of the Farallones

Monterey Bay

Channel Islands

Papahānaumokuākea
Marine National Monument

Hawaiian Islands
Humpback Whale

Fagatele Bay
(American Samoa)

Thunder Bay

Stellwagen Bank

Monitor

Gray's Reef

Florida Keys

Flower Garden Banks

b. Map of the National Oceanic and Atmospheric Administration (NOAA) national marine sanctuary system, which includes 13 national marine sanctuaries and one national monument.

that rather than focus on a single, narrow goal such as reviving a specific fish population, ocean management should focus on preserving the health and function of the entire marine ecosystem.

One proposed approach that would enhance ecosystem-based management would be to establish networks of fully protected marine reserves, within which no habitat destruction or resource extraction would be allowed. Currently less than 5 percent of U.S. marine environments are set aside as fully protected marine reserves, yet these areas have successfully preserved threatened habitats and increased populations of exploited organisms (**Figure 11.15a**).

The United States has designated **national marine sanctuaries** along the Atlantic, Pacific, and Gulf of Mexico coasts to minimize human impacts and protect unique natural resources and historic sites. These sanctuaries include kelp forests off the coast of California, coral reefs in the Florida Keys, fishing grounds along the continental shelf, and deep submarine canyons, as well as shipwrecks and other sites of historic value (**Figure 11.15b**). NOAA's *National Marine Sanctuary Program* administers the sanctuaries, which, like many federal lands, are managed for multiple purposes, including conservation,

recreation, education, fishing, mining of some resources, scientific research, and ship salvaging.

In 2006, President George W. Bush established the world's largest protected marine area when he designated the northwestern Hawaiian Islands and surrounding waters—an area almost as large as California—as a national monument. Now named the Papahānaumokuākea Marine National Monument, this protected area is home to more than 7000 species, including seabirds, fishes, marine mammals, coral reef colonies, and other organisms, approximately one-quarter of which are found only there.

Like most other countries, the United States recognizes the importance of the ocean to life on this planet. However, it remains to be seen if the United States and other countries will make a strong commitment to protecting and managing the global ocean.

CONCEPT CHECK

1. **Which** international treaties aim to protect ocean resources?

2. **What** three strategies does the U.S. Commission on Ocean Policy recommend?

CASE **STUDY**

The Dead Zone in the Gulf of Mexico

Nitrogen and phosphorus from the Mississippi River—products of fertilizer and manure runoff from midwestern fields and livestock operations—are deemed largely responsible for a huge **dead zone** in the Gulf of Mexico. Except for bacteria that thrive in oxygen-free environments, no life exists in the dead zone because the water there—where these nutrients have been deposited—does not contain enough dissolved oxygen to support fishes or other aquatic organisms (see **Figure a**). Dead zones form seasonally worldwide; more than 405 occur along global coastlines. The Gulf of Mexico dead zone, one of the largest in the ocean, extends from the seafloor up into the water column, sometimes to within a few meters of the surface. In 2011 it covered about 21,000 km² (8000 mi²), an area the size of New Jersey. It generally persists from March to September. In March and April, snowmelt and spring rains flow from the Mississippi River into the Gulf, and the dead zone is most severe during June–August.

The low-oxygen condition in a dead zone, known as **hypoxia**, occurs when algae (phytoplankton) grow rapidly because of the presence of nutrients in the water (see **Figure b**). Dead algae sink to the bottom and are decomposed by bacteria, which deplete the water of dissolved oxygen, leaving too little for other sea life.

Scientists are now seeing evidence that ocean warming induced by global climate change may be exacerbating dead zones. Dead zones, including the one in the Gulf of Mexico, are expanding, they are emerging closer to shore than ever before, and they are forming even in areas of the ocean that don't receive agricultural runoff.

Increased frequency and size of dead zones threaten biodiversity and harm coastal fisheries. The EPA has taken some measures to control nitrogen and phosphorus inputs to the Mississippi River but recognizes that the dead zone problem is immense in scope and will take billions of dollars and decades of effort to fix.

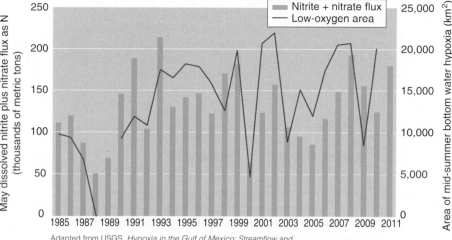

Adapted from USGS. *Hypoxia in the Gulf of Mexico: Streamflow and Nutrient Delivery to the Gulf of Mexico for October 2010 to May 2011* (2011).

a. The graph presents dissolved nitrite plus nitrate flux to the Gulf of Mexico (bars), in the month of May 1985–2011, and (line) the mid summer area of the northern Gulf of Mexico in which bottom waters exhibit very low oxygen levels (less than 2 mg/L), 1985–2010. This area defines the size of the Gulf of Mexico dead zone (USGS 2011).

b. Enhanced NASA satellite image illustrates summer phytoplankton (algae) activity along the Gulf of Mexico coastline. Reds and oranges indicate high concentrations of phytoplankton and river sediments, and corresponding low-oxygen levels.

Summary

1 The Global Ocean 274

1. The global ocean is a huge body of salt water that surrounds the continents. It affects the hydrologic cycle and other cycles of matter, influences climate and weather, and provides food to millions.

2. Prevailing winds over the ocean generate **gyres**, large, circular ocean current systems that often encompass an entire ocean basin. The **Coriolis effect** is a force resulting from Earth's rotation that influences the paths of surface ocean currents, which move in a circular pattern, clockwise in the Northern Hemisphere and counterclockwise in the Southern Hemisphere.

3. The ocean and the atmosphere are strongly linked. The **El Niño–Southern Oscillation (ENSO)** event, which is responsible for much of Earth's interannual climate variability, is a periodic, large-scale warming of surface waters of the tropical eastern Pacific Ocean that temporarily alters both ocean and atmospheric circulation patterns. A **La Niña** event occurs when surface water in the eastern Pacific Ocean becomes unusually cool. Its effects on weather patterns are less predictable than an ENSO event's effects.

pollute the adjacent water and also causes a net loss of wild fish because many of the fishes farmed are carnivorous.

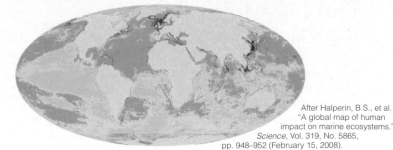

After Halperin, B.S., et al. "A global map of human impact on marine ecosystems." *Science*, Vol. 319, No. 5865, pp. 948–952 (February 15, 2008).

Mapping Human Impacts.

2 Major Ocean Life Zones 278

1. The vast ocean is subdivided into major life zones. The biologically productive **intertidal zone** is the area of shoreline between low and high tides. The **benthic environment** is the ocean floor, which extends from the intertidal zone to the deep-ocean trenches. Most of the benthic environment consists of sediments where many animals burrow. Shallow benthic habitats include sea grass beds, kelp forests, and coral reefs. The **pelagic environment** is divided into two provinces. The **neritic province** is the part of the pelagic environment from the shore to where the water reaches a depth of 200 m (650 ft). Organisms that live in the neritic province are all floaters or swimmers. The **oceanic province**, "the deep sea," is the part of the pelagic environment where the water depth is greater than 200 m. The oceanic province is the largest marine environment, comprising about 75 percent of the ocean's water.

2. **Marine pollution** is generated by many human activities, including the release of trash and contaminants through commercial shipping, ocean dumping of sludge and industrial wastes, and discarding of plastics that are potentially harmful to marine organisms. Marine environments are also deteriorated by coastal development and the extraction of offshore minerals.

3. The **ocean conveyor belt** moves cold, salty, deep-sea water from higher to lower latitudes, affecting regional and possibly global climate. Global climate change associated with human activities may alter the link between the ocean conveyor belt and global climate.

3 Human Impacts on the Ocean 284

1. The most serious problem for marine fisheries is the overharvesting of many species to the point that their numbers are severely depleted. Fishermen usually concentrate on a few fish species with high commercial value. In doing so, they also catch **bycatch**: fishes, marine mammals, sea turtles, seabirds, and other animals caught unintentionally in a commercial fishing catch and then discarded. **Aquaculture** is the growing of aquatic organisms (fishes, shellfish, and seaweeds) for human consumption. Aquaculture is common in developing nations with abundant cheap labor, and it is limited by the size of the space dedicated to cultivation. Aquaculture produces wastes that

4 Addressing Ocean Problems 291

1. International initiatives aimed at protecting the global ocean include the **U.N. Convention on the Law of the Sea (UNCLOS)**, a "constitution for the ocean" that protects ocean resources, and the **U.N. Fish Stocks Agreement**, the first international treaty to regulate marine fishing. In the United States, marine fisheries are regulated by the **Magnuson-Stevens Fishery Conservation and Management Act**.

2. Long-term goals for halting and reversing destruction of the ocean focus on adopting an ecosystem-based approach to management of ocean environments. Consolidating ocean programs, funding research on marine ecosystems, and enhancing ocean education to instill in citizens a stewardship ethic could improve U.S. ocean policy.

Key Terms

- aquaculture 287
- benthic environment 278
- bycatch 286
- El Niño–Southern Oscillation (ENSO) 276
- gyres 274
- intertidal zone 278
- neritic province 281
- oceanic province 283

What is happening in this picture?

- This biochemist is studying cuttings of bleached coral in the Indian Ocean.

- Why are bleached corals pale?

- What sort of human impacts cause coral bleaching?

- Is the occurrence of coral bleaching likely to increase or decrease? Why?

Critical and Creative Thinking Questions

1. How do ocean currents affect climate on land? In particular, describe the role of the ocean conveyor belt.

2. Compare the different global effects of El Niño with those of La Niña. How are the two events similar? How are they different?

3. Identify which of the ocean life zones at right would be home to each of the following organisms: giant squid, kelp, tuna, and mussels. Explain your answers.

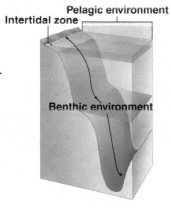

4. The use of plastic shopping bags has been banned in many U.S. cities, including Los Angeles and Washington, DC, and statewide in Hawaii. How might these bans influence human impacts on the ocean?

5. Describe the global character of the ocean and its importance to life on Earth in terms of the effects of mismanagement of the bluefin tuna fishery in the Mediterranean or the expansion of the Gulf of Mexico dead zone.

6. Imagine that you live in a small Atlantic coast community where a company wants to set up an aquaculture facility in a salt marsh. What are its benefits and its environmental drawbacks? Would you support or oppose this proposal? Explain your answer.

7. If global climate change trends continue, why might Italy grow cooler, and how would impacts on the ocean trigger that change?

8. Which U.N. treaty might impose limits on the number of fishing vessels allowed to catch tuna in international waters?

Use graph to answer questions 9–10.

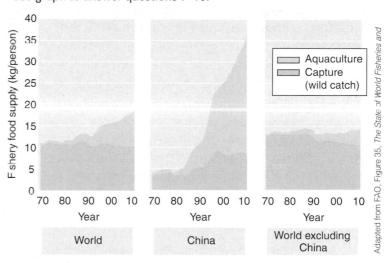

9. In 2010, what proportion of the fishery food supply was provided by aquaculture in the entire world? in China? in the world excluding China?

10. What overall trends in fishery food supply, from both aquaculture and capture, were exhibited by China, and by the rest of the world, between 1970 and 1990? between 1990 and 2010? What role do regional differences in aquaculture production play in these trends?

Sustainable Citizen Question

11. Assess your personal seafood consumption, or that of a friend or family member. Which species do you eat? Where and how are they obtained? Research these details by investigating online the Monterey Bay Aquarium Seafood Watch© Program. List possible impacts to the ocean from your seafood consumption, and identify choices you might make to reduce them.

Mineral and Soil Resources

COPPER BASIN, TENNESSEE

The United States—like countries worldwide—is growing increasingly more dependent on nonfuel mineral resources (see graph), but this dependence can trigger high environmental costs. Copper Basin, Tennessee, provides an example of environmental degradation caused by smelting, a stage of mineral processing that removes impurities from metals. During the 19th century mining companies in southeastern Tennessee extracted copper ore—rock containing copper—from the ground and dug vast pits to serve as open-air smelters. They cut down the surrounding trees to fire the smelters, producing the high temperatures needed for separating copper from other substances in the ore. One of these substances, sulfur, reacted with oxygen in the air to form sulfur dioxide. The sulfur dioxide entered the atmosphere, reacted with water vapor there, and became sulfuric acid that returned to Copper Basin as acid precipitation.

Ecological ruin took only a few short years (see larger photograph, taken in 1973). Acid precipitation killed plants. Without plants to hold the soil in place, erosion cut gullies in the rolling hills. The forest animals disappeared along with the plants, their food and shelter destroyed.

State and federal reclamation efforts were only marginally successful until the 1970s, when specialists began using new replanting techniques. The new plants had a greater survival rate, and as they became established, their roots held the soil in place (see inset photo of a reforested section, in 2008). Birds and field mice slowly began to return.

Today, reclamation of Copper Basin continues; the goal is to have the entire area under plant cover by the middle of the 21st century. The return of the original forest ecosystem will take at least a century or two. Plant scientists and land reclamation specialists have learned a lot from Copper Basin, and they will put this knowledge to use in future reclamation projects around the world.

WileyPLUS

U.S. Raw Nonfuel Mineral Materials Put into Use Annually, 1900–2006

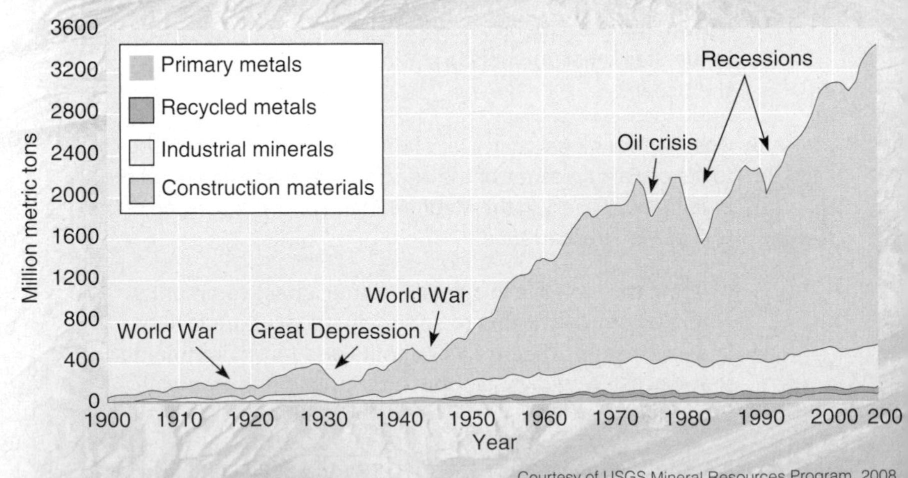

Legend:
- Primary metals
- Recycled metals
- Industrial minerals
- Construction materials

Annotations on graph: World War, Great Depression, World War, Oil crisis, Recessions. Y-axis: Million metric tons (0–3600). X-axis: Year (1900–2000).

Courtesy of USGS Mineral Resources Program, 2008.

Interpreting Data

Explain the differences in use of various mineral categories—why do you suppose usage has increased so much more in some groups than in others?

© Andre Jenny/Alamy

Emory Kristof/NG Image Collection

CHAPTER OUTLINE

Plate Tectonics and the Rock Cycle   298
- Volcanoes
- Earthquakes
- The Rock Cycle

Economic Geology: Useful Minerals   302
- Minerals: An Economic Perspective
- How Minerals Are Extracted and Processed

Environmental Implications of Mineral Use   306
- Mining and the Environment
- ◼ EnviroDiscovery: Not-so-Precious Gold
- Environmental Impacts of Refining Minerals
- Restoration of Mining Lands

Soil Properties and Processes   309
- Soil Formation and Composition
- ◼ What a Scientist Sees: Soil Profile
- Soil Organisms

Soil Problems and Conservation   312
- Soil Erosion
- Soil Pollution
- Soil Conservation and Regeneration
- ◼ Environmental InSight: Soil Conservation
- ◼ Case Study: Coping with "Conflict Minerals"

CHAPTER PLANNER ✓

☐ Study the picture and read the opening story.

☐ Scan the Learning Objectives in each section:
   p. 298 ☐   p. 302 ☐   p. 306 ☐   p. 309 ☐   p. 312 ☐

☐ Read the text and study all figures and visuals.
   Answer any questions.

Analyze key features

☐ Process Diagram, p. 301

☐ EnviroDiscovery, p. 307

☐ What a Scientist Sees, p. 310

☐ Environmental InSight, p. 315

☐ Case Study, p. 317

☐ Stop: Answer the Concept Checks before
   you go on:
   p. 301 ☐   p. 305 ☐   p. 308 ☐   p. 312 ☐   p. 316 ☐

End of Chapter

☐ Review the Summary and Key Terms.

☐ Answer What is happening in this picture?

☐ Answer the Critical and Creative Thinking Questions.

Plate Tectonics and the Rock Cycle

LEARNING OBJECTIVES

1. **Define** *plate tectonics* and explain its relationship to earthquakes and volcanic eruptions.

2. **Diagram** a simplified version of the rock cycle.

Earth's layers and surface structure • Figure 12.1

a. The Main Layers of Planet Earth.

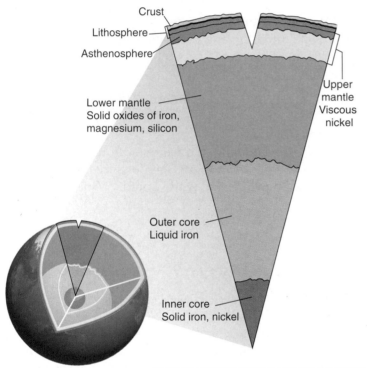

Crust

Lithosphere

Asthenosphere

Lower mantle
Solid oxides of iron, magnesium, silicon

Upper mantle
Viscous nickel

Outer core
Liquid iron

Inner core
Solid iron, nickel

Geology is an essential part of environmental science. To better understand the environmental effects of humans on mineral and soil resources, you must first know something about the geologic properties of Earth's outer layers.

Earth's outermost rigid rock layer, the *lithosphere*, consists of Earth's crust—the outermost layer—and the uppermost part of the mantle. It is composed of seven large plates, plus a few smaller ones, that float on the *asthenosphere*, the region of the mantle where rocks become hot and soft (**Figure 12.1**). Continents and landmasses are situated on some of these plates. As the plates move across Earth's surface, the continents change their relative positions. **Plate tectonics**, the study of the movement of these plates, explains how most features on Earth's surface originate.

> **plate tectonics**
> The study of the processes by which the lithospheric plates move over the asthenosphere.

An area where two plates meet—a **plate boundary**—is a site of intense geologic activity (**Figure 12.2**). Three types of plate boundaries—divergent, convergent, and transform—exist; all three types occur both in the ocean and on land. Two plates move apart at a **divergent plate boundary**. When two plates move apart, a ridge of molten rock from the mantle wells up between them; the ridge continually expands as the plates move farther apart. The Atlantic Ocean is growing as a result of the buildup of

b. Plates and Plate Boundary Locations.
There are seven major independent plates that move horizontally across Earth's surface. Arrows show the directions of plate movements. The three types of plate boundaries are explained in Figure 12.2.

GLOBAL LOCAL
Which plate do you live on? Are you near a plate boundary?

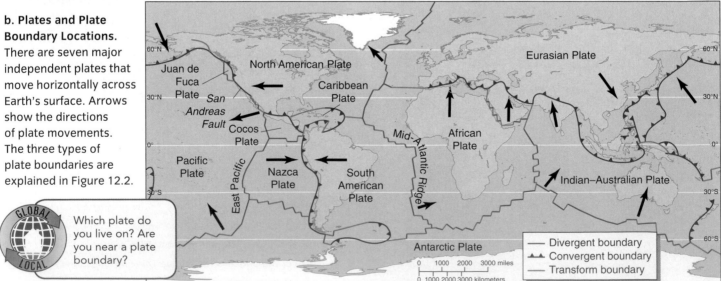

60°N

North American Plate

Juan de Fuca Plate

San Andreas Fault

Cocos Plate

30°N

Pacific Plate

East Pacific

Nazca Plate

South American Plate

Caribbean Plate

Mid-Atlantic Ridge

African Plate

Eurasian Plate

0°

30°S

Indian–Australian Plate

60°S

Antarctic Plate

0 1000 2000 3000 miles
0 1000 2000 3000 kilometers

— Divergent boundary
▲▲ Convergent boundary
— Transform boundary

Plate boundaries • Figure 12.2

All three types of plate boundaries occur both in the ocean and on land.

a. Two plates move apart at a divergent plate boundary.

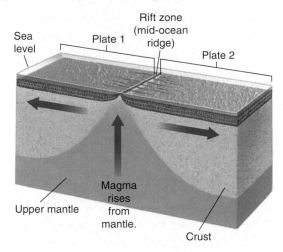

b. When two plates collide at a convergent plate boundary in the seafloor, subduction may occur. Convergent collision can also form a mountain range (not shown).

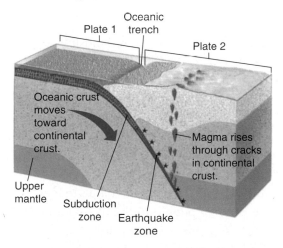

c. At a transform plate boundary, plates move horizontally in opposite but parallel directions. On land, such a boundary is often evident as a long, thin valley due to erosion along the fault line.

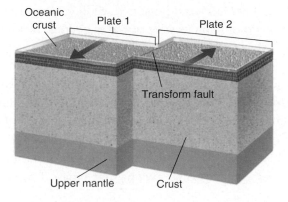

lava along the Mid-Atlantic Ridge, where two plates are diverging.

When two plates collide at a **convergent plate boundary**, one of the plates sometimes descends under the other in the process of **subduction**. Convergent collision can also form a mountain range; the Himalayas formed when the plate carrying India converged into the plate carrying Asia. At a **transform plate boundary**, plates move horizontally in opposite but parallel directions. On land, such a boundary is often evident as a long, thin valley due to erosion along the fault line.

Earthquakes and volcanoes are common at plate boundaries. San Francisco, California (noted for its earthquakes), and the volcano Mount Saint Helens in Washington State are both situated on plate boundaries.

Volcanoes

The movement of tectonic plates on the hot, soft rock of the asthenosphere causes most volcanic activity. In places where the asthenosphere is close to the surface, heat from this part of Earth's mantle melts the surrounding rock, forming pockets of **magma**. When one plate slides under or away from another, this magma may rise to the surface, often forming volcanoes. Magma that reaches the surface is called **lava**.

Volcanoes occur at three kinds of locations: at subduction zones, at spreading centers, and above hot spots. Subduction zones around the Pacific Basin have given rise to hundreds of volcanoes around Asia and the Americas, in a region known as the "ring of fire." Iceland is a volcanic island that formed along the Mid-Atlantic Rift Zone as the adjoining plates there spread apart. The volcanic Hawaiian Islands formed as the Pacific Plate moved over a **hot spot**, a rising plume of magma that flows from an opening in Earth's crust beneath the ocean or continents.

The largest volcanic eruption in the 20th century occurred in 1991, when Mount Pinatubo in the Philippines exploded (see Figure 9.9). Despite the evacuation of more than 200,000 people, several hundred deaths occurred, mostly from the collapse of buildings under the thick layer of wet ash that blanketed the area. The volcanic cloud produced when Mount Pinatubo erupted extended upward some 48 km (30 mi). We are used to hearing about human activities affecting climate, but many significant natural phenomena, including volcanoes, also affect global climate. The lava and ash ejected into the atmosphere by the eruption blocked much of the sun's warmth and caused a slight cooling of global temperatures (0.2–0.5°C) for a year or so.

Earthquakes

Forces inside Earth sometimes push and stretch rocks in the lithosphere. The rocks absorb this energy for a time, but eventually, as the energy accumulates, the stress is too great, and the rocks suddenly shift or break. The energy—released as **seismic waves**, vibrations that rapidly spread through rock in all directions—causes one of the most powerful events in nature, an earthquake. Most earthquakes occur along **faults**, fractures in the crust where rock moves forward and backward, up and down, or from side to side. Fault zones are often found at plate boundaries.

The site where an earthquake begins, often far below the surface, is the **focus** (**Figure 12.3**). Directly above the focus, at Earth's surface, is the earthquake's **epicenter**. When seismic waves reach the surface, they cause the ground to shake. Buildings and bridges may collapse, and roads may break. One of the instruments used to measure seismic waves is a seismograph, which helps seismologists (scientists who study earthquakes) determine where an earthquake started, how strong it was, and how long it lasted.

Seismologists record more than 1 million earthquakes each year. Some of these are major, but most are too small for humans to feel, equivalent to readings of about 2 on the *Richter scale*, a measure of the magnitude of energy released by an earthquake. In populated areas, a magnitude 5 earthquake usually causes some property damage, and quakes of 8 or higher cause massive property destruction and kill large numbers of people. Although the public is familiar with the Richter scale, most seismologists prefer to use a more accurate scale, the *moment magnitude scale*, to measure earthquakes, especially those larger than magnitude 6.5 on the Richter scale. The moment magnitude scale calculates the total energy that a quake releases.

In January 2010, an earthquake with a moment magnitude of 7.0 struck in an area approximately 25 km (16 mi) from the capital of Haiti. About 230,000 people were killed, making it one of the deadliest earthquakes on record. Most of these people died due to the structural collapse of poorly constructed buildings. About one million people whose homes were destroyed became refugees. The Caribbean region is prone to earthquakes due to movements between the North American, South American, and Caribbean plates. Puerto Rico, Jamaica, Dominican Republic, Martinique, and Guadeloupe have had earthquakes greater than magnitude 7 in the past.

Side effects of earthquakes include landslides and tsunamis. A *landslide* is an avalanche of rock, soil, and other debris that slides swiftly down a mountainside. In 2008 a powerful earthquake in a mountainous area in Sichuan Province, China, and its aftershocks, triggered massive landslides and the structural collapse of many buildings. About 70,000 people were killed, and more than 1.5 million were left homeless.

A *tsunami*, a giant sea wave caused by an underwater earthquake, volcanic eruption, or landslide, sweeps across the ocean at more than 750 km (450 mi) per hour. Although a tsunami may be only about 1 m (3 ft) high in deep-ocean water, it can build to a wall of water 30 m (about 100 ft)—as high as a 10-story building—when it comes ashore, often far from where the original earthquake triggered it. Tsunamis have caused thousands of deaths, particularly along the Pacific coast. Colliding tectonic plates in the Indian Ocean triggered tsunamis in 2004 that killed more than 230,000 people in South Asia and Africa. Not only did the tsunamis cause catastrophic loss of life and destruction of property, but they resulted in widespread environmental damage. Salt water that moved inland as far as 3 km (1.9 mi) polluted soil and groundwater. Oil and gasoline from overturned cars, trucks, and boats contaminated the land and poisoned wildlife. Coral reefs and other offshore habitats were also damaged or destroyed.

In March 2011, an earthquake with a magnitude of 8.9 hit Japan. This earthquake, which was the strongest ever recorded in Japan, also triggered a disastrous tsunami. Other tsunamis generated by the earthquake hit coastal areas of several Pacific Rim countries, although they caused substantially less damage than in Japan. The death toll in Japan numbered in the thousands and would have been worse except that Japan is one of the

Earthquakes • Figure 12.3

Earthquakes occur when plates along a fault suddenly move in opposite directions relative to one another. This movement triggers seismic waves that radiate through the crust.

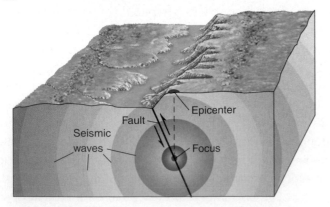

The rock cycle • Figure 12.4

Rocks do not remain in their original form forever. This highly simplified diagram shows how rock cycles from one form to another.

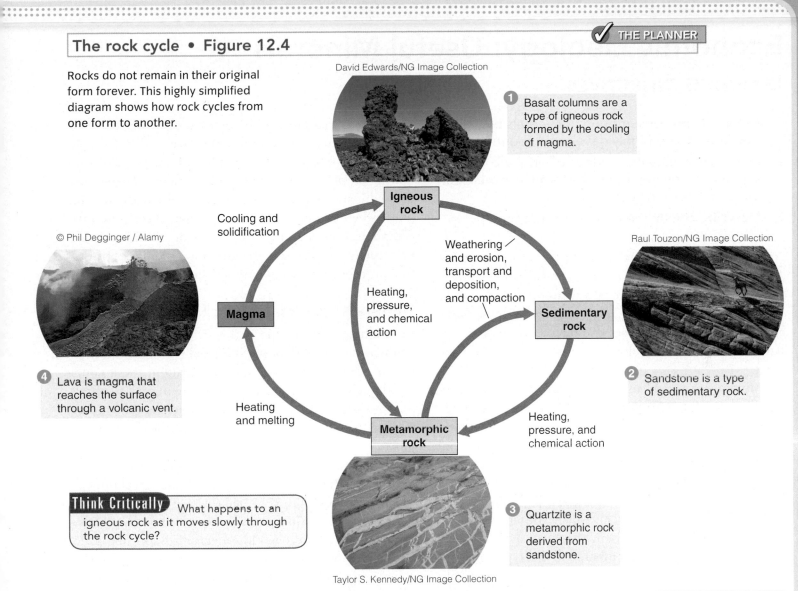

David Edwards/NG Image Collection

1 Basalt columns are a type of igneous rock formed by the cooling of magma.

© Phil Degginger / Alamy

Raul Touzon/NG Image Collection

Cooling and solidification

Igneous rock

Weathering and erosion, transport and deposition, and compaction

Heating, pressure, and chemical action

Magma

Sedimentary rock

4 Lava is magma that reaches the surface through a volcanic vent.

Heating and melting

Metamorphic rock

Heating, pressure, and chemical action

2 Sandstone is a type of sedimentary rock.

Think Critically What happens to an igneous rock as it moves slowly through the rock cycle?

3 Quartzite is a metamorphic rock derived from sandstone.

Taylor S. Kennedy/NG Image Collection

best prepared of all nations when it comes to earthquake disasters. Japan lies on a seismically active junction of several tectonic plates. (See Chapter 17 for a discussion of damage to Japan's nuclear power plants from the quake.)

The Rock Cycle

Rocks, which are aggregates of one or more **minerals**, fall into three categories, based on how they formed: igneous, sedimentary, and metamorphic. **Igneous rocks** form when magma rises from the mantle and cools. **Sedimentary rocks** form when small fragments of weathered, eroded rocks (or marine organisms) are deposited, compacted, and cemented together. **Metamorphic rocks** form when intense heat and pressure alter igneous, sedimentary, or other metamorphic rocks.

minerals Elements or compounds of elements that occur naturally in Earth's crust.

Earth's internal structure and the basic geologic processes that we have presented in this chapter result in a **rock cycle**, in which rock moves from one physical state or location to another (**Figure 12.4**). The rock cycle is similar to the other cycles of matter, such as the carbon and hydrologic cycles (see Chapter 5). However, rocks are formed and move through the environment over geologic time, much more slowly than the elements of the other cycles.

CONCEPT CHECK   **STOP**

1. **What** are tectonic plates and plate boundaries? Where do earthquakes and volcanoes usually occur? Why?

2. **What** are the three types of rock? How are the three types of rock interconnected in the rock cycle?

Economic Geology: Useful Minerals

LEARNING OBJECTIVES

1. **Contrast** the consumption of minerals by developing and highly developed countries.
2. **Distinguish** between surface mining and subsurface mining, using the terms *overburden* and *spoil bank* in your answer.
3. **Describe** briefly the process of smelting.

Earth's outermost layer, the crust, contains many kinds of minerals that are of economic importance. We now focus on the economic and environmental impacts of extracting and using mineral resources. We then consider soil, the part of the crust where biological and physical processes meet.

Minerals are such an integral part of our lives that we often take them for granted. Steel, an essential building material, is a blend of iron and other metals. Beverage cans, aircraft, automobiles, and buildings all contain aluminum. Copper, which readily conducts electricity, is used for electrical and communications wiring. The concrete used in buildings and roads is made from sand and gravel, as well as cement, which contains crushed limestone. Sulfur, a component of sulfuric acid, is an indispensable industrial mineral used to make plastics and fertilizers, and to refine oil. Other important minerals include platinum, mercury, manganese, and titanium. Tantalum, a rare hard metal that is resistant to corrosion, has recently become important to the production of capacitors in a range of electronic devices. (See the *Case Study* at the end of the chapter for more on this mineral, including its ties to conflict in the Democratic Republic of the Congo.)

Earth's minerals are elements or (usually) compounds of elements and have precise chemical compositions. **Sulfides** are mineral compounds in which certain elements are combined chemically with sulfur, and **oxides** are mineral compounds in which elements are combined chemically with oxygen. Minerals are metallic or nonmetallic (**Figure 12.5**). **Metals** are minerals such as iron, aluminum, and copper, which are malleable, lustrous, and good conductors of heat and electricity. **Nonmetallic minerals**, such as sand, stone, salt, and phosphates, lack these characteristics.

Rocks are naturally formed mixtures of minerals that have varied chemical compositions. **Ore** is rock that contains a large enough concentration of a particular mineral to be profitably mined and extracted. **High-grade ores** contain relatively large amounts of particular minerals, whereas **low-grade ores** contain lesser amounts. Although some minerals are abundant, all minerals are nonrenewable resources that are not replenished by natural processes on a human timescale.

Minerals: An Economic Perspective

At one time, most of the highly developed nations had abundant mineral deposits that enabled them to industrialize. In the process of industrialization, these countries largely depleted their domestic reserves of minerals so that they must increasingly turn to developing countries. This is particularly true for countries in Europe, Japan, and, to a lesser extent, the United States.

The level of mineral consumption varies widely between highly developed and developing countries. The United States and Canada, which have about 5.1 percent of the world's population, consume about one-fourth of many of the world's metals. It is too simplistic, however, to divide the world into two groups, the mineral consumers (highly developed countries) and the mineral producers (developing countries). Many of the world's top mineral producers are highly developed countries and top mineral consumers (for example, the United States, Canada, Australia, and Russia). Many developing countries, however, lack any significant mineral deposits.

Mineral production and consumption in China is increasing dramatically as the country industrializes. For example, China smelted more than 34 percent of the world's primary aluminum (obtained from ores, not recycling). China also consumes almost all of this aluminum, making it the world's largest producer and largest consumer of primary aluminum. As China's mineral production and economic growth expand, the natural capital on which its economic growth is based is being degraded. Recall from Chapter 3 that **natural capital** is Earth's resources and processes that sustain humans and other living organisms (see Figure 3.14). Natural capital includes minerals, soils, fresh water, clear air, forests, wildlife, and fisheries.

Some important minerals and their uses • Figure 12.5

Gypsum, silicon, and sulfur are nonmetals. All other minerals shown are metals.

Aluminum
Mikhail Pozhenko/iStockphoto
Aircraft, motor vehicles, packaging (cans, foil), water treatment

Chromium
iStockphoto
Chrome plate, dyes and paints, steel alloys (cutlery)

Cobalt
Matthew Ragen/iStockphoto
Corrosion and wear-resistant alloys, pigments (cobalt blue)

Gold
iStockphoto
Jewelry, money, restorative dentistry

Iron
Mike Clarke/iStockphoto
Steel (alloy of iron) buildings and machinery

Magnesium
Christoph Ermel/iStockphoto
Beverage cans, electronic devices, firecrackers, flares

Mercury
© gmnicholas/iStockphoto
Industrial chemicals, electric and electronic applications, batteries

Molybdenum
Dave White/iStockphoto
High-temperature alloys for aircraft, industrial motors

Nickel
Jack Cobben/iStockphoto
Coins, metal plating, alloys with various uses

Potassium
Matt Meadows/Alamy
Fertilizers, photography

Silver
Vladimir Melnik/iStockphoto
Jewelry, silverware, photography, electronics

Titanium
Wesley VanDinter/iStockphoto
Alloy in steel and other industrial alloys, pigment in paints, plastics

Zinc
Predrag Novakovic/iStockphoto
Galvanizing steel, alloys (brass), anode in alkaline batteries

Gypsum (CaSO$_4$—2H$_2$O)
Nikki Lowry/iStockphoto
Drywall, plaster of Paris, soil conditioner

Silicon
Don Wilkie/iStockphoto
Electronic devices, semiconductors, natural stone, glass, concrete

Sulfur
Jozsef Szasz-Fabian/iStockphoto
Industrial chemicals, insecticides, gunpowder, vulcanized tires

Types of mining operations • Figure 12.6

a. Surface mining This open-pit copper mine in Utah is the largest human-made excavation in the world.

b. Surface mining Strip mining removes overburden along narrow strips to reach the ore beneath.

Because industrialization increases the demand for minerals, developing countries that at one time met their mineral needs with domestic supplies become increasingly reliant on foreign supplies as development occurs. Many industrialized nations, including the United States, have stockpiled strategically important minerals to reduce their dependence on potentially unstable suppliers. Recently, China has begun to stockpile *rare earth metals*, which are 17 elements such as dysprosium and terbium that are important in high-technology applications like hybrid-car batteries, wind turbines, and laser-guided missiles. China, which controls more than 90 percent of the global supply of rare earth metals, has also reduced its exports to other countries, in a move that could affect market prices.

How Minerals Are Extracted and Processed

The process of making mineral deposits available for human consumption occurs in several steps. First, a particular mineral deposit is located. Geologic knowledge of Earth's crust and how minerals are formed is used to estimate locations of possible mineral deposits. Once these sites are identified, geologists drill or tunnel for mineral samples and analyze their composition. Second, mining extracts the mineral from the ground. Third, the mineral is processed, or refined, by concentrating it and removing impurities. Finally, the purified mineral is used to make a product.

Extracting Minerals The depth of a particular mineral deposit determines whether surface or subsurface

mining will be used. In **surface mining**, minerals are extracted near the surface. Surface mining is more common because it is less expensive than subsurface mining. Because even surface mineral deposits occur in rock layers beneath Earth's surface, the overlying soil and rock layers, called **overburden**, must first be removed, along with the vegetation growing in the soil. Then giant power shovels scoop out the minerals.

There are two kinds of surface mining: open-pit surface mining and strip mining. Iron, copper, stone, and gravel are usually extracted through **open-pit surface mining**, in which a giant hole, called a quarry, is dug in the ground to extract the minerals (**Figure 12.6a**). In **strip mining**, a trench is dug to extract the minerals (**Figure 12.6b**). Then a new trench is dug parallel to the old one, and the overburden from the new trench is put into the old one, creating a hill of loose rock called a **spoil bank**.

Subsurface mining extracts minerals too deep in the ground to be removed by surface mining. It disturbs the land less than surface mining, but it is more expensive and more hazardous for miners. Miners face risk of death or injury from explosions

surface mining The extraction of mineral and energy resources near Earth's surface by first removing the soil, subsoil, and overlying rock strata.

overburden Soil and rock overlying a useful mineral deposit.

spoil bank A hill of loose rock created when the overburden from a new trench is put into the already excavated trench during strip mining.

subsurface mining The extraction of mineral and energy resources from deep underground deposits.

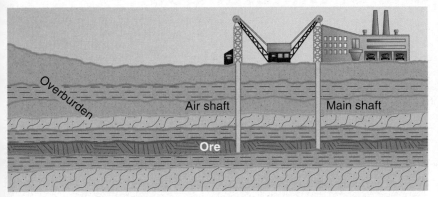

c. Subsurface mining In a shaft mine, a hole is dug straight through the overburden to the ore, which is removed up through the shaft in buckets.

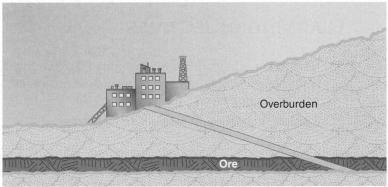

d. Subsurface mining In a slope mine, an entry to the ore is dug at an angle so that the ore can be hauled out in carts.

or collapsing walls, and prolonged breathing of dust in subsurface mines can result in lung disease.

Subsurface mining may be done with underground shaft mines or slope mines. A **shaft mine**, often used for mining coal, is a direct vertical shaft to the vein of ore (**Figure 12.6c**). The ore is broken up underground and then hoisted through the shaft to the surface in buckets. A **slope mine** has a slanting passage that makes it possible to haul the broken ore out of the mine in cars rather than to hoist it up in buckets (**Figure 12.6d**). Sump pumps keep a subsurface mine dry, and a second shaft is usually installed for ventilation.

Processing Minerals Processing minerals often involves smelting. Purified copper, tin, lead, iron, manganese, cobalt, or nickel smelting is done in a blast furnace.

> **smelting** The process in which ore is melted at high temperatures to separate impurities from the molten metal.

Figure 12.7 shows a blast furnace used to smelt iron. Iron ore, limestone rock, and coke (modified coal used as an industrial fuel) are added at the top of the furnace, while heated air or oxygen is added at the bottom. Chemical reactions take place throughout the furnace as the ore moves downward: The iron ore reacts with coke to form molten iron and carbon dioxide, whereas the limestone reacts with impurities in the ore to form a molten mixture called **slag**. Both molten iron and slag collect at the bottom, but slag floats on molten iron because it is less dense than iron. The slag is cooled and then disposed of. Note the vent near the top of the iron smelter for exhaust gases. If air pollution control devices are not installed, many dangerous gases are emitted during smelting.

Blast furnace • Figure 12.7

Towerlike furnaces separate metal from impurities in the ore. The energy for smelting comes from a blast of heated air.

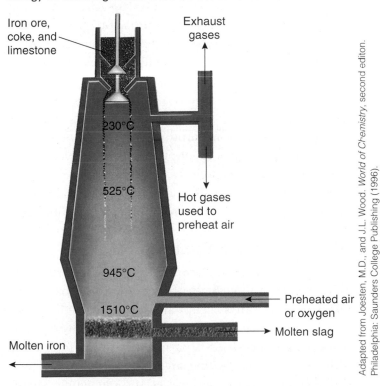

Adapted from Joesten, M.D., and J.L. Wood. *World of Chemistry*, second edition. Philadelphia: Saunders College Publishing (1996).

CONCEPT CHECK **STOP**

1. **How** does mineral consumption differ between highly developed and developing countries?

2. **What** are the steps involved in surface mining? in subsurface mining?

3. **Why** and how are minerals smelted?

Environmental Implications of Mineral Use

LEARNING OBJECTIVES

1. **Relate** the environmental impacts of mining and refining minerals. Include a brief description of acid mine drainage.
2. **Explain** how mining lands can be restored.

There is no question that mineral use harms the environment. The extraction, processing, and disposal of minerals has **external costs** (see Chapter 3). Mining disturbs and damages the land, and the processing and disposal of minerals pollute the air, soil, and water. Although pollution can be controlled and damaged lands can be restored, these remedies are costly. Historically, the environmental cost of extracting, processing, and disposing of minerals has not been incorporated into the actual price of mineral products to consumers (the environmental impacts are similar when these processes are applied to fossil fuel resources, as described in Chapter 17).

Most highly developed countries have regulatory mechanisms in place to minimize environmental damage from mineral consumption, and many developing nations are in the process of putting them in place. Such mechanisms include policies to prevent or reduce pollution, restore mining sites, and exclude certain recreational and wilderness sites from mineral development.

Mining and the Environment

Mining, particularly surface mining, disturbs large areas of land. In the United States, functioning and abandoned metal and coal mines occupy an estimated 9 million hectares (22 million acres). Because mining destroys existing vegetation, mined land is particularly prone to erosion, with wind erosion causing air pollution and water erosion polluting nearby waterways and damaging aquatic habitats.

Open-pit mining of gold and other minerals uses huge quantities of water. As miners dig deeper, they eventually hit the water table and must pump out the water to keep the pit dry. Farmers and ranchers in open-pit mining areas are concerned about depletion of the groundwater they need for irrigation. Environmentalists and others would like the mining operations to reinject the water into the ground after pumping it out.

Acid mine drainage • Figure 12.8

The characteristic orange acid runoff contains sulfuric acid contaminated with heavy metals. Photographed in Preston County, West Virginia.

© Thomas R. Fletcher/Alamy

Mining affects water quality. According to the Worldwatch Institute, mining has contributed to the contamination of at least 19,000 km (11,800 mi) of streams and rivers in the United States. Rocks rich in minerals often contain high concentrations of heavy metals such as arsenic and lead. Rainwater seeping through the sulfide minerals in mine waste produces sulfuric acid, which dissolves the heavy metals and other toxic substances in the spoil banks. These acids, called acid mine drainage, are highly toxic and are washed into soil and water, including groundwater, by precipitation runoff (**Figure 12.8**). When such acids and toxic compounds make their way into nearby lakes and streams, particularly through "toxic pulses" of thunderstorms or spring snowmelt, they are particularly harmful to waterfowl, fish, and other wildlife in the watershed. Although some acid drainage occurs naturally, mining exposes large areas of dissolved toxic substances to precipitation, greatly accelerating polluted runoff.

> **acid mine drainage**
> Pollution caused when sulfuric acid and dangerous dissolved materials such as lead, arsenic, and cadmium wash from mines into nearby lakes and streams.

EnviroDiscovery

Not-so-Precious Gold

Gold is a precious metal used primarily for jewelry and as a medium of exchange in many countries (see figure). Worldwide demand for gold is increasing, and the environment is suffering from the increased mining. The waste from mining and processing ore is enormous: 6 tons of wastes are produced to yield enough gold to make two wedding rings. The world's largest gold mine, located in Indonesia but owned by a U.S. company, dumps more than 200,000 metric tons of contaminated tailings into the local river each day, where they threaten waterfowl and fishes, as well as underground drinking water supplies.

Small-scale miners use other extraction techniques with destructive side effects: soil erosion, production of silt that clogs streams and threatens aquatic organisms, and contamination from mercury used to extract the gold. The environmental hazards of gold mining do not end when the gold is carried away: If not disposed of properly, mining wastes cause long-term problems such as acid mine drainage and heavy-metal contamination. Additionally, gold mining operations of all scales use huge amounts of energy—mostly from burning fossil fuels—to obtain and process the ore.

Relative importance of the many uses of gold

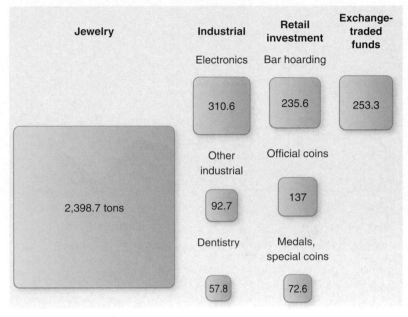

Adapted from Larmer, B. "The Price of Gold." *National Geographic* (January 2009).

Environmental Impacts of Refining Minerals

Approximately 80 percent or more of mined ore consists of impurities that become wastes after processing. These wastes, called **tailings**, are usually left in giant piles on the ground or in ponds near the processing plants (**Figure 12.9a**). The tailings contain toxic materials such as cyanide, mercury, uranium, and sulfuric acid. When left exposed, they contaminate the air, soil, and water (**Figure 12.9b**).

Smelting plants may emit large quantities of air pollutants, particularly sulfur, during mineral processing. Unless expensive pollution control devices are added to smelters, the sulfur escapes into the atmosphere, where it forms sulfuric acid. (The environmental implications

Environmental impact of tailings • Figure 12.9

© Ron Chapple Stock/Alamy

a. Tailings—dumped here in rural Utah—cause air, soil, and water pollution and have serious effects on land use.

b. Dissolved uranium contaminates a creek downstream of a former uranium and copper mine in southeastern Utah.

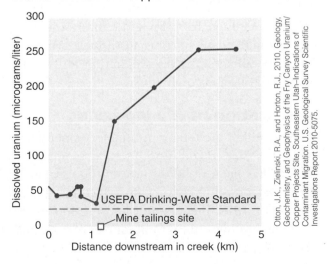

Otton, J.K., Zielinski, R.A., and Horton, R.J., 2010, Geology, Geochemistry, and Geophysics of the Fry Canyon Uranium/Copper Projects Site, Southeastern Utah–Indications of Contaminant Migration. U.S. Geological Survey Scientific Investigations Report 2010-5075.

Restoration of mining lands • Figure 12.10

Bull elk forage on a reclaimed surface coal mine in Hindman, Kentucky. Restoration of mining lands makes them usable once again, or at least stabilizes them so that further degradation does not occur.

NewsCom

of the resulting acid precipitation are discussed in Chapter 9.) Pollution control devices for smelters are the same as the devices used for the burning of sulfur-containing coal—scrubbers and electrostatic precipitators.

Contaminants in ores include the heavy metals lead, cadmium, arsenic, and zinc. These toxic elements may pollute the atmosphere during the smelting process and cause harm to humans. Smelters emit airborne pollutants as well as hazardous liquid and solid wastes that can pollute the soil and water.

One of the most significant environmental impacts of mineral production is the large amount of energy required to mine and refine minerals, particularly if they are being refined from low-grade ore. Most of this energy is obtained by burning fossil fuels, which depletes non-renewable energy reserves and produces carbon dioxide and other air pollutants.

Restoration of Mining Lands

When a mine is no longer profitable to operate, the land can be reclaimed, or restored to a seminatural condition, as has been done to most of the Copper Basin in Tennessee (see the chapter introduction). Reclamation prevents further degradation and erosion of the land, eliminates or neutralizes local sources of toxic pollutants, and makes the land productive for purposes other than

mining (**Figure 12.10**). Restoration also makes such areas visually attractive.

A great deal of research is available on techniques of restoring land degraded by mining, called **derelict land**. Restoration involves filling in and grading the area to the shape of its natural contours and then planting vegetation to hold the soil in place. The establishment of plant cover is not as simple as throwing a few seeds on the ground. Often the topsoil is completely gone or contains toxic levels of metals, so special types of plants that tolerate such a challenging environment must be used, such as acid-tolerant species. According to experts, the main limitation on the restoration of derelict lands is not a lack of knowledge but the lack of funding.

The **Surface Mining Control and Reclamation Act** of 1977 requires reclamation of areas that were surface mined for coal. However, no federal law is in place to require restoration of derelict lands produced by other kinds of mines. As a result, restoration of mining lands often does not occur.

CONCEPT CHECK

1. **What** are three harmful environmental effects of mining and processing minerals?

2. **How** are mining lands restored?

Soil Properties and Processes

LEARNING OBJECTIVES

1. **Define** *soil* and identify the factors involved in soil formation.
2. **Describe** the composition of soil and the organization of soil into horizons.
3. **Relate** at least two ecosystem services performed by soil organisms and briefly discuss nutrient cycling.

The relatively thin surface layer of Earth's crust is **soil**, which consists of mineral and organic matter modified by the natural actions of agents such as weather, wind, water, and organisms. It is easy to take soil for granted. We walk on and over it throughout our lives but rarely stop to think about how important it is to our survival.

soil The uppermost layer of Earth's crust, which supports terrestrial plants, animals, and microorganisms.

Soil supports virtually all terrestrial food webs. Vast numbers and kinds of organisms, mainly microorganisms, inhabit soil and depend on it for shelter, food, and water. Plants anchor themselves in soil, and from it they receive essential minerals and water. Terrestrial plants could not survive without soil, and because we depend on plants for our food, humans could not exist without soil either (**Figure 12.11**).

Soil Formation and Composition

Soil is formed from *parent material,* rock that is slowly broken down, or fragmented, into smaller and smaller particles by biological, chemical, and physical **weathering processes**. It takes a long time, sometimes thousands of years, for rock to disintegrate into finer and finer mineral particles. Time is also required for organic material to accumulate in the soil. Soil formation is a continuous process that involves interactions between Earth's solid crust and the biosphere. The weathering of parent material beneath already formed soil continues to add new soil.

Organisms and climate both play essential roles in weathering, sometimes working together. Carbon dioxide released when soil organisms respire diffuses into the soil and reacts with soil water to form carbonic acid; lichens and other organisms produce other kinds of acids. These acids etch tiny cracks in the rock, where water collects.

In a temperate climate, the alternate freezing and thawing of the water during the winter causes the cracks to enlarge, breaking off small pieces of rocks. Small plants then become established and send their roots into the larger cracks, fracturing the rock further.

Topography, a region's surface features (such as the presence or absence of mountains and valleys), is also involved in soil formation. Steep slopes often have little or no soil on them because soil and rock are continually transported down the slopes by gravity. Runoff from precipitation tends to amplify erosion on steep slopes. Moderate slopes and valleys, on the other hand, may encourage the formation of deep soils.

Soil is composed of four distinct parts: mineral particles, organic matter, water, and air. The mineral portion, which comes from weathered rock (parent material), is the

Cut-away view of prairie soil in Kansas • Figure 12.11

Soil is an important natural resource that humans and countless soil organisms rely on.

Jim Richardson/NG Image Collection

of the soil. One of the best ways to reduce the effects of wind erosion on soil is to plant shelterbelts that lessen the impact of wind (**Figure 12.18**).

Restoration of soil fertility to its original level is a slow process. The land cannot be farmed or grazed until the soil has completely recovered. Disaster is likely if the land is put back to use before the soil has completely recovered. But the restriction of land use for an indefinite period may be difficult to accomplish. How can land use be restricted when people's livelihoods, and maybe even their lives, depend on it?

Soil Conservation Policies in the United States

The Food Security Act (Farm Bill) of 1985 contains provisions for two main soil conservation programs: a conservation compliance program and the Conservation Reserve Program. The conservation compliance program requires farmers with highly erodible land to develop and adopt a 5-year conservation plan for their farms that includes erosion-control measures. If they do not comply, they lose federal agricultural subsidies such as price supports.

The **Conservation Reserve Program (CRP)** is a voluntary subsidy program that pays U.S. farmers to stop producing crops on highly erodible farmland. It requires planting native grasses or trees on such land and then "retiring" it from further use for 10 to 15 years. During that time the land may not be grazed, nor may the grass be harvested for hay. The CRP has benefited the environment. Since its inception in 1985, annual loss of soil on CRP lands planted with grasses or trees has been reduced more than 90 percent, from an average of 7.7 metric tons of soil per hectare (8.5 tons per acre) to 0.6 metric ton per hectare (0.7 ton per acre). Because the vegetation is not disturbed once it is established, it provides biological habitat. Small and large mammals, birds of prey, and ground-nesting birds such as ducks have increased in number and kind on CRP lands. The reduction in soil erosion has improved water quality and enhanced fish populations in surrounding rivers and streams.

The future of the Conservation Reserve Program is unclear. Historically, farmers are more likely to practice soil conservation and participate in the CRP during hard financial times and periods of agricultural surpluses (when food prices are low). Because food prices have risen in recent years, and federal mandates support converting corn crops into ethanol fuel, some farmers are beginning to grow crops on former CRP lands, removing them from the program.

Shelterbelts surrounding kiwi orchards • Figure 12.18

Trees protect the delicate fruits from the wind and reduce wind erosion of farmland soil. Photographed on the North Island, New Zealand.

Global Locator

NEW ZEALAND

NG Maps

NATIONAL GEOGRAPHIC

© Kevin Fleming/CORBIS

CONCEPT CHECK   STOP

1. **What** is sustainable soil use?

2. **How** does soil erosion affect plants growing in the soil?

3. **How** do conservation tillage, contour plowing, and shelterbelts contribute to soil conservation?

CASE **STUDY**

Coping with "Conflict Minerals"

Tantalum is a dark-gray, very hard metal that plays an important part in our high-tech society (**Figure a**). Because of its extreme abilities to resist heat and corrosion and conduct electricity, tantalum is prized as the source material of electronic capacitors. It is found in cell phones, computer circuits, digital cameras, gaming hardware, many other electronic gadgets, and even in weapon systems. Tantalum has been mined primarily in Australia (see **Figure b**), but its presence in the Democratic Republic of the Congo (DRC) made it a controversial mineral. In the DRC tantalum is mined in a form known as "coltan," an ore containing both tantalite—an oxide of tantalum—and a mineral known as columbite. During the recent devastating war in the DRC, which began in 1998 and included conflicts among nine African nations resulting in the deaths of millions, various militias took over the mining and selling of coltan and other "blood" minerals, including diamonds. These military groups forced DRC residents into unsafe labor and created environmental havoc, primarily by harvesting native species—most notably gorillas—for bushmeat. The sale of coltan supported the conflict.

Although peace accords were negotiated in 2003, fighting continues in the eastern DRC, and coltan remains a black-market commodity. The prized ore is also present in the border region of Venezuela and Colombia, where it has again triggered illegal activity, luring in drug smugglers and paramilitary groups.

The developed world's dependence on tantalum creates dilemmas for manufacturers of high-tech goods: How do they avoid conflict sources of this metal? And how do consumers know that the manufacturers are avoiding these sources? In 2010 the U.S. Congress passed legislation as part of the Dodd-Frank Act to prevent electronics manufacturers from supporting wars by purchasing conflict minerals such as coltan from the DRC and neighboring countries. Unfortunately, legal mining operations are often near the illicit ones controlled by militia groups. U.S. researchers are pursuing laser-induced breakdown spectroscopy (LIBS) technology that can generate specific geochemical "fingerprints" for distinct mineral samples. The LIBS technique offers promise for manufacturers and consumers hoping to avoid conflict sources of minerals such as coltan.

a. Coltan, a form of tantalum mined from the Democratic Republic of the Congo's Masisi Territory.

© Lucas Oleniuk/Zuma Press/Corbis

b. International sources of tantalum, 1990–2009.

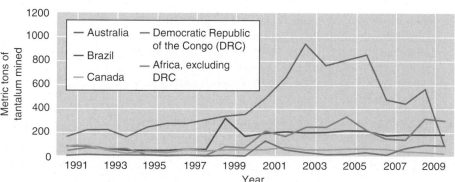

Adapted from United States Geological Survey (USGS) *Minerals Yearbook* annual data. 1990–2009.

Think Critically How do you think poverty levels affect a nation's potential to exploit conflict minerals? How might the war in the DRC have been different if the nation lacked such minerals?

Soil Problems and Conservation   317

Summary

1 Plate Tectonics and the Rock Cycle 298

1. The lithosphere, Earth's outermost rigid rock layer, is composed of plates that float on the asthenosphere, the region of the mantle where rocks become hot and soft. **Plate tectonics** is the study of the processes by which the lithospheric plates move over the asthenosphere. **Plate boundaries** are often sites of intense geologic activity: earthquakes, volcanoes, and mountain building.

2. Rocks are aggregates of one or more **minerals**. The **rock cycle** shows how rock slowly cycles from one form to another. The three categories of rock are **igneous, metamorphic,** and **sedimentary rock.**

© Phil Degginger / Alamy

2 Economic Geology: Useful Minerals 302

1. Minerals are metallic or nonmetallic elements or compounds of elements that occur naturally in Earth's crust. Highly developed nations consume a disproportionate share of the world's minerals, but as developing countries become industrialized, their needs for minerals increase.

2. Minerals are extracted through surface or subsurface mining. **Surface mining** removes the **overburden:** the overlying soil, subsoil, and rock strata. Strip mining, a type of surface mining, produces a **spoil bank** when the overburden from a new trench is put into an excavated trench. **Subsurface mining** extracts resources from deep underground deposits.

3. Processing minerals often involves **smelting,** melting the ore in a blast furnace to separate impurities from the metal.

3 Environmenal Implications of Mineral Use 306

1. Surface mining destroys vegetation across large areas, increasing erosion. **Open-pit mining** uses huge quantities of water. Mining also affects water quality. **Acid mine drainage** is pollution caused when dissolved toxic materials wash from mines into nearby lakes and streams.

2. **Derelict lands** degraded by mining can be restored by filling in and grading the land to its natural contours and then planting vegetation to hold the soil in place.

4 Soil Properties and Processes 309

1. **Soil** is the uppermost layer of Earth's crust and supports terrestrial plants, animals, and microorganisms. Soil is formed from parent material—rock that is slowly fragmented into small particles through biological, chemical, and physical **weathering processes.**

2. Soil is composed of mineral particles, organic matter, water, and air. **Soil horizons** are the horizontal layers into which many soils are organized, from the surface to the underlying parent material.

3. Soil organisms provide **ecosystem services** such as maintaining soil fertility and preventing soil erosion. Soil organisms carry out **nutrient cycling,** the pathway of nutrient minerals or elements from the environment through organisms and back to the environment.

5 Soil Problems and Conservation 312

1. **Sustainable soil use** is the wise use of soil resources, without a reduction in the amount or fertility of soil, so soil is productive for future generations. Soil used in a sustainable way renews itself by natural processes year after year.

2. Water, wind, ice, and other agents cause **soil erosion,** the wearing away or removal of soil from the land. Soil erosion reduces fertility because essential minerals and organic matter are removed. Erosion causes sediments and pesticide and fertilizer residues to pollute nearby waterways.

3. Good soil conservation practices promote sustainable soil use. In **conservation tillage,** residues from previous crops partially cover the soil to help hold it in place until newly planted seeds are established. **Crop rotation,** the planting of different crops in a field over a period of years, decreases the insect damage, disease, and mineral depletion that occur when one crop is grown continuously. **Contour plowing,** which matches the natural contour of the land, helps control erosion of land with variable topography. **Strip cropping** produces alternating strips of different crops along natural contours. **Terracing** reduces soil erosion on steep slopes. A **shelterbelt** is a row of trees planted as a windbreak to reduce soil erosion.

Key Terms

- acid mine drainage 306
- conservation tillage 314
- contour plowing 314
- crop rotation 314
- minerals 301
- nutrient cycling 312

- overburden 304
- plate tectonics 298
- shelterbelt 316
- smelting 305
- soil 309
- soil erosion 313

- soil horizons 311
- spoil bank 304
- subsurface mining 304
- surface mining 304
- sustainable soil use 312

What is happening in this picture?

- On March 11, 2011, a tsunami breaches the seawall and rushes into the city of Miyako, Japan, following a strong earthquake in the region.

- Why is Japan more likely than the U.S. East Coast to experience a tsunami?

- What do tsunamis and volcanic eruptions have in common?

- Go online to compile a summary of the human and environmental damage that resulted from the Japan 2011 earthquake.

AFP/Getty Images

Critical and Creative Thinking Questions

1. Why is Mount Everest getting taller? Why is the Pacific Ocean shrinking?

2. Outline the process through which rocks are recycled over geologic time. How are they connected?

3. How many minerals have you come in contact with today? Which were metals; which were nonmetals?

4. What is the difference between surface and subsurface mining? open-pit and strip mines? shaft and slope mines? When is each most likely to be used?

5. Why is pollution control equipment needed for blast furnaces?

6. How does acid mine drainage damage nearby streams and groundwater?

7. Describe how Copper Basin, Tennessee, became an environmental disaster. Are reclamation efforts making a difference there?

8. What are the roles of weathering, organisms, and topography in soil formation?

9. What are the different soil horizons in a soil profile? Do soil horizons provide any information about a soil's ability to support plant growth? Explain.

10. How does the developed world's insatiable need for electronics and other goods contribute to violent conflicts in some developing regions?

11. Why should environmentally sustainable societies protect their soil resources? How might they do so?

12. This graph relates spring wheat production (measured in kg of wheat per hectare of cultivated land) to soil erosion (measured in cm of topsoil lost) in the northern Great Plains. How does soil erosion affect wheat yield?

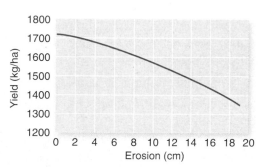

Williams, J.R. and D.L. Tanaka. "Economic Evolution of Topsoil Loss in Spring Wheat Production in Northern Great Plains, U.S." *Soil and Tillage Research*, Vol. 37 (1996).

Sustainable Citizen Question

13. Describe two soil conservation methods that assist in sustainable food production. As a food consumer, where might you shop, what questions might you ask, or what sort of information on labels might you look for to encourage soil conservation?

Land Resources

U.S. NATIONAL PARKS

In 2003 and 2004 two new U.S. national parks were created, Congaree National Park in South Carolina and Great Sand Dunes National Park in Colorado. These parks bring the total number of U.S. national parks to 58. The number of people visiting national parks has increased dramatically since parks were first established, now numbering nearly 300 million visitors annually (see graph).

One of the smallest national parks, Congaree preserves the largest remaining intact hardwood bottomland (swamp) forest in the United States (see inset). It provides crucial terrestrial and aquatic wildlife habitat for many species.

Designating a national park does not happen overnight. The Sierra Club began campaigning to preserve the Congaree Swamp in 1969, and Congress established the Congaree Swamp National Monument in 1976. More than 25 years elapsed before the Congaree was designated a national park, after its growing recognition as a unique and important land resource.

Great Sand Dunes National Park features the highest sand dunes in North America, some as tall as 230 m (750 ft) (see large photograph, with the Sangre de Cristo Mountains in the background), which formed as rain and wind eroded the surrounding mountains. As with Congaree, establishing Great Sand Dunes National Park took several years and included its designation as a national monument. Land purchases involved the Nature Conservancy, as well as federal, state, and private donors.

National parks represent much more than wildlife sanctuaries. Parks preserve the land for future generations to enjoy the beauty of fast-disappearing natural areas.

Interpreting Data
During what decade were the most parks established? When did visitation change the most? How might you explain the trend in number of visitors between 2000 and 2010?

WileyPLUS

Number of U.S. National Parks and Annual Park Visitation, 1900–2010

Based on data from the National Park Service.

© Jerry Downs/SuperStock

CHAPTER OUTLINE

Land Use in the United States   322

Forests   324
- Forest Management
  - What a Scientist Sees: Harvesting Trees
  - EnviroDiscovery: Ecologically Certified Wood
- Deforestation
  - Environmental InSight: Tropical Deforestation
- Forests in the United States

Rangelands   333
- Rangeland Degradation and Desertification
- Rangeland Trends in the United States

National Parks and Wilderness Areas   336
- National Parks
  Environmental InSight: National Parks
- Wilderness Areas
- Management of Federal Lands

Conservation of Land Resources   341
  Case Study: The Tongass Debate over
  Clear-Cutting

Jane Leaman/Alamy

CHAPTER PLANNER

- ❑ Study the picture and read the opening story.
- ❑ Scan the Learning Objectives in each section:
  p. 322 ❑   p. 324 ❑   p. 333 ❑   p. 336 ❑   p. 341 ❑
- ❑ Read the text and study all figures and visuals.
  Answer any questions.

Analyze key features

- ❑ Process Diagram, p. 324
- ❑ What a Scientist Sees, p. 327
- ❑ EnviroDiscovery, p. 328
- ❑ Environmental InSight, p. 331 ❑   p. 337 ❑
- ❑ National Geographic Map, pp. 342–343 ❑
- ❑ Case Study, p. 344
- ❑ Stop: Answer the Concept Checks before
  you go on:
  p. 323 ❑   p. 332 ❑   p. 335 ❑   p. 340 ❑   p. 341 ❑

End of Chapter

- ❑ Review the Summary and Key Terms.
- ❑ Answer What is happening in this picture?
- ❑ Answer the Critical and Creative Thinking Questions.

321

Land Use in the United States

LEARNING OBJECTIVE

1. **Summarize** current land ownership in the United States.

Private citizens, corporations, and nonprofit organizations own more than 60 percent of the land in the United States, and Native American tribes own more than 2 percent. State and local governments own another 9 percent. The federal government owns the rest (nearly 28 percent).

Government-owned land encompasses all types of ecosystems, from tundra to desert, and includes land that contains important resources such as minerals and fossil fuels, land that possesses historical or cultural significance, and land that provides critical biological habitat. Most federally owned land is in Alaska and 11 western states (**Figure 13.1**).

Federal land is managed primarily by four agencies, three in the U.S. Department of the Interior—the Bureau of Land Management (BLM), the Fish and Wildlife Service (FWS), and the National Park Service (NPS)—and one in the U.S. Department of Agriculture—the U.S. Forest Service (USFS) (**Table 13.1**).

Government-owned lands and other rural areas provide vital **ecosystem services** that benefit humans living far from public forests, grasslands, deserts, and wetlands. These services include wildlife habitat, flood and erosion

Selected federal lands • Figure 13.1

Shown are national parks and preserves; national wildlife refuges; national forests, grasslands, and wilderness; and national marine sanctuaries in the United States. Note the preponderance of federal lands in western states and Alaska. Other federal lands, such as military installations and research facilities, aren't shown.

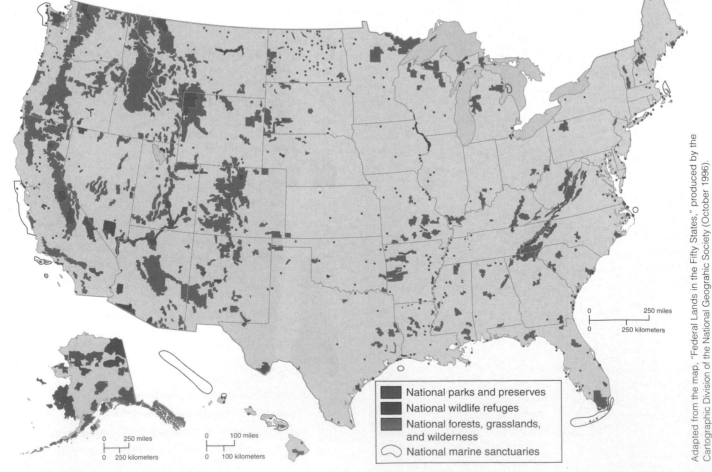

Legend:
- National parks and preserves
- National wildlife refuges
- National forests, grasslands, and wilderness
- National marine sanctuaries

Adapted from the map, "Federal Lands in the Fifty States," produced by the Cartographic Division of the National Geograhic Society (October 1996).

Administration of federal lands • Table 13.1

Agency	Land held	Primary uses	Area in millions of hectares (acres)
Bureau of Land Management (Dept. of the Interior)	National resource lands	Mining, livestock grazing, oil and natural gas extraction	102 (253)
U.S. Forest Service (Dept. of Agriculture)	National forests	Logging, recreation, conservation of watersheds, wildlife habitat, mining, livestock grazing, oil and natural gas extraction	78 (193)
U.S. Fish and Wildlife Service (Dept. of the Interior)	National wildlife refuges	Wildlife habitat; also logging, hunting, fishing, mining, livestock grazing, oil and natural gas extraction	38 (93)
National Park Service (Dept. of the Interior)	National Park Service	Recreation, wildlife habitat	34 (84)
Other—includes Department of Defense, Corps of Engineers (Dept. of the Army), and Bureau of Reclamation (Dept. of the Interior)	Remaining federal lands	Military uses, wildlife habitat	23 (57)

Source: U.S. Department of Interior, U.S. Department of Agriculture, and U.S. Department of Defense.

control, and groundwater recharge. Undisturbed land breaks down pollutants and recycles wastes. Natural environments provide homes for organisms. One of the best ways to maintain biological diversity and to protect endangered and threatened species is by preserving or restoring the natural areas to which these organisms are adapted.

Undisturbed public lands are ecosystems that scientists use as a benchmark, or point of reference, to determine the impact of human activity. Geologists, zoologists, botanists, ecologists, and soil scientists are some of the scientists who use government-owned lands for scientific inquiry. These areas provide perfect settings for educational experiences not only in science but also in history, because they can be used to demonstrate the condition of the land when humans originally settled it (**Figure 13.2**).

Public lands are important for their recreational value, providing places for hiking, swimming, boating, rafting, sport hunting, and fishing. Wild areas are important to the human spirit. Forest-covered mountains, rolling prairies, barren deserts, and other undeveloped areas are aesthetically pleasing and also help us recover from the stresses of urban and suburban living. We can escape the tensions of the civilized world by retreating, even temporarily, to the solitude of natural areas.

Not all public lands remain undeveloped. As you will see throughout this chapter, many public lands are developed for uses ranging from logging to cattle grazing to mineral extraction.

El Capitan and Bridalveil Fall in Yosemite National Park, California • Figure 13.2

Yosemite National Park, 1 of 58 national parks and nearly 400 total sites in the National Park Service system, includes rugged granite peaks and waterfalls, as well as valleys, meadows, and wilderness areas.

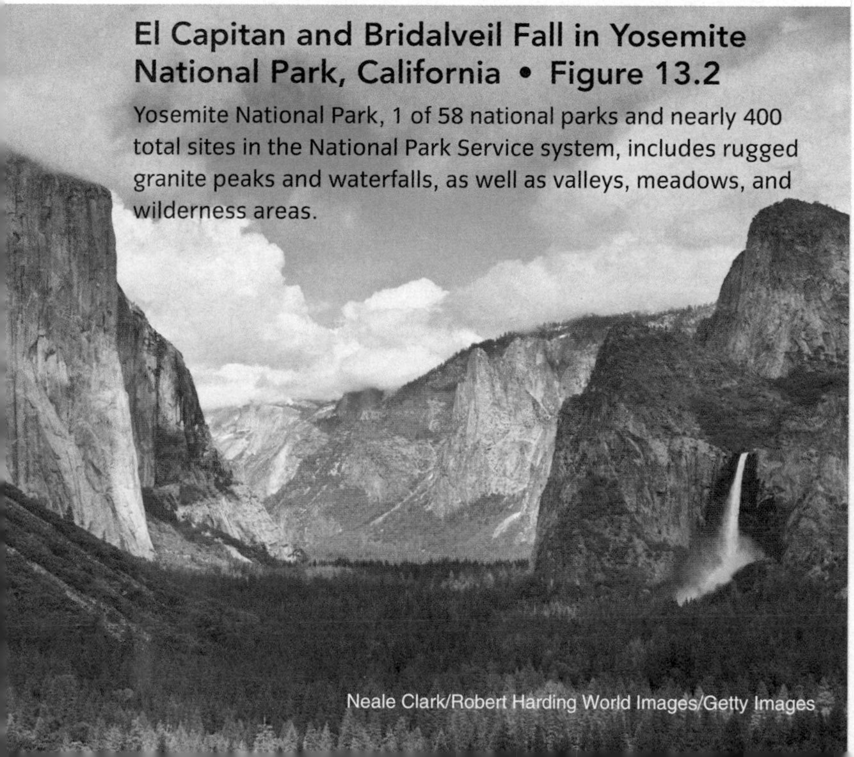

Neale Clark/Robert Harding World Images/Getty Images

CONCEPT CHECK STOP

1. **What** percentage of land in the United States is privately owned? What percentage is owned by the federal government?

Forests

LEARNING OBJECTIVES

1. **Define** *sustainable forestry* and explain how monocultures and wildlife corridors are related to it.

2. **Define** *deforestation*, including *clear-cutting*, and list the main causes of tropical deforestation.

3. **Describe** national forests and state which government agencies administer them and current issues of concern.

Forests, important ecosystems that provide many goods and services to support human society, occupy about one-fourth of Earth's total land area. Timber harvested from forests is used for fuel, construction materials, and paper products. Forests supply nuts, mushrooms, fruits, and medicines. Forests provide employment for millions of people worldwide and offer recreation and spiritual sustenance in an increasingly crowded world.

Forests also provide a variety of beneficial ecosystem services, such as influencing climate conditions. If you walk into a forest on a hot summer day, you will notice that the air is cooler and moister than it is outside the forest. This is the result of a biological cooling process called *transpiration*, in which water from the soil is absorbed by roots, transported through plants, and then evaporated from their leaves and stems. Transpiration provides moisture for clouds, eventually resulting in precipitation (**Figure 13.3**). Thus, forests help maintain local and regional precipitation.

PROCESS DIAGRAM

Role of forests in the hydrologic cycle • Figure 13.3

THE PLANNER

Forests return most of the water that falls as precipitation to the atmosphere by transpiration. When an area is deforested, almost all precipitation is lost as runoff.

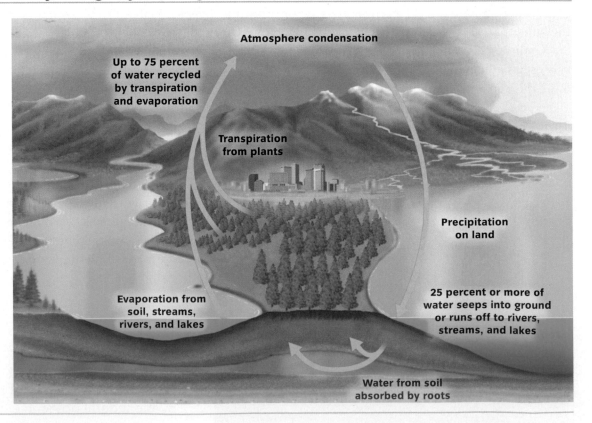

Atmosphere condensation

Up to 75 percent of water recycled by transpiration and evaporation

Transpiration from plants

Precipitation on land

Evaporation from soil, streams, rivers, and lakes

25 percent or more of water seeps into ground or runs off to rivers, streams, and lakes

Water from soil absorbed by roots

Think Critically

In which part of the hydrologic cycle do trees and other plants play a huge role? How would clearing the trees on a stretch of land disrupt this process? What would happen to water in soil there?

Forests play an essential role in regulating global biogeochemical cycles, such as those for carbon and nitrogen. Photosynthesis by Earth's approximately 1 trillion canopy trees removes large quantities of heat-trapping carbon dioxide from the atmosphere and fixes it into carbon compounds, while releasing oxygen back into the atmosphere. Forests thus act as carbon "sinks," which may help mitigate climate warming, and they produce oxygen, which almost all organisms require for cellular respiration.

Tree roots hold vast tracts of soil in place, reducing erosion and mudslides. Forests protect watersheds because they absorb, hold, and slowly release water; this moderation of water flow provides a more regulated flow of water downstream, even during dry periods, and helps control floods and droughts. Forest soils remove impurities from water, improving its quality. In addition, forests provide a variety of essential habitats for many organisms, such as mammals, reptiles, amphibians, fishes, insects, lichens and fungi, mosses, ferns, conifers, and numerous kinds of flowering plants.

Forest Management

Management for timber production disrupts a forest's natural condition and alters its species composition and other characteristics. Specific varieties of commercially important trees are planted, and those trees not as commercially desirable as others are thinned out or removed. Traditional forest management often results in low-diversity forests. In the southeastern United States, many tree plantations of young pine grown for timber and paper production are all the same age and are planted in rows a fixed distance apart (**Figure 13.4**). In the U.S. Northwest and southwestern Canada, tree farms of hybrid poplars are becoming increasingly common, primarily because of the tree's viability as a paper source. All of these "forests" are essentially monocultures—areas uniformly covered by one crop, like a field of corn. Herbicides are sprayed to kill shrubs and herbaceous plants between the rows. One of the disadvantages of monocultures is that they are at increased risk of damage from insect pests and disease-causing microorganisms. Consequently, pests and diseases must be controlled in managed forests, usually by applying insecticides and fungicides. Also, because managed forests contain few

> **monoculture** Ecological simplification in which only one type of plant is cultivated over a large area.

Wesley Hitt/Photographer's Choice/Getty Images

Tree plantation • Figure 13.4

This intensively managed pine plantation is a monoculture, with trees of uniform size and age. Such plantations have little biodiversity and provide limited habitat, but they supplement harvesting of trees in wild forests to provide the United States with the timber it requires. Photographed in Arkansas.

kinds of food, they can't support the variety of organisms typically found in natural forests.

In recognition of the many ecosystem services performed by natural forests, a newer method of forest management, known as **ecologically sustainable forest management**, or simply **sustainable forestry**, is evolving. Sustainable forestry maintains a mix of forest trees, by age and species, rather than imposing a monoculture. This broader approach seeks to conserve forests for the

> **sustainable forestry** The use and management of forest ecosystems in an environmentally balanced and enduring way.

long-term commercial harvest of timber and nontimber forest products. Sustainable forestry also attempts to sustain biological diversity by providing improved habitats for a variety of species, to prevent soil erosion and improve soil conditions, and to preserve watersheds that produce clean water. Effective sustainable forest management involves cooperation among environmentalists, loggers, farmers, indigenous peoples, and local, state, and federal governments.

When logging adheres to sustainable forestry principles, unlogged areas and **habitat corridors** are set aside

---

habitat corridor
A protected zone that connects isolated unlogged or undeveloped areas.

---

as sanctuaries for organisms. The purpose of habitat corridors is to provide animals with escape routes, should they be needed, and to allow them to migrate so they can interbreed. (Small, isolated, inbred populations may have an increased risk of extinction.) Habitat corridors may also allow large animals such as the Florida panther to maintain large territories. Some scientists question the effectiveness of habitat corridors, although recent research in fragmented landscapes suggests that habitat corridors help certain wildlife populations persist. Additional research is needed to determine the effectiveness of habitat corridors for all endangered species.

The actual methods for ecologically sustainable forest management that distinguish it from traditional forest management are under development. Such practices vary from one forest ecosystem to another, in response to different environmental, cultural, and economic conditions. In Mexico, many sustainable forestry projects involve communities that are economically dependent on forests. Because trees have such long life spans, scientists and forest managers of the future will judge the results of today's efforts.

Harvesting Trees According to the U.N. Food and Agriculture Organization (FAO), about 3.4 million m³ (120 million ft³) of wood are harvested annually (for fuelwood, timber, and other products). The five countries with the greatest tree harvests are the United States, Canada, Russia, Brazil, and China; these countries currently produce more than half the world's timber. About 50 percent of harvested wood is burned directly as fuelwood or used to make charcoal. (Partially burning wood in a large kiln from which air is excluded converts the wood into charcoal.) Most fuelwood and charcoal are used in developing countries (see Chapter 18). Highly developed countries consume more than three-fourths of the remaining 50 percent of harvested wood for paper and wood products.

Loggers harvest trees in several ways—selective cutting, shelterwood cutting, seed tree cutting, and clear-cutting (see *What a Scientist Sees*). **Selective cutting**, in which mature trees are cut individually or in small clusters while the rest of the forest remains intact, allows the forest to regenerate naturally. Selective cutting has fewer negative effects on the forest environment than other methods of tree harvest, but it is not as profitable in the short term because timber is not removed in great enough quantities.

The removal of all mature trees in an area over an extended period is **shelterwood cutting**. In the first year of harvest, undesirable tree species and dead or diseased trees are removed. Subsequent harvests occur at intervals of several years, allowing time for remaining trees to grow. Little soil erosion occurs with this method of tree removal, even though more trees are removed than in selective cutting.

In **seed tree cutting**, almost all trees are harvested from an area; a scattering of desirable trees is left behind to provide seeds for the regeneration of the forest.

Clear-cutting is harvesting timber by removing all trees from an area and then either allowing the area to reseed and regenerate itself naturally or planting the area with one or more specific varieties of trees. Timber companies prefer clear-cutting because it is the most cost-effective way to harvest trees. Clear-cutting in small patches

---

clear-cutting A logging practice in which all the trees in a stand of forest are cut, leaving just the stumps.

---

actually benefits some wildlife species, such as deer and certain songbirds. These species thrive in the regrowth of trees and shrubs that follows removal of the overhead canopy. However, clear-cutting over wide areas is ecologically unsound. It destroys biological habitats and increases soil erosion, particularly on sloping land, sometimes degrading land so much that reforestation doesn't take place. Mudslides on steep hillsides that were clear-cut can follow heavy rains, damaging properties and roads and killing people. Sometimes the land is so degraded from clear-cutting that reforestation does not take place. Clear-cut areas at lower elevations are usually regenerated successfully, whereas those at high elevations are often difficult to regenerate. Obviously, most recreational benefits of forests are lost when clear-cutting occurs.

WHAT A SCIENTIST SEES

Harvesting Trees

a. Aerial view of a large patch of clear-cut forest in British Columbia, Canada. Clear-cutting is the most common but most controversial type of logging. The obvious line is a road built to haul away the logs.

Stephen Sharnoff/NG Image Collection

b. As a forest scientist looks at a clear-cut forest, he or she may think about the various kinds of tree harvesting (1 to 3) that are less environmentally destructive than clear-cutting (4).

(1) In **selective cutting**, the older, mature trees are selectively harvested from time to time, and the forest regenerates itself naturally.

(2) In **shelterwood cutting**, less desirable and dead trees are harvested. As younger trees mature, they produce seedlings, which continue to grow as the now-mature trees are harvested.

(3) **Seed tree cutting** involves the removal of all but a few trees, which are allowed to remain to provide seeds for natural regeneration.

(4) In **clear-cutting**, all trees are removed from a particular site. Clear-cut areas may be reseeded or allowed to regenerate naturally.

EnviroDiscovery
Ecologically Certified Wood

Many homebuilders and homeowners are interested in "green" wood for flooring and other building materials (see photograph). Such wood is ecologically certified by a legitimate third party, such as the German-based Forest Stewardship Council (FSC), to have come from a forest managed with environmentally sound and socially responsible practices. Although these areas remain a small percentage of world forests, by mid 2012 the FSC had certified, as well managed, more than 150 million hectares (more than 371 million acres) in 80 countries. Certification is based on sustainability of timber resources, socioeconomic benefits provided to local people, and forest ecosystem health, which includes such considerations as preservation of wildlife habitat and watershed stability.

Green forestry has its detractors. Traditional forestry organizations are skeptical about the reliability of FSC investigations and the economic viability of this type of forestry. Trade experts caution that government efforts to specify the purchase of certified timber could violate global free-trade agreements. Still, green timber is gaining market share, pleasing business owners and consumers alike, and offering the promise of better conservation practices in managed forests.

Oliver Berg Deutsch Presse Agentur/NewsCom

The Forest Stewardship Council ecologically certifies "green" wood. Often, the consumer pays no additional premium, or only slightly more, for ecologically certified wood, which has become so popular that demand threatens to exceed supply.

Deforestation

The most serious problem facing the world's forests is deforestation. According to latest FAO estimates, world forests shrank by more than 13 million hectares (32 million acres) *annually* between 2000 and 2010. This amounts to a net 10-year loss equivalent to an area the size of Costa Rica. This estimate of forest loss does not take into account remaining forests that have been thinned or degraded by overharvesting, declining biological diversity, and reduced soil fertility. Causes of the decades-long trend of deforestation include fires triggered by drought and land-clearing practices, expansion of agriculture, construction of roads, tree harvests, insects, disease, and mining.

Most of the world's deforestation is currently taking place in South America and Africa, according to the FAO. Between 2000 and 2010, South America lost about 4.0 million hectares (9.9 million acres) of forest per year, and Africa lost 3.4 million hectares (8.4 million acres) annually. In contrast, estimated forested area in North

> **deforestation**
> The temporary or permanent clearance of large expanses of forest for agriculture or other uses.

America and Central America did not change between 2000 and 2010, and Europe and Asia actually gained forested areas, through either natural regrowth or increasing forest plantations.

Results of Deforestation Deforestation results in decreased soil fertility, as the essential mineral nutrients found in most forest soils leach away rapidly without trees to absorb them. Uncontrolled soil erosion, particularly on steep deforested slopes, affects the production of hydroelectric power as silt builds up behind dams. Increased sedimentation of waterways caused by soil erosion harms downstream fisheries. In drier areas, deforestation contributes to the formation of deserts (discussed shortly). Regulation of water flow is disrupted when a forest is removed, so that the affected region experiences alternating periods of flood and drought.

Deforestation contributes to the extinction of many species. (See Chapter 15 for a discussion of the importance of tropical forests as repositories of biological diversity.) Many tropical species, in particular, have limited ranges within a forest, so they are especially vulnerable to habitat modification and destruction. Migratory species, including birds and butterflies, also suffer because of deforestation.

Deforestation contributes to regional and global climate changes. Trees release substantial amounts of moisture into the air; in the hydrologic cycle, about 97 percent of the water that roots absorb from the soil is evaporated directly into the atmosphere and then falls back to Earth. When a large forested area is removed, local rainfall may decline, droughts may become more common in that region, and temperatures may rise slightly. Studies suggest that the local climate has become warmer and drier in parts of Brazil where huge tracts of the rain forest were burned. As the forest continues to shrink, it is possible that the changing regional climate will no longer be able to support the forest. Thus, large parts of what had once been tropical rain forest could become savanna.

Deforestation also contributes to an increase in global temperature by releasing carbon originally stored in the trees into the atmosphere as carbon dioxide, which enables the air to retain heat. When an old-growth forest is harvested, researchers estimate that it takes about 200 years for the replacement forest to accumulate the equivalent amount of carbon stored in the original trees.

Boreal Forests and Deforestation Extensive deforestation in boreal forests due to logging began in the late 1980s and continues today. **Boreal forests** occur in Alaska, Canada, Scandinavia, and northern Russia and are dominated by coniferous evergreen trees such as spruce, fir, cedar, and hemlock. The boreal forest biome is the world's largest, covering about 11 percent of Earth's land. Harvested primarily through clear-cut logging, boreal forests are the primary source of the world's industrial wood and wood fiber. The annual loss of boreal forests is estimated to encompass an area twice as large as the Amazonian rain forests of Brazil.

About 1 million hectares (2.5 million acres) of forest in Canada—currently the world's biggest timber exporter—are logged annually (**Figure 13.5**). Most of Canada's forests are subject to logging contracts, known as tenures, between provinces and companies. On the basis of current harvest quotas, logging was widely considered unsustainable in Canada until recently. In 2010 logging companies and environmental groups brokered an agreement in which logging was suspended in more than 300,000 km^2 (about 116,000 mi^2) of boreal forest, an area the size of Great Britain. An additional 385,000 km^2 (about 149,000 mi^2) were designated for restricted logging using sustainable guidelines.

Extensive tracts of Siberian forests in Russia are harvested, although estimates are unavailable. Alaska's boreal forests are at risk because the U.S. government may increase logging on public lands in the future.

Logging in Canada's boreal forest • Figure 13.5
About 80 percent of Canada's forest products are exported to the United States.

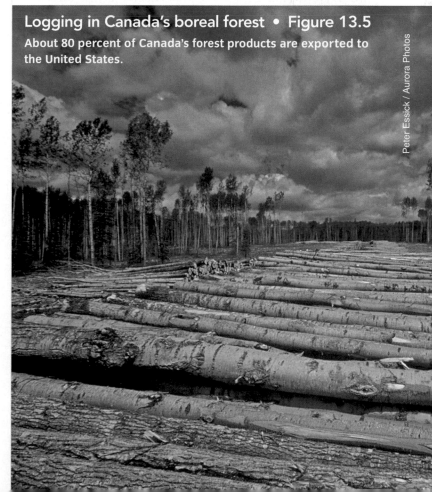

Peter Essick / Aurora Photos

Forests in the United States

In recent years, most temperate forests in the Rocky Mountains, Great Lakes region, and New England and other eastern states have been holding steady or even expanding. In Vermont, the amount of land covered by forests increased from 35 percent in 1850 to 75 percent in 2010. Expanding forests are the result of *secondary succession* on abandoned farms (see Figure 6.21), the commercial planting of tree plantations on both private and public lands, and government protection. Although these second- and third-growth forests generally don't have the biological diversity of virgin stands, many organisms have successfully become reestablished in the regenerated areas. The good news about these returning forests is tempered, however, by the fact that we are contributing to deforestation elsewhere by importing more timber to meet the increased demand for lumber, paper, and other wood products; we also import beef raised on former forest lands.

Slightly less than one-half of U.S. forests are privately owned (**Figure 13.7**); three-fourths of these private lands are in the Northeast and Midwest. Many private owners are under economic pressure to subdivide the land and develop tracts for housing or shopping malls, as they seek ways to recoup their high property taxes. Projected conversion of forests to agricultural, urban, and suburban lands over the next 40 years will have the greatest potential impact in the South, where more than 85 percent of forest is privately owned and logging is largely unregulated.

U.S. National Forests According to the USFS, the United States has 155 national forests encompassing 78 million hectares (193 million acres) of land, mostly in Alaska and western states. The USFS manages most national forests, and the BLM oversees the remainder. National forests have been established to provide U.S. citizens with the maximum benefits of natural resources such as fish, wildlife, and timber. Multiple uses include timber harvesting; mining; livestock foraging; hunting, fishing, and other forms of outdoor recreation; water resources and watershed protection; and habitat for fishes and wildlife. Recreation, which increased dramatically in national forests during the 1990s and early 2000s, ranges from camping at designated campsites to backpacking in the wilderness. Visitors to national forests swim, boat, picnic, and observe nature. With so many possible uses of national forests, conflicts inevitably arise, particularly between timber interests and those who wish to preserve the trees for other purposes.

Road building is a particularly contentious issue in national forests, in part because the USFS builds taxpayer-funded roads to allow private logging companies access to forests to remove timber. (See the *Case Study* at the end of the chapter for an example.) Road building in national forests is environmentally destructive when improper construction accelerates soil erosion and mudslides (particularly on steep terrain) and causes water pollution in streams. Biologists are concerned that the many roads that are built fragment wildlife habitat and provide entries for disease organisms and invasive species.

Forest ownership in the United States
• Figure 13.7

Much U.S. forest is privately owned.

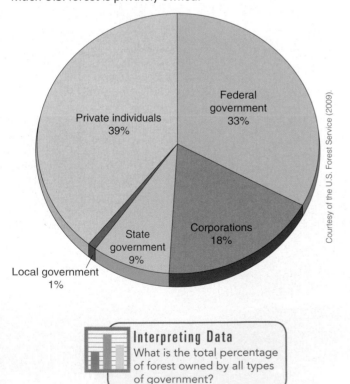

Courtesy of the U.S. Forest Service (2009).

Interpreting Data
What is the total percentage of forest owned by all types of government?

CONCEPT CHECK STOP

1. **What** is sustainable forestry?
2. **What** is deforestation? What are four important causes of tropical deforestation?
3. **Why** does the fact that U.S. national forests have been created for multiple uses result in both positive and negative outcomes? Give examples.

Rangelands

LEARNING OBJECTIVES

1. **Describe** rangelands and their general uses.
2. **Define** *desertification* and explain its relationship to overgrazing.
3. **Describe** how conservation easements help protect privately owned rangelands.
4. **Identify** the government agencies that administer public rangelands and describe current issues of concern.

Rangelands are grasslands, in both temperate and tropical climates, that serve as important areas of food production for humans by providing fodder for livestock such as cattle, sheep, and goats (**Figure 13.8**). Rangelands may be mined for minerals and energy resources, used for recreation, and preserved for biological habitat and for soil and water resources. The predominant vegetation of rangelands includes grasses, *forbs* (small plants other than grasses), and shrubs.

> **rangeland**
> Grassland area that is not intensively managed and is used for grazing livestock.

Rangeland Degradation and Desertification

Grasses, the predominant vegetation of rangelands, have *fibrous root systems*, in which many roots form diffuse networks in the soil to anchor the plants. Plants with fibrous roots hold the soil in place quite well, thereby reducing soil erosion. Grazing animals eat the leafy shoots of the grass, and the fibrous roots continue to develop, allowing the plants to recover and regrow to their original size.

Carefully managed grazing is beneficial for grasslands. Because rangeland vegetation is naturally adapted to grazing, when grazing animals remove mature vegetation, the activity stimulates rapid regrowth. At the same time, the hooves of grazing animals disturb the soil surface enough to allow rainfall to more effectively reach the root systems of grazing plants. Several studies have reported that moderate levels of grazing encourage greater plant diversity.

The **carrying capacity** of a rangeland is the maximum number of animals the natural vegetation can sustain over an indefinite period without deterioration of the ecosystem. When the carrying capacity of a rangeland

Rangeland • Figure 13.8

Rangeland is considered a renewable resource when its carrying capacity—the number of animals it can sustain without suffering deterioration—is not exceeded. Photographed along the Salmon River in Idaho.

Steve Smith/Superstock

is exceeded, **overgrazing** of grasses and other plants occurs. When plants die, the ground is left barren, and the exposed soil is susceptible to erosion. Sometimes plants that do not naturally grow in a rangeland but that can tolerate the depleted soil invade an overgrazed area. In parts of the Texas Hill Country that were overgrazed, junipers—which are not good forage food—replaced the lush grasses.

Most of the world's rangelands lie in semiarid areas that have natural extended droughts. During dry periods, the carrying capacity of the rangeland is considerably lower because the lack of precipitation reduces plant productivity.

Native grasses in these dry lands can survive severe drought: The aboveground portion of the plant dies back, but the extensive root system remains alive and holds the soil in place. When the rains return, the roots develop new shoots. But when an extended drought

overgrazing
A situation that occurs when too many grazing animals consume the plants in a particular area, leaving the vegetation destroyed and unable to recover.

desertification
Degradation of once-fertile rangeland or tropical dry forest into nonproductive desert.

occurs in conjunction with overgrazing, once-fertile rangeland may be converted to desert as reduced grass cover allows winds to erode the soil. Even when the rains return, the degradation may be so extensive that the land cannot recover. Water erosion removes the little remaining topsoil, and the sand left behind forms dunes.

Land degradation is both a natural and a human-induced process that decreases the future ability of the land to support crops or livestock. This progressive degradation, which induces unproductive desert-like conditions on formerly productive rangeland (or tropical dry forest), is **desertification** (**Figure 13.9**). It reduces the agricultural productivity of economically valuable land, forces out many organisms, and threatens endangered species. Worldwide, desertification seems to be on the increase. The United Nations estimates that

Desertification in the African Sahel region, Mali • Figure 13.9

Livestock in Mali, Africa, have eaten all the ground cover. The dead trees were stripped of branches to feed livestock and provide firewood. Overgrazing, in combination with extended drought, as well as with overexploitation by desperately poor people, is increasing the amount of unproductive desert area in the Sahel.

Global Locator

NG Maps

SAHEL

AFRICA

NATIONAL GEOGRAPHIC

K. Tumanowicz/Science Source

each year since the mid-1990s, 3,560 km² (1,374 mi²)—an area about the size of Rhode Island—has turned into desert.

Rangeland Trends in the United States

Rangelands make up approximately 30 percent of the total land area in the United States, mostly in the western states. Of this, approximately one-third is publicly owned

| conservation easement A legal agreement that protects privately owned forest, rangeland, or other property from development for a specified number of years. |

and two-thirds is privately owned. Much of the private rangeland is under increasing pressure from developers, who want to subdivide the land into lots for homes and condominiums. To preserve the open land, conservation groups often pay ranchers for **conservation easements** that prevent future owners from developing the land. An estimated 400,000 hectares (1 million acres) of private rangelands are protected by conservation easements.

Excluding Alaska, there are at least 89 million hectares (220 million acres) of public rangelands in the United States. The BLM manages approximately 69 million hectares (170 million acres) of public rangelands, and the USFS manages an additional 20 million hectares (50 million acres).

Overall, the condition of public rangelands in the United States has slowly improved since the low point of the Dust Bowl in the 1930s, when the combined effects of poor agricultural practices, severe winds, and extended drought led to devastating soil erosion and dramatic declines in soil productivity. Much of this improvement is attributed to fewer livestock being permitted to graze the rangelands after passage of the **Taylor Grazing Act** of 1934, the **Federal Land Policy and Management Act** of 1976, and the **Public Rangelands Improvement Act** of 1978. Better livestock management practices, such as controlling the distribution of animals on a range through fencing or herding, as well as scientific monitoring, have also contributed to rangeland recovery.

But restoration is slow and costly, and more is needed. Rangeland management includes seeding in places where plant cover is sparse or absent, conducting controlled burns to suppress shrubby plants, constructing fences to allow rotational grazing, controlling invasive weeds, and protecting habitats of endangered species. Most livestock operators use public rangelands in a way that results in their overall improvement.

Issues Involving Public Rangelands The federal government distributes permits that allow private livestock operators to use public rangelands for grazing in exchange for a fee that is much lower than the cost of grazing on private land. The permits are held for many years and are not open to free-market bidding by the general public—that is, only ranchers who live in the local area are allowed to obtain grazing permits. As of late 2011, nearly 18,000 federal (BLM) grazing permits and leases were in force, all in western states. Montana, Wyoming, and New Mexico are the states holding the most federal grazing permits, together accounting for almost half of the national total. Most grazing permits are issued for cattle; however, they also cover rangeland use for horses, burros, sheep, and goats.

Some environmental groups are concerned about the ecological damage caused by overgrazing of public rangelands and want to reduce the number of livestock animals allowed to graze. They want public rangelands managed for other uses, such as biological habitat, recreation, and scenic value, rather than exclusively for livestock grazing. To accomplish this goal, they would like to purchase grazing permits and set aside the land for nongrazing purposes.

Conservative economists have joined environmentalists in criticizing the management of federal rangelands. According to policy analysts at Taxpayers for Common Sense, in 2010 taxpayers contributed at least $115 million more than the grazing fees collected in order to support grazing on public rangelands. This money is used to manage and maintain the rangelands, including installing water tanks and fences, and to repair damage caused by overgrazing. Taxpayers for Common Sense and other free-market groups want grazing fees increased to cover all costs of maintaining herds on publicly owned rangelands.

CONCEPT CHECK STOP

1. **What** are rangelands?
2. **How** can overgrazing of rangeland lead to desertification?
3. **How** do conservation easements help protect privately owned rangelands?
4. **Which** agencies manage public rangelands? What management issues do they face?

National Parks and Wilderness Areas

LEARNING OBJECTIVES

1. **State** which government agency administers U.S. national parks and describe current issues of concern.

2. **Define** *wilderness* and discuss the administration and goals of the National Wilderness Preservation System and the problems faced by wilderness areas.

3. **Explain** the differences between the wise-use and environmental movements' views on the use of public lands.

Many acres of federal land are set aside either as national park property or as wilderness areas. Both types of land were established to encourage the protection of the natural environment, and both experience conflicts associated with how best to use and manage these protected areas.

National Parks

In 1872 Congress established the world's first national park, Yellowstone National Park, in federal lands in the territories of Montana and Wyoming. The purpose of the park was to protect great scenic beauty and biological diversity in an unimpaired condition for present and future generations. Created in 1916 as a federal bureau in the Department of the Interior, the **National Park Service** (NPS) was originally composed of large, scenic areas in the West such as Yellowstone, Yosemite Valley, and the Grand Canyon (**Figure 13.10a**). Today the NPS has more cultural and historical sites—battlefields and historically important buildings and towns—than places of scenic wilderness. The NPS currently administers 397 sites, 58 of which are national parks (see the introduction to this chapter), encompassing 34.2 million hectares (84.4 million acres).

Because the NPS believes that knowledge and understanding increase enjoyment, one of its primary roles is to teach people about the natural environment, management of natural resources, and history of a site by providing nature walks and guided tours of its parks. Exhibits along roads and trails, evening campfire programs, museum displays, and lectures are other common educational tools.

The popularity and success of U.S. national parks (**Figure 13.10b**) have encouraged many other nations to establish national parks. Today the U.N. Environment Programme identifies more than 3500 national parks, as defined by the International Union for the Conservation of Nature (IUCN), in nearly 100 countries. As in the United States, parks in other countries usually have multiple roles, ranging from providing biological habitat to facilitating human recreation.

Threats to U.S. Parks Some national parks are overcrowded (**Figure 13.10c**). Problems plaguing urban areas are also found in popular national parks during peak seasonal use, including crime, vandalism, litter, traffic jams, and pollution of the soil, water, and air. In addition, thousands of resource violations, from cutting live trees and collecting plants, minerals, and fossils, to defacing historical structures with graffiti and setting fires, are investigated in national parks each year. Park managers have had to reduce visitor access to park areas that have become degraded from overuse, and in some cases to restrict vehicle traffic (**Figure 13.10d**).

Many people think more funding is needed to maintain and repair existing parks. Facilities at some of the largest, most popular parks, such as Yosemite, the Grand Canyon, and Yellowstone, were last upgraded some 30 years ago. Recreation fees account for less than $200 million of the approximately $3.1 billion a year the NPS spends. Although steps were taken to make parks more self-sufficient, they still depend on general tax revenues to pay for their operations.

Some national parks have imbalances in wildlife populations. Populations of many mammal species are in decline, including bears, white-tailed jackrabbits, and red foxes. For example, grizzly bear populations in national parks of the western United States are threatened. Grizzlies are territorial and require large areas of wilderness as habitat, and the presence of humans in national parks may adversely affect them. Most importantly, the parks may be too small to support grizzlies. Fortunately, so far grizzly bears have survived in sustainable numbers in Alaska and Canada.

Other mammal populations—notably elk—have proliferated. Elk in Yellowstone National Park's northern range increased from a population of 3100 in 1968 to a record high of 19,000 in 1994. Ecologists documented that elk reduced the abundance of native vegetation, such as willow and aspen, and seriously eroded stream banks.

a. Grand Canyon National Park in Arizona.
This view from Mohave Point shows sunset over the south rim of the canyon.

© Ivan Kuzmin/Alamy

BWAC Images/Alamy

c. Heavy Traffic at Cades Cove, Great Smoky Mountains National Park in Tennessee.
Great Smoky Mountains National Park receives the most visitors of any U.S. national park. The popularity of certain national parks threatens to overwhelm them.

Jim Parkin/Alamy

d. Shuttle Bus at Zion National Park in Utah.
In operation since 2000, the Zion National Park shuttle system eliminates traffic congestion. During peak months, personal vehicles are banned from popular areas; visitors instead ride the propane-powered shuttle.

b. The 10 Most Popular National Parks.

Courtesy of the National Park Service.

National park	Number of recreational visitors in 2011 (in millions)
Great Smoky Mountains (North Carolina, Tennessee)	9.0
Grand Canyon (Arizona)	4.3
Yosemite (California)	3.9
Yellowstone (Wyoming, Montana, Idaho)	3.4
Rocky Mountain (Colorado)	3.2
Olympic (Washington)	3.0
Zion (Utah)	2.8
Grand Teton (Wyoming)	2.6
Acadia (Maine)	2.4
Cuyahoga Valley (Ohio)	2.2
Total visitors to all parks in the national park system	278.9

National Parks and Wilderness Areas   337

Tom Murphy/NG Image Collection

Gray wolves prey on elk in Yellowstone National Park • Figure 13.11

Since their reintroduction, gray wolf populations have gained a secure foothold in Yellowstone National Park. Early studies of the effects of these predators support scientists' predictions that wolves will help reduce the burgeoning elk population.

The reintroduction of gray wolves to Yellowstone, which began in 1995, has helped reduce the elk population (**Figure 13.11**), which in turn has led to improved aspen and willow growth and growing numbers of herbivores.

National parks are increasingly becoming islands of natural habitat surrounded by human development. Development on the borders of national parks limits the areas in which wild animals may range, forcing them into isolated populations. Ecologists have found that when environmental stressors occur, several small "island" populations are more likely to become threatened than a single large population occupying a sizable range (see Chapter 15).

Wilderness Areas

Wilderness encompasses regions where the land and its community of organisms are not greatly disturbed by human activities, where humans may visit but don't live permanently. The U.S. Congress recognized that increased

> **wilderness**
> A protected area of land in which no human development is permitted.

human population and expansion into wilderness areas might result in a future where no lands exist in their natural condition. Accordingly, the **Wilderness Act** of 1964

authorized the U.S. government to set aside federally owned land that retains its primeval character and lacks permanent improvements or human habitation, as part of the **National Wilderness Preservation System** (NWPS). These federal lands range in size from tiny islands of uninhabited land to portions of national parks, national forests, and national wildlife refuges (**Figure 13.12**). Although mountains are the most common wilderness areas, portions of other ecosystems have been set aside, including tundra, desert, and wetlands.

Areas designated as wilderness are given the highest protection of any federal land. These areas are to remain natural and unchanged so they will be unimpaired for future generations to enjoy. The same four government agencies that regulate all publicly owned land—the NPS, USFS, FWS, and BLM—oversee 757 wilderness areas comprising 44.1 million hectares (109 million acres) of land. More than one-half of the lands in the NWPS lie in Alaska, and western states are home to much of the remainder. Because few sites untouched by humans exist in the eastern states, requirements were modified in 1975 so that the wilderness designation could be applied to certain federally owned lands where forests are recovering from logging.

Millions of people visit U.S. wilderness areas each year, and some areas are overwhelmed by this traffic: Eroded trails, soil and water pollution, litter and trash, and human congestion predominate over quiet, unspoiled land. Government agencies now restrict the number of people allowed into each wilderness area at one time so that human use doesn't seriously affect the wilderness.

Some of the most popular wilderness areas may require more intensive future management, such as the development of trails, outhouses, cabins, and campsites. These amenities are not encountered in true wilderness, posing a dilemma between wilderness preservation and human use and enjoyment of wild lands.

Limiting the number of human guests in a wilderness area doesn't control all the factors that threaten wilderness, however. **Invasive species** have the potential to upset the natural balance among native species. For example, white pine blister rust, a foreign (non-native) fungus that kills white pine trees, has invaded the wilderness in the northern Rocky Mountains. Wilderness managers are concerned that declining white pine populations could affect the population of grizzly bears in the region because pine seeds are a major part of the grizzlies' diet. The Wilderness Act specifies both the preservation of natural conditions and the avoidance of intentional

Diverse wilderness areas protected in the National Wilderness Preservation System • Figure 13.12

a. At more than 3.7 million hectares (9.1 million acres), the Wrangell–Saint Elias Wilderness in Alaska is the nation's largest designated wilderness area.

Rich Reid /NG Image Collection

Kevin Schafer/Alamy

James Schwabel/Alamy

b. The Cabeza Prieta Wilderness, Arizona's largest wilderness area, includes 325,000 hectares (803,000 acres) of isolated desert landscapes.

c. The Pelican Island National Wildlife Refuge in Florida is home to the 2.4-hectare (6-acre) Pelican Wilderness, the smallest protected area in the NWPS.

ecological management. In this example, the only way to preserve as much as possible of the original wilderness may be to intentionally manipulate the white pine population by breeding and planting fungus-resistant trees.

Large tracts of wilderness, most of it in Alaska, have been added to the NWPS since passage of the Wilderness Act in 1964. People who view wilderness as a nonrenewable resource support the designation of additional wilderness areas, particularly in the lower 48 states. In March

2009 President Barack Obama signed the Omnibus Public Land Management Act of 2009, which designated 52 new U.S. wilderness areas and added almost 1 million hectares (more than 2 million acres) to the NWPS, most of it outside Alaska. Increasing the amount of federal land designated as wilderness in the NWPS is opposed by groups who operate businesses on public lands (such as timber, mining, ranching, and energy companies) and by their political representatives.

Management of Federal Lands

How do we best manage the legacy of federal lands? Should federal lands be managed under multiple uses, or should they be preserved so that they benefit U.S. citizens for generations to come? These questions have divided many Americans into two groups, each a coalition of several hundred grassroots organizations. Those who wish to exploit resources on federal lands are known collectively as the *wise-use movement*; this group also includes many corporations. Those who wish to preserve the resources on federally owned lands are known collectively as the *environmental movement*. The descriptions of the two movements that follow are mainstream; any given group may not support all of the listed goals of a movement.

In general, people who support the wise-use movement think that the government overregulates environmental protection and that property owners should have more flexibility to use natural resources. They believe that the primary purpose of federal lands is to enhance economic growth (**Figure 13.13**).

A logging site in Gifford Pinchot National Forest, Cascade Range, Washington
• Figure 13.13

The wise-use movement favors opening federal lands to logging and other types of economic development.

James P. Blair/NG Image Collection

Some of the goals of the wise-use movement include the following:

1. Put all national forests, including (and especially) old-growth forests, under timber management.

2. Permit mining and commercial development of wilderness areas, wildlife refuges, and national parks, where appropriate.

3. Allow unrestricted development of wetlands.

4. Sell parts of resource-rich federal lands to private interests, such as mining, oil, coal, ranching, and timber groups, for resource extraction.

Many organizations that embrace the wise-use movement have environmentally friendly names. The National Wetlands Coalition, for example, consists primarily of real estate developers and energy companies that want to drain and develop wetlands. Similarly, logging companies support the American Forest Resource Alliance.

In contrast to the wise-use movement, the environmental movement views federal lands as a legacy of U.S. citizens. They think that:

1. The primary purpose of public lands is to protect biological diversity and ecosystem integrity.

2. Those who extract resources from public lands should pay U.S. citizens compensation equal to the fair market value of the resource and not be subsidized by taxpayers.

3. Those who use public lands should be held acountable for any environmental damage they cause.

CONCEPT CHECK STOP

1. **What** government agency administers the National Park System? What problems does it face?

2. **What** is wilderness, and how does the U.S. National Wilderness Preservation System seek to protect it? What problems do protected wilderness areas face?

3. **How** do the wise-use and environmental movements differ in their views on the use of public lands? What conflicts might arise as a result of these differences?

Conservation of Land Resources

LEARNING OBJECTIVES

1. **Name** at least three of the most endangered ecosystems in the United States.

2. **Describe** several of the criteria used to evaluate whether an ecosystem is endangered.

O ur ancestors considered natural areas an unlimited resource to exploit. They appreciated prairies as valuable agricultural land and forests as immediate sources of lumber and eventual farmland. This outlook was practical as long as there was more land than people needed. But as the population increased and the amount of available land decreased, people began to view land as a limited resource. Thus, exploitation has increasingly shifted to preservation of the remaining natural areas in the United States and elsewhere around the world (**Figure 13.14**).

Although all types of ecosystems must be conserved, several are in particular need of protection. The U.S. Geological Survey and the Defenders of Wildlife commissioned studies that ranked the most endangered ecosystems in the United States. They used four criteria:

1. The area lost or degraded since Europeans colonized North America

2. The number of present examples of a particular ecosystem, or the total area

3. An estimate of the likelihood that a given ecosystem will lose a significant area or be degraded during the next 10 years

4. The number of threatened and endangered species living in that ecosystem

Table 13.2 lists the 15 most endangered U.S. ecosystems based on these criteria. Examples include the South Florida landscape, southern Appalachian spruce-fir forests, and longleaf pine forests and savannas. As these ecosystems are lost and degraded, the organisms that compose them decline in number and in genetic diversity. Researchers have also found that some rare types of soils are threatened, endangering individual species that rely on them. Implementing conservation strategies

The 15 most endangered ecosystems in the United States (in order of priority) • Table 13.2

South Florida landscape

Southern Appalachian spruce-fir forests

Longleaf pine forests and savannas

Eastern grasslands, savannas, and barrens

Northwestern grasslands and savannas

California native grasslands

Coastal communities in the lower 48 states and Hawaii

Southwestern riparian communities

Southern California coastal sage scrub

Hawaiian dry forest

Large streams and rivers in the lower 48 states and Hawaii

Cave systems

Tallgrass prairie

California river- and stream-bank communities and wetlands

Florida scrub

Source: From Box 1.1 in Noss, R.F., M.A. O'Connoll, and D.D. Murphy. *The Science of Conservation Planning: Habitat Conservation Under the Endangered Species Act,* Island Press. World Wildlife Fund (1997).

that set aside ecosystems is the best way to preserve an area's biodiversity.

As you have seen in this chapter, government agencies, private conservation groups, and private citizens have begun to set aside natural areas for permanent preservation. Such activities ensure that our children and grandchildren will inherit a world with wild places and other natural ecosystems.

CONCEPT CHECK STOP

1. **What** are three U.S. ecosystems that need protection? How do you think humans have caused these ecosystems to become endangered?

2. **What** are three criteria used to evaluate whether an ecosystem is endangered?

GeoBytes

LARGEST NATIONAL PARK
North East Greenland National Park, Greenland, 972,000 km² (375,000 mi²)

LARGEST MARINE PARK
Northwestern Hawaiian Islands Marine National Monument, U.S., 360,000 km² (140,000 mi²)

LARGEST TROPICAL FOREST PARK
Tumucumaque National Park in the Brazilian Amazon 24,135 km² (9319 mi²)

BIODIVERSITY HOTSPOTS
Conservation International identifies world regions suffering from a severe loss of biodiversity.

WORLD HERITAGE SITES
The United Nations Educational, Scientific, and Cultural Organization (UNESCO) recognizes natural and cultural sites of "universal value."

UNPROTECTED AREA
88% of Earth's land surface

PROTECTED AREA
12%

NG Maps

Protected areas worldwide represent 12 percent of the Ear land surface, according to the U Environmental Programme Wor Conservation Monitoring Centr Only 0.5 percent of the marine environment is within protected areas—an amount considered inadequate by conservationists because of the increasing threa of overfishing and coral reef los worldwide.

HAWAII VOLCANOES NATIONAL PARK, HAWAII
The park includes two of the world's most active volcanoes, Kilauea and Mauna Loa. The landscape shows the results of 70 million years of volcanism, including calderas, lava flows, and black sand beaches. Lava spreads out to build the island, and seawater vaporizes as lava hits the ocean at 1149°C (2100°F). The national park, created in 1916, covers 10 percent of the island of Hawaii and is a refuge for endangered species such as the hawksbill turtle and Hawaiian goose. It was made a World Heritage site in 1987.

GALÁPAGOS NATIONAL PARK, ECUADOR
Galápago means "tortoise" in Spanish, and at one time 250,000 giant tortoises roamed the islands. Today about 15,000 remain, and 3 of the original 14 subspecies are extinct—and the Pinta Island tortoise may be extinct soon. In 1959, Ecuador made the volcanic Galápagos Islands a national park, protecting the giant tortoises and other endemic species. The archipelago became a World Heritage site in 1978, and a marine reserve surrounding the islands was added in 2001.

WESTERN UNITED STATES
An intricate mix of public lands—including national forests, wilderness areas, wildlife refuges, and national parks such as Arches (above)—embraces nearly half the surface area of 11 western states. Ten out of 19 World Heritage sites in the United States are found here. It was in the West that the modern national park movement was born in the 19th century, with the establishment of Yellowstone and Yosemite National Parks.

MADIDI NATIONAL PARK, BOLIVIA
Macaws may outnumber humans in Madidi, Bolivia's second largest national park, established in 1995. A complex community of plants, animals, and native Indian groups share this 18,900–km² (7300–mi²) reserve, part of the Tropical Andes biodiversity hotspot. Indigenous communities benefit from ecotourism.

AMAZON BASIN, BRAZIL
Indigenous peoples help manage reserves in Brazil that are linked with Jaú National Park. The park and reserves are part of the Central Amazon Conservation Complex, a World Heritage site covering more than 60,000 km² (23,000 mi²). It is the largest protected area in the Amazon Basin and one of the most biologically rich regions on the planet.

ARCTIC REGIONS
Polar bears find safe havens in Canadian parks, such as on Ellesmere Island, and in Greenland's huge protected area—Earth's largest—that preserves the island's frigid northeast. In 1996 countries with Arctic lands adopted the Circumpolar Protected Areas Network Strategy and Action Plan to help conserve ecosystems. Today 15 percent of Arctic land area is protected.

GLOBAL LOCAL

What do protected areas around the world have in common? How are their concerns shared by state and local governments and organizations?

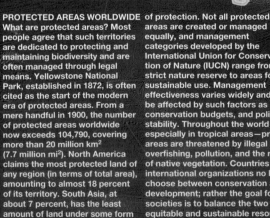

PROTECTED AREAS WORLDWIDE
What are protected areas? Most people agree that such territories are dedicated to protecting and maintaining biodiversity and are often managed through legal means. Yellowstone National Park, established in 1872, is often cited as the start of the modern era of protected areas. From a mere handful in 1900, the number of protected areas worldwide now exceeds 104,790, covering more than 20 million km² (7.7 million mi²). North America claims the most protected land of any region (in terms of total area), amounting to almost 18 percent of its territory. South Asia, at about 7 percent, has the least amount of land under some form of protection. Not all protected areas are created or managed equally, and management categories developed by the International Union for Conservation of Nature (IUCN) range from strict nature reserve to areas for sustainable use. Management effectiveness varies widely and can be affected by such factors as conservation budgets, and political stability. Throughout the world—but especially in tropical areas—protected areas are threatened by illegal hunting, overfishing, pollution, and the removal of native vegetation. Countries and international organizations no longer choose between conservation and development; rather the goal for societies is to balance the two for equitable and sustainable resource use.

◀ **WILDEST AREAS**
Although generally far from cities, the world's remaining wild places play a vital role in a healthy global ecosystem. The boreal forests of Canada and Russia, for instance, help clean the air we breathe by absorbing carbon dioxide and providing oxygen. With the human population increasing by an estimated 1 billion over the next 15 years, many wild places could fall within reach of the plow or under a cloud of smog.

For millennia, lands have been set aside as sacred ground or as hunting reserves for the powerful. Today, great swaths are protected for recreation, habitat conservation, biodiversity preservation, and resource management. Some groups may oppose protected spaces because they want access to resources now. Yet local inhabitants and governments are beginning to see the benefits of conservation efforts and sustainable use for human health and future generations.

Wildest biomes
■ Wildlands
□ Ice or snow cover

NG Maps

SAREKS NATIONAL PARK, SWEDEN
This remote 1970–km² (760–mi²) park, established in 1909 to protect the alpine landscape, is a favorite of backcountry hikers. It boasts some 200 mountains more than 1800 m (5900 ft) high, narrow valleys, and about 100 glaciers. Sareks forms part of the Laponian Area World Heritage site and has been a home to the Saami (or Lapp) people since prehistoric times.

AFRICAN RESERVES
Some 120,000 elephants roam Chobe National Park in northern Botswana. Africa has more than 7500 national parks, wildlife reserves, and other protected areas, covering about 9 percent of the continent. Protected areas are under enormous pressure from expanding populations, civil unrest and war, and environmental disasters.

WOLONG NATURE RESERVE, CHINA
Giant pandas freely chomp bamboo in this 2000–km² (772–mi²) reserve in Sichuan Province, near the city of Chengdu. Misty bamboo forests host a number of endangered species, but the critically endangered giant panda—among the rarest mammals in the world—is the most famous resident. Only about 1600 giant pandas exist in the wild.

KAMCHATKA, RUSSIA
Crater lakes, ash-capped cones, and diverse plant and animal species mark the Kamchatka Peninsula—a World Heritage site—located between the icy Bering Sea and Sea of Okhotsk. The active volcanoes and glaciers form a dynamic landscape of great beauty, known as "The Land of Fire and Ice." Kamchatka's remoteness and rugged landscape help fauna flourish, producing record numbers of salmon species and half of the Steller's sea-eagles on Earth.

GUNUNG PALUNG NATIONAL PARK, INDONESIA
A tree frog's perch could be precarious in this 900-km² (347–mi²) park on the island of Borneo, in the heart of the Sundaland biodiversity hotspot. The biggest threat to trees and animals in the park and region is illegal logging. Gunung Palung contains a wider range of habitats than any other protected area on Borneo, from mangroves to lowland and cloud forests. A number of endangered species, such as orangutans and sun bears, depend on the dense forests.

AUSTRALIA & NEW ZEALAND
Uluru, a red sandstone monolith (formerly known as Ayers Rock), and the vast Great Barrier Reef, one of the largest marine parks in the world, are outstanding examples of Australia's protected areas—which make up more than 10 percent of the country's area and conserve a diverse range of unique ecosystems. About a third of New Zealand is protected, and it is a biodiversity hotspot because of threats to flightless native birds, such as the kakapo and kiwi. Cats, stoats, and other predators, introduced to New Zealand by settlers, kill thousands of birds each year.

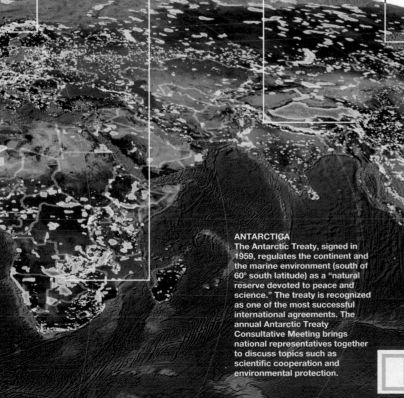

ANTARCTICA
The Antarctic Treaty, signed in 1959, regulates the continent and the marine environment (south of 60° south latitude) as a "natural reserve devoted to peace and science." The treaty is recognized as one of the most successful international agreements. The annual Antarctic Treaty Consultative Meeting brings national representatives together to discuss topics such as scientific cooperation and environmental protection.

NG Maps

NATIONAL GEOGRAPHIC

CASE STUDY

The Tongass Debate over Clear-Cutting

Despite its northern location along Alaska's southeastern coast, the Tongass National Forest is one of the world's few temperate rain forests (**Figure a**; also see Chapter 6 for a description of the temperate rainforest biome). It is one of the wettest places in the United States. This moisture supports old-growth forest of giant Sitka spruce, yellow cedar, and western hemlock, some of which are 700 years old. This 6.9-million-hectare (17-million-acre) forest, the largest in the National Forest System, provides habitat for a wealth of wildlife, such as grizzly bears and bald eagles.

The Tongass is a prime logging area because a single large Sitka spruce may yield as much as 23.6 m² (10,000 board ft) of high-quality timber (**Figure b**). The logging industry forms the basis of much of the local economy but conflicts with environmental interests seeking to avoid overharvesting. Regeneration of mature forest after it is clear-cut can take several centuries.

As in most other national forests, it is expensive to log in the Tongass. To cover high operating costs, timber interests such as pulp mills rely on obtaining the timber from the federal government at below-market prices. This right was granted in 1954 by a contract that expired in the 1990s. In 1990, congressional efforts to pass the Tongass Timber Reform Act, which would force timber interests to pay market prices, were bitterly opposed. The compromise agreement, reached in 1997, provided timber to the mills at market prices. As a result of this legislation, clear-cut logging continued in the Tongass, but at lower rates than in the past.

In 1999, the Tongass Land Management Plan of 1997 was modified after several dozen appeals were filed against it. The modified plan protected an additional 40,500 hectares (100,000 acres) of old-growth forest from logging, bringing the total protected area in the Tongass to 95,000 hectares (234,000 acres),

and increased timber harvest rotations from 100 years to 200 years in specially designated wildlife areas. This change reduces the impact of forest fragmentation and protects the Sitka black-tailed deer population, used for food by native tribes.

In 2008 a new amendment to the Forest Plan for the Tongass Forest was announced. This plan replaces the 1997 Tongass Land Management Plan and provides direction for managing the land and resources of the Tongass National Forest based on current laws.

Alaska's Tongass National Forest

a. This temperate rain forest (light green area) is in southeastern Alaska along the Pacific Ocean.

b. A view of vast stretches of old-growth forest in the Tongass, as well as swatches that have already been clear-cut.

Summary

1 Land Use in the United States 322

1. More than one-half of U.S. land is privately owned. Approximately one-third—including many types of ecosystems and land uses—is owned by the federal government. Nine percent belongs to state and local governments, and more than 2 percent to Native American tribes.

2 Forests 324

1. **Sustainable forestry** is the use and management of forest ecosystems in an environmentally balanced and enduring way. Sustainable forestry maintains a mix of forest trees, by age and species, rather than a **monoculture**, in which only one type of plant is cultivated over a large area. Adopting sustainable forestry principles requires setting aside sanctuaries and **habitat corridors**, protected zones that connect isolated unlogged or undeveloped areas.

2. **Deforestation** is the temporary or permanent clearing of large expanses of forest for agriculture or other uses. **Clear-cutting** is a logging practice in which all the trees in a stand of forest are cut, leaving just the stumps; clear-cutting over a wide area is ecologically unsound. The major causes of tropical deforestation are subsistence farming, commercial logging, and cattle ranching, all accelerated by growing human populations. Increased needs for fuelwood drive deforestation of tropical dry forests.

3. Most U.S. national forests are managed by the U.S. Forest Service (USFS); the rest are overseen by the Bureau of Land Management (BLM). National forests face conflicts associated with supporting multiple uses: timber harvest; livestock forage; water resources and watershed protection; mining; hunting, fishing, and other forms of recreation; and habitat for fishes and wildlife.

3 Rangelands 333

1. **Rangelands** are grasslands that aren't intensively managed and are used for grazing livestock. Rangelands are also mined for mineral and energy resources, used for recreation, and preserved for biological habitat and for soil and water resources.

2. **Overgrazing** is the destruction of vegetation caused by too many grazing animals consuming the plants in a particular area, leaving them unable to recover. Overgrazing accelerates **land degradation**, which decreases the future ability of the land to support crops or livestock. **Desertification** is the degradation of once-fertile rangeland or tropical dry forest into nonproductive desert.

3. **A conservation easement** is a legal agreement that protects privately owned forest or other property from development for a specified number of years. Conservation groups often pay for conservation easements to preserve open rangeland.

4. The BLM manages more than three-fourths of U.S. public rangelands, excluding Alaska; the USFS manages the remainder. Current issues on public rangelands include conflicts between environmental groups and ranchers over the number of livestock allowed to graze and the potential to manage the areas for uses such as biological habitat, recreation, and scenic value. Conflicts also arise over whether grazing fees paid by livestock operators on public lands should be high enough to cover all costs of maintaining herds, removing taxpayer burden.

Seed tree cutting

Clear-cutting

4 National Parks and Wilderness Areas 336

1. **The National Park Service** administers 397 sites in the United States, including 58 national parks. Some problems the sites encounter include overcrowding, pollution, crime, resource violations, and imbalanced wildlife populations.

2. **Wilderness** is a protected area of land in which no human development is permitted. The **National Wilderness Preservation System** consists of four U.S. government agencies—the NPS, USFS, FWS, and BLM—that oversee 757 wilderness areas. Some problems these areas face include overuse and overcrowding by visitors, pollution, erosion, and the introduction of invasive species.

3. Those who support the **wise-use movement** believe a primary purpose of federal lands is to enhance economic growth. They think that the government overregulates environmental protection and that property owners should have more flexibility to use natural resources. Those who support the **environmental movement** view federal lands as a legacy of U.S. citizens and thus want to preserve resources on federally owned lands.

5 Conservation of Land Resources 341

1. Endangered U.S. ecosystems include the south Florida landscape, southern Appalachian spruce–fir forests, and longleaf pine forests and savannas.

2. Criteria used to evaluate whether an ecosystem is endangered and to what degree it is threatened include its history of land loss and degradation, its prospects for future loss or degradation, the area the ecosystem occupies, and the number of threatened and endangered species living in that ecosystem.

Key Terms

- clear-cutting 326
- conservation easement 335
- deforestation 328
- desertification 334
- habitat corridor 326
- monoculture 325
- overgrazing 334
- rangeland 333
- sustainable forestry 325
- wilderness 338

What is happening in this picture?

Julia "Butterfly" Hill lived in this 600–1000-year-old, 180-foot-tall California redwood for more than 2 years in the late 1990s, to keep a lumber company from cutting down the tree.

- Would Hill's perspective on wilderness better fit the wise-use movement or the environmental movement?

- Explain the likely differences in the perspectives of Hill and the lumber company, especially given that the tree is on the company's land.

- Based on issues faced in Tongass National Forest, why do ecologists and environmentalists think that the logging of old-growth trees causes particular damage?

Critical and Creative Thinking Questions

1. Why is deforestation a serious global environmental problem?

2. What are the environmental effects of clear-cutting on steep mountain slopes? on tropical rainforest land?

3. Distinguish between rangeland degradation and desertification. Why is moderate grazing beneficial to rangelands, yet overgrazing leads to erosion?

4. Debate conflicts over logging in Tongass National Forest from two points of view: that of the wise-use movement and that of the environmental movement.

5. Do you think additional federal lands should be added to the wilderness system? Why or why not?

6. Consider the message that the cartoon shown here sends about the popularity of U.S. national parks. What might park managers do to protect parks while encouraging their use? What are some of the uses that must be considered in managing a national park?

Sustainable Citizen Question

7. Should private landowners have control over what they wish to do to their land? If you were a landowner, what are some land-use decisions you might have to make that could affect both the public and the environment? Identify a ranked list of five or more priorities you would follow.

8. Explain how economic growth and sustainable use of natural resources can be compatible goals.

9. How does an ecosystem's population growth rate affect its likelihood of meeting the criteria for endangered ecosystems? Describe the four criteria in your answer.

10. Which federal agencies are responsible for managing public rangelands? What environmental and economic issues do they face?

11. Given the important contributions of forests in providing both timber and ecosystem services, how would you manage U.S. public forests if you were in charge? Whose interests would you have to balance?

12. For three decades, Forest A has been harvested using clear-cutting, and Forest B has been managed through sustainable forestry. Describe five differences you would expect to see between the two forests, including sustainable forest practices in your description.

For questions 13 and 14, examine the following graph of world land use that was assembled by the World Resources Institute and the U.N. Food and Agriculture Organization.

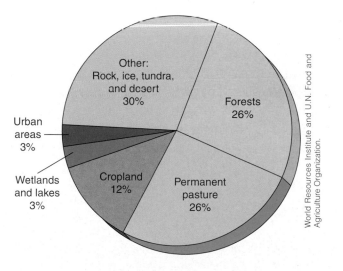

13. How much of the world's total land area is used for agriculture? How might conservation easements protect some of these areas?

14. Based on types of habitats represented, which of the categories shown are most likely to include the most endangered U.S. ecosystems?

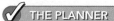

Agriculture and Food Resources

MAINTAINING GRAIN STOCKS

When people have access at all times to adequate amounts and kinds of food needed for healthy, active lives, they are said to have *food security*. World grain stocks provide a measure of food security (see larger photograph). World grain stocks are the amounts of rice, wheat, corn, and other grains remaining from previous harvests as a cushion against poor harvests and rising costs. Grain stocks have been decreasing since their all-time highs in the mid-1980s and late 1990s (see graph). The amount of grain stockpiled in 2010 would have fed the world's people for only 72 days. According to the United Nations, world grain stocks should not fall below a minimum of 70 days' supply in a given year.

World grain stocks have dropped in the past few years for several reasons. Many severe weather events have occurred—record heat waves, severe droughts, and numerous wildfires—suggesting that the climate is warming, and environmental conditions such as rising temperatures and falling water tables have caused poor harvests. Also, as the United States and other countries search for gasoline substitutes to reduce dependency on foreign oil, corn yields will be increasingly diverted to ethanol production (to blend with gasoline) instead of to food and animal feed.

World grain stocks have also fallen because consumption of beef, pork, poultry, and eggs has increased in developing countries such as China (see inset photo), where growing affluence has led some people to diversify their diets. This trend represents a global pattern: In highly developed countries, animal products account for 40 percent of the calories people consume, compared to only 5 percent of the calories people in developing countries consume. Increased consumption of meat and meat products has prompted a surge in the amount of grain used to feed the world's billions of livestock animals: More than one-third of the world's grain is now used to feed livestock. Thus, the global trend of eating more meat and other animal products is linked to increased use of grains and other feed crops for livestock.

WileyPLUS

⊕

H. Donnezan/Photo Researchers Inc.

World Grain Stocks, 1960–2010

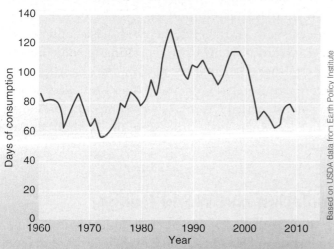

Based on USDA data from Earth Policy Institute

Interpreting Data
Identify which years, if any, world grain stocks have fallen below U.N.-recommended levels.

James A. Sugar/NG image Collection

CHAPTER OUTLINE

World Food Problems   350
- Population and World Hunger
- ■ Environmental InSight: World Hunger
- Poverty and Food

The Principal Types of Agriculture   353

Challenges of Agriculture   355
- Loss of Agricultural Land
- Global Decline in Domesticated Plant and Animal Varieties
- Increasing Crop Yields
- Increasing Livestock Yields
- Environmental Impacts
- ■ Environmental InSight: Impacts of Industrialized Agriculture

Solutions to Agricultural Problems   360
- Moving to Sustainable Agriculture
- ■ EnviroDiscovery: A New Weapon against Locust Swarms
- Genetic Engineering: A Solution or a Problem?

Controlling Agricultural Pests   364
- Benefits of Pesticides
- Problems with Pesticides
- ■ What a Scientist Sees: Pesticide Use and New Pest Species
- Alternatives to Pesticides
- ■ Case Study: Organic Agriculture

CHAPTER PLANNER ✓

- ❏ Study the picture and read the opening story.
- ❏ Scan the Learning Objectives in each section:
  p. 350 ❏   p. 353 ❏   p. 355 ❏   p. 360 ❏   p. 364 ❏
- ❏ Read the text and study all figures and visuals. Answer any questions.

Analyze key features

- ❏ Environmental InSight, p. 351 ❏   p. 359 ❏
- ❏ Process Diagram, p. 353 ❏   p. 362 ❏
- ❏ EnviroDiscovery, p. 361
- ❏ What a Scientist Sees, p. 366
- ❏ Case Study, p. 368
- ❏ Stop: Answer the Concept Checks before you go on:
  p. 352 ❏   p. 354 ❏   p. 358 ❏   p. 363 ❏   p. 367 ❏

End of Chapter

- ❏ Review the Summary and Key Terms.
- ❏ Answer What is happening in this picture?
- ❏ Answer the Critical and Creative Thinking Questions.

World Food Problems

LEARNING OBJECTIVES

1. **Differentiate** between undernutrition and overnutrition.

2. **Define** *food insecurity* and relate it to human population, poverty, and world hunger.

The U.N. Food and Agriculture Organization (FAO) reported at the 2009 World Summit on Food Security that at least 1 billion people—more than three times the population of the United States—lack the food needed for healthy, productive lives. Most of these people live in rural areas of the poorest developing countries (**Figure 14.1a**). Two-thirds of the world's hungry live in just seven countries—Bangladesh, China, the Democratic Republic of the Congo, Ethiopia, India, Indonesia, and Pakistan—with 40 percent in India and China alone.

The average adult human must consume enough food to get approximately 2600 *kilocalories*, or simply *Calories*, per day. People who receive fewer calories than needed are undernourished. Over an extended period of undernourishment, their health and stamina decline, even to the point of death. Worldwide, an estimated 182 million children under age 5 suffer from **undernutrition** and are seriously underweight, according to the World Health Organization (WHO).

> **undernutrition**
> A type of malnutrition in which underconsumption of calories or nutrients leaves the body weakened and susceptible to disease.

People might receive enough calories in their diets but still be malnourished because they do not receive enough essential nutrients, such as proteins, vitamin A, iodine, or iron. Adults suffering from malnutrition are more susceptible to disease and have less strength to function productively than those who are well fed. In addition to being more susceptible to disease, malnourished children do not grow or develop normally. Because malnutrition affects cognitive development, malnourished children typically do not perform well in school. The FAO estimates that 2 billion people worldwide suffer from micronutrient deficiencies (**Figure 14.1 b, c, d**). Also, more than half the deaths in children younger than 5 years old in developing countries are associated with malnutrition (**Figure 14.1e**).

People who eat more food than necessary are overnourished. Generally, a person suffering from **overnutrition** has a diet high in saturated (animal) fats, sugar, and salt. Overnutrition, which is most common among people in highly developed nations such as the United States, results in obesity, high blood pressure, and an increased likelihood of disorders such as diabetes, heart disease, and some cancers. Overnutrition is also emerging in some developing countries, particularly in urban areas, where, as people earn more money, their diets shift from consumption of cereal grains to consumption of more livestock products and processed foods high in fat and sugar.

> **overnutrition**
> A type of malnutrition in which an overconsumption of calories leaves the body susceptible to disease.

Population and World Hunger

According to the FAO, 66 countries are considered low income and food deficient, which means they cannot produce enough food or afford to import enough food to feed the entire population. South Asia, with an estimated 330 million hungry people, and sub-Saharan Africa, with an estimated 217.5 million who are hungry, are the regions of the world with the greatest **food insecurity**. In 2010, the FAO identified 22 countries—17 of them in sub-Saharan Africa—as being in a state of protracted crisis, "those environments in which a significant proportion of the population is acutely vulnerable to death, disease and disruption of livelihoods over a prolonged period of time." Such countries commonly experience food insecurity.

> **food insecurity**
> The condition in which people live with chronic hunger and malnutrition.

Today, sub-Saharan Africa produces less food per person than it did in 1950. Several factors contribute to the food shortages in Africa; these include civil wars and military actions, HIV/AIDS (which has killed or incapacitated much of the agricultural workforce in some

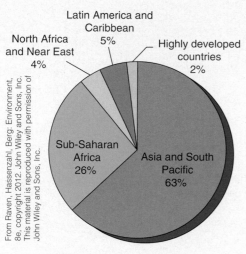

From Raven, Hassenzahl, Berg: Environment, 8e, copyright 2012. John Wiley and Sons, Inc. This material is reproduced with permission of John Wiley and Sons, Inc.

Latin America and Caribbean 5%

North Africa and Near East 4%

Highly developed countries 2%

Sub-Saharan Africa 26%

Asia and South Pacific 63%

a. Most of the world's undernourished people live in Asia and sub-Saharan Africa.

Paul Souders/©Corbis

b. Millions of children suffer from *kwashiorkor*, caused by severe protein deficiency. Note the characteristic swollen belly, which results from fluid retention. Photographed in Haiti.

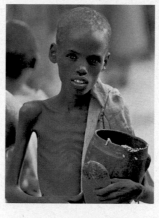

© Reuters/Corbis Images

c. *Marasmus* is progressive emaciation caused by a diet low in both total calories and protein. Symptoms include a pronounced slowing of growth and extreme wasting of muscles. Photographed in Somalia.

Paul Lovichi Photography/Alamy

d. Globally, millions of adult men and women are hungry. This homeless man is suffering from severe malnutrition and starvation. Photographed in New Delhi, India.

GLOBAL LOCAL — What can you do to reduce hunger where you live? to reduce hunger globally?

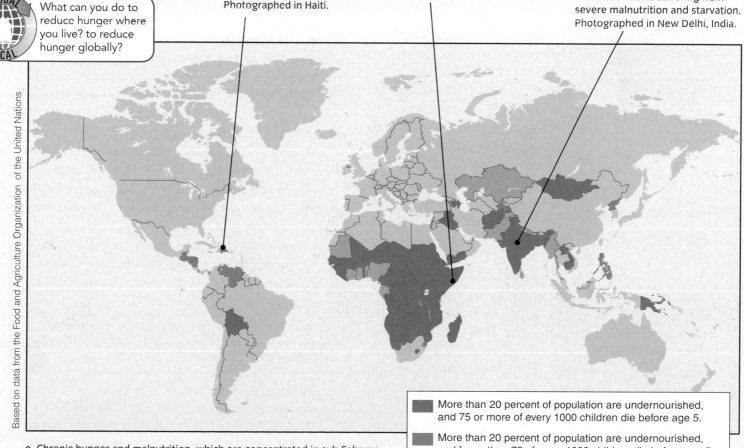

Based on data from the Food and Agriculture Organization of the United Nations

More than 20 percent of population are undernourished, and 75 or more of every 1000 children die before age 5.

More than 20 percent of population are undernourished, and fewer than 75 of every 1000 children die before age 5.

Less than 20 percent of population are undernourished, and 75 or more of every 1000 children die before age 5.

e. Chronic hunger and malnutrition, which are concentrated in sub-Saharan Africa and South Asia, are linked to increased mortality for infants and children.

Total world grain production and grain production per person, 1970 to 2011
• Figure 14.2

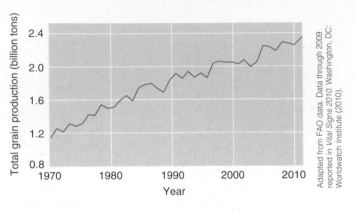

a. Total world grain production increased from 1.1 billion tons in 1970 to more than 2.3 billion tons in 2011.

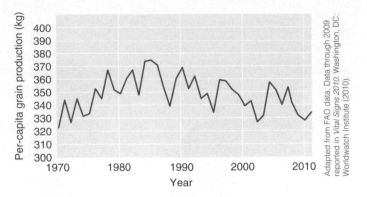

b. The amount of grain produced per person has not changed significantly in the past 40 years.

countries), floods, droughts, and soil erosion from hilly, marginal farmlands. People with food insecurity always live under the threat of starvation.

Experts agree that world hunger, population, poverty, and environmental problems are interrelated, but they disagree on the relative importance of each factor.

> **economic development**
> An expansion in a region's or country's economy, viewed by many as the best way to raise the standard of living.

Different groups propose different solutions for resolving the world's food problems, including controlling population growth, promoting the **economic development** of countries that do not produce adequate food, and correcting the inequitable distribution of resources. All experts agree that population pressures exacerbate world food problems.

Although annual grain production almost doubled from 1970 to 2011 (**Figure 14.2**), the world population increased so rapidly during that period that the amount of grain *per person* has not changed appreciably. By 2050, farmers will have to grow almost 30 percent more grain than they do now so that the 9.3 billion people living then—the median projection—can be fed.

Currently, the amount of available grain per person varies greatly from one country to another; in the United States, it is about 1.2 metric tons (1.3 tons) per person, most of which is fed to livestock, whereas in Zimbabwe, it is about 90 kg (200 lb) per person.

Poverty and Food

The main cause of undernutrition is poverty. Infants, children, and the elderly are most susceptible to poverty and chronic hunger. The world's poorest people—those living in developing countries in Asia, Africa, and Latin America—do not own land on which to grow food and do not have sufficient money to purchase food. Poverty and hunger are not restricted to developing nations, however; poor hungry people are also found in the United States, Europe, and Australia.

World food problems are many, as are their solutions. We must increase the sustainable production of food (discussed later in the chapter) and improve food distribution. Highly developed nations can provide economic assistance and technical aid to help farmers in developing countries produce more food. Globally, chronic hunger will persist so long as the human population remains above the level that the environment can support.

| CONCEPT CHECK | **STOP** |

1. **What** is the difference between undernutrition and overnutrition? Where is each type of malnutrition most prevalent in the world?

2. **What** is food insecurity and how does it relate to human population, poverty, and world hunger?

The Principal Types of Agriculture

LEARNING OBJECTIVES

1. **Contrast** industrialized agriculture with subsistence agriculture.

2. **Describe** three kinds of subsistence agriculture.

Agriculture can be roughly divided into two types: industrialized agriculture and subsistence agriculture. Most farmers in highly developed countries and some in developing countries practice *high-input agriculture*, or **industrialized agriculture**. It relies on large inputs of capital and energy (in the form of fossil fuels) to make and

> **industrialized agriculture** Modern agricultural methods that require large capital inputs and less land and labor than traditional methods.

run machinery, purchase seed, irrigate crops, and produce agrochemicals such as commercial inorganic fertilizers and pesticides (**Figure 14.3**). Industrialized agriculture produces high **yields** (the amount of food produced per unit of land), which allows forests and other natural areas to remain wild instead of being converted to agricultural land. However, the productivity of industrialized agriculture comes with costs, such as soil degradation and increased pesticide resistance in agricultural pests; we discuss these and other problems later in the chapter (and in Chapters 4 and 12).

Energy inputs in industrialized agriculture • Figure 14.3

 THE PLANNER

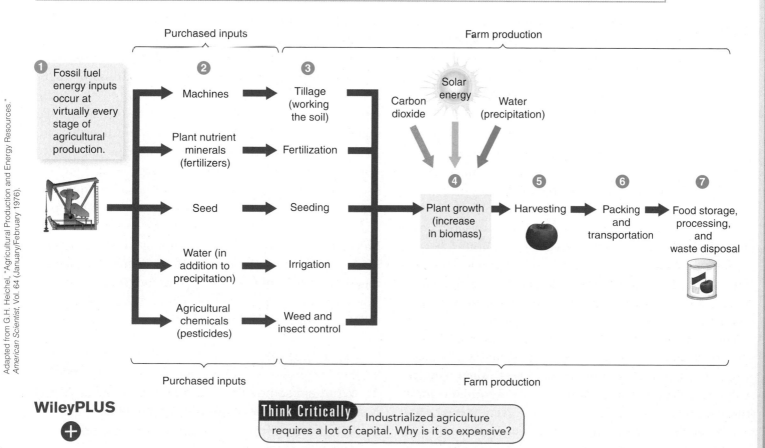

Purchased inputs

Farm production

1 Fossil fuel energy inputs occur at virtually every stage of agricultural production.

Machines → Tillage (working the soil)

Plant nutrient minerals (fertilizers) → Fertilization

Solar energy

Carbon dioxide   Water (precipitation)

Seed → Seeding

4 Plant growth (increase in biomass) → **5** Harvesting → **6** Packing and transportation → **7** Food storage, processing, and waste disposal

Water (in addition to precipitation) → Irrigation

Agricultural chemicals (pesticides) → Weed and insect control

Purchased inputs

Farm production

WileyPLUS ⊕

Think Critically Industrialized agriculture requires a lot of capital. Why is it so expensive?

Global Locator

AFRICA

KENYA

NG Maps

NATIONAL GEOGRAPHIC

Nomadic sheep herders in Kenya • Figure 14.4

---

subsistence agriculture
Traditional agricultural methods that are dependent on labor and a large amount of land to produce enough food to feed oneself and one's family.

Most farmers in developing countries practice **subsistence agriculture**, the production of enough food to feed oneself and one's family, with little left over to sell or reserve for hard times. Subsistence agriculture also requires large inputs of energy, but from humans and draft animals rather than from fossil fuels.

Some types of subsistence agriculture require large tracts of land. **Shifting cultivation** is a form of subsistence agriculture in which short periods of cultivation are followed by longer periods of fallow (land being left uncultivated), during which the land reverts to forest. Shifting cultivation supports relatively small populations. **Slash-and-burn agriculture** is a type of shifting cultivation that involves clearing small patches of tropical forest to plant crops (see Chapter 13). Slash-and-burn agriculture is land-intensive; because tropical soils lose their productivity quickly when they are cultivated, farmers using slash-and-burn agriculture must move from one area of forest to another every 3 years or so.

Nomadic herding, in which livestock is supported by land too arid for successful crop growth, is a

similarly land-intensive form of subsistence agriculture (**Figure 14.4**). Nomadic herders must continually move their livestock to find adequate food for the animals.

Intercropping is a form of intensive subsistence agriculture that involves growing a variety of plants on the same field simultaneously. When certain crops are grown together, they produce higher yields than when they are grown as **monocultures**. (A monoculture is the cultivation of only one type of plant over a large area.) One reason for higher yields is that different pests are found on each crop, and intercropping discourages the buildup of any single pest species to economically destructive levels. **Polyculture** is a type of intercropping in which several kinds of plants that mature at different times are planted together. In polyculture practiced in the tropics, fast- and slow-maturing crops are often planted together so that different crops can be harvested throughout the year.

CONCEPT CHECK STOP

1. **What** are some differences between industrialized agriculture and subsistence agriculture?

2. **What** are shifting cultivation, nomadic herding, and intercropping?

Challenges of Agriculture

LEARNING OBJECTIVES

1. **Discuss** recent trends in loss of U.S. agricultural land, global declines in domesticated plant and animal varieties, and efforts to increase crop and livestock yields.

2. **Relate** the benefits and problems associated with the green revolution.

3. **Describe** the environmental impacts of industrialized agriculture, including land degradation and habitat fragmentation.

The United States has more than 120 million hectares (300 million acres) of **prime farmland**, land that has the soil type, growing conditions, and available water to produce food, forage, fiber, and oilseed crops. U.S. agriculture faces a decline in prime farmland. Other challenges include coping with declining numbers of domesticated varieties, improving crop and livestock yields, and addressing agriculture's environmental impacts.

Loss of Agricultural Land

There is considerable concern that much of the nation's prime agricultural land is falling victim to urbanization and suburban sprawl or spread by being converted to parking lots, housing developments, and shopping malls (**Figure 14.5**). In certain areas of the United States, loss of rural land is a significant problem. According to the American Farmland Trust, the top five U.S. farm areas threatened by population growth and urban/suburban spread are California's Central Valley, south Florida, California's coastal region, the Mid-Atlantic Chesapeake region (Maryland to New Jersey), and the North Carolina Piedmont. More than 160,000 hectares (400,000 acres) of prime U.S. farmland are lost each year.

The 1996 Farm Bill included funding for the establishment of a national *Farmland Protection Program*. (Many states and local jurisdictions also have farmland protection programs.) This voluntary program lets farmers sell **conservation easements** that prevent their farmland from being converted to nonagricultural uses. The easements are in effect from a minimum of 30 years to forever. As with other conservation easements, the farmers retain full rights to use their property—in this case, for agricultural purposes.

Suburban spread onto agricultural land • Figure 14.5

Homes and businesses occupy land that was once cornfields in York County, Pennsylvania.

AP Photo/York Daily Record, Bil Bowden

White leghorn hen • Figure 14.6

If you eat eggs, they are probably laid by white leghorn hens, which produce almost all of the eggs in the United States as well as those in many other countries. This breed is prolific at egg laying, but it is not particularly good for meat production.

Stephen Aumus/US Dept. of Agriculture/SPL

Global Decline in Domesticated Plant and Animal Varieties

A global trend is currently under way to replace the many local varieties of a particular crop or domesticated farm animal with just a few kinds (**Figure 14.6**). A traditional variety is adapted to the climate where it was bred and contains a unique combination of traits conferred by its unique combination of genes. Modern varieties, which are bred for uniformity and maximum production, are generally more susceptible to insect pests and disease and less able to adapt to environmental changes, including climate change.

When farmers abandon traditional varieties in favor of more modern ones, the traditional varieties frequently face extinction. This represents a great loss in genetic diversity because each variety's characteristic combination of genes gives it distinctive nutritional value, size, color, flavor, resistance to disease, and adaptability to different climates and soil types.

To preserve older, more diverse varieties of plants and animals, many countries, including the United States, are collecting **germplasm**: seeds, plants, and plant tissues of traditional crop varieties and the sperm and eggs of traditional livestock breeds.

> **germplasm** Any plant or animal material that may be used in breeding.

Increasing Crop Yields

Until the 1940s, agricultural yields among various countries, whether highly developed or developing, were generally equal. Advances by research scientists since then have dramatically increased food production in highly developed countries (**Figure 14.7**). Greater knowledge of plant nutrition has resulted in production of fertilizers that promote high yields. The use of pesticides to control insects, weeds, and disease-causing organisms has also improved crop yields.

The Green Revolution By the middle of the 20th century, serious food shortages occurred in many developing countries coping with growing populations. The development and introduction during the 1960s of high-yield varieties of wheat and rice to Asian and Latin American countries gave these nations the chance to provide their people with adequate supplies of food (**Figure 14.8**). But the high-yield varieties required intensive industrial cultivation methods, including the use of commercial inorganic fertilizers, pesticides, and mechanized machinery, to realize their potential. These agricultural technologies were passed from highly developed nations to developing nations.

Using modern cultivation methods and the high-yield varieties of certain staple crops to produce more food per acre of cropland is known as the **green revolution**. Some of the success stories of the green revolution are remarkable. During the 1920s, Mexico produced

Average U.S. wheat yields, 1950 to 2010 • Figure 14.7

Each year shown is actually an average of 3 years to minimize the effects that poor weather conditions might have in a single year. Similar increases in yield have occurred in other grain crops.

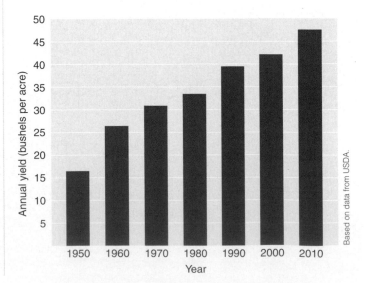

Based on data from USDA.

Development of high-yield rice varieties
• Figure 14.8

Traditional rice plants on the left are taller and do not yield as much grain (clusters at top of plant) as the more modern varieties shown in the middle and on the right. The rice plant in the middle was developed during the 1960s by crossing a high-yield, disease-resistant variety with a dwarf variety to prevent the grain-heavy plants from falling over. Improvements since the green revolution have been modest, as shown in the rice variety developed during the 1990s (right). Some researchers think rice and certain other genetically improved crops are near their physical limits of productivity.

Tall conventional plant   Improved high-yielding plant   Newer plant with fewer leaves

Increased meat consumption, 1980–2005
• Figure 14.9

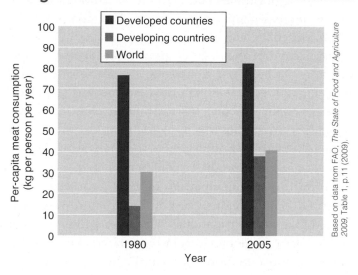

Based on data from FAO, *The State of Food and Agriculture 2009*, Table 1, p. 11 (2009).

less than 0.7 metric tons (0.8 ton) of wheat per hectare annually. During the green revolution years that began in 1965, Mexico's annual wheat production rose to more than 2.4 metric tons (2.7 tons) per hectare. Indonesia, another green revolution success story, formerly imported more rice than any other country in the world. Today Indonesia produces enough rice to feed its people and export some. The outcome of the green revolution was not just greater food production: Now that developing countries produced enough food, they could focus more on economic development.

Critics argue that the green revolution has made developing countries dependent on imported technologies, such as agrochemicals and tractors, at the expense of traditional agriculture. The two most important problems associated with higher crop production are the high energy costs built into this type of agriculture and the serious environmental problems caused by the intensive use of fertilizers and pesticides.

Increasing Crop Yields in the Post–Green Revolution Era The world demand for rice, wheat,

and corn will certainly increase in the coming years. This rise in demand will require a corresponding rise in grain production to feed the human population, as well as the livestock needed to satisfy the appetites of the increasing numbers of affluent people who can afford to buy meat (**Figure 14.9**). Also, the increased use of crop plants to produce **biofuels**, such as ethyl alcohol, rather than food will increase global demand.

This challenge cannot be met by increasing the amount of land under cultivation, as the best arable lands are already being cultivated. Projected freshwater shortages, rising costs of agricultural chemicals, and deteriorating soil quality caused by intensive agricultural techniques may further constrain productivity. As Figure 14.8 demonstrates, recent progress in coaxing more grain out of crops genetically improved during the green revolution has resulted in diminishing returns. Grain yields have continued to rise since the 1960s, but in recent years the rates of increase have not been as great as they were previously.

Despite these problems, most plant geneticists think we can produce enough food in the 21st century to meet demand if countries spend more money in support of a concerted scientific effort to improve crops. Many scientists regard genetic engineering as one of the keys to breeding more productive varieties. In addition, modern agricultural methods, such as water-efficient irrigation, will need to be introduced to developing countries that do not currently have them if crop yields are to continue increasing.

Increasing Livestock Yields

The use of hormones and antibiotics, although controversial, increases animal production. **Hormones**, usually administered by ear implants, regulate livestock bodily functions and promote faster growth. Although U.S. and Canadian farmers use hormones, the European Union (EU) currently restricts imports of hormone-treated beef because of health concerns for human consumers. EU regulators cite studies which suggest that these hormones or their breakdown products, both found in trace amounts in meat and meat products, could cause cancer or affect the growth of young children. In 1999 an international scientific committee organized by the FAO and WHO concluded that the trace amounts of hormones found in beef are safe because they are very low compared to the normal hormone concentrations found in the human body.

Modern agriculture has also embraced the routine addition of low doses of **antibiotics** to feed for healthy pigs, chickens, and cattle. These animals gain 4 to 5 percent more weight than untreated animals, presumably because they expend less energy fighting infections.

Several studies link the indiscriminate use of antibiotics in humans and livestock to the evolution of bacterial strains that are resistant to antibiotics (**Figure 14.10a**). When an antibiotic is used to treat a bacterial infection, a few bacteria may survive because they are genetically resistant to the antibiotic, and they pass these genes to future generations. As a result, the bacterial population contains a larger percentage of antibiotic-resistant bacteria than before.

Because there is increasing evidence that the use of antibiotics in agriculture reduces their medical effectiveness for humans, WHO recommended in 2003 that routine use of antibiotics in livestock be eliminated. Many European countries have complied, but the United States and many other countries continue the practice.

Environmental Impacts

Industrialized agriculture has many environmental effects (**Figure 14.10b**). The agricultural use of fossil fuels and pesticides produces air pollution. Untreated animal wastes and agricultural chemicals such as fertilizers and pesticides cause water pollution, which reduces biological diversity, harms fisheries, and leads to outbreaks of nuisance species. According to the Environmental Protection Agency, agricultural practices are the single largest cause of surface water pollution in the United States.

Industrialized agriculture has favored the replacement of traditional family farms with large agribusiness conglomerates. In the United States, most cattle, hogs, and poultry are now grown in feedlots and livestock factories (**Figure 14.10c**). The large concentrations of animals in livestock factories create many environmental problems, including air and water pollution (**Figure 14.10d**).

Many insects, weeds, and disease-causing organisms have developed or are developing resistance to **pesticides**. Pesticide resistance forces farmers to apply progressively larger quantities of pesticides (**Figure 14.10e**). Pesticide residues contaminate our food supply and reduce the number and diversity of beneficial microorganisms in the soil. Fishes and other aquatic organisms are sometimes killed by pesticide runoff into lakes, rivers, and estuaries.

Land degradation is a reduction in the potential productivity of land. Soil erosion, which is exacerbated by large-scale mechanized operations, causes a decline in soil fertility, and the eroded sediments damage water quality. Other types of degradation are compaction of soil by heavy farm machinery and waterlogging and **salinization** (salting) of soil from improper irrigation methods.

Clearing grasslands and forests and draining wetlands to grow crops result in **habitat fragmentation** that reduces biological diversity. Many species are endangered or threatened as a result of habitat loss to agriculture. The most dramatic example of habitat loss in North America is tallgrass prairie, more than 90 percent of which has been converted to agriculture.

> **pesticide** A toxic chemical used to kill pests.
>
> **degradation (of land)** Natural or human-induced reduction in the potential ability of the land to support crops or livestock.
>
> **habitat fragmentation** The breakup of large areas of habitat into small, isolated patches.

CONCEPT CHECK ⬡ STOP

1. **What** is happening to the number of domesticated plant and animal varieties? Why?

2. **What** is the green revolution? What are some of its benefits and problems?

3. **What** are the major environmental problems associated with industrialized agriculture?

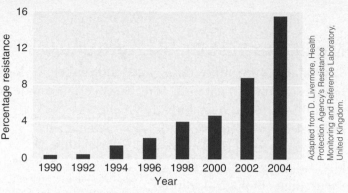

Adapted from D. Livermore, Health Protection Agency's Resistance Monitoring and Reference Laboratory, United Kingdom.

a. Evolution of Antibiotic Resistance. Shown is the increasing resistance of *E. coli* in blood and cerebrospinal infections to the antibiotic ciprofloxacin.

b. Some Common Problems of Industrialized Agriculture.

Air pollution

• Pesticide sprays
• Soil particles from wind erosion
• Odors from livestock factories
• Greenhouse gases from combustion of fossil fuels
• Other air pollutants from combustion of fossil fuels

Water issues

• Groundwater depletion from irrigation
• Pollution from fertilizer and pesticide runoff
• Sediment pollution from eroding soil particles
• Pollution from animal wastes (livestock factories)
• Enrichment of surface water from fertilizer runoff and livestock wastes

Loss of biological diversity

• Habitat fragmentation (clearing land and draining wetlands)
• Monocultures (lack of diversity in croplands)
• Stress from air and water pollution
• Stress from pesticides
• Replacement of many traditional crop and livestock varieties with just a few

Land degradation

• Soil erosion
• Loss of soil fertility
• Soil salinization
• Soil pollution (pesticide residues)
• Waterlogged soil from improper irrigation

© Macduff Everton/CORBIS

c. Hog Factory. The hogs remain indoors and are fed and watered by machines at timed intervals throughout the day.

USDA/NRCS/Jeff Vanuga

d. Waste Lagoon in an Automated Hog Farm. Some livestock factories produce as much sewage as small cities.

e. Colorado Potato Beetle on Potato leaf. As a result of exposure to heavy pesticides over the years, these beetles are resistant to most insecticides registered for use on potatoes.

Courtesy Peggy Greb/USDA

Solutions to Agricultural Problems

LEARNING OBJECTIVES

1. **Define** *sustainable agriculture* and contrast it with industrialized agriculture.

2. **Identify** the potential benefits and problems of genetic engineering.

ood production poses an environmental quandary. The green revolution and industrialized agriculture have unquestionably met the food requirements of most of the human population, even as that population has more than doubled since 1960. But we pay for these food gains with serious environmental problems, and we do not know if industrialized agriculture is sustainable for more than a few decades. To compound the issue, we must continue to increase food production to feed the growing human population—but the resulting damage to the environment may reduce our chances of continuing those food increases into the future.

Fortunately, farming practices and techniques exist that ensure a sustainable output at yields comparable to those of industrialized agriculture. Farmers who practice industrialized agriculture can adopt these alternative agricultural methods, which cost less and are less damaging to the environment. Advances are also being made in sustainable subsistence agriculture.

Moving to Sustainable Agriculture

Sustainable agriculture (also called **alternative**, or **low-input**, **agriculture**) carefully integrates certain modern agricultural techniques with traditional farming methods. Sustainable agriculture is modeled after natural ecosystems, with their high biological diversity, biodegradation of materials, and maintenance of soil fertility. To this end, sustainable agriculture relies on beneficial biological processes and environmentally friendly chemicals that disintegrate quickly and do not persist as residues in the environment. A sustainable farm consists of field crops, trees that bear fruits and nuts, small herds of livestock, and even tracts of forest (**Figure 14.11**). Such diversification protects a sustainable farmer against unexpected changes in the marketplace.

The breeding of disease-resistant crop plants and the maintenance of animal health rather than the continual use of antibiotics to prevent disease are important parts of sustainable agriculture. Water and energy conservation are also practiced.

> **sustainable agriculture**
> Agricultural methods that maintain soil productivity and a healthy ecological balance while having minimal long-term impacts.

Some goals of sustainable agriculture • Figure 14.11

Sustainable agriculture, which is modeled on natural ecosystems, protects the soil so that it does not become depleted.

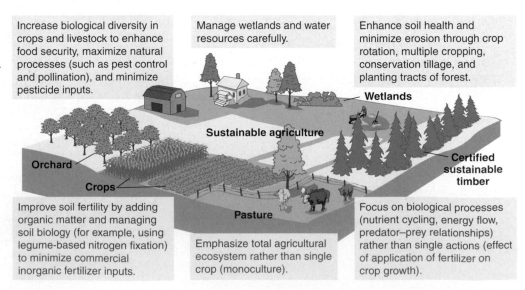

Increase biological diversity in crops and livestock to enhance food security, maximize natural processes (such as pest control and pollination), and minimize pesticide inputs.

Manage wetlands and water resources carefully.

Enhance soil health and minimize erosion through crop rotation, multiple cropping, conservation tillage, and planting tracts of forest.

Wetlands

Sustainable agriculture

Orchard

Certified sustainable timber

Crops

Improve soil fertility by adding organic matter and managing soil biology (for example, using legume-based nitrogen fixation) to minimize commercial inorganic fertilizer inputs.

Pasture

Emphasize total agricultural ecosystem rather than single crop (monoculture).

Focus on biological processes (nutrient cycling, energy flow, predator–prey relationships) rather than single actions (effect of application of fertilizer on crop growth).

Instead of using large quantities of chemical pesticides, sustainable agriculture controls pests by enhancing natural predator–prey relationships. Biological diversity is maintained to minimize pest problems. For example, providing hedgerows (rows of shrubs) between fields provides a habitat for birds and other insect predators.

An important goal of sustainable agriculture is to preserve the quality of agricultural soil. Crop rotation, conservation tillage, and contour plowing help control erosion and maintain soil fertility (see Figures 12.17c and 3.1a). Sloping hills converted to mixed-grass pastures erode less than hills planted with field crops, thereby conserving the soil and supporting livestock.

Animal manure added to soil not only cuts costs but decreases the need for high levels of commercial inorganic fertilizers. Legumes such as soybeans, clover, and alfalfa convert atmospheric nitrogen into a form that plants can use in a process called *biological nitrogen fixation*. Thus legumes also reduce the need for nitrogen fertilizers.

Sustainable agriculture is not a single program but a series of programs adapted for specific soils, climates, and farming requirements. Some sustainable farmers—those who practice **organic agriculture**—use no commercial inorganic fertilizers or pesticides; others use a system of *integrated pest management (IPM)*, which incorporates the limited use of pesticides with pest-controlling biological and cultivation practices. (See the *Case Study* at the end of the chapter for more on organic agriculture.)

In growing recognition of the environmental problems associated with industrialized agriculture, more and more mainstream farmers are trying some methods of sustainable agriculture. These methods cause fewer environmental problems to the agricultural ecosystem, or **agroecosystem**, than industrialized agriculture. This trend away from using intensive techniques that produce high yields and toward methods that focus on long-term sustainability of the soil is sometimes referred to as the **second green revolution**.

EnviroDiscovery

A New Weapon against Locust Swarms

One particular non-insect biological control agent targets desert locusts, which periodically increase in number and swarm across the African Sahel, threatening crops across 12 million hectares (30 million acres) located south of the Sahara Desert (see photo). Extensive swarms can move well beyond the Sahel, including into southern Africa. During a major swarm in 1988, African farmers sprayed massive quantities of pesticides into the environment to bring the locust population under control. However, many people were concerned about the adverse ecological and health effects of using such large quantities of pesticides. Economists were also concerned about high short-term costs of using pesticides ($300 million for the 1988 swarm) as well as long-term economic costs.

After the 1988 swarm, an international team of scientists worked for 15 years to develop a biological control for locusts that consists of fungal spores. Called Green Muscle, it is environmentally safe and very effective. The use of Green Muscle in Tanzania—the first major application of the biological control agent—averted much of the threat from a 2009 locust infestation that could have impacted crops that feed as many as 15 million people.

A boy runs through a cloud of locusts during a 2004 swarm in Senegal, in the Sahel.

© Pierre Holtz/Reuters/Corbis

NG Maps

Global Locator

SAHEL

AFRICA

TANZANIA

NATIONAL GEOGRAPHIC

Genetic engineering • Figure 14.12

✓ THE PLANNER

This example of genetic engineering uses a plasmid, a small circular molecule of DNA (genetic material) found in many bacteria. The plasmid of the bacterium *Agrobacterium* introduces into a plant desirable genes from another organism.

WileyPLUS ➕

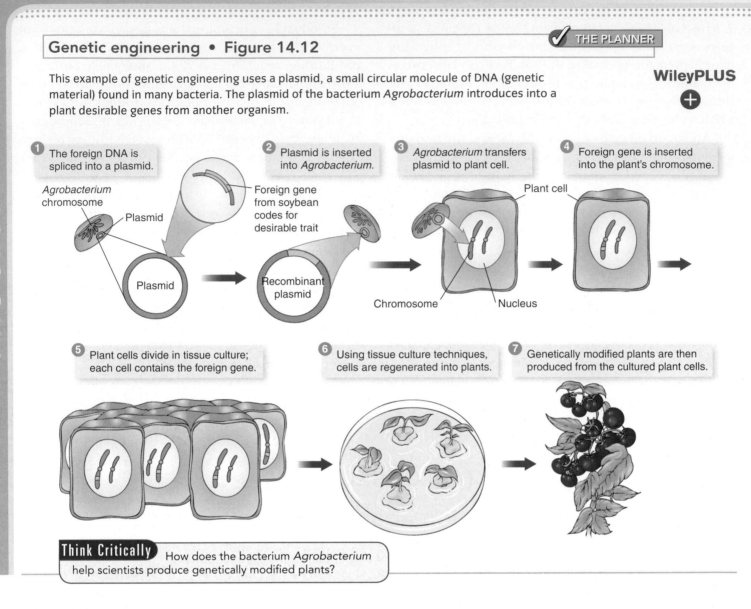

1 The foreign DNA is spliced into a plasmid.

Agrobacterium chromosome

Plasmid

Plasmid

Foreign gene from soybean codes for desirable trait

2 Plasmid is inserted into *Agrobacterium*.

Recombinant plasmid

3 *Agrobacterium* transfers plasmid to plant cell.

Plant cell

Chromosome

Nucleus

4 Foreign gene is inserted into the plant's chromosome.

5 Plant cells divide in tissue culture; each cell contains the foreign gene.

6 Using tissue culture techniques, cells are regenerated into plants.

7 Genetically modified plants are then produced from the cultured plant cells.

Think Critically How does the bacterium *Agrobacterium* help scientists produce genetically modified plants?

Genetic Engineering: A Solution or a Problem?

Genetic engineering is a controversial technology that has begun to revolutionize medicine and agriculture. The agricultural goals of genetic engineering are not new. Using traditional breeding methods, farmers and scientists have developed desirable characteristics in crop plants and agricultural animals for centuries. It takes time—15 years or more—to develop such genetically improved organisms using traditional breeding methods. Genetic engineering has the potential to accomplish the same goal in a fraction of that time.

genetic engineering The manipulation of genes (for example, taking a specific gene from one species and placing it into an unrelated species) to produce a particular trait.

Genetic engineering differs from traditional breeding methods in that desirable genes from any organism can be used, not just those from the species of the plant or animal being improved. If a gene for disease resistance found in soybeans would be beneficial to tomatoes, a genetic engineer can splice the soybean gene into the tomato plant's DNA (**Figure 14.12**). Traditional breeding methods could not do this because soybeans and tomatoes belong to separate groups of plants and do not interbreed.

Genetic engineering has the potential to produce more nutritious food plants that contain all the essential amino acids (which no single food crop currently does) or that would be rich in necessary vitamins. Crop plants resistant to viral diseases, drought, heat, cold, herbicides, salty or acidic soils, and insect pests are also being developed.

World production of GM crops • Figure 14.13

a. The world's top producers of GM crops.

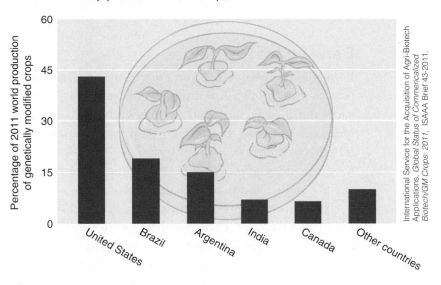

International Service for the Acquisition of Agri-Biotech Applications, Global Status of Commericalized Biotech/GM Crops: 2011, ISAAA Brief 43-2011.

b. The production of GM crops has increased rapidly.

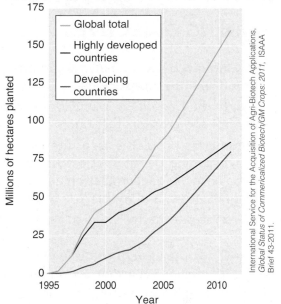

International Service for the Acquisition of Agri-Biotech Applications, Global Status of Commericalized Biotech/GM Crops: 2011, ISAAA Brief 43-2011.

The first **genetically modified (GM)** crops were approved for commercial planting in the United States in the early 1990s. The United States is the world's top producer of GM crops (**Figure 14.13a**). Since 2000, however, GM production in developing nations has increased faster than in highly developed countries (**Figure 14.13b**). Developing nations produced half of the world's GM crops in 2011.

Genetic engineering has been used to develop more productive farm animals, including rapidly growing hogs and fishes. Perhaps the greatest potential contribution of animal genetic engineering is the production of vaccines against disease organisms that harm agricultural animals. For example, genetically engineered vaccines have been developed to protect cattle against the deadly viral disease rinderpest, which is economically devastating in parts of Asia and Africa.

Concerns about Genetically Modified Foods

During the late 1990s and early 2000s, opposition to genetically engineered crops increased in many countries in Europe and Africa. In 1999 the EU placed a 5-year moratorium on virtually all approvals of GM crops (the moratorium has since been lifted), and subsequently refused to buy U.S. corn because it might be genetically modified. (Currently, 80 percent or more of the U.S. corn crop is genetically modified, mostly to protect against insect pests and provide resistance to herbicides [weed killers].) One concern is that the inserted genes could spread from GM crops

to weeds or wild relatives of crop plants and possibly harm natural ecosystems in the process. Scientists recognize this concern as legitimate and must take special precautions to avoid this possibility. Critics also worry that some consumers might develop food allergies to GM foods, although scientists routinely screen new GM crops for allergenicity.

A growing body of evidence indicates that current GM crop plants are as safe for human consumption as crops grown by conventional or organic agriculture. According to the FAO, GM crops widely planted in North America and elsewhere appear to have posed little threat to the environment. However, a complete analysis of the costs and benefits of *long-term* planting of GM crops remains to be done, in part because more research on the environmental impacts of GM crops is needed. To that end, strict guidelines exist in areas of genetic engineering research that could possibly affect the environment. Much research is currently being conducted to assess the effects of introducing GM crops whose foreign genes might spread to non-GM plants and be incorporated into their genetic makeup. It is important to assess the biology of each GM organism to determine if it has characteristics that might cause an environmental hazard under certain conditions.

CONCEPT CHECK **STOP**

1. **What** is sustainable agriculture? What are some features of a sustainable farm?

2. **Why** are some people opposed to GM crops?

Controlling Agricultural Pests

LEARNING OBJECTIVES

1. **Distinguish** between narrow-spectrum and broad-spectrum pesticides.

2. **Relate** the benefits of pesticides in disease control and crop protection.

3. **Summarize** problems associated with pesticide use, including genetic resistance, ecosystem imbalances, bioaccumulation and biological magnification, and mobility in the environment.

4. **Describe** alternative ways to control pests.

Any organism that interferes in some way with human welfare or activities is a **pest**. Some weeds, insects, rodents, bacteria, fungi, nematodes (microscopic worms), and other pest organisms compete with humans for food; other pests cause or spread disease. People try to control pests, usually by reducing the size of the pest population. Using *pesticides* is the most common way of doing this, particularly in agriculture. Pesticides can be grouped by their target organisms—that is, by the pests they are supposed to eliminate. **Insecticides** kill insects, **herbicides** kill plants, **rodenticides** kill rodents such as rats and mice, and **fungicides** kill fungi (**Figure 14.14**).

Bananas being sprayed with fungicide before shipment to grocery stores • Figure 14.14

Mayela Lopez/AFP/Getty Images

Benefits of Pesticides

Pesticides can effectively control organisms, such as insects, that transmit devastating human diseases. Fleas and lice carry the microorganism that causes typhus in humans. Malaria, also caused by a microorganism, is transmitted to millions of humans each year by female *Anopheles* mosquitoes. Pesticides help control the population of mosquitoes, thereby reducing the incidence of malaria.

Pesticides also protect crops. It is estimated that pests eat or destroy more than one-third of the world's crops. Given our expanding population and world hunger, it is easy to see why control of agricultural pests is desirable. Pesticides—applied heavily primarily in the United States and other highly developed countries—reduce the amount of a crop lost through competition with weeds, consumption by insects, and diseases caused by plant **pathogens** (microorganisms, such as fungi and bacteria, that cause disease).

Why are agricultural pests found in such great numbers in our fields? Part of the reason is that agriculture is usually a **monoculture**, in which the field cultivated with a single species represents a simple ecosystem. In contrast, forests, wetlands, and other natural ecosystems are complex and contain many species, including predators and parasites that control pest populations, as well as plant species that pests do not use for food. A monoculture reduces the dangers and accidents that might befall a pest as it searches for food. In the absence of many natural predators and in the presence of plenty of food, a pest population thrives and grows, damaging more of the crop.

Problems with Pesticides

The ideal pesticide is a **narrow-spectrum pesticide** that kills only the intended organism and does not harm other species. The perfect pesticide would readily break down into safe materials such as water, carbon dioxide, and oxygen. The ideal pesticide would stay exactly where it was put and would not move around in the environment. Unfortunately, no pesticide is perfect. Most pesticides are **broad-spectrum pesticides**. Some pesticides

> **broad-spectrum pesticide** A pesticide that kills a variety of organisms, including beneficial organisms, in addition to the target pest.

do not degrade readily, or they break down into compounds as dangerous as—if not more dangerous than—the original pesticide. And most pesticides move around the environment quite a bit.

The prolonged use of a particular pesticide can cause a pest population to develop genetic resistance to the pesticide. In the 50 years during which pesticides have been widely used, at least 520 species of insects and mites and at least 84 weed species have evolved genetic resistance to certain pesticides (**Figure 14.15**).

> **genetic resistance**
> An inherited characteristic that decreases the effect of a given agent (such as a pesticide) on an organism (such as a pest).

An organism exposed to a chemically stable pesticide that takes years to break down may accumulate high concentrations of the toxin, a phenomenon known as **bioaccumulation**. Organisms at higher levels on food webs tend to have greater concentrations of bioaccumulated pesticide stored in their bodies than those lower on food webs, through a process known as **biological magnification** (**Figure 14.16**; also see Figure 4.8).

One of the worst problems associated with pesticide use is that pesticides affect more species than the pests for which they are intended. Beneficial insects and other organisms are harmed and killed as effectively as pest insects. Quite often the stress of carrying pesticides in their tissues makes organisms more vulnerable to predators, diseases, or other stressors in their environments. Because the natural enemies of pests often starve or migrate in search of food after a pesticide is sprayed in an area, pesticides are indirectly responsible for a large reduction in the populations of these natural enemies. Pesticides also kill natural enemies directly because predators may consume lethal amounts of a pesticide while consuming the pests. After a brief period, the pest population rebounds and gets larger than ever, partly because no natural predators are left to keep its numbers in check. In some instances, the

Genetic resistance • Figure 14.15

The number of pest species evolving genetic resistance to pesticides has increased.

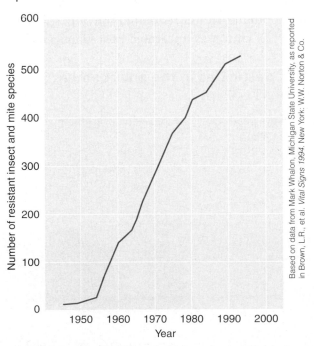

Based on data from Mark Whalon, Michigan State University, as reported in Brown, L.R., et al. *Vital Signs 1994.* New York: W.W. Norton & Co.

Interpreting Data
How is this figure similar to Figure 14.10a, which shows the evolution of resistance to antibiotics?

Peregrine falcons make a comeback • Figure 14.16

Peregrine falcons were once an endangered species in the United States. Bioaccumulation of pesticides such as DDT caused the birds to lay eggs with extremely thin shells, causing the chicks' deaths. With the ban of DDT and similar pesticides, the falcons recovered.

George Grall/NGS Image Collection

WHAT A SCIENTIST SEES

Pesticide Use and New Pest Species

An infestation of red scale insects on lemons occurred in California after DDT was sprayed to control a different pest. Prior to DDT treatment, red scale did not cause significant economic injury to citrus crops.

a. Red scale infestation—shown here on oranges—makes the fruit unappealing and unfit for market.

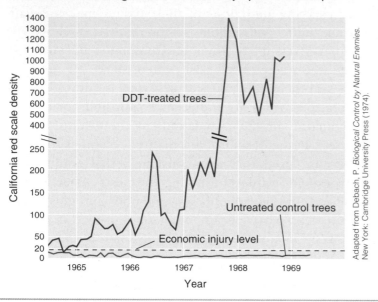

Adapted from Debach, P. *Biological Control by Natural Enemies.* New York: Cambridge University Press (1974).

© Graphic Science / Alamy

b. Scientists understand that the red scale infestation is a direct result of DDT spraying. This graph compares red scale populations on DDT-treated trees (red line) and untreated trees (blue line).

use of a pesticide has resulted in a pest problem that did not exist before (see *What a Scientist Sees*).

Another problem associated with pesticides is that they do not stay where they are applied but tend to move through the soil, water, and air, sometimes traveling long distances (**Figure 14.17**). Pesticides applied to agricultural lands wash into rivers and streams, where they can harm fishes. Pesticide mobility is also a problem for humans. About 14 million U.S. residents drink water containing traces of five widely used herbicides, and some people living where the herbicides are commonly used face a slightly elevated cancer risk because of their exposure.

Alternatives to Pesticides

Given their many problems, pesticides are clearly not the final solution to pest control. Fortunately, pesticides are not the only weapons in our arsenal. Alternative ways to control pests include cultivation methods, **biological controls**, **pheromones** and hormones, reproductive controls, genetic controls, quarantine, and irradiation (**Table 14.1**). A combination of these methods

in agriculture, often including a limited use of pesticides as a last resort, is known as **integrated pest management**

Mobility of pesticides in the environment • Figure 14.17

A helicopter sprays pesticides on a crop—and everything else in its pathway—in California.

Lawrence Migdale/Photo Researchers, Inc.

Alternative methods of controlling agricultural pests • Table 14.1

Pest control method	How it works	Disadvantages
Cultivation methods	Interplanting mixes different plants, as by alternating rows; strip cutting alternates crop harvest by portion—remaining portions protect natural predators and parasites of pests	No appreciable disadvantages; more care must be taken in harvest
Biological controls	Naturally occurring predators, parasites, or disease organisms are used to reduce pest populations	Organism introduced for biological control can unexpectedly affect the environment or other organisms
Pheromones and hormones	Sexual attractants (pheromones) lure pest species to traps; synthetic regulatory chemicals (hormones) disrupt pests' growth and development	Hormones might affect beneficial species
Reproductive controls	Sterilizing some members of pest population reduces population size	Expensive; must be carried out continually
Genetic controls	Selective breeding or genetic engineering develops pest-resistant crops	Plant pathogens evolve rapidly, adapting to disease-resistant host plant; plant breeders forced to constantly develop new strains
Quarantine	Governments restrict importation of foreign pests, diseases	Not foolproof; pests are accidentally introduced
Irradiating foods	Harvested foods are exposed to ionizing radiation that kills potentially harmful microorganisms	Consumers concerned about potential radioactivity; irradiation forms traces of potentially carcinogenic chemicals (free radicals)

(**IPM**; **Figure 14.18**). IPM is an important part of **sustainable agriculture**.

Because conventional pesticides are used sparingly in IPM, and the least toxic pesticides are applied, IPM allows farmers to control pests with a minimum of environmental disturbance and often at minimal cost. IPM is based on two fundamental premises. First, pests are *managed* rather than eradicated. Farmers who adopt IPM allow a low level of pests in their fields and accept a certain amount of economic damage from the pests. They periodically sample the pest population in the field to determine when the benefit of using pesticides exceeds the cost of that action. Second, IPM requires that farmers be educated to understand what strategies will work best in their particular situations. To be effective, IPM requires a thorough knowledge of the life cycles and feeding habits of the pests as well as all their interactions with their hosts and other organisms.

Tools of Integrated Pest Management (IPM) • Figure 14.18

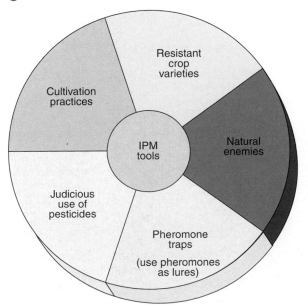

CONCEPT CHECK STOP

1. **What** is a broad-spectrum pesticide?
2. **What** are two important benefits of pesticide use?
3. **Discuss** two problems associated with the use of pesticides.
4. **How** can pests be controlled without pesticides?

✓ THE PLANNER

Organic Agriculture

Farmers practicing **organic agriculture** use no commercial inorganic fertilizers or pesticides (**Figure a**). According to guidelines established by the **Organic Food Production Act** in 1990, *organic foods* are crops grown in soil free of commercial inorganic fertilizers and pesticides for at least three years. If the land that the crops are grown on has been inspected, private or state agencies label it *certified organic*, which specifies the highest standards. Cattle and other livestock labeled *USDA organic* are not treated with antibiotics or hormones and are fed organic feed grown without commercial inorganic fertilizers or pesticides. Animals certified *humanely raised and handled* get fresh air and exercise and are not raised in cramped pens. Federal standards for certification of organically grown food went into effect in 2002, replacing standards that varied from state to state.

Internationally, Australia is the leading country by far in the amount of land dedicated to organic agriculture (**Figure b**). The United States ranks third worldwide, with organic agriculture

covering nearly 2 million hectares in 2008 (the most recent year for which data are available). Approximately 54 percent of this area was devoted to organic crops, and the remaining 46 percent to organic pasture and rangeland.

Consumer demand for a safe, pesticide-free food supply translated to total U.S. sales of $3.16 billion in 2008 (crops and animal products). A great deal of organic products are sold locally; 44 percent in 2008 were marketed within 100 miles of the farm. Although organic agriculture still makes up a small percentage of total U.S. farms, the practice has grown rapidly in recent years, in both the amount of land farmed and the number of operations certified as organic (**Figure c**). Organic farms and ranches operate in all 50 states, but California is home to almost 20 percent of U.S. organic operations.

a. Organic farms do not use pesticides; operators are often concerned that their crops will be exposed to windborne pesticides from other farms.

Inga Spence/Photolibrary/Getty Images

Think Critically Many organic producers and consumers avoid GM organisms, yet GM material has been inadvertently introduced into some otherwise organic U.S. crops and animals. Suggest some ways this could have taken place without the deliberate action of the farmers involved.

b. The ten countries with the most organic agricultural land (in 2010, except where noted).

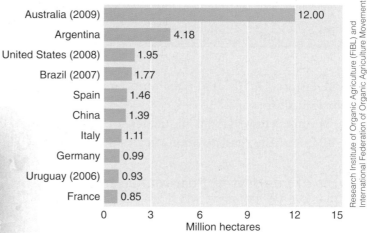

Research Institute of Organic Agriculture (FIBL) and International Federation of Organic Agriculture Movements (iFoam) 2012. *The World of Organic Agriculture*, 2012.

c. Growth in U.S. organic agriculture, 1995–2008

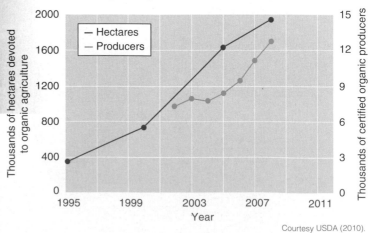

Courtesy USDA (2010).

Summary

1 World Food Problems 350

1. **Undernutrition** is a health-threatening underconsumption of calories or nutrients that leaves the body weakened and susceptible to disease. **Overnutrition** is a health-threatening overconsumption of calories or nutrients.

2. **Food insecurity**, in which people live with chronic hunger, is exacerbated by population growth, environmental problems, and poverty.

2 The Principal Types of Agriculture 353

1. **Industrialized agriculture** uses modern methods requiring large capital input and less land and labor than traditional methods. **Subsistence agriculture** requires labor and a large amount of land to produce enough food to feed a family.

2. There are three types of subsistence agriculture. In **shifting cultivation**, short periods of cultivation are followed by longer periods of fallow. In **nomadic herding**, carried out on arid land, herders move livestock continually to find food for them. **Intercropping** involves simultaneously growing a variety of plants on the same field.

© Images of Africa Photobank/David Keith Jones/Alamy

3 Challenges of Agriculture 355

1. **Prime farmland** in the United States is being lost to urbanization and urban sprawl. Global declines in plant and animal varieties have led many countries to collect **germplasm**, plant and animal material that may be used in breeding. Farmers and ranchers strive to increase yields by administering **hormones** and **antibiotics** to livestock.

2. The **green revolution** introduced modern cultivation methods and high-yield crop varieties to Asia and Latin America. These methods require developing nations to import energy-intensive technologies and to face environmental problems caused by the use of inorganic fertilizers and pesticides.

3. Environmental problems caused by industrialized agriculture include air pollution from the use of fossil fuels and pesticides, water pollution from untreated animal wastes and agricultural chemicals, pesticide-contaminated foods and soils, and increased resistance of pests to pesticides. **Land degradation** decreases the future ability of the land to support crops or livestock. Clearing grasslands and forests and draining wetlands to grow crops have resulted in **habitat fragmentation**.

4 Solutions to Agricultural Problems 360

1. **Sustainable agriculture** uses methods that maintain soil productivity and a healthy ecological balance while minimizing long-term impacts.

2. **Genetic engineering**, the manipulation of genes to produce a particular trait, can produce more nutritious crops or crop plants that are resistant to pests, diseases, or drought. Concerns about genetic engineering include unknown environmental effects.

5 Controlling Agricultural Pests 364

1. A **pesticide** is any toxic chemical used to kill pests. Most pesticides are **broad-spectrum pesticides** that kill other organisms in addition to the intended pest.

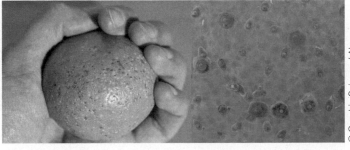

© Graphic Science / Alamy

2. Pesticides can effectively control disease-carrying organisms and crop pests.

3. Pesticide use leads to several problems: Pests develop **genetic resistance**, an inherited characteristic that decreases the effect of a given agent (such as a pesticide) on an organism; ecosystem imbalances occur when pesticides affect species other than the intended pests; and some pesticides exhibit persistence, degrading very slowly. **Bioaccumulation** is the buildup of a persistent pesticide in an organism's body. **Biological magnification** is the increased concentration of pesticides in organisms at higher levels in food webs. Pesticides also show mobility, moving to places other than where they were applied.

4. Alternatives to pesticides include **biological controls**, which use disease organisms, parasites, or predators to control pests. **Pheromones**, produced by animals to stimulate a response in other members of the same species, attract and trap pest species. **Integrated pest management** is a combination of pest control methods that keep a pest population small enough to prevent substantial economic loss.

Key Terms

- broad-spectrum pesticide 364
- degradation (of land) 358
- economic development 352
- food insecurity 350
- genetic engineering 362
- genetic resistance 365
- germplasm 356
- habitat fragmentation 358
- industrialized agriculture 353
- overnutrition 350
- pesticide 358
- subsistence agriculture 354
- sustainable agriculture 360
- undernutrition 350

What is happening in this picture?

This trap, which was placed in the middle of a pea crop, contains a pheromone to attract pea moth males. Note the dead males in the trap.

- What would be some advantages and disadvantages of pheromone traps?

- What would be the more conventional method for getting rid of these moth pests?

- Pheromone traps are one tool of integrated pest management. What other tools must also be employed by farmers using this approach?

- How does this approach fit into the practice of sustainable agriculture?

- Why isn't IPM in widespread use by farmers?

Nigel Cattlin/Alamy

Critical and Creative Thinking Questions

1. Compare undernutrition and overnutrition. Describe the types of people and the regions of the world most likely to be affected by each.

2. How and why do (a) population growth and (b) the rising consumption of meat affect world grain carryover stocks and world food security?

Use the following graph to answer questions 3 and 4.

School attainment and undernourishment by region, 2000

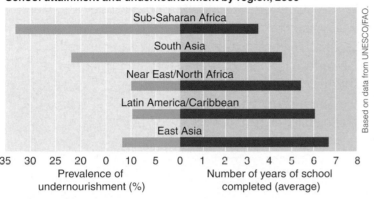

Prevalence of undernourishment (%)

Number of years of school completed (average)

Based on data from UNESCO/FAO.

3. Based on the graph, what conclusion can you make about how the level of education people attain in a region is related to that region's prevalence of undernutrition? Give examples of data from specific regions to support your conclusion.

4. How do these results fit in with our understanding of the role that poverty plays in the world's hunger problems?

5. Distinguish between shifting cultivation and slash-and-burn agriculture.

6. What are the environmental impacts of urban/suburban sprawl on agriculture land?

7. What was the green revolution? Describe a few of its successes and shortcomings and compare those to changes that have occurred in the recent "second green revolution."

8. Name two environmental problems associated with industrialized agriculture and give at least three examples of ways that industrialized agriculture could be made more sustainable.

9. Contrast the use of the following in industrialized and sustainable agriculture: water resources, fertilizers, and chemical pesticides.

10. How is developing new genetically modified crops and livestock promising in feeding the world's people? What are some concerns associated with growing genetically modified crops?

11. These charts compare the 2011 global acreage of genetically modified crops (screened areas) to the total production. How many hectares of genetically modified soybeans were planted? cotton? corn?

Soybeans

75% of 100 million hectares

Mark Thiessen/ NGS Staff/ NG Image Collection

Cotton

82% of 30 million hectares

Mark Thiessen/ NGS Staff/ NG Image Collection

Corn

32% of 159 million hectares

Rebecca Hale/ NGS Staff/ NG Image Collection

Based on data from International Service for the Acquisition of Agri-Biotech Applications (ISAAA), 2011.

12. Which type of pesticide has fewer deleterious environmental effects, broad-spectrum or narrow-spectrum? Explain your answer.

13. Overall, do you think the benefits of pesticide use outweigh the disadvantages, or vice versa? Give at least two reasons for your answer.

14. How and why is a pesticide's effectiveness altered by its heavy, long-term use?

Sustainable Citizen Question

15. How might your choosing to consume only organic foods affect environmental problems associated with food production? What about choosing vegetarianism? Compare the environmental benefits of organic and vegetarian diets.

✓ THE PLANNER

Biodiversity and Conservation

DISAPPEARING FROGS

Since the 1970s, many of the world's frog populations have dwindled or disappeared. At least 42 percent of the more than 6000 known amphibian species are declining, and about one-third are threatened or already extinct. Approximately165 species are believed to be extinct. Amphibian declines are occurring worldwide, even in remote, pristine locations, such as the high reaches of California's Sierra Nevada Mountains (see large photograph).

Biologists cannot pinpoint a single cause of the massive die-offs and suspect they reflect a combination of threats. According to data accumulated by the International Union for Conservation of Nature (IUCN), habitat loss likely causes the greatest harm to these species, but chemical pollutants, infectious diseases, increased levels of UV-B radiation, and climate change are among other factors threatening amphibian populations (see graph).

graphingactivity

Chytridiomycosis, a rapidly spreading disease caused by a chytrid fungus, had been recorded in 43 countries and in 36 U.S. states. Responsible for the disappearance of more than 100 amphibian species worldwide, and afflicting almost 300 species, this disease is thought to exacerbate the harm caused by other threats.

Rescue efforts around the globe include testing and treating animals affected by chytrids (see inset photo). Large-scale conservation programs, which include the captive breeding of vulnerable amphibian species, resulted in the establishment of the Amphibian Ark, an international group affiliated with the IUCN that seeks to preserve at least 500 species for reintroduction.

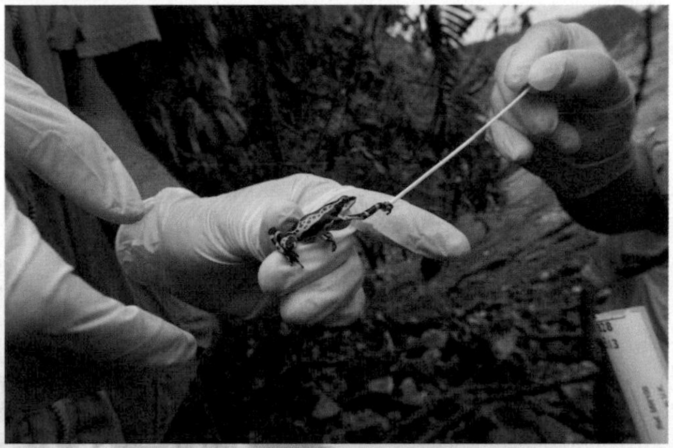

Joel Sartore/NG Image Collection

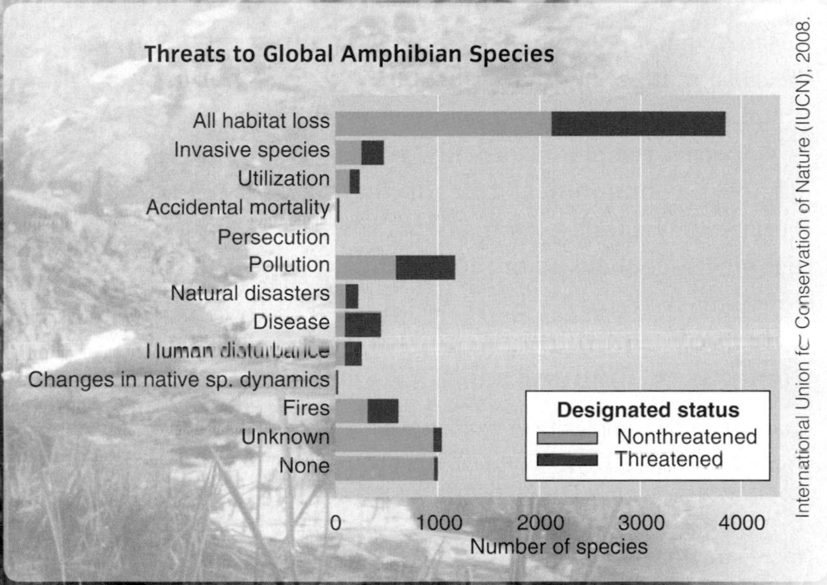

Threats to Global Amphibian Species
All habitat loss
Invasive species
Utilization
Accidental mortality
Persecution
Pollution
Natural disasters
Disease
Human disturbance
Changes in native sp. dynamics
Fires
Unknown
None
Designated status
Nonthreatened
Threatened
0
1000
2000
3000
4000
Number of species

International Union for Conservation of Nature (IUCN), 2008.

Interpreting Data

Which threats are having the greatest impact on threatened species? on nonthreatened ones?

Joel Sartore/NG Image Collection

CHAPTER OUTLINE

Species Richness and Biological Diversity 374
- How Many Species Are There?
- Why We Need Biodiversity
- Importance of Genetic Diversity

Endangered and Extinct Species 378
- Endangered and Threatened Species
- Areas of Declining Biological Diversity
- EnviroDiscovery: Is Your Coffee Bird Friendly®?
- Earth's Biodiversity Hotspots
- What a Scientist Sees: Where Is Declining Biological Diversity the Most Serious?
- Human Causes of Species Endangerment
- Environmental InSight: Threats to Biodiversity

Conservation Biology 386
- Protecting Habitats
- Restoring Damaged or Destroyed Habitats
- Conserving Species
- Environmental InSight: Efforts to Conserve Species

Conservation Policies and Laws 390
- The Endangered Species Act
- International Conservation Policies and Laws
- Case Study: The Challenges of Protecting Rare Species

CHAPTER PLANNER

- ❑ Study the picture and read the opening story.
- ❑ Scan the Learning Objectives in each section: p. 374 ❑ p. 378 ❑ p. 386 ❑ p. 390 ❑
- ❑ Read the text and study all figures and visuals. Answer any questions.

Analyze key features

- ❑ EnviroDiscovery, p. 380
- ❑ What a Scientist Sees, p. 381
- ❑ Environmental InSight, p. 383 ❑ p. 389 ❑
- ❑ Case Study, p. 393
- ❑ Stop: Answer the Concept Checks before you go on: p. 377 ❑ p. 386 ❑ p. 390 ❑ p. 392 ❑

End of Chapter

- ❑ Review the Summary and Key Terms.
- ❑ Answer What is happening in this picture?
- ❑ Answer the Critical and Creative Thinking Questions.

Species Richness and Biological Diversity

LEARNING OBJECTIVES

1. **Describe** factors associated with species richness.
2. **Define** *biological diversity* and distinguish among species richness, genetic diversity, and ecosystem diversity.
3. **Relate** several important ecosystem services provided by biological diversity.

A **species** is a group of more or less distinct organisms that are capable of interbreeding with one another in the wild but that do not interbreed with organisms outside their group. We do not know exactly how many species exist. In fact, biologists now realize how little we know about Earth's diverse organisms.

How Many Species Are There?

Scientists estimate that there may be as few as 5 million or as many as 100 million different species inhabiting Earth. According to the International Union for Conservation of Nature (IUCN), about 1.7 million species have been scientifically named and described to date, including nearly 310,000 plant species, more than 64,000 vertebrate animal species, and 1 million insect species. About 10,000 new species are identified each year.

Species richness varies greatly from one community

> **species richness**
> The number of different species in a community.

to another. It is determined by several factors: the abundance of potential ecological niches, geographic isolation, habitat stress, dominance of one species over others, closeness to the margins of adjacent communities, and geologic history.

A complex community, such as a tropical rain forest or a coral reef, offers a greater variety of potential ecological niches than does a simple community such as a mountain chaparral. A similar effect is seen in comparisons between more structurally complex and less structurally complex areas within the same type of habitat (**Figure 15.1**).

Species richness is inversely related to the geographic isolation of a community. Isolated island communities are much less diverse than communities in similar environments found on continents. This is partly the result of the difficulty many species have in reaching and successfully colonizing an island, and locally extinct species are not readily replaced in isolated environments such as islands or mountaintops. Isolated areas are often small and possess fewer potential ecological niches.

Generally, species richness is also inversely related to the environmental stress of a habitat. Only species capable of tolerating extreme environmental conditions can live in an environmentally stressed community, such as a polluted stream or a polar region exposed to a harsh climate. Species richness is also reduced when one species occupies a decided position of dominance within a community because that species may appropriate a disproportionate share of resources, thus crowding out other species.

Species richness is usually greater at the edges of adjacent communities than in their centers. This is because an **ecotone**—a transitional zone where communities meet—contains all or most of the ecological niches of

Effect of community complexity on species richness • Figure 15.1

The structural complexity of chaparral vegetation in California (*x*-axis) is based on vegetation height and density, from low complexity (very dry scrub) to high complexity (woodland). Note that species richness in birds increases as vegetation becomes more structurally complex.

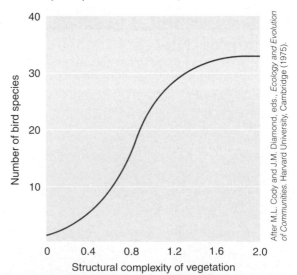

After M.L. Cody and J.M. Diamond, eds., *Ecology and Evolution of Communities*. Harvard University, Cambridge (1975).

Interpreting Data
What is the percentage difference between the number of bird species supported by the least structurally complex areas and the number supported by the most structurally complex areas?

the adjacent communities as well as some niches unique to the ecotone. The change in species composition produced at ecotones is known as the **edge effect**.

Geologic history greatly affects species richness. Tropical rain forests are probably old, stable communities that have undergone few climate changes in Earth's entire history. During this time, myriad species evolved in tropical rain forests, having experienced few or no abrupt climate changes that might have led to their extinction. In contrast, glaciers have repeatedly altered temperate and arctic regions during Earth's history, as the climate alternately cooled and warmed. An area recently vacated by glaciers will have a low species richness because few species will as yet have had a chance to enter it and become established. Scientists expect that future climate change will alter species richness, although it is difficult to predict the impact in any given area.

Why We Need Biodiversity

The variation among organisms is referred to as biological diversity, or **biodiversity**. Biological diversity occurs at

> **biological diversity** The number and variety of Earth's organisms; consists of three components: genetic diversity, species richness, and ecosystem diversity.

all levels of biological organization, from populations to ecosystems (Chapter 5). It takes into account three components: **genetic diversity**, the genetic variety *within* all populations of that species; species richness; and **ecosystem diversity**, the variety of ecosystems found on Earth and the variety of interactions among organisms in natural communities (**Figure 15.2**). For example, a forest community—with its trees, shrubs, vines, herbs, insects, worms, vertebrate animals, fungi, bacteria, and other microorganisms—has greater ecosystem diversity than a cornfield.

Humans depend on the contributions of thousands of species for their survival. For example, insects are instrumental in several ecological and agricultural processes, including pollination of crops, weed control, and insect pest control. Bacteria and fungi provide us with foods, antibiotics and other medicines, and biological processes such as nitrogen fixation (see Chapter 5). However, relatively few species have been evaluated for their potential usefulness to humans. At least 250,000 of the approximately 310,000 known plant species have yet to be assessed for industrial, medicinal, or agricultural potential. The same is true for most of the millions of microorganisms, fungi, and animals.

Levels of biodiversity • Figure 15.2

John Prior Images/Alamy

a. The variation in color patterns is evidence of the genetic diversity in the land snail species *Cepaea nemoralis*, common in western and central Europe.

© blickwinkel/Alamy

b. Species richness in a European grassland or woodland is represented by the land snail species, the grasshopper, and the grass.

Annie Griffiths Belt/NG Image Collection

c. A scenic view in Switzerland reflects ecosystem diversity across central Europe.

© Yan Gluzberg/Alamy

Alligators in the environment • Figure 15.3

The American alligator plays an integral role in its natural ecosystem. Alligators help maintain populations of smaller fishes by eating gar, a fish that preys on those smaller fishes. Alligators dig underwater holes that other aquatic organisms use during droughts when the water level is low. Their nest mounds eventually form small islands colonized by trees and other plants. The trees on these islands support heron and egret populations. The alligator habitat is maintained in part by underwater "gator trails," which help clear out aquatic vegetation that might eventually form a marsh.

Ecosystem Services and Species Richness The living world is a complex system. Each ecosystem is composed of many parts that are organized and integrated to maintain the ecosystem's overall performance. The activities of all organisms are interrelated; we depend on one another and on the physical environment, often in subtle ways (**Figure 15.3**). When one species declines, other species linked to it may either decline or increase in number.

Ecosystems supply human societies with many environmental benefits, or **ecosystem services** (**Table 15.1**). Forests are not just a source of lumber; they provide watersheds from which we obtain fresh water, limit the number and severity of local floods, and reduce soil erosion. Many flowering plant species depend on insects to transfer pollen for reproduction. Soil dwellers, from earthworms to bacteria, develop and maintain soil fertility for plants. Bacteria and fungi perform the crucial task of decomposition, which allows nutrients to cycle in the ecosystem. Conservationists maintain that ecosystems with greater species richness supply ecosystem services better than ecosystems with lower species richness.

ecosystem services Important environmental benefits such as clean air, clean water, and fertile soil, that the natural environment provides.

You might think that the loss of some species from an ecosystem would not endanger the rest of the organisms, but this is far from true. If enough species are removed, the entire ecosystem will change. Species richness within a community may provide the community with resistance, the ability to withstand environmental disturbances, natural or human events that disrupt a community, and with resilience, the ability to recover quickly to its former state following an environmental disturbance.

Importance of Genetic Diversity

The maintenance of a broad genetic base is critical for the long-term health and survival of each species. Consider economically important crop plants. During the 20th century, plant scientists developed genetically uniform, high-yielding varieties of important food crops such as wheat. However, genetic uniformity resulted in increased

Ecosystem services • Table 15.1

Ecosystem	Services provided
Forests	Purify air and water
	Produce and maintain soil
	Absorb carbon dioxide (carbon storage)
	Provide wildlife habitat
	Provide humans with wood and recreation
Freshwater systems (rivers and streams, lakes, and groundwater)	Moderate water flow and mitigate floods
	Dilute and remove pollutants
	Provide wildlife habitat
	Provide humans with drinking and irrigation water, food, transportation corridors, electricity, and recreation
Grasslands	Purify air and water
	Produce and maintain soil
	Absorb carbon dioxide (carbon storage)
	Provide wildlife habitat
	Provide humans with livestock and recreation
Coasts	Provide a buffer against storms
	Dilute and remove pollutants
	Provide wildlife habitat, including food and shelter for young marine species
	Provide humans with food, harbors, transportation routes, and recreation

Source: Adapted from p. 527 of *Climate Change Impacts on the United States,* a report of the National Assessment Synthesis Team, U.S. Global Change Research Program, Cambridge University Press (2001).

Interpreting Data
Are any of the ecosystem services provided by all of the listed ecosystems? If yes, identify. How does each ecosystem contribute to pollution control?

susceptibility to pests and disease. Crossing the "super strains" with more genetically diverse relatives can allow disease and pest resistance to be reintroduced into such plants.

Genetic engineering, the incorporation of genes from one organism into a different species (see Chapter 14), makes it possible to use organisms' genetic resources on a wide scale. Genetic engineering has provided new vaccines, more productive farm animals, and disease-resistant agricultural plants.

Although we have the skills to transfer genes from one organism to another, we do not have the ability to *make* genes that encode for specific traits. Genetic engineering depends on the broad base of genetic diversity from which it obtains genes. It has taken hundreds of millions of years for **evolution** to produce the genetic diversity found on our planet today. This diversity may hold solutions to today's problems and to future problems we have not begun to imagine. It would be unwise to allow such an important part of our heritage to disappear.

Medicinal, Agricultural, and Industrial Importance of Organisms

The genetic resources of organisms are vitally important to the pharmaceutical industry, which incorporates hundreds of chemicals derived from plants and other organisms into its medicines. From extracts of cherry and horehound for cough medicines to certain ingredients of periwinkle and mayapple for cancer therapy, derivatives of plants play important roles in the treatment of illness and disease (**Figure 15.4**). Many of the natural products taken directly from marine organisms are promising anticancer or antiviral drugs. The AIDS (acquired immunodeficiency syndrome) drug AZT (azidothymidine), for example, is a synthetic derivative of a compound from a sponge. The 20 best-selling prescription drugs in the United States are either natural products, natural products that are slightly modified chemically, or synthetic drugs whose chemical structures were obtained from organisms.

The agricultural importance of plants and animals is indisputable because we must eat to survive. However, the number of different kinds of foods we eat is limited compared to the total number of edible species available in any given region. Many species are probably nutritionally superior to our common foods.

Modern industrial technology depends on a broad range of products from organisms. Plants supply oils and lubricants, perfumes and fragrances, dyes, paper, lumber, waxes, rubber and other elastic latexes, resins,

Cyril Ruoso / Biosphoto

Medicinal value of the rosy periwinkle • Figure 15.4

The rosy periwinkle produces chemicals that are effective against certain cancers. Drugs from the rosy periwinkle have increased the chance of surviving childhood leukemia from about 5 percent to more than 95 percent. These leaves and flowers are being harvested in the Berenty Private Reserve, Madagascar.

poisons, cork, and fibers. Animals provide wool, silk, fur, leather, lubricants, and waxes, and they are important in medical research.

Insects secrete a large assortment of chemicals that represent a wealth of potential products. Certain beetles produce steroids with birth-control potential, fireflies produce a compound that may be useful in treating viral infections, and centipedes secrete a fungicide over the eggs of their young that could help control the fungi that attack crops. Because biologists estimate that perhaps 90 percent of all insects have yet to be identified, insects represent an important potential biological resource.

Aesthetic, Ethical, and Spiritual Value of Organisms

Organisms not only contribute to human survival and physical comfort, they provide recreation, inspiration, and spiritual solace. Our natural world is a thing of beauty largely because of the diversity of living forms found in it. Artists have attempted to capture this beauty in drawings, paintings, sculpture, and photography, and poets, writers, architects, and musicians have created works reflecting and celebrating the natural world.

CONCEPT CHECK

1. **What** are two determinants of species richness? Give an example of each.

2. **What** is biological diversity?

3. **What** are ecosystem services? How do ecosystem services provided by forests compare to those provided by coasts?

Endangered and Extinct Species

LEARNING OBJECTIVES

1. **Define** *extinction* and distinguish between background extinction and mass extinction.

2. **Contrast** threatened and endangered species.

3. **Describe** four human causes of species endangerment and extinction.

Extinction, the death of a life form, occurs when the last member of a species dies. When a species is extinct, it will never reappear. Biological extinction is the fate of all species, much as death is the fate of all individuals. According to one estimate, for every 2000 species that have ever lived, 1999 of them are extinct today.

> **extinction** The elimination of a species from Earth.

During the time in which organisms have occupied Earth, a continuous, low-level extinction of species, or **background extinction**, has occurred. A second kind of extinction, commonly referred to as **mass extinction**, has occurred perhaps five or six times throughout Earth's history. During a mass extinction event, a large number of species disappear during a relatively short period of geologic time.

The causes of past mass extinctions are not well understood, but biological and environmental factors were probably involved. A major climate change could have triggered a mass extinction. Marine organisms are particularly vulnerable to temperature changes; many would likely have become extinct if Earth's temperature changed just a few degrees. Mass extinctions might also have been triggered by catastrophes, such as the collision of Earth with a large asteroid or comet. The impact could have forced massive quantities of dust into the atmosphere, blocking the sun's rays and cooling the planet.

Although extinction is a natural biological process, it is greatly accelerated by human activities. The burgeoning human population has spread into almost all areas of Earth. Whenever humans invade an area, the habitats of many organisms are disrupted or destroyed, which contributes to their extinction.

Earth's biological diversity is disappearing at an unprecedented rate (**Figure 15.5**). Conservation biologists estimate that species are now becoming extinct at a rate of 100 to 1000 times the natural rate of background extinctions. The IUCN listed almost 20,000 plant and animal species as threatened with extinction in 2012, including an estimated 25 percent of mammals, 13 percent of birds, and 41 percent of amphibians. Approximately 9000 plant species are listed as threatened.

Endangered and Threatened Species

The Endangered Species Act legally defines an **endangered species** as a species in imminent danger of extinction throughout all or a significant portion of its **range**. (The area in which a particular species is found is its range.) A species is endangered when its numbers are so severely reduced that it is in danger of becoming extinct without human intervention. Many endangered species share certain characteristics that seem to have made them more vulnerable to extinction: having an extremely small (localized) range; requiring a large territory; living on islands; having low reproductive success, often the result of a small population size or low reproductive rates; needing specialized breeding areas; and having specialized feeding habits. A species is legally defined as **threatened** when extinction is less imminent but its population is quite low and the species is likely to become endangered in the foreseeable future.

> **endangered species** A species that faces threats that may cause it to become extinct within a short period.

> **threatened species** A species whose population has declined to the point that it may be at risk of extinction.

Endangered and threatened species represent a decline in biological diversity because as their numbers decrease, their genetic variability is severely diminished. Long-term survival and evolution depend on genetic diversity, so a decline in genetic diversity heightens the risk of extinction for endangered and threatened species compared to species that have greater genetic variability.

Representative endangered or extinct species from around the world • Figure 15.5

Officials at the U.S. Fish and Wildlife Service estimate that more than 500 U.S. species have become extinct during the past 200 years. Of these, roughly half have become extinct since 1980.

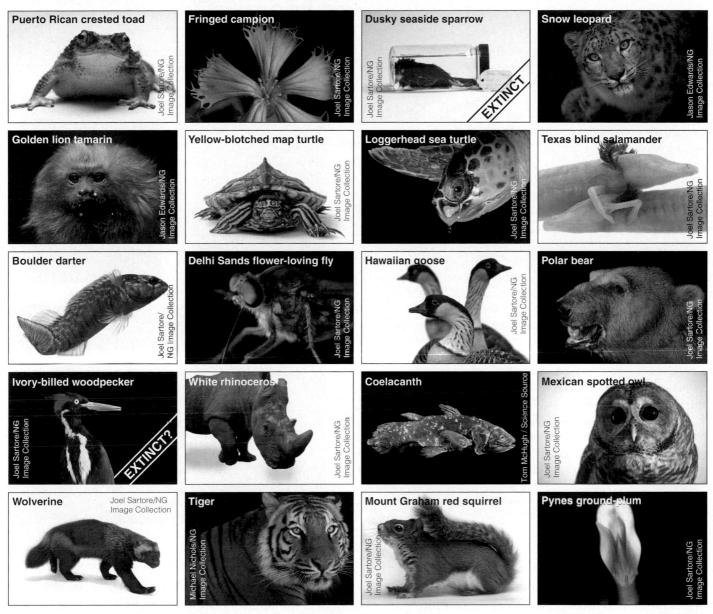

Areas of Declining Biological Diversity

Declining biological diversity is a concern throughout the United States but is most serious in the states of Hawaii (where 63 percent of species are at risk) and California (where about 29 percent of species are at risk). At least two-thirds of Hawaii's native forests are gone.

As serious as declining biological diversity is in the United States, it is even more serious abroad, particularly in tropical rain forests. Tropical rain forests are being destroyed faster than almost all other ecosystems; approximately 1 percent of these ecosystems are being cleared or severely degraded each year. The forests are making way for human settlements, banana plantations, oil and mineral explorations, and other human activities. (For further discussion of tropical rain forests, see Chapters 6 and 13.)

Tropical rain forests are home to thousands or even millions of species. Many species in tropical rain forests are **endemic** (that is, they are not found anywhere else in the world), and the clearing of tropical rain forests contributes to their extinction.

> **endemic species**
> Organisms that are native to or confined to a particular region.

EnviroDiscovery

Is Your Coffee Bird Friendly®?

Many species of migratory songbirds, favorites among North American bird lovers, are in decline, and Americans' coffee habits may play a role. In the tropics, high-yield farms cultivating coffee in full sunlight—known as *sun plantations*—are rapidly replacing traditional *shade plantations*. This switch is affecting wintering birds common to southern Mexico, the Caribbean, Costa Rica, and Colombia. Shade plantations grow coffee plants in the shade of tropical rainforest trees. These trees support a vast diversity of songbird species that winter in the tropics (one study counted 150 species in 5 hectares [12.4 acres]), as well as large numbers of other vertebrates and insects. In contrast, sun plantations provide poor bird habitat. Sun-grown varieties of coffee, treated with large inputs of chemical pesticides and fertilizers, outproduce the shade-grown varieties but lack the diverse products that come from the shade trees. A substantial portion of the region's shade plantations have been converted to sun plantations since the 1970s: 17 percent in Mexico, 40 percent in Costa Rica, and 69 percent in Colombia.

Songbird populations have declined alarmingly during this period. Researchers counted 94 to 97 percent fewer bird species on sun plantations in Colombia and Mexico than on shade-grown coffee plantations. Various conservation organizations and development agencies, such as the Smithsonian Migratory Bird Center (SMBC) and the U.S. Agency for International Development have initiated programs to certify coffee as "shade grown," which allows consumers the chance to support the preservation of tropical rain forests. Shade-grown coffee typically costs more than sun-grown coffee because it is hand picked, involves more care in selecting only ripe beans, and is often certified organic.

The SMBC certifies shade-grown and organic coffee.

Available bird habitat is much greater in shade coffee plantations (left) than in sun plantations (right).

© John Mitchell/Alamy

© David Evans/National Geographic Society/Corbis

The scarlet tanager spends its summers in eastern North America and its winters in Central and South America.

George Grall/NG Image Collection

Perhaps the most unsettling outcome of tropical **deforestation** is its disruptive effect on evolution. In Earth's past, mass extinctions were followed over millions of years by the evolution of new species as replacements for those that died out. In the past, tropical rain forests may have supplied ancestral organisms from which other organisms evolved. Destroying tropical rain forests may reduce nature's ability to replace its species.

Earth's Biodiversity Hotspots

In the 1980s ecologist Norman Myers of Oxford University coined the term **biodiversity hotspots**. In 2000, using plants as their criteria,

> **biodiversity hotspots** Relatively small areas of land that contain an exceptional number of endemic species and are at high risk from human activities.

Myers and ecologists at Conservation International identified 25 biological hotspots around the world (see *What a Scientist Sees*). Interestingly, these 25 hotspots for plants contain 29 percent of the world's endemic bird species, 27 percent of endemic mammal species, 38 percent of endemic reptile species, and 53 percent of endemic amphibian species. Many humans—nearly 20 percent of the world's population—live in the hotspots. Of the 25 hotspots, 15 are tropical, and 9 are mostly or solely islands. Many biologists recommend that conservation planners focus on preserving land in these hotspots to reduce the mass extinction of species that is currently under way.

WHAT A SCIENTIST SEES

Where Is Declining Biological Diversity the Most Serious?

a. If you've visited a zoo and seen a black-and-white ruffed lemur, you may know that lemurs are found in nature only on Madagascar and the small neighboring Comoro Islands. However, most people do not realize that the ruffed lemur comes from a biological hotspot. (Photographed at Folsom Childrens' Zoo in Lincoln, Nebraska.)

Joel Sartore/NG Image Collection

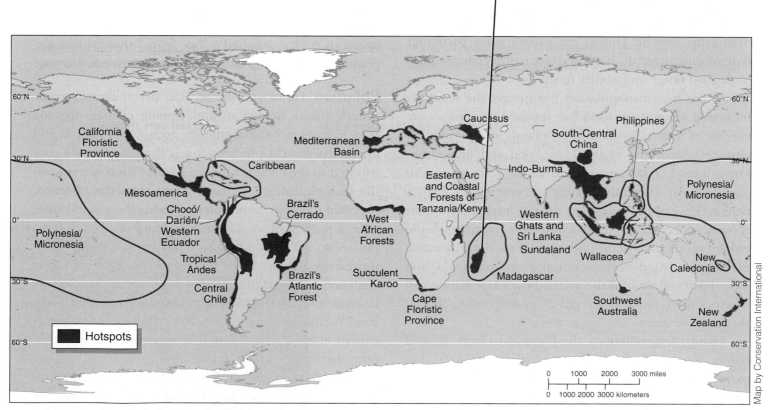

Map by Conservation International

b. Ecologists have identified Madagascar and neighboring islands as one of the world's biological hotspots. These hotspots, which are rich in endemic species, are at great risk from human activities. All of Madagascar's 33 lemur species are currently in danger of extinction.

Selected examples of the many established foreign species accidentally or deliberately introduced into the United States. Invasive species cause great ecological and economic harm.

Invasive Species The introduction of a non-native or foreign species into an ecosystem in which it did not evolve often upsets the balance among the organisms living in that area and interferes with the ecosystem's normal functioning. The foreign species, which typically do not have any natural predators in their new habitat, may compete with native species for food or habitat or may prey on them. Generally, an introduced competitor or predator has a greater negative effect on local organisms than do native competitors or predators. Foreign species whose introduction causes economic or environmental harm are called **invasive species** (**Figure 15.7**). Although invasive species may be introduced into new areas by natural means, humans are usually responsible for such introductions, either knowingly or unknowingly.

invasive species
Foreign species that spread rapidly in a new area if free of predators, parasites, or resource limitations that may have controlled their population in their native habitat.

Cargo-carrying ocean vessels carry *ballast water* from their ports of origin to increase their stability in the ocean. When they reach their destinations, they discharge this water into local bays, rivers, or lakes. Ballast water may contain clams, mussels, worms, small fishes, and crabs, along with millions of microscopic aquatic organisms. These organisms, if they establish themselves, may threaten the area's aquatic environment and contribute to the extinction of native organisms.

One of North America's greatest biological threats—and its most costly aquatic invader—is the zebra mussel, a native of the Caspian Sea, probably introduced through ballast water flushed into the Great Lakes by a foreign ship in 1985 or 1986. Since then, the tiny freshwater mussel has clustered in extraordinary densities on hulls of boats, piers, buoys, and, most damaging of all, water intake systems (**Figure 15.8**).

The zebra mussel's strong appetite for algae and zooplankton reduces the food supply of native fishes,

Zebra mussels • Figure 15.8

Zebra mussels, here clogging a section of water pipe, have caused billions of dollars in damage, in addition to displacing native clams and mussels.

mussels, and clams, threatening their survival. The U.S. Coast Guard estimates that economic losses and control efforts associated with the zebra mussel cost the United States about $5 billion each year.

Two other examples further illustrate the harm caused by invasive species. Two species of Asian carp were introduced by aquaculture facilities flooding into the Mississippi River in the 1990s; they spread rapidly and now threaten the multimillion dollar Great Lakes commercial and sport fisheries. The Burmese python, released into Florida's Everglades National Park by pet owners (see Chapter 6 opener), has multiplied and since 2000 has decimated small mammal populations, essentially wiping out some species there. Estimates of the number of non-native species now in the United States vary widely, but there may be as many as 50,000. Of these, approximately 4300 are considered invasive species. Worldwide, most regions are estimated to contain 10 to 30 percent foreign species.

Overexploitation Sometimes species become endangered or extinct as a result of deliberate efforts to eradicate or control their numbers. Many of these species prey on game animals or livestock. Ranchers, hunters, and government agents have reduced populations of large predators such as wolves and grizzly bears. Some animals are killed because they cause problems for humans. The Carolina parakeet, a beautiful green, red, and yellow bird endemic to the southern United States, was extinct by 1920, exterminated by farmers because it ate fruit and grain crops.

Prairie dogs and pocket gophers were poisoned and trapped so extensively by ranchers and farmers that between 1900 and 1960 they disappeared from most of their original geographic range. As a result of sharply decreased numbers of prairie dogs, the black-footed ferret, the natural predator of these animals, became endangered. A successful captive-breeding program has allowed black-footed ferrets to be reintroduced into the wild and reproduce successfully, though some populations have been decimated by disease.

Unregulated hunting, or overhunting, was a factor contributing to the extinction of certain species in the past but is now strictly controlled in most countries. The passenger pigeon was one of the most common birds in North America in the early 1800s, but a century of overhunting resulted in its extinction in the early 1900s. Unregulated hunting was one of several factors that caused the near extinction of the American bison.

Illegal commercial hunting, or **poaching**, endangers many larger animals, such as the tiger, cheetah, and snow leopard, whose beautiful furs are quite valuable. Rhinoceroses are slaughtered primarily for their horns, used for ceremonial dagger handles in the Middle East and for purported medicinal purposes in Asian medicine. Bears are killed for their gallbladders, used in Asian medicine to treat ailments ranging from indigestion to heart problems. Endangered American turtles are captured and exported illegally to China, where they are killed for food. Caimans (reptiles similar to crocodiles) are killed for their skins and made into shoes and handbags. Although all these animals are legally protected, the demand for their products on the black market has led to their being hunted illegally.

In West Africa, poaching has contributed to the decline in lowland gorilla and chimpanzee populations. The meat (called *bushmeat*) of these rare primates and other protected species, such as anteaters, elephants, and mandrill baboons, provides an important source of protein for indigenous people. Bushmeat is also sold to urban restaurants. This demand for a meat source increases the incidence of poaching.

Live organisms collected through **commercial harvest** end up in zoos, aquaria, biomedical research laboratories, circuses, and pet stores. Several million birds are commercially harvested each year for the pet trade, but unfortunately many of them die in transit, and many more die from improper treatment after they are in their owners' homes. Although it is illegal to capture endangered animals from nature, there is a thriving black market, mainly because collectors in the United States, Canada, Europe, and Japan are willing to pay large

World Conservation Strategy seeks to preserve the vital ecosystem services on which all life depends for survival and to develop sustainable uses of organisms and their ecosystems.

The Convention on Biological Diversity was produced by the 1992 Earth Summit to decrease the rate of extinction of the world's endangered species. This treaty requires that each signatory nation inventory its own biodiversity and develop a **national conservation strategy**, a detailed plan for managing and preserving the biological

Illegal trade in products made from endangered species • Figure 15.14

A merchant in Myanmar (Burma) deals in wildlife products.

Terry Whitaker/Alamy

diversity of that specific country. Despite an increase in conservation efforts since the Earth Summit, however, the loss of biological diversity is not diminishing.

The exploitation of endangered species is somewhat controlled through legislation at the international level: 175 countries participate in the **Convention on International Trade in Endangered Species of Wild Flora and Fauna (CITES)**, which went into effect in 1975. Originally drawn up to protect endangered animals and plants considered valuable in the highly lucrative international wildlife trade, CITES bans the hunting, capturing, and selling of endangered or threatened species and regulates trade of organisms listed as potentially threatened. CITES protects at various levels more than 30,000 species of plants and animals. Unfortunately, enforcement of this treaty varies from country to country. Even where enforcement exists, the penalties are not severe. As a result, illegal trade continues in rare, commercially valuable species (**Figure 15.14**).

The goals of CITES often stir up controversy over issues such as who actually owns the world's wildlife and whether global conservation concerns take precedence over competing local interests. These conflicts often highlight socioeconomic differences between wealthy consumers of CITES products and poor people who trade in endangered organisms.

The case of the African elephant, discussed earlier in the chapter, is a good example of these controversies. Listed as an endangered species since 1989 to halt the slaughter of elephants driven by the ivory trade, the species seems to have recovered in southern Africa. However, poaching rebounded in the 2000s. According to wildlife specialists, by 2008 about 8 percent of the African elephant population was being killed for ivory each year. This level of slaughter is greater than the level in 1989, when the ban went into effect.

CONCEPT CHECK

1. **What** are the goals of the Endangered Species Act? Why is the ESA considered controversial?

2. **What** is the World Conservation Strategy?

The Challenges of Protecting Rare Species

Although a large, powerful predator, the tiger—found in India, Indonesia, Thailand, Russia, and a few other Asian countries—is in danger of becoming extinct (**Figure a**). About 3200 tigers remain in the wild, down from an estimated 100,000 a century ago, and the areas in which they are found have declined to only 7 percent of their historical range.

Tigers are protected by law, yet they are illegally hunted to meet a growing demand for tiger skins, bones, organs, and meat; some body parts are used in traditional Asian medicine. Many of the laws passed to protect tigers are not enforced. Also, as the human population continues to grow, more wild land is altered for human needs, leaving tiger habitat increasingly fragmented, and humans hunt the same animals that tigers do.

Clearly, a concerted commitment is needed at local, national, and international levels if wild tiger populations are to be saved. Science, economics, social standards, cultural beliefs and traditions, political factors, human population growth, and national and international policies can all influence the fate of wild tiger populations.

A very different species, the California condor, has hovered on the brink of extinction for some time (**Figure b**). A huge bird that scavenges carrion and requires a large, undisturbed territory, the California condor once ranged across North America. The species' low birth rate and particular mating habits, combined with habitat destruction, poaching, DDT poisoning, contamination of food supplies with lead shot, and power line hazards, rapidly decimated condor populations. By 1982, only 22 condors remained, all in California, and the species faced imminent extinction.

Recovery efforts took the controversial step of capturing all remaining wild condors to initiate captive breeding (**Figure c**). The first captively bred condors were released into the wild in 1992. Care and training of young birds has evolved to address reintroduction challenges; lead poisoning in particular remains a major threat. No condor population is yet self-sustaining (able to replace its numbers without captive breeding), but their numbers are climbing slowly. As of March 2012, condors totaled 405—still short of the recovery program's goal of 450 individuals—with more than half living wild at four sites in California, one in Mexico, and one in Arizona.

Steve Winter/NG Image Collection

a. A female tiger with her cub, Bandhavgarh National Park, India.

Michael Nichols/NG Image Collection

c. To avoid attachment to humans, handlers use puppets to rear captive condor chicks for release into the wild.

b. The critically endangered California condor.

Tom McHugh/Science Source

Summary

1 Species Richness and Biological Diversity 374

1. **Species richness** is the number of different species in a community. High species richness is associated with communities that are ecologically complex, not isolated, geologically old and stable, and not subject to environmental stress. Species richness is also higher when no one species dominates the community.

2. **Biological diversity** is the number and variety of Earth's organisms; it consists of three components: genetic diversity, species richness, and ecosystem diversity. **Genetic diversity** is the genetic variety within all populations of a given species. **Ecosystem diversity** is the variety of interactions among organisms in natural communities.

3. Ecosystems with greater species richness are better able to supply **ecosystem services**: environmental benefits such as clean air to breathe, clean water to drink, and fertile soil in which to grow crops.

2 Endangered and Extinct Species 378

1. **Extinction** is the elimination of a species from Earth. **Background extinction**, a continuous, low-level extinction of species, has occurred throughout Earth's history. **Mass extinction**, in which many species disappear during a relatively short period of geologic time, has occurred only a few times in Earth's history.

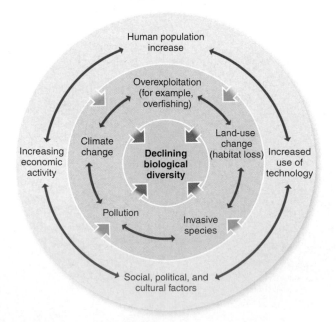

2. An **endangered species** is a species that faces threats that may cause it to become extinct within a short period. A species is defined as **threatened** when extinction is less imminent but its population is quite low.

3. Humans cause species endangerment through habitat destruction, fragmentation, and degradation; pollution; the spread of invasive species; and the overexploitation of biological resources. **Endemic species** are organisms that are native to or confined to a specific place. **Biodiversity hotspots** are areas that contain particularly high numbers of endemic species. **Invasive species** are foreign species, usually introduced by humans, that spread rapidly in a new area where they are free of predators, parasites, or resource limitations that may have controlled their population in their native habitat.

3 Conservation Biology 386

1. **Conservation biology** is the scientific study of how humans affect organisms and of the development of ways to protect biological diversity. **In situ conservation** includes the establishment of parks and reserves to preserve biological diversity in nature; **ex situ conservation** involves conservation of biological diversity in human-controlled settings such as zoos and seed banks.

2. **Restoration ecology** is the study of the historical condition of a human-damaged ecosystem, with the goal of returning it as closely as possible to its former state.

4 Conservation Policies and Laws 390

1. **The Endangered Species Act (ESA)** authorizes the U.S. Fish and Wildlife Service (FWS) to protect endangered and threatened species in the United States and abroad. The ESA requires the FWS to select critical habitats and design a detailed recovery plan for each species listed. Species are designated as endangered or threatened entirely on biological grounds, not economic factors. The act does not compensate private property owners who suffer financial losses related to its enforcement.

2. The **World Conservation Strategy**, formulated by the International Union for the Conservation of Nature, the World Wildlife Fund, and the U.N. Environment Programme, seeks to conserve biological diversity worldwide, to preserve vital ecosystem services, and to develop sustainable uses of organisms and their ecosystems.

Key Terms

- biodiversity hotspots 380
- biological diversity 375
- conservation biology 386
- ecosystem services 376
- endangered species 378
- endemic species 379
- extinction 378
- invasive species 384
- restoration ecology 387
- species richness 374
- threatened species 378

What is happening in this picture?

This little brown bat is suffering from white-nose syndrome—characterized by the frosting of fungus on its nose—a disease that has killed millions of bats in the United States and Canada, wiping out entire populations at many sites.

- How might bats' living habits affect the spread of disease?

- Bats are voracious predators of insects. In areas where bat populations have vanished, what could be some effects on the ecosystem? on the local economy?

- What do affected bats have in common with many bees and frogs?

Ryan von Linden/NY Dept of Environmental Conservation/AP Photo

Critical and Creative Thinking Questions

1. Is biological diversity a renewable or nonrenewable resource? Could it be seen both ways? Explain.

2. List at least five important ecosystem services provided by living organisms.

3. Describe factors affecting species richness and explain how and why the species richness of a wheat field might differ from that of a coral reef.

4. What are the four main causes of species endangerment and extinction? Which do biologists consider most important?

5. What are invasive species? How might their presence particularly affect biodiversity hotspots?

6. Incorporate what you have learned about disappearing frogs and other declining species to compare threatened and endangered species and define *extinction*.

8. How would stabilizing the human population affect biological diversity?

9. In *A Sand County Almanac* and *Sketches Here and There*, Aldo Leopold wrote, "To keep every cog and wheel is the first precaution of intelligent tinkering." How does his statement relate to this chapter?

10. Why do you suppose the Svalbard Seed Vault in Norway is called the "Doomsday vault"? How is this name reflected in the vault's location, buried in a frozen hillside?

11. What are the goals of restoration ecology? Are there any disadvantages associated with restoring disturbed areas?

The Nature Conservancy evaluated the extent of human-caused habitat disturbance in the world's various biomes. Use the data in the graph below to answer questions 12 and 13.

12. On which global region have humans had the greatest impact—polar, temperate, or tropical? Suggest why human impact has been greatest in this region, and describe specifically how this impact likely affects biodiversity there.

13. Which biome has had the lowest percentage of habitat disturbance? Suggest two possible reasons why human impacts on this biome may increase greatly in the future.

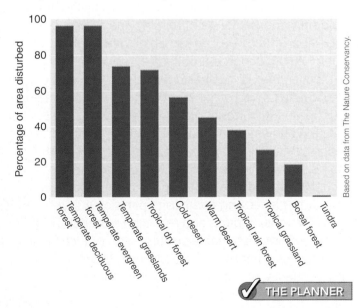

Based on data from The Nature Conservancy.

✓ THE PLANNER

Solid and Hazardous Waste

REUSING AND RECYCLING OLD AUTOMOBILES

In the United States, about 35 million motor vehicles leave service each year out of roughly 250 million in use (see graph). Most are exported to developing countries, but about 11 million cars and trucks are discarded (see photo of a Virginia salvage yard). Although 75 percent of a car can be reused as secondhand parts or recycled, the remaining 25 percent—glass, metals, plastics, fabrics, rubber, foam, and leather—usually ends up in landfills. Economics is an important aspect of the problem, because companies must make money as they reuse and recycle auto components.

About 37 percent of the iron and steel scrap reprocessed in the United States comes from old cars. According to the Environmental Protection Agency, recycling scrap iron and steel produces 86 percent less air pollution and 76 percent less water pollution than mining and refining an equivalent amount of iron ore.

Recycling plastic, which automakers use because it is lightweight and improves fuel efficiency, is one of the biggest challenges in auto recycling. No industry standards currently exist for plastic parts, so the kinds and amounts used in cars vary a great deal. As many as 15 plastics comprise some dashboards, and because many of these plastics are chemically incompatible, they cannot be melted together for recycling.

Auto manufacturers around the world have begun to address the challenge of reusing and recycling old cars. Japan and the European Union have mandated that by 2015, 95 percent of each discarded car must be recoverable. Toyota has developed a way to recover urethane foam and other shredded materials to make soundproofing material. Honda, Mercedes-Benz, Peugeot, Toyota, Volkswagen, Volvo, and other auto manufacturers design cars so that major components from old authomobiles can be reused in new vehicles. Other parts are designed such that they can be separated and recycled (see inset of recyclable plastic and composite parts on a Toyota vehicle).

While reusing parts and recycling materials reduce waste, a more sustainable strategy is to reduce reliance on automobiles. Shifting to public transit, reducing trip lengths and frequency, and other behavioral changes decrease the need to dispose of old automobiles.

WileyPLUS

Motor vehicles in the United States, 2000 to 2011

After decades of growth in the automobile industry, the number of vehicles in the United States leveled off arround 2008.

Based on data from Polk Automotive Research, "Automotive Industry Dashboard, July 2012."

Interpreting Data Questions

In what year did the number of cars and trucks peak? Do you think the number will go up or down over the next decade?

CHAPTER OUTLINE

Solid Waste   398
- Types of Solid Waste
- Disposal of Solid Waste
- ■ What a Scientist Sees: Sanitary Landfills
- ■ EnviroDiscovery: The U.S.–China Recycling Connection

Reducing Solid Waste   404
- Source Reduction
- Reusing Products
- Recycling Materials
- ■ Environmental InSight: Recycling in the United States
- Integrated Waste Management

Hazardous Waste   409
- Types of Hazardous Waste
- ■ EnviroDiscovery: Handling Nanotechnology Safely

Managing Hazardous Waste   412
- Chemical Accidents
- Public Policy and Toxic Waste Cleanup
- Managing Toxic Waste Production
- ■ Case Study: High-Tech Waste

Joel Sartore/NG Image Collection

CHAPTER PLANNER ✓

❑ Study the picture and read the opening story.

❑ Scan the Learning Objectives in each section:
   p. 398 ❑   p. 404 ❑   p. 409 ❑   p. 412 ❑

❑ Read the text and study all figures and visuals. Answer any questions.

Analyze key features

❑ What a Scientist Sees, p. 401

❑ Process Diagram, p. 403 ❑   p. 408 ❑

❑ EnviroDiscovery, p. 404 ❑   p. 410 ❑

❑ Environmental InSight, p. 407

❑ Case Study, p. 415

❑ Stop: Answer the Concept Checks before you go on:
   p. 404 ❑   p. 408 ❑   p. 411 ❑   p. 414 ❑

End of Chapter

❑ Review the Summary and Key Terms.

❑ Answer What is happening in this picture?

❑ Answer the Critical and Creative Thinking Questions.

Solid Waste

LEARNING OBJECTIVES

1. **Distinguish** between municipal and nonmunicipal solid waste.

2. **Describe** the features of a modern sanitary landfill and relate some of the problems associated with sanitary landfills.

3. **Describe** the features of a mass burn incinerator and relate some of the problems associated with incinerators.

4. **Explain** the composting process.

The United States generates more solid waste per capita than any other country. (Canada is a close second.) According to the EPA, in 2010, the average person in the United States produced 2.01 kg (4.43 lb) of solid waste per day. This amount corresponded to a total of 227 million metric tons. This represents a small decrease from 2005, a change attributable to the economic downturn that began in 2008.

Waste generation is an unavoidable consequence of the prosperous, high-technology, industrial economies of the United States and other highly developed nations. Many products that would be repaired, reused, or recycled in less affluent nations are simply thrown away. Nobody likes to think about solid waste, but it is certainly a concern of modern society—we keep producing it, and places to dispose of it safely are dwindling in number (**Figure 16.1**).

municipal solid waste Solid materials discarded by homes, offices, stores, restaurants, schools, hospitals, prisons, libraries, and other commercial and institutional facilities.

Types of Solid Waste

Municipal solid waste consists of the combined residential and commercial waste produced in a municipal area. Municipal solid

Drinking water options • Figure 16.1

Global demand for bottled water is increasing, but most of these empty bottles are not recycled. In 2011 the average American consumed 110 L (117 qt) of bottled water, which translates into 60 million plastic water bottles per day (**a**). A return to water fountains (**b**) or a shift to refillable bottles (**c**) could reduce some of this waste.

David McGlynn/Photographer's Choice/Getty Images

© D A Barnes/Alamy

© Radius Images/Alamy

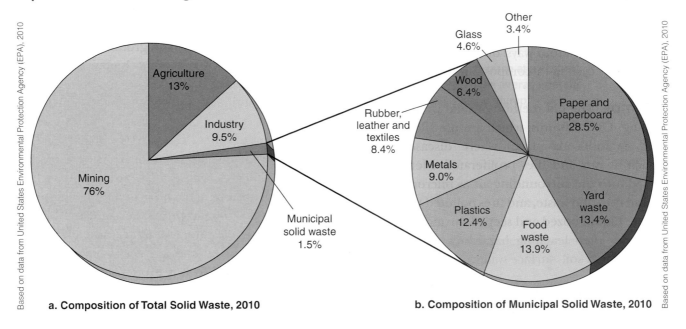

Based on data from United States Environmental Protection Agency (EPA), 2010

a. Composition of Total Solid Waste, 2010

b. Composition of Municipal Solid Waste, 2010

Based on data from United States Environmental Protection Agency (EPA), 2010

waste is a heterogeneous mixture composed primarily of paper and paperboard; yard waste; plastics; food waste; metals; rubber, leather, and textiles; wood; and glass (**Figure 16.2**). The proportions of the major types of solid waste in this mixture change over time. Today's solid waste contains more paper and plastics than in the past, whereas the amounts of glass and steel have declined. The total amount of municipal solid waste produced in the United States in 2010 was 2.8 times as much as in 1960, while per-person waste generation increased by a factor of only 1.7 (**Figure 16.3**). This shows how both population growth and behavioral changes affect our impacts on the environment.

Municipal solid waste makes up only a small proportion—less than 2 percent—of the total solid waste produced each day. **Nonmunicipal solid waste**, which includes mining, agricultural, and industrial wastes, is produced in substantially larger amounts. Most solid waste generated in the United States is from nonmunicipal sources.

> **nonmunicipal solid waste** Solid waste generated by industry, agriculture, and mining.

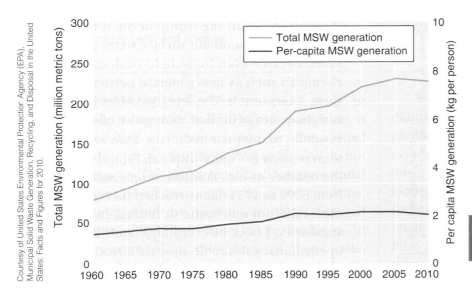

Courtesy of United States Environmental Protection Agency (EPA), Municipal Solid Waste Generation, Recycling, and Disposal in the United States: Facts and Figures for 2010.

Total and per-capita municipal solid waste in the United States, 1960–2010 • Figure 16.3

Until the past few years, total production of municipal solid waste in the United States increased every five years. Individual waste production increased as well, reaching a plateau around 1990. The recent decrease in total and per-capita production are attributable to the economic downturn that began in 2008.

Interpreting Data

In what year did total waste reach a maximum? What about per-capita waste? What accounts for the difference?

EnviroDiscovery
Handling Nanotechnology Safely

Nanotechnology is in the news a lot these days. **Nanomaterials**, which are unique materials and devices designed on the ultrasmall scale of atoms or molecules, have numerous possible applications. For example, nanoparticles of cadmium selenide might be injected into cancerous tissue, where they would accumulate inside cancer cells; when exposed to ultraviolet radiation, these nanoparticles glow, and surgeons could more easily excise the cancerous tissues and leave the healthy tissues intact. Nanocrystals have the potential to be used in thin-film solar panels to convert solar energy to electricity. Silica nanoparticles embedded in glass make a heat-resistant glass capable of withstanding temperatures of up to 1800°C for several hours (see photograph).

Despite the potential of nanotechnology, particles on the nanometer scale (a nanometer is one-billionth of a meter) might pose health, safety, or environmental risks. No one knows for sure. The EPA has adopted a precautionary approach (see the section on the precautionary principle in Chapter 4) and decided to regulate nanomaterials that might adversely affect the environment. This means that the burden of proof about product safety will fall on companies that sell nanotechnology. Similarly, the Food and Drug Administration will have to oversee regulation of nanotechnology that has potential health and safety risks.

Nanotechnology
The glass between the man and the fire has fire-resistant properties because of the addition of silica nanoparticles.

Here we discuss dioxins and PCBs because they are some of the most persistent hazardous compounds that contaminate our environment.

Dioxins **Dioxins** are a group of related organic chemical compounds. They become toxic when chemically combined with chlorine or (less commonly) bromine. Incineration of medical and municipal wastes accounts for 70 to 95 percent of known human emissions of dioxins. Some other known sources of dioxins are iron ore mills, copper smelters, cement kilns, metal recycling, coal combustion, and chemical accidents. Pulp and paper plants that use chlorine for bleaching release dioxin in sludge and wastewater. Newer bleaching techniques avoid chlorine, thereby reducing or eliminating dioxins (**Figure 16.12**). Motor vehicles, outdoor grills, and cigarette smoke emit minor amounts of dioxins. Forest fires and volcanic eruptions are natural sources of dioxins. Dioxins also form during the production of some pesticides.

Dioxins emitted to the air settle on plants, soil, and bodies of water; from there they are incorporated into the food web. When humans and other animals ingest

Air pollution from a paper mill
• Figure 16.12 _____

Simpson paper pulp mill near Tacoma, Washington. This plant was an early adopter of a chlorine free bleaching process. Consequently, it no longer produces measurable amounts of chlorinated dioxins.

Dioxin concentrations in guillemot eggs, 1967–2010 • Figure 16.13

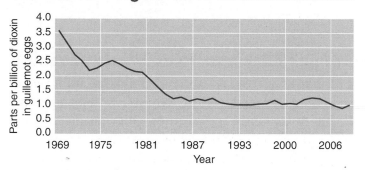

Adapted from the Swedish Environmental Protection Agency (2011). National Environmental Monitoring, Seas and Costal Areas Programme Area.

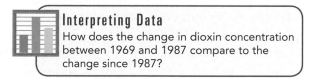

Interpreting Data

How does the change in dioxin concentration between 1969 and 1987 compare to the change since 1987?

dioxins—primarily in contaminated meat, dairy products, and fish—they store and accumulate the dioxins in their fatty tissues (see the bioaccumulation and biomagnification discussion in Chapter 4). Because dioxins are so widely distributed in the environment, virtually everyone has dioxins in their body fat.

Dioxins cause several kinds of cancer in laboratory animals, but the data conflict on their cancer-causing ability in humans. A 2001 EPA report suggests that dioxins probably cause several kinds of cancer in humans and likely affect the human reproductive, immune, and nervous systems. Because dioxins are passed through human milk, nursing infants are considered particularly at risk.

Measures to reduce dioxin emissions have led to some success. A study supported by the Swedish Environmental Protection Agency showed that dioxins in guillemot eggs (a guillemot is a kind of a seabird) decreased from 3.5 parts per billion (ppb) to around 1 ppb between 1969 and 2010 (**Figure 16.13**).

PCBs **Polychlorinated biphenyls (PCBs)** are a group of 209 industrial chemicals composed of carbon, hydrogen, and chlorine. PCBs were manufactured in the United States between 1929 and 1979 for a wide variety of uses: as cooling fluids in electrical transformers, electrical capacitors, vacuum pumps, and gas-transmission turbines; and in hydraulic fluids, fire retardants, adhesives, lubricants, pesticide extenders, inks, and other materials. Prior to the EPA ban in the 1970s, PCBs were dumped in large quantities into landfills, sewers, and fields. Such improper disposal is one of the reasons PCBs are still a threat today.

Studying bacteria that break down PCBs in contaminated soil • Figure 16.14

A microbiologist adds soil to a "bioreactor" to test the ability of certain bacteria to treat contaminated soil.

Photo Researchers, Inc.

The dangers of PCBs first became evident in Japan in 1968, where hundreds of people ate rice bran oil accidentally contaminated with PCBs and consequently experienced serious health problems, including liver and kidney damage. A similar mass poisoning tied to PCBs occurred in Taiwan in 1979. Since then, toxicity tests conducted on animals indicate that PCBs harm the skin, eyes, reproductive organs, and gastrointestinal system. PCBs are endocrine disrupters: They interfere with hormones released by the thyroid gland. Several studies have demonstrated that in utero exposure to PCBs can lead to certain intellectual impairments in children. PCBs may be carcinogenic; they are known to cause liver cancer in rats, and studies in Sweden and the United States have shown a correlation between high PCB concentrations in the body and incidences of certain cancers.

Although high-temperature incineration is one of the most effective ways to destroy PCBs in most solid waste, it is too costly to be used for the removal of PCBs that have leached into soil and water. One way to remove PCBs from soil and water is to extract them using solvents. This method is undesirable because the solvents themselves are hazardous chemicals, and these extraction methods are also costly.

Recently, researchers have discovered several bacteria that degrade PCBs at a fraction of the cost of incineration. Additional research is needed to make the biological degradation of PCBs practical (**Figure 16.14**).

CONCEPT CHECK STOP

1. **What** is hazardous waste?

2. **What** are two sources of dioxins? of PCBs?

Managing Hazardous Waste

LEARNING OBJECTIVES

1. **Compare** the Resource Conservation and Recovery Act and the Comprehensive Environmental Response, Compensation, and Liability Act (the Superfund Act).
2. **Explain** how green chemistry is related to source reduction.

We have the technology to manage toxic waste in an environmentally responsible way, but it can be extremely expensive. Although great strides have been made in educating the public about the problems of hazardous waste, we have only begun to address many issues of hazardous waste disposal. Hazardous wastes, once released into the environment, are extremely difficult to remove. Consequently, any approach that avoids the production of hazardous wastes, such as chlorine-free blending at pulp and paper mills, are the most strategic for minimizing contamination. No country currently has an effective hazardous waste management program, but several European countries have led the way by producing smaller amounts of hazardous waste and by using fewer hazardous substances.

Chemical Accidents

When a chemical accident occurs in the United States, whether at a factory or during the transport of hazardous chemicals, the National Response Center (NRC) is notified. The NRC assigns an on-scene coordinator to determine the size and chemical nature of the accident. Most chemical accidents reported to the NRC involve oil, gasoline, or other petroleum spills. The remaining accidents involve more than 1000 other hazardous chemicals, such as ammonia, sulfuric acid, and chlorine.

Chemical safety programs have traditionally stressed accident mitigation and adding safety systems to existing procedures. More recently, industry and government agencies have stressed accident prevention through the **principle of inherent safety**, in which industrial processes are redesigned to involve less toxic materials so that dangerous accidents are less likely to occur in the first place. The principle of inherent safety is an important aspect of source reduction.

Public Policy and Toxic Waste Cleanup

Currently, two federal laws dictate how hazardous waste should be managed: (1) the Resource Conservation and Recovery Act, which is concerned with managing hazardous waste being produced now, and (2) the Superfund Act, which provides for the cleanup of abandoned and inactive hazardous waste sites.

The **Resource Conservation and Recovery Act (RCRA)** was passed in 1976 and amended in 1984. Among other things, RCRA instructs the EPA to identify which wastes are hazardous and to provide guidelines and standards to states for hazardous waste management programs. RCRA bans hazardous waste from land disposal unless it is treated to meet the EPA's standards of reduced toxicity. In 1992 the EPA initiated a major reform of RCRA to expedite cleanups and streamline the permit system to encourage hazardous waste recycling.

In 1980 the **Comprehensive Environmental Response, Compensation, and Liability Act (CERCLA)**, commonly known as the **Superfund Act**, established a program to tackle the huge challenge of cleaning up abandoned and illegal toxic waste sites across the United States. At many of these sites, hazardous chemicals have migrated deep into the soil and have polluted groundwater. The greatest threat to human health from toxic waste sites comes from drinking water laced with such contaminants.

Cleaning Up Existing Toxic Waste: The Superfund Program

The federal government estimates that the United States has more than 400,000 hazardous waste sites with leaking chemical storage tanks and drums (both above and below ground), pesticide dumps, and piles of mining waste. This estimate does not include the hundreds or thousands of toxic waste sites at military bases and nuclear weapons facilities.

By 2010, over 10,000 sites were in the CERCLA inventory, which means the EPA had identified them as qualifying for cleanup (**Figure 16.15**). (This count does not include more than 1000 sites that had been cleaned up and removed from the CERCLA inventory since 1980.)

The sites posing the greatest threat to public health and the environment are placed on the **Superfund National Priorities List**, and the federal government will assist in their cleanup. As of 2010, 1146 sites were on the National Priorities List. The five states that led the list in 2010 were New Jersey (112 sites), California (98 sites), Pennsylvania (96 sites), New York (87 sites), and Michigan (66 sites). The average cost of cleaning up a site is $20 million.

One reason for the urgency over cleaning up the sites on the National Priorities List is their locations. With the growth of cities and their suburbs, residential developments now surround many of the dumps. Because the federal government cannot clean up every old dump in the United States, the current landowner, prior owners, and anyone who has dumped waste on or transported waste to a particular site may be liable for cleanup costs.

Although critics decry the slow pace and high cost of cleaning up Superfund sites, the existence of CERCLA is a deterrent to further polluting. Companies that produce hazardous waste are now fully aware of the costs of liability and cleanup and are more likely to properly dispose of their hazardous wastes.

Managing Toxic Waste Production

The Superfund Act deals only with hazardous waste produced in the past, not the large amount of toxic waste produced today. There are three ways to manage hazardous waste: (1) source reduction, (2) conversion to less hazardous materials, and (3) long-term storage.

As with municipal solid waste, the most effective approach is source reduction—that is, using less hazardous or nonhazardous materials in industrial processes. Source reduction relies on the increasingly important field of green chemistry.

> **green chemistry**
> A subdiscipline of chemistry in which commercially important chemical processes are redesigned to significantly reduce environmental harm.

Important principles of green chemistry include preventing the creation of hazardous materials, choosing less hazardous materials over more hazardous ones, and minimizing the amounts of resources used to make products. The latter includes not just chemicals, but also equipment and energy needs.

For example, chlorinated solvents are widely used in electronics, dry cleaning, foam insulation, and industrial cleaning. To accomplish source reduction, it is sometimes possible to substitute a less hazardous water-based solvent for the toxic chlorinated one. Substantial source reduction of chlorinated solvents can also be accomplished by reducing solvent emissions. Installing solvent-saving devices benefits the environment and also saves money because smaller amounts of chlorinated solvents must be purchased. No matter how efficient source reduction becomes, however, it may never entirely eliminate hazardous waste.

The second-best way to deal with hazardous waste is to reduce its toxicity by chemical, physical, or biological means, depending on the nature of the waste. High-temperature incineration, for example, reduces dangerous compounds such as pesticides, PCBs, and organic solvents to safe products such as water and carbon dioxide. The resulting ash is hazardous and must be disposed of in a landfill designed for hazardous materials. Incineration using a *plasma torch* produces such high temperatures (up to 10,000°C, five times higher than temperatures of conventional incinerators) that hazardous waste is almost completely converted to nontoxic gases, such as carbon dioxide and nitrogen.

Cleaning up hazardous waste • Figure 16.15

a. Toxic waste in deteriorating drums at a site near Washington, DC. The metal drums in which much of the waste is stored have corroded and started to leak. Old toxic waste dumps are commonplace around the United States.

Courtesy USDA

b. Cleanup of a hazardous waste site near Minneapolis, Minnesota. Removal and destruction of the wastes are complicated by the fact that usually nobody knows what chemicals are present.

© Greg Smith/© Corbis

Cutaway view through a hazardous waste landfill • Figure 16.16

The bottom of this hazardous waste landfill has two layers of compacted clay, each covered by a high-density plastic liner. (Some hazardous waste landfills have three layers of compacted clay.) A drain system located above the plastic and clay liners allows liquid leachate to collect in a basin where it can be treated, and a leak detection system is installed between the clay liners. Barrels of hazardous waste are placed above the liners and covered with soil.

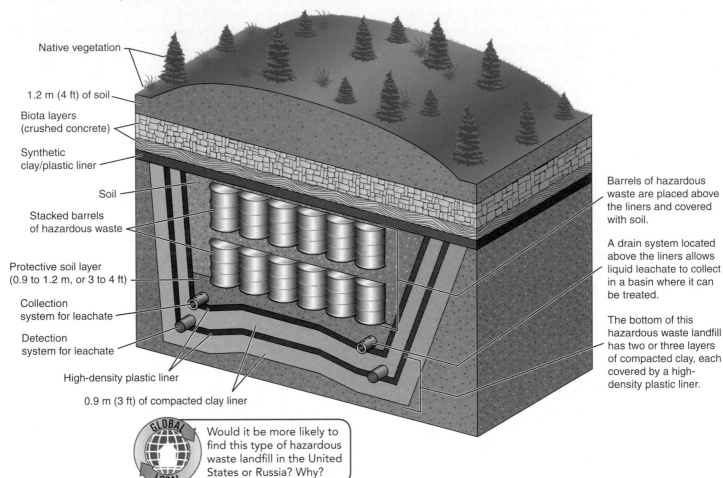

Native vegetation

1.2 m (4 ft) of soil

Biota layers
(crushed concrete)

Synthetic
clay/plastic liner

Soil

Stacked barrels
of hazardous waste

Protective soil layer
(0.9 to 1.2 m, or 3 to 4 ft)

Collection
system for leachate

Detection
system for leachate

High-density plastic liner

0.9 m (3 ft) of compacted clay liner

Barrels of hazardous
waste are placed above
the liners and covered
with soil.

A drain system located
above the liners allows
liquid leachate to collect
in a basin where it can
be treated.

The bottom of this
hazardous waste landfill
has two or three layers
of compacted clay, each
covered by a high-
density plastic liner.

GLOBAL LOCAL — Would it be more likely to find this type of hazardous waste landfill in the United States or Russia? Why?

Adapted from Rocky Mountain Arsenal Remediation Venture Office.

Hazardous waste that is not completely detoxified must be placed in long-term storage. Hazardous waste landfills are subject to strict environmental criteria and design features. They are located as far as possible from aquifers, streams, wetlands, and residences. Design components include several layers of compacted clay and high-density plastic liners at the bottom of the landfill to prevent leaching of hazardous substances into surface water and groundwater (**Figure 16.16**). Leachate is collected and treated to remove contaminants. The entire facility and nearby groundwater deposits are carefully monitored to make sure there is no leakage.

Some toxic liquid waste, such as explosives and pesticides, can be disposed of by *deep-well injection*. In this technique, the liquid waste is injected through pipes that extend hundreds of meters into an injection zone located between two impermeable areas. In such geologic formations, the waste is not likely to migrate into aquifers that could be used for drinking or irrigation.

CONCEPT CHECK STOP

1. **How** are the Resource Conservation and Recovery Act and the Comprehensive Environmental Response, Compensation, and Liability Act alike, and what is the focus of each act?

2. **How** is green chemistry applied to reducing sources of hazardous waste?

CASE STUDY

High-Tech Waste

In the United States and other highly developed countries, the average computer is replaced every 18 to 24 months, not because it is broken but because rapid technological developments and new generations of software make it obsolete. Old computers may still be in working order, but they have no resale value and are even difficult to give away. As a result, they sit in warehouses, garages, and basements—or are frequently thrown away with the trash. According to the EPA, more than 300 million electronic devices were discarded in 2006 (latest data available), and 80 percent of these were disposed of in sanitary landfills.

This disposal represents a huge waste of the high-quality plastics and metals (aluminum, copper, tin, nickel, palladium, silver, and gold) that make up computers. Computers also contain the toxic heavy metals lead, cadmium, mercury, and chromium, which could potentially leach from landfills into soil and groundwater. Not long ago, many computers contained as much as 3.5 kg (8 lb) of lead in their monitors and circuit boards. The recent transition to liquid crystal display (LCD) monitors, as well as more efficient circuitry, has reduced much of the lead content (an example of dematerialization). At the same time, smart phones, tablets, mp3 players, and other small, short-lived personal electronic products continue to proliferate. Several states have passed legislation requiring businesses and residents to **e-cycle** consumer electronics—that is, recycle PCs, monitors, cell phones, and televisions.

In 2009, about 25 percent of all electronics and about 38 percent of all computers in the United States were e-cycled. Although some companies handle obsolete computers in the United States (see photograph), many U.S. computers are shipped overseas to be recycled in developing countries such as India, Pakistan, and China. There the computers are disassembled, often using methods that are potentially dangerous to the workers taking them apart. For example, circuit boards are often burned to obtain the small amount of gold in them, and burning releases hazardous fumes into the air.

Some highly developed countries have been more progressive than the United States in dealing with their computer waste. The European Union implemented a Waste Electrical and Electronic Equipment plan in 2005 to recover, recycle, and dispose of electronic waste and remove some of the most hazardous chemicals. Japan and other industrialized nations have put in place similar policies.

Peter Essick / Aurora Photos

Obsolete computer equipment

The computer monitors at this Texas electronic recycler are being disassembled. Functional tubes will be exported to Thailand, where they will be used to manufacture inexpensive televisions. Broken tubes will be recycled or disposed of in the United States.

Summary

1 Solid Waste 398

1. **Municipal solid waste** consists of solid materials discarded by homes, office buildings, stores, restaurants, schools, hospitals, prisons, libraries, and other facilities. **Nonmunicipal solid waste** consists of solid waste generated by industry, agriculture, and mining.

2. Use of **sanitary landfills** is the most common method of solid waste disposal, involving compacting and burying waste under a shallow layer of soil. Layers of compacted clay and plastic sheets prevent **leachate** (liquid waste) from seeping into groundwater. Problems with sanitary landfills include the potential for methane gas to seep out and cause explosions, the accidental leaking of toxic leachate, a lack of existing landfill space, and resistance to new landfills near homes and businesses.

3. A **mass burn incinerator** is a large furnace that burns all solid waste except for unburnable items such as refrigerators. Problems associated with incineration of solid waste include the potential for air pollution, difficulties in disposing of the toxic ash produced, the high costs of the process, and difficulties in choosing incinerator sites.

4. In composting, yard waste, food scraps, and other organic wastes are transformed by microbial action into a material that, when added to soil, improves its condition.

© D A Barnes/Alamy

© Radius Images/Alamy

2 Reducing Solid Waste 404

1. Reducing consumption is the surest way to reduce waste production. In **source reduction**, products are designed and manufactured in ways that decrease the volume of solid waste and the amount of hazardous waste in the solid waste stream.

2. The volume of solid waste produced can be decreased through source reduction, reuse of products, and recycling of materials. Recycling conserves natural resources and is more environmentally benign than landfill disposal but requires a market for the recycled goods.

3. **Integrated waste management** is a combination of the best waste management techniques into a consolidated program to deal effectively with solid waste.

3 Hazardous Waste 409

1. **Hazardous waste** is a discarded chemical that threatens human health or the environment. Hazardous chemicals may be solids, liquids, or gases and include a variety of acids, dioxins, abandoned explosives, heavy metals, infectious wastes, nerve gas, organic solvents, PCBs, pesticides, and radioactive substances.

2. **Dioxins** are hazardous chemicals formed as unwanted by-products in industrial processes and incineration that include organic compounds and chlorine or, less commonly, bromine. **Polychlorinated biphenyls (PCBs)** are hazardous, oily, industrial chemicals composed of carbon, hydrogen, and chlorine.

4 Managing Hazardous Waste 412

1. The **Resource Conservation and Recovery Act (RCRA)** instructs the EPA to identify hazardous waste and to provide guidelines and standards for states' hazardous waste management programs. The **Comprehensive Environmental Response, Compensation, and Liability Act (CERCLA)**, or **Superfund Act**, established a program whose goal is to clean up abandoned and illegal toxic waste sites across the United States.

2. The most effective approach to managing hazardous waste is source reduction, reducing the amount and toxicity of hazardous materials used in industrial processes. Source reduction relies on **green chemistry**, a subdiscipline of chemistry in which commercially important chemical processes are redesigned to reduce environmental harm.

Key Terms

- green chemistry 413
- hazardous waste 409
- integrated waste management 408
- mass burn incinerator 402
- municipal solid waste 398
- nonmunicipal solid waste 399
- sanitary landfill 400
- source reduction 405

What is happening in this picture?

Ulrich Doering/Alamy

This clothing market in Tanzania sells used clothing from Europe and America.

• How does this practice affect the volume of solid waste produced?

• These clothes, which come from nonprofit charities such as Goodwill, are sold in local marketplaces. How does this practice affect African clothing industries?

• Africa now exports about $1.6 billion of its textiles and clothing to the United States each year. However, Africa's main competition for the U.S. market is China and other Asian countries. Explain this global connection.

Critical and Creative Thinking Questions

1. How could source reduction efforts reduce the volume of waste that arises from abandoned automobiles?

2. Explain why reducing consumption is the surest way to reduce production of wastes.

3. Compare the advantages and disadvantages of disposing of municipal solid waste in sanitary landfills and by incineration.

4. List what you think are the best ways to treat each of the following types of solid waste and explain the benefits of the processes you recommend: paper, plastic, glass, metals, food waste, and yard waste.

5. What are dioxins, and how are they produced? What harm do they cause?

6. Suppose hazardous chemicals were suspected to be leaking from an old dump near your home. Outline the steps you would take to (1) have the site evaluated to determine whether there is a danger and (2) mobilize the local community to get the site cleaned up.

7. What are the goals, strengths, and weaknesses of the Superfund Act?

8. What is integrated waste management? Why must a sanitary landfill always be included in any integrated waste management plan?

9. What is hazardous waste, and how does green chemistry help address the problem of hazardous waste?

In an effort to reduce municipal solid waste, many communities have required customers to pay for garbage collection according to the amount of garbage they generate, an approach termed "unit pricing," or "pay as you throw." The following chart applies to questions 10 and 11 and illustrates the effects of unit pricing in San Jose, California, on garbage sent to landfills and on wastes diverted through recycling and through separation of yard wastes.

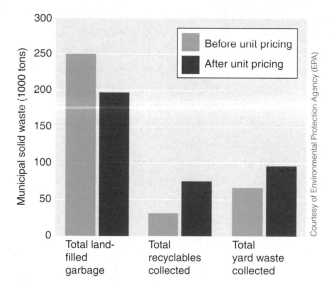

10. How did the implementation of unit pricing in San Jose affect the amount of garbage sent to landfills?

11. How did the implementation of unit pricing affect the quantity of materials recycled or of yard wastes collected?

Sustainable Citizen Question

12. Estimate the total amount of material collected before and after unit pricing went into effect. Did the policy change consumption, or just encourage citizens to sort their waste? What other policies might encourage people to reduce consumption? What would be the pros and cons of such a policy?

✓ THE PLANNER

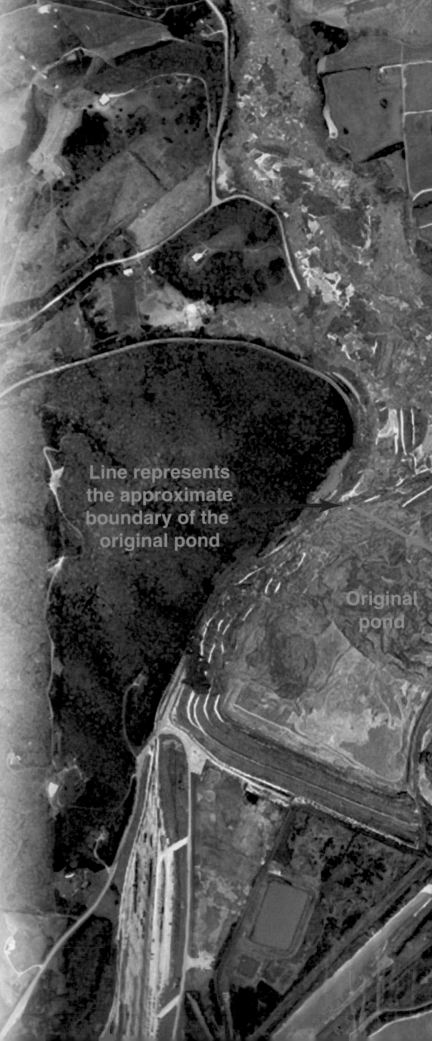

Nonrenewable Energy Resources

COAL AND THE ENVIRONMENT

About 300 years ago, industrialists discovered that burning coal could heat water to drive steam turbines. That discovery allowed the Industrial Revolution to occur, transforming almost all aspects of life for people, first in Europe, then in the Americas, and now around the word. The downsides of coal—mining dangers, air and water pollution, and solid waste—have always been apparent. Through the centuries, as coal use has increased, people have developed technologies to manage these problems. Nonetheless, as long as coal remains a major energy source, we will constantly struggle with its many environmental consequences.

In many parts of the world, coal is burned using old technology, leading to air, water, and ground contamination. Modern coal burning technology can eliminate most particulate matter, sulfur, and mercury from the exhaust of coal-fired power plants. However, technology to remove carbon dioxide, one of the biggest environmental threats we face today, is just now being developed.

Even where technology is highly advanced, large-scale use of coal can have devastating consequences. As coal consumption has increased in the United States over the past five decades, so has the production of fly ash, a major waste by-product (see inset). On December 22, 2008, a pond containing about 4 million m^3 (140 ft^3) of fly ash slurry—water mixed with the ash left over after coal is burned—broke open in Kingston, Tennessee, covering 1.2 km^2 (0.5 mi^2) of the surrounding area (see photograph; the bluish-gray material is fly ash slurry). The spill destroyed houses and roads, and it contaminated the Emory and Clinch rivers, which feed the Tennessee River, the Ohio River, and eventually the Mississippi River before reaching the ocean.

Line represents the approximate boundary of the original pond

Original pond

graphingactivity

C

LEA

1.

2.

3.

Dist

Data
recov
be re
techn

Coal consumption and fly ash production in the United States, 1966 to 2009

Based on data from the U.S. Geological Survey and U.S. Department of Energy.

— Fly ash production
— Coal consumption

Interpreting Data Question

Did the amount of fly ash produced for each ton of coal consumed remain constant over this time span?

Percentage of proved recoverable

Emory River

Tennessee Valley Authority

CHAPTER OUTLINE

Energy Consumption   420

Coal   421
- Coal Mining
- Environmental Impacts of Coal
- Making Coal Cleaner

Oil and Natural Gas   423
- Reserves of Oil and Natural Gas
- Environmental Impacts of Oil and Natural Gas
- ▪ Environmental InSight: The *Exxon Valdez* and *Deepwater Horizon* Oil Spills

Nuclear Energy   430
- Conventional Nuclear Fission
- Nuclear Energy and Fossil Fuels
- Safety and Accidents in Nuclear Power Plants
- The Link Between Nuclear Energy and Nuclear Weapons
- Radioactive Wastes
- ▪ What a Scientist Sees: Yucca Mountain
- ▪ EnviroDiscovery: A Nuclear Waste Nightmare
- Decommissioning Nuclear Power Plants
- ▪ Case Study: The Arctic National Wildlife Refuge

CHAPTER PLANNER ✓

❑ Study the picture and read the opening story.

❑ Scan the Learning Objectives in each section:
   p. 420 ❑   p. 421 ❑   p. 423 ❑   p. 430 ❑

❑ Read the text and study all figures and visuals. Answer any questions.

Analyze key features

❑ Process Diagram, p. 424 ❑   p. 431 ❑   p. 433 ❑

❑ National Geographic Map, pp. 426–427

❑ Environmental InSight, p. 429

❑ What a Scientist Sees, p. 437

❑ EnviroDiscovery, p. 438

❑ Case Study, p. 439

❑ Stop: Answer the Concept Checks before you go on:
   p. 420 ❑   p. 423 ❑   p. 430 ❑   p. 438 ❑

End of Chapter

❑ Review the Summary and Key Terms.

❑ Answer What is happening in this picture?

❑ Answer the Critical and Creative Thinking Questions.

Summary

1 Energy Consumption 420

1. The per person energy consumption in highly developed nations is eight times higher than that in developing nations. However, highly developed countries consume less total energy than do developing countries. Experts expect total energy use in developing countries to grow rapidly in the coming decades, but to plateau in highly developed countries.

2 Coal 421

1. **Surface mining** is the extraction of mineral and energy resources near Earth's surface by first removing the soil, subsoil, and overlying rock strata. **Subsurface mining** is the extraction of mineral and energy resources from deep underground deposits. Surface mining is less expensive and safer but causes more serious environmental problems than subsurface mining.

2. Coal mining can lead to landslides and can pollute streams with sediment and **acid mine drainage**, when sulfuric acid and dangerous dissolved materials such as lead, arsenic, and cadmium wash from coal and metal mines into nearby lakes and streams. In **mountaintop removal** a huge shovel removes an entire mountaintop to reach coal located below. Burning coal releases more CO_2 and contributes more extensively to global climate warming than does producing an equivalent amount of energy by burning other fossil fuels. The combustion of coal contributes to **acid deposition**, in which acid falls from the atmosphere to the surface as precipitation or as dry acid particles.

3. Power plants can make coal a cleaner fuel by installing **scrubbers** to clean the power plants' exhaust. **Fluidized-bed combustion** is a clean-coal technology in which crushed coal is mixed with limestone to neutralize the acidic sulfur compounds produced during combustion.

3 Oil and Natural Gas 423

1. More than half of the world's total estimated oil and natural gas reserves are located in the Persian Gulf region.

2. A serious spill along an oil transportation route creates an environmental crisis, particularly in aquatic ecosystems. The burning of oil and natural gas produces CO_2 that can contribute to global climate change. Burning oil also leads to acid deposition by producing nitrogen oxides. Natural gas contains almost no sulfur and produces less CO_2 and other pollutants compared to oil and coal.

© Christopher Morris/VII/Corbis

4 Nuclear Energy 430

1. **Nuclear energy** is the energy released by nuclear **fission** or fusion. A **nuclear reactor** is a device that initiates and maintains a controlled nuclear fission chain reaction to produce energy for electricity. A typical reactor contains a **reactor core**, where nuclear fission occurs; a **steam generator**; a **turbine**; and a **condenser**.

2. Generating electric power through nuclear energy emits few pollutants (such as CO_2) into the atmosphere compared to the combustion of coal but generates highly dangerous radioactive waste, such as **spent fuel**, the used fuel elements that were irradiated in a nuclear reactor.

3. As evidenced at Chernobyl and Fukushima Daiichi, accidents at nuclear power plants can release dangerous levels of radiation into the environment and result in human casualties. The safe storage of radioactive wastes is another concern associated with nuclear energy. **Low-level radioactive wastes** are radioactive solids, liquids, or gases that give off small amounts of ionizing radiation. **High-level radioactive wastes** are radioactive solids, liquids, or gases that initially give off large amounts of ionizing radiation. Radioactive wastes must be isolated securely for thousands of years. One option for the retirement of an aging nuclear power plant is to decommission it by dismantling it after it closes.

© AIR PHOTO SERVICE/AFLO/ Nippon News/Corbis

Key Terms

- acid mine drainage 422
- enrichment 431
- fission 431
- fluidized-bed combustion 423
- high-level radioactive wastes 436
- hydraulic fracturing 428
- low-level radioactive wastes 436
- nuclear energy 430
- nuclear reactor 432
- spent fuel 434
- subsurface mining 422
- surface mining 422

What is happening in this picture?

© Kurt Rogers/San Francisco Chronicle/Corbis

In November 2007, a container ship called the *Cosco Busan* struck a bridge tower in the San Francisco Bay, spilling about 200,000 L (54,000 gal) of heavy fuel oil. While this is a much smaller amount than was spilled by the *Exxon Valdez* or *Deepwater Horizon*, it occurred near a densely populated and environmentally productive area.

- Research this spill—what were the impacts, and how was it managed?

- What are the lessons from this spill about the relative importance of the size of the spill and its location?

Critical and Creative Thinking Questions

1. Distinguish among coal, oil, natural gas, and nuclear energy, and compare the environmental impacts of each.

2. What economic priorities and environmental concerns might be shared by coal-mining regions in West Virginia and in China?

3. Some environmental analysts think that the latest war in Iraq was related in part to gaining control over the supply of Iraqi oil. Do you think this is plausible? Explain why or why not.

4. Do you think oil drilling should be permitted in the Arctic National Wildlife Refuge? Why or why not?

5. Which major consumer of oil is most vulnerable to disruption in the event of another energy crisis: electric power generation, motor vehicles, heating and air conditioning, or industry? Why?

6. Which of the following produces the most CO_2 per unit of energy produced: oil, natural gas, or coal?

7. India, China, and the United States all have large coal reserves. Do you think any or all of them should build more coal-fired power plants over the next 20 years? If so, what technologies will help them minimize environmental impacts?

8. Describe the major impacts of the accidents at Three Mile Island, Chernobyl, and Fukushima Daiichi. Do additional research if necessary.

9. Some scholars think the Industrial Revolution may have been concentrated in Europe and North America because coal is located there. Explain the connection between coal and the Industrial Revolution.

10. Examine the bar graph, which shows the six countries with the greatest natural gas deposits. Is most of the world's natural gas located in North and South America, or in Europe and the Persian Gulf countries?

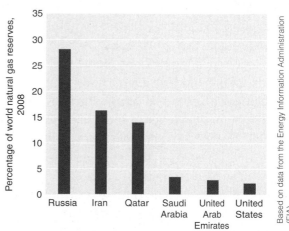

Based on data from the Energy Information Administration (EIA).

Sustainable Citizen Question

11. When major oil spills like the *Exxon Valdez* and the *Deepwater Horizon* occur, many people place the blame on the owners or operators. Consider whether people who benefit from petroleum products (gasoline-powered automobiles, plastic water bottles, diesel farm equipment) are also to blame. Should knowing that oil spills will occur matter as we make consumer choices? Provide some arguments on both sides of this issue.

THE PLANNER

Renewable Energy Resources

REDUCING HEAT LOSS

Two strategies can decrease our reliance on fossil fuels: finding alternative energy sources and reducing energy use. While energy reduction can sometimes require infrastructure investments or lifestyle changes, it can often be relatively simple and inexpensive. Older technologies and buildings often use much more energy than do newer models, and upgrading or replacing them can save energy. The more expensive energy is, the shorter the payback time on investments in insulation, efficient appliances, and other energy-saving technologies.

The picture to the right is a thermal image of a house in New Haven, Connecticut. In contrast to typical photographs, thermal images capture the heat escaping from an object. The dark blue and green represent cold, while red and white are hottest. Thermal images allow homeowners to identify where their homes are leaking heat and reduce those losses by insulating, caulking cracks, and replacing older windows.

Most of the windows in this house appear to be very cool (blue/green color), although the first-floor window on the right may be older or broken, as suggested by its red color. The roof (red and white) may be the biggest energy loser; attic insulation could save considerable energy for this household.

Cars, appliances, and other energy-intensive technologies can also be evaluated with thermal imaging. Heat from a light bulb represents wasted energy (see inset). Of three types of light bulbs, an LED (bottom left) is the coolest, while a traditional incandescent (top) emits much more heat. A compact fluorescent (bottom right) is intermediate—more efficient than the incandescent but less efficient than the LED (see graph).

Tyrone Turner/NG Image Collection

graphingactivity

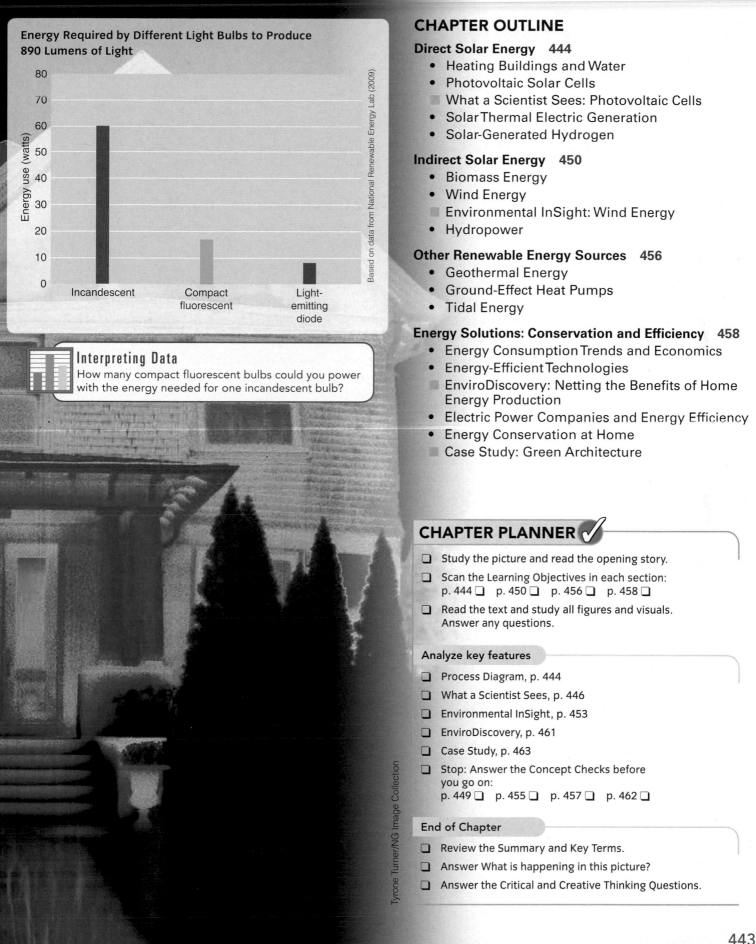

Energy Required by Different Light Bulbs to Produce 890 Lumens of Light

Based on data from National Renewable Energy Lab (2009).

Interpreting Data

How many compact fluorescent bulbs could you power with the energy needed for one incandescent bulb?

Tyrone Turner/NG Image Collection

CHAPTER OUTLINE

Direct Solar Energy   444
- Heating Buildings and Water
- Photovoltaic Solar Cells
- What a Scientist Sees: Photovoltaic Cells
- Solar Thermal Electric Generation
- Solar-Generated Hydrogen

Indirect Solar Energy   450
- Biomass Energy
- Wind Energy
- Environmental InSight: Wind Energy
- Hydropower

Other Renewable Energy Sources   456
- Geothermal Energy
- Ground-Effect Heat Pumps
- Tidal Energy

Energy Solutions: Conservation and Efficiency   458
- Energy Consumption Trends and Economics
- Energy-Efficient Technologies
- EnviroDiscovery: Netting the Benefits of Home Energy Production
- Electric Power Companies and Energy Efficiency
- Energy Conservation at Home
- Case Study: Green Architecture

CHAPTER PLANNER ✓

- ❏ Study the picture and read the opening story.
- ❏ Scan the Learning Objectives in each section:
  p. 444 ❏   p. 450 ❏   p. 456 ❏   p. 458 ❏
- ❏ Read the text and study all figures and visuals. Answer any questions.

Analyze key features

- ❏ Process Diagram, p. 444
- ❏ What a Scientist Sees, p. 446
- ❏ Environmental InSight, p. 453
- ❏ EnviroDiscovery, p. 461
- ❏ Case Study, p. 463
- ❏ Stop: Answer the Concept Checks before you go on:
  p. 449 ❏   p. 455 ❏   p. 457 ❏   p. 462 ❏

End of Chapter

- ❏ Review the Summary and Key Terms.
- ❏ Answer What is happening in this picture?
- ❏ Answer the Critical and Creative Thinking Questions.

Direct Solar Energy

LEARNING OBJECTIVES

1. **Distinguish** between active and passive solar heating and describe how each is used.

2. **Contrast** the advantages and disadvantages of photovoltaic solar cells and solar thermal electric generation in converting solar energy into electricity.

3. **Explain** how fuel cells work.

The sun produces a tremendous amount of energy, most of which dissipates into space. Only a small portion is radiated to Earth. Solar energy differs from fossil and nuclear fuels in that it is always available; we will run out of solar energy only when the sun's nuclear fire burns out. To make solar energy useful, we must collect it.

In most cases, we must also transform solar energy from light and heat into other useful forms, usually electricity or biomass.

Heating Buildings and Water

In **active solar heating**, a series of collection devices mounted on a roof or in a field gather solar energy. The most common solar collection device is a panel or plate of black metal, which absorbs the sun's energy (**Figure 18.1**). Active solar heating is used primarily for heating water, either for

active solar heating A system of putting the sun's energy to use in which collectors absorb solar energy and pumps or fans distribute the collected heat.

Active solar water heating • Figure 18.1

✓ THE PLANNER

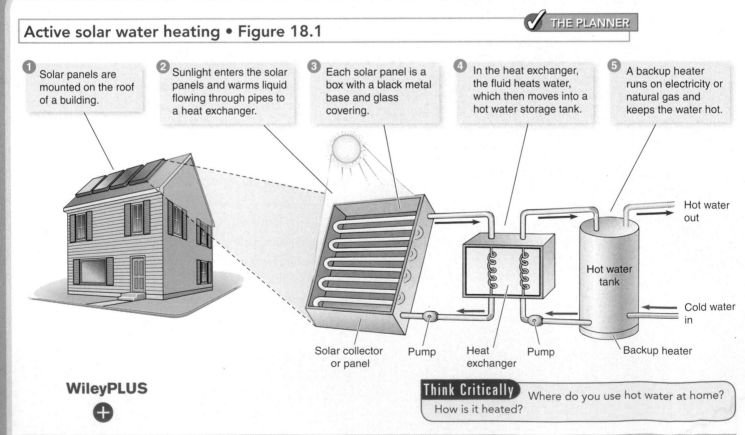

① Solar panels are mounted on the roof of a building.

② Sunlight enters the solar panels and warms liquid flowing through pipes to a heat exchanger.

③ Each solar panel is a box with a black metal base and glass covering.

④ In the heat exchanger, the fluid heats water, which then moves into a hot water storage tank.

⑤ A backup heater runs on electricity or natural gas and keeps the water hot.

Hot water out

Hot water tank

Cold water in

Solar collector or panel   Pump   Heat exchanger   Pump   Backup heater

WileyPLUS
⊕

Think Critically Where do you use hot water at home? How is it heated?

household use or for swimming pools. Heat absorbed by a solar collector is transferred to a fluid inside the panel, which is then pumped to the heat exchanger, where the heat is transferred to water that will be stored in the hot water tank. Solar domestic water heating can provide a family's hot water needs year-round. Because more than 8 percent of energy consumed in the United States goes toward heating water, active solar heating could potentially supply a significant amount of the nation's energy demand.

Active solar energy is not used for space heating as commonly as it is used for heating water, but it may become more important as natural gas, oil, and electricity prices continue to rise.

In **passive solar heating**, solar energy heats buildings without the need for pumps or fans to distribute the

> **passive solar heating** A system of putting the sun's energy to use that does not require mechanical devices to distribute the collected heat.

heat. Certain design features are incorporated into a passive solar heating system to warm buildings in winter and help them remain cool in summer (**Figure 18.2**).

In the Northern Hemisphere, large south-facing windows receive more total sunlight during the day than windows facing other directions. Sunlight entering through the windows provides heat, which is then stored in floors and walls made of concrete, packed earth, or stone, or in containers of water. This stored heat is transmitted throughout the building naturally by **convection**, the circulation that occurs because warm air rises and cooler air sinks.

Buildings with passive solar heating systems must be well insulated so that accumulated heat doesn't escape. Depending on a building's design and location, passive heating can save as much as 80 percent of heating costs. Currently, about 7 percent of new homes built in the United States have passive solar heating features.

Photovoltaic Solar Cells

Photovoltaic (PV) solar cells can convert sunlight directly into electricity (see *What a Scientist Sees*). Individual PV cells can be used to power small devices like wrist watches. However, in order to generate energy to power a building or supply electricity for industrial or commercial uses, PV cells are usually arranged on large panels that absorb sunlight even on cloudy or rainy days.

> **photovoltaic (PV) solar cell** A wafer or thin film of solid-state materials, such as silicon or gallium arsenide, that is treated with certain metals in such a way that the film generates electricity when solar energy is absorbed.

Passive solar heating • Figure 18.2

James P. Blair/NG Image Collection

a. This home in Santa Fe, New Mexico, requires no heating in winter.

b. Several passive design features are incorporated into this home.

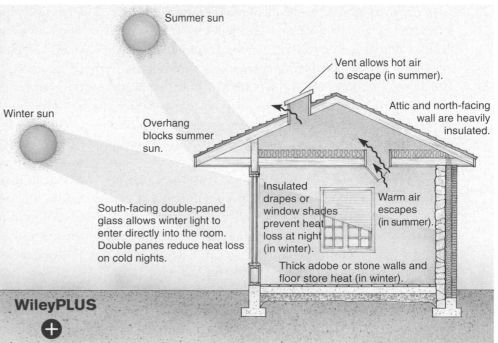

Summer sun

Winter sun

Vent allows hot air to escape (in summer).

Attic and north-facing wall are heavily insulated.

Overhang blocks summer sun.

South-facing double-paned glass allows winter light to enter directly into the room. Double panes reduce heat loss on cold nights.

Insulated drapes or window shades prevent heat loss at night (in winter).

Warm air escapes (in summer).

Thick adobe or stone walls and floor store heat (in winter).

WileyPLUS
⊕

WHAT A SCIENTIST SEES

Photovoltaic Cells

Courtesy Solar Design Associates

a. A student seeing the roof of the Intercultural Center of Georgetown University probably knows that it has arrays of photovoltaic (PV) cells to collect solar energy. The PV system supplies about 10 percent of the school's electricity.

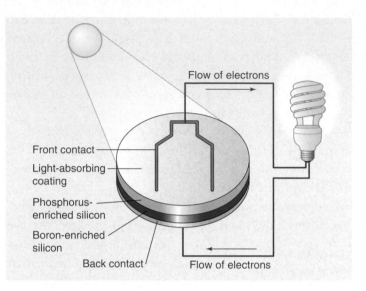

b. A scientist looking at those arrays knows that photovoltaic cells contain silicon and other materials. Sunlight excites electrons, which are ejected from silicon atoms. Useful electricity is generated when the ejected electrons flow out of the PV cells through a wire.

PV cells generate electricity with no pollution and minimal maintenance. They can be used on any scale, from small portable modules attached to camping lanterns to large, multimegawatt power plants, and can power satellites, uncrewed airplanes, highway signals, wristwatches, and calculators. The widespread use of PV cells to generate electricity is currently limited by their low efficiency at converting solar energy to electricity and by the amount of land needed to hold the number of solar panels required for large-scale use. Several thousand acres of today's PV panels would be required to produce the electricity generated by a single large conventional power plant.

In remote areas not served by electric power plants, such as the rural areas of developing countries, it can be more economical to use PV cells for electricity than to extend powerlines. Photovoltaics generate energy that can pump water, refrigerate vaccines, grind grain, charge batteries, and supply rural homes with lighting. According to the Institute for Sustainable Power, more than 1 million households in developing countries have installed rooftop PV solar cells. A PV panel the size of two pizza boxes supplies a rural household with enough electricity for five lights, a radio, and a television.

Utility companies can purchase PV devices in modular units, which can become operational in a short period, allowing them to increase generating capacity in small increments. The PV units can provide the additional energy, for example, to power irrigation pumps on hot, sunny days.

The cost of manufacturing PV modules has steadily declined over the past 35 years, from an average factory price of almost $90 per watt in 1975 to about $4.00 per watt in 2010. The cost of producing electricity from PVs has steadily declined from 1970 to the present. As a result of this progress, in 2010 the cost was about $0.15 per kilowatt-hour (**Figure 18.3**). **Table 18.1** compares the costs of generating electricity using different energy sources, including photovoltaics.

Future technological progress may make PVs economically competitive with electricity produced using conventional energy sources. The production of "thin-film" solar cells (**Figure 18.4**), which are much less expensive to

Cost of electricity from solar photovoltaic cells, 1980–2010 • Figure 18.3

The cost of generating electricity using photovoltaic cells has dropped from over one dollar per kilowatt-hour three decades ago to about $0.15 in 2010.

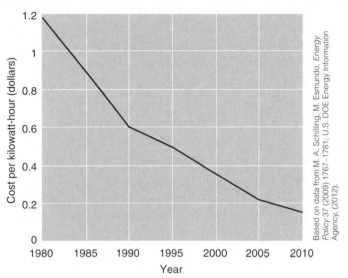

Based on data from M. A. Schilling, M. Esmundo, *Energy Policy* 37 (2009) 1767–1781; U.S. DOE Energy Information Agency, (2012).

Think Critically The average cost of electricity to consumers in the United States is currently about 12 cents per kilowatt-hour. Based on this graph, in what year do you expect power from photovoltaics to cost less than the average cost?

Generating costs of electric power plants, 2009 • Table 18.1

Energy source	Generating costs (cents per kilowatt-hour)*
Hydropower	2–10
Biomass	6–9
Geothermal	3–8
Wind	4–7
Solar thermal	5–13
Photovoltaics	15–25
Natural gas	5–7
Coal	6–8
Nuclear power	6–8

*Electricity production and consumption are measured in kilowatt-hours (kWh). As an example, one 50-watt light bulb that is on for 20 hours uses 1 kWh of electricity (50 × 20 = 1000 watt-hours = 1 kWh).

Solar shingles • Figure 18.4

These thin-film solar cells look much like conventional roofing materials.

© Kyoungil Jeon/Getty; inset, © Fabio Bianchini/iStockphoto

manufacture than standard PVs, has decreased costs for PVs across the board. Japan is considering a national policy that all new buildings incorporate PV on rooftops by 2030. Through the Million Solar Roofs initiative, California expects to have solar roofing on 1 million buildings by 2016. By late 2012, Californians had installed over 1000 MW of solar capacity on their roofs. This is equivalent to the power generated by a typical nuclear power plant. Another technological advance that shows promise is dye-sensitized solar cells, which can be produced at about one-fifth the cost of conventional silicon panels.

While operating PVs creates no air or water pollution, the manufacturing process requires industrial chemicals, many of which are toxic. Finding cleaner ways to produce solar panels will be important as more are produced.

Solar Thermal Electric Generation

Systems that concentrate solar energy to heat fluids have long been used for buildings and industrial processes. In **solar thermal electric generation**, electricity is produced by systems that collect

solar thermal electric generation A means of producing electricity in which the sun's energy is concentrated using mirrors or lenses onto a fluid-filled pipe; the heated fluid is used to generate electricity.

sunlight and concentrate it using mirrors or lenses to heat a fluid to high temperatures.

In one such system, computer-guided trough-shaped mirrors track the sun for optimum efficiency, center sunlight on nearby oil-filled pipes, and heat the oil to 390°C (735°F) (**Figure 18.5**). The hot oil is circulated to a water storage system and used to boil water into superheated steam, which turns a turbine to generate electricity. The plant produces about 370 MW of electricity. Worldwide, concentrated solar thermal plants produce over 1 gigawatt (GW). In 2011, an additional 17 GW of capacity was planned or under construction.

Solar thermal systems often have a backup—usually natural gas—to generate electricity at night and during cloudy days when solar power isn't operating. The world's largest solar thermal system of this type currently operates in the Mojave Desert in southern California.

Solar thermal energy systems are inherently more efficient than other solar technologies because they concentrate the sun's energy. With improved engineering, manufacturing, and construction methods, solar thermal energy is becoming cost-competitive with fossil fuels (see Table 18.1). In addition, the environmental benefits of solar thermal plants are significant: These plants don't produce air pollution or contribute to acid rain or global climate change.

Solar-Generated Hydrogen

Increasingly, people think of hydrogen as the fuel of the future, as it is abundant as well as easily produced. Electricity generated by any energy source can split water into the gases oxygen and hydrogen. Consequently, producing hydrogen results in the same environmental and security problems associated with the underlying energy source, including those discussed in Chapter 17.

Hydrogen is a clean fuel; it produces water and heat as it burns, but it produces no sulfur oxides, carbon monoxide, hydrocarbon particulates, or CO_2 emissions. It does produce some nitrogen oxides, though in amounts fairly easy to control. Hydrogen has the potential to provide energy for transportation (in the form of hydrogen-powered automobiles) as well as for heating buildings and producing electricity.

It may seem wasteful to use electricity generated from solar energy to make hydrogen that will then be used to generate electricity. However, the electricity generated by existing photovoltaic cells must be used immediately,

Solar thermal electric generation • Figure 18.5

a

Pasquale Sorrentino/Science Source Images

a. A solar thermal plant in California uses troughs to focus sunlight on a fluid-filled tube, as shown in **b.** The heated oil is pumped to a water tank, where it generates steam used to produce electricity. For simplicity, arrows show sunlight converging on several points; sunlight actually converges on the pipe throughout its length.

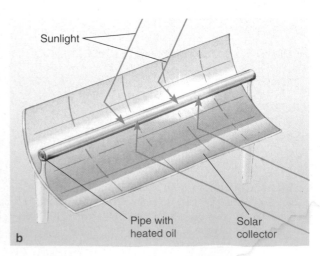

Sunlight

Pipe with heated oil

Solar collector

b

Fuel cells • Figure 18.6

a. These experimental fuel cells combine hydrogen and oxygen to create electricity.

© Kim Kulish/Corbis

b. Cross section of a fuel cell.

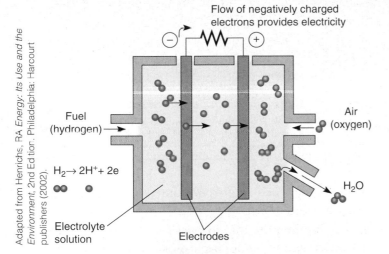

Adapted from Henrichs, RA *Energy: Its Use and the Environment*, 2nd Edition. Philadelphia: Harcourt publishers (2002).

Flow of negatively charged electrons provides electricity

Fuel (hydrogen)

$H_2 \rightarrow 2H^+ + 2e$

Air (oxygen)

H_2O

Electrolyte solution

Electrodes

c. Fuel cells can be adapted to many applications.

© Kimimasa Mayama/©Corbis

whereas hydrogen offers a convenient way to store solar energy as chemical energy. It can be transported by pipeline, possibly less expensively than electricity is transported by wire.

Production of hydrogen from PV electricity currently has a relatively low efficiency (perhaps 10 percent), which means that very little of the solar energy absorbed by the PV cells is actually converted into the chemical energy of hydrogen fuel. Low efficiency translates into high costs. Scientists are working to improve this efficiency and make solar-generated hydrogen fuel commercially viable.

We face other challenges besides high costs if we are to replace gasoline with hydrogen as a transportation fuel. First, we would need to develop a complex infrastructure (such as hydrogen pipelines) to provide hydrogen to service stations. Because hydrogen is extremely volatile, it must be stored, handled, and transported very carefully. Another challenge is developing **fuel cells** for motor vehicles that are inexpensive, safe, and allow the vehicle to drive a long distance without the need to refuel.

> **fuel cell** A device that directly converts chemical energy into electricity. A fuel cell requires hydrogen and oxygen from the air.

A fuel cell is an electrochemical cell similar to a battery (**Figure 18.6**). Fuel cells represent the most promising way to use hydrogen.

Whereas batteries store a fixed amount of energy, fuel cells produce power as long as they are supplied with fuel. Fuel cells are available to power everything from cell phones to city buses. While fuel cells are currently more expensive than other energy sources, the cost has dropped from $275 per kW capacity in 2002 to just $51 per kW in 2010. For most applications (including automobiles) however, they remain very expensive.

CONCEPT CHECK STOP

1. **What** is active solar energy? passive solar energy?

2. **What** are the advantages of producing electricity by solar thermal energy? using hydrogen and photovoltaic (PV) solar cells?

3. **How** do fuel cells work?

Indirect Solar Energy

LEARNING OBJECTIVES

1. **Define** *biomass* and outline its use as a source of energy.
2. **Compare** the potential of wind energy and hydropower.

Some renewable energy sources indirectly use the sun's energy. Combustion of *biomass* (organic matter) is an example of indirect solar energy because plants use solar energy for photosynthesis and store the energy in biomass. Windmills, or wind turbines, use *wind energy* to generate electricity. The damming of rivers and streams to generate electricity is a type of *hydropower*—the energy of flowing water.

Biomass Energy

Biomass, one of the oldest fuels known to humans, consists of materials such as wood, fast-growing plant and algal crops, crop wastes, sawdust and wood chips, and animal wastes. Biomass contains chemical energy that comes from the sun's radiant energy, which photosynthetic organisms use to form organic molecules. Biomass is a renewable form of energy if managed properly.

> **biomass** Plant and animal material used as fuel.

Biomass fuel, which may be a solid, liquid, or gas, is burned to release its energy. Solid biomass fuels such as wood, charcoal (wood turned into coal by partial burning), animal dung, and peat (partly decayed plant matter found in bogs and swamps) supply a substantial portion of the world's energy. At least half of the human population relies on biomass as their main source of energy. In many developing countries, wood is the primary fuel for cooking and heat (**Figure 18.7**).

It is possible to convert biomass, particularly animal wastes, into **biogas**. Biogas, which is usually composed of a mixture of gases (mostly methane), is like natural gas. It is a clean fuel—its combustion produces fewer pollutants than either coal or solid biomass. In India and China, several million family-sized **biogas digesters** use microbial decomposition of household and agricultural wastes to produce biogas for cooking and lighting (**Figure 18.8**). When biogas conversion is complete,

Biomass • Figure 18.7

Firewood is the major energy source for most of the developing world. Photographed in Garadawa, Niger.

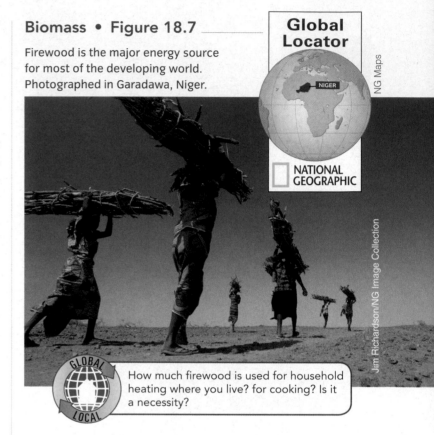

Global Locator

NIGER

NATIONAL GEOGRAPHIC

NG Maps

Jim Richardson/NG Image Collection

GLOBAL LOCAL

How much firewood is used for household heating where you live? for cooking? Is it a necessity?

the solid remains are removed from the digester and used as fertilizer.

Biogas has the potential to power fuel cells to generate electricity. A pilot program at Boston's main sewage treatment plant began producing electricity from biogas in 1997. Sewage sludge in large biogas digesters produces methane, which is then burned in a methane fuel cell to produce enough electricity for 150 homes. Like the hydrogen fuel cells discussed earlier in the chapter, methane fuel cells produce relatively few pollutants.

Biomass can also be converted into liquid fuels, especially **methanol** (methyl alcohol) and **ethanol** (ethyl alcohol), which can replace gasoline in internal combustion engines. In many parts of the world, automotive fuels must contain 10 percent or more ethanol. *Biodiesel*, made from plant or animal oils, is becoming popular as an alternative fuel for diesel engines in trucks, farm equipment, and boats. The oil is often refined from waste oil produced at restaurants (such as the oil used to make french fries); biodiesel burns much cleaner than diesel fuel.

Global Locator

INDIA

NATIONAL GEOGRAPHIC

NG Maps

WileyPLUS

Biogas digester in India • Figure 18.8

This small-scale biogas digester is being evaluated at a research center. Animal manure placed in the digester decomposes, releasing methane gas that can be used as cooking fuel.

Some U.S. energy companies convert corn, sugarcane, or wood crops to alcohol; others are interested in the commercial conversion of agricultural and municipal wastes into ethanol. Currently, the profitability of ethanol is possible only because of government **subsidies** that reduce ethanol's cost.

In 2010, just over 1.4 million barrels of ethanol were consumed each day worldwide. Of this, almost 60 percent was used in the United States, 27 percent was used in Brazil, and just 7 percent in all of Europe. By comparison, the United State consumed 22 percent of the world's gasoline, while Europe consumed 17 percent and Brazil just 2.5 percent.

Biomass is attractive as a source of energy—and popular with U.S. politicians—because it reduces dependence on fossil fuels. It is popular with consumers because they can easily shift from gasoline and diesel without disruptive lifestyle changes. Also, biomass often makes use of waste products, thereby reducing our waste disposal problem. Although it is not completely free of the pollution problems of fossil fuels, biomass combustion produces

levels of sulfur and ash that are lower than those that coal produces.

Some problems associated with obtaining energy from biomass include the use of land and water that might otherwise be dedicated to agriculture. This shift toward energy production might decrease food production, contributing to higher food prices and reducing food supplies even as the population is growing.

Also, as mentioned earlier, at least half of the world's population relies on biomass as its main source of energy. Unfortunately, in many areas people burn wood faster than they replant trees. Intensive use of wood for energy has resulted in severe damage to the environment, including soil erosion, deforestation and desertification, air pollution, and degradation of water supplies.

Excessive use of crop biomass can also harm soil quality. Crop residues, such as cornstalks, are increasingly being used for energy. Crop residues left in and on the ground prevent erosion by holding the soil in place; their removal would eventually deplete the soil of minerals and reduce its productivity.

Wind Energy

Wind results from the sun warming the atmosphere. **Wind energy** is an indirect form of solar energy: The radiant energy of the sun is transformed into mechanical energy through the movement of air molecules. Wind is sporadic over much of Earth's surface, varying in direction and magnitude. Like direct solar energy, wind energy is a highly dispersed form of energy. Harnessing wind energy to generate electricity has great potential.

> **wind energy**
> Electric energy obtained from surface air currents caused by the solar warming of air.

New wind turbines are huge—100 m (328 ft) tall—and have long blades designed to harness wind energy efficiently (**Figure 18.9a, b**). As turbines have become larger and more efficient, costs for wind power have declined rapidly—from $.40 per kilowatt-hour in 1980 to a current cost of $.04 to $.07 per kilowatt-hour. Wind power is cost-competitive with many forms of conventional energy. Advances such as turbines that use variable-speed operation may make wind energy an important global source of electricity during the first half of the 21st century.

During the 1990s and early 2000s, wind became the world's fastest-growing source of energy (**Figure 18.9c**). According to the Global Wind Energy Commission, in 2012, China, Germany, and the United States led the world as the top producers of wind energy. China is rapidly expanding its wind energy capacity, with plans to produce at least 100 GW by 2015. This is more than the total global wind energy production in 2005. Denmark currently generates 21 percent of its electricity using wind energy, much of it offshore where ocean winds are strong. Other leading wind energy countries include Spain and India.

Harnessing wind energy is most profitable in areas with consistent winds, such as islands, coastal areas, mountain passes, and grasslands. The world's largest concentration of wind turbines is currently located in the Tehachapi Pass at the southern end of the Sierra Nevada mountain range in California.

In the continental United States, some of the best locations for large-scale electricity generation from wind energy are on the Great Plains. The 10 states with the greatest wind energy potential, according to the American Wind Energy Association, are North Dakota, Texas, Kansas, South Dakota, Montana, Nebraska, Wyoming, Oklahoma, Minnesota, and Iowa. In fact, if we developed the wind energy in North Dakota, Texas, and Kansas to their full potential, we could supply enough electricity to meet the current needs of the entire United States! Wind power projects are under way in these and many other states.

Currently, wind energy is captured and placed into regional electricity grids. Deploying wind energy on a national scale—for example, wind energy produced in Texas and used in New York City—requires the development of new technologies for storing and distributing energy.

Wind produces no waste and is a clean source of energy. It produces no emissions of sulfur dioxide, carbon dioxide, or nitrogen oxides. Every kilowatt-hour of electricity generated by wind power rather than fossil fuels prevents as much as 1 kg (2.2 lb) of the greenhouse gas CO_2 from entering the atmosphere.

Although the use of wind power doesn't cause major environmental problems, one concern is the deaths of birds and bats. The California Energy Commission estimated that several hundred birds turned up dead in the vicinity of the 7000 turbines at Altamont Pass in California during a two-year study; most had collided with the turbines. Studies later determined that Altamont Pass is a major bird migration pathway. Wind power sites have implemented technical "fixes" such as painted blades and anti-perching devices, or have shut down operations during peak bird migration periods. Developers of future wind farm sites currently conduct voluntary wildlife studies and try to locate sites away from bird and bat routes.

Not all people welcome wind power projects with open arms. The Maple Ridge Wind Farm in upstate New York has almost 200 windmills. Many local residents appreciate the extra money that wind-farm leases provide to the local economy. However, others think the wind turbines ruin their view of the Adirondack Mountains. A similar dispute is occurring over the proposed Nantucket Sound wind farm off the coast of Massachusetts. This 130-turbine wind farm, if built, will be the first offshore wind project in the United States. Massachusetts has some of the highest electricity costs in the nation, and proponents of the wind farm point out that wind energy will help lower energy costs. However, many people are concerned that the wind farm will adversely affect the local economy because tourists will find it offensive.

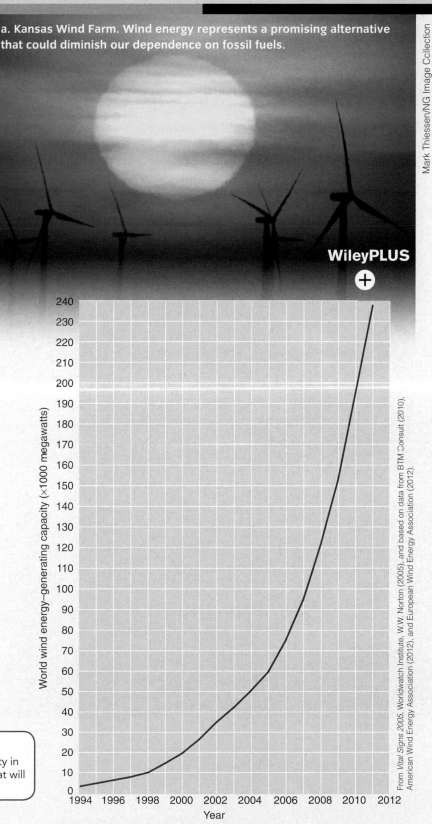

a. Kansas Wind Farm. Wind energy represents a promising alternative that could diminish our dependence on fossil fuels.

Mark Thiessen/NG Image Collection

WileyPLUS +

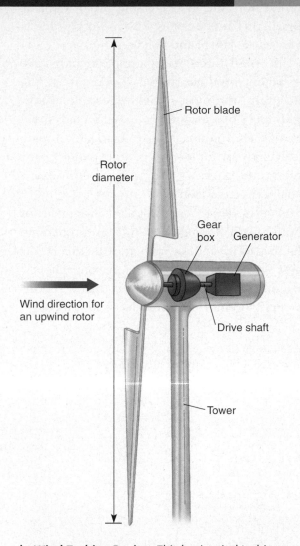

Rotor blade

Rotor diameter

Gear box

Generator

Wind direction for an upwind rotor

Drive shaft

Tower

b. Wind Turbine Design. This basic wind turbine design has a horizontal axis (horizontal refers to the orientation of the drive shaft). Airflow causes the turbine's blades to turn 15 to 60 revolutions per minute (rpm). As the blades turn, gears within the turbine spin the drive shaft. This spinning powers the generator, which sends electricity through underground cables to a nearby utility. (The tower isn't drawn to scale and is much taller than depicted.)

Interpreting Data

What was the world wind energy–generating capacity in 1998? 2008? If the trend in this figure continues, what will it be in 2018?

World wind energy–generating capacity (×1000 megawatts)

240
230
220
210
200
190
180
170
160
150
140
130
120
110
100
90
80
70
60
50
40
30
20
10
0

1994 1996 1998 2000 2002 2004 2006 2008 2010 2012

Year

From *Vital Signs 2005*, Worldwatch Institute, W.W. Norton (2005), and based on data from BTM Consult (2010), American Wind Energy Association (2012); and European Wind Energy Association (2012).

c. Trends in Wind Energy Use. Global wind energy has shown record growth in recent years.

Hydropower

The sun's energy drives the hydrologic cycle, which includes precipitation, evaporation, transpiration, and drainage and runoff (see Figure 5.9). As water flows from higher elevations back to sea level through rivers and streams, dams can harness and make use of its energy. The potential energy of water held back by a dam is converted to kinetic energy as the water turns turbines to generate electricity (**Figure 18.10**). Hydropower, which is more concentrated than solar energy, is more efficient than any other energy source for producing electricity; about 90 percent of available hydropower energy is converted into consumable electricity.

Hydropower generates approximately 19 percent of the world's electricity, making it the form of solar energy in greatest use. China is the world's largest producer of hydroelectric power, generating over 200 GW. Besides China, the

> **hydropower**
> A form of renewable energy that relies on flowing or falling water to generate electricity.

nine countries with the greatest hydroelectric production are, in decreasing order, Canada, Brazil, the United States, Russia, Norway, India, Venezuela, Japan, and Sweden.

In the United States, approximately 2200 hydropower plants produce about 7 percent of the country's electricity, the most of any renewable energy source. Highly developed countries have already built dams at most of their potential sites. In many developing nations—particularly in undeveloped, unexploited parts of Africa and South America—hydropower represents a great potential source of electricity.

Traditional hydropower technology is only suited for large dams with rapidly flowing water, such as those found on the Columbia River and its tributaries in the Pacific Northwest and British Columbia. Therefore, only about 3 percent of existing U.S. dams generate electricity. New designs allow modern turbines to harness electricity from large,

Hydroelectric power • Figure 18.10

© TPG/Top Photo/Corbis

a. Water generates electricity as it moves through the Three Gorges Dam, which spans the Yangtse River in China.

b. A controlled flow of water released down the penstock turns a turbine, which generates electricity.

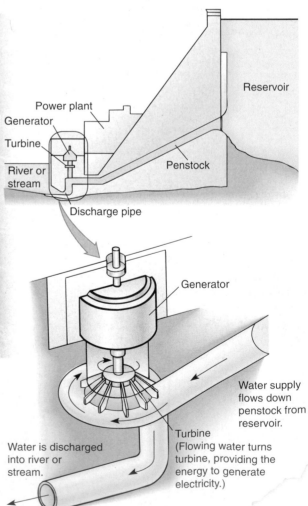

Reservoir

Power plant

Generator

Turbine

River or stream

Penstock

Discharge pipe

Generator

Water supply flows down penstock from reservoir.

Turbine (Flowing water turns turbine, providing the energy to generate electricity.)

Water is discharged into river or stream.

slow-moving rivers or from streams with small flow capacities. These new technologies have the potential to increase the amount of hydroelectric power generated by existing dams.

Building a dam changes the natural flow of a river: Water backs up, flooding large areas of land and forming a reservoir, which destroys plant and animal habitats. Native fishes are particularly being harmed by dams because the original river ecosystem is so altered. The migration of spawning fish is also altered (see Chapter 10). Below the dam, the once-powerful river is reduced to a relative trickle. The natural beauty of the countryside is affected, and certain forms of wilderness recreation are made impossible or less enjoyable, although the dams permit water sports in the reservoir.

With at least 200 large dams around the world, earthquakes may occur during and after the filling of the reservoir behind the dam. The larger the reservoir and the faster it is filled, the greater the intensity of seismic activity. An area need not be seismically active to have earthquakes induced by reservoirs.

If a dam breaks, people and property downstream may be affected. In addition, waterborne diseases may spread through the population. **Schistosomiasis** is a tropical disease caused by a parasitic worm. As much as half the population of Egypt suffers from this disease, which resulted when Egypt dammed the Nile River. First, the Low Aswan Dam was built on the Nile River in 1902 to control flooding. Lake Nasser was created in 1971 when the High Aswan Dam was completed (**Figure 18.11**). The large

Aswan High Dam construction, 1964
• Figure 18.11 _____

The Aswan High Dam was completed in 1971, creating Lake Nasser.

© Bettmann/Corbis

reservoir behind the dam provides habitat for the worm, which spends part of its life cycle in the water. Humans are infected by the worm through bathing, swimming, walking barefoot along water banks, and drinking infected water.

In arid regions, the creation of a reservoir results in greater evaporation because it has a larger surface area in contact with the air than did the stream or river. As a result, serious water loss and increased salinity of the remaining water may occur.

When an area behind a dam is flooded, the trees and other vegetation die and are decomposed, releasing the large quantities of carbon that were tied up in organic molecules in the plant bodies as carbon dioxide and methane. These gases absorb infrared radiation and are therefore associated with global climate change.

The construction of large dams involuntarily displaces people from lands flooded by reservoirs. The *Three Gorges Project*, the largest dam ever built, was recently completed on the Yangtze River in China (see Chapter 10). The reservoir behind this dam is 632 km (412 mi) long. The tops of as many as 100 mountains have become small islands, fragmenting habitat and threatening 57 endangered species. The reservoir has displaced almost 2 million people, the largest number for any dam project.

The environmental and social impacts of a dam may not be acceptable to the people living in a particular area. Laws prevent or restrict the building of dams in certain locations. In the United States, the Wild and Scenic Rivers Act prevents the hydroelectric development of certain rivers, although this law protects less than 1 percent of the nation's total river system. Other countries, such as Norway and Sweden, have similar laws.

Dams cost a great deal to build but are relatively inexpensive to operate. A dam has a limited life span, usually 50 to 100 years, because over time the reservoir fills in with silt until it cannot hold enough water to generate electricity. This trapped silt, which is rich in nutrients, is prevented from enriching agricultural lands downstream. For example, Egypt must rely on heavy applications of chemical fertilizer downstream from the Aswan Dam to maintain the fertility of the Nile River Valley.

CONCEPT CHECK

1. **What** is biomass, and how is it used?

2. **What** are the advantages and disadvantages of using wind to produce electricity? of using hydropower to produce electricity?

Other Renewable Energy Sources

LEARNING OBJECTIVE

1. **Describe** geothermal energy, ground-effect heat pumps, and tidal energy.

Geothermal energy and tidal energy are renewable energy sources that are not derived from solar energy. *Geothermal energy* is the naturally occurring heat within Earth. This heat is used for space heating and to generate electricity. *Ground-effect heat pumps use stable temperatures just below Earth's surface to heat buildings in winter and cool them in summer. Tidal energy,* caused by the changes in water level between high and low tides, is exploited to generate electricity on a limited scale.

Geothermal Energy

Earth's subsurface can be accessed for energy in two very different ways. Geothermal energy relies on high-temperature heat from Earth's interior. Ground-effect heat pumps, discussed later, take advantage of the constant but much lower temperature a few meters below Earth's surface. **Geothermal energy**, the natural heat within Earth, arises from Earth's core, from friction along continental plate boundaries, and from the decay of radioactive elements. The amount of geothermal energy is enormous. Scientists estimate that just 1 percent of the heat contained in the uppermost 10 km (6 mi) of Earth's crust is equivalent to 500 times the energy contained in all of Earth's oil and natural gas resources.

> **geothermal energy** Energy from Earth's hot interior, used for space heating or generation of electricity.

Geothermal energy is typically associated with volcanism. Large underground reservoirs of heat exist in areas of geologically recent volcanism. As groundwater in these areas travels downward and is heated, it becomes buoyant and then rises until it is trapped by an impermeable layer in Earth's crust, forming a **hydrothermal reservoir**. Hydrothermal reservoirs contain hot water and possibly steam, depending on the temperature and pressure of the fluid. Some of the hot water or steam may escape to the surface, creating hot springs or geysers. Hot springs have been used for thousands of years for bathing, cooking, and heating buildings. Drilling a well brings the hot fluid from a hydrothermal reservoir to the surface, where a power station may use it to supply heat directly to consumers or to generate electricity (**Figure 18.12a**). The electricity these power stations generate is inexpensive and reliable.

The United States is the world's largest producer of geothermal electricity. Electric power is currently produced at 17 different geothermal fields in California, Nevada, Utah, and Hawaii. The world's largest geothermal power plant—The Geysers in northern California—provides electricity for 1.7 million homes. Other important producers of geothermal energy include the Philippines, Italy, Japan, Mexico, Indonesia, and Iceland (**Figure 18.12b**).

Iceland, a country with minimal oil and natural gas resources, is located on the mid-Atlantic ridge, a boundary between two continental plates. Iceland is therefore an island of intense volcanic activity with considerable geothermal resources. Iceland uses geothermal energy to generate electricity to heat two-thirds of its homes. In addition, most of the fruits and vegetables required by the people of Iceland are grown in geothermally heated greenhouses.

Is geothermal energy renewable? As a source of heat for geothermal energy, the planet is inexhaustible on a human timescale. However, the water used to transfer the heat to the surface isn't inexhaustible. Some geothermal applications recirculate all the water back into the underground reservoir, ensuring many decades of heat extraction from a given reservoir.

Geothermal energy is considered environmentally benign because it emits only a fraction of the air pollutants released by conventional fossil fuel–based energy technologies. The most common environmental hazard is the emission of hydrogen sulfide (H_2S) gas, which comes from the very low levels of dissolved minerals and salts found in the steam or hot water. Hydrogen sulfide smells like rotten eggs and is toxic to humans in high concentrations. A lesser concern is that the surrounding land may subside, or sink, as the water from hot springs and their connecting underground reservoirs is removed.

Scientists are studying how to economically extract some of the vast amount of geothermal energy stored in hot, dry rock. Such a technology could greatly expand the extent and use of geothermal resources.

Deep geothermal power plant • Figure 18.12

a. Steam separated from hot water pumped from underground turns a turbine and generates electricity.

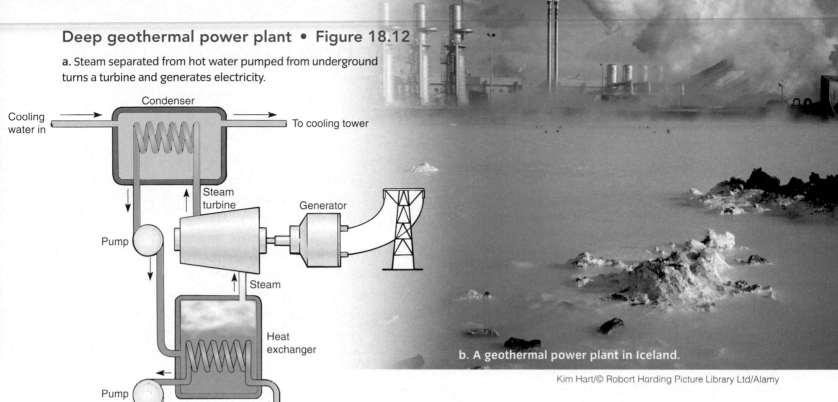

b. A geothermal power plant in Iceland.

Kim Hart/© Robert Harding Picture Library Ltd/Alamy

Ground-Effect Heat Pumps

Ground-effect heat pumps (GHPs), also called ambient heat systems, take advantage of the difference in temperature between Earth's surface and subsurface (at depths from 1 m to about 100 m or about 3 to 300 ft). In an underground arrangement of pipes containing circulating fluids, GHPs extract natural heat in winter, when Earth acts as a heat source, and transfer excess heat underground in summer, when Earth acts as a heat sink. Ground-effect heating systems can be modified to provide supplemental hot water.

Though GHPs have been available for many years, they aren't widely used because their installation is expensive. However, with the growth of *green architecture* (see the *Case Study* at the end of this chapter) and rising fuel costs, commercial and residential use of GHPs is on the rise. The system's benefits include low operating costs—which may be half those of conventional systems—and high efficiency.

Tidal Energy

Tides, or the rise and fall of the surface waters of the ocean and seas that occur twice each day, are the result of the gravitational pull of the moon and the sun. A dam built across a bay can harness the energy of large tides to generate electricity. As the tide falls, water flowing back to the ocean over the dam's spillway turns a turbine and generates electricity through **tidal energy**.

Power plants using tidal power are in operation in France, Russia, China, and Canada. Tidal energy can't become a significant resource worldwide because few areas experience large enough differences in water level between high and low tides to make power generation feasible. The most promising locations for tidal power in North America include the Bay of Fundy in Nova Scotia, Passamaquoddy Bay in Maine, Puget Sound in Washington, and Cook Inlet in Alaska.

Other problems associated with tidal energy include the high cost of building a tidal power station and potential environmental problems associated with tidal energy in **estuaries**, coastal areas where river currents meet ocean tides. Fishes and countless invertebrates migrate to estuaries to spawn. Building a dam across the mouth of an estuary would prevent these animals from reaching their breeding habitats.

CONCEPT CHECK   STOP

1. **What** are the pros and cons of using geothermal energy to produce electricity, and what are the pros and cons of using tidal power to produce electricity?

Energy Solutions: Conservation and Efficiency

LEARNING OBJECTIVES

1. **Distinguish** between energy conservation and energy efficiency and give examples of each.
2. **Summarize** options to conserve energy at home.

Human requirements for energy will continue to increase, if only because the human population is growing. In addition, energy consumption continues to increase as developing countries raise their standards of living. We must therefore place a high priority not only on developing alternative sources of energy but on **energy conservation** and **energy efficiency**.

To illustrate the difference between energy conservation and energy efficiency, let's consider automobile gasoline consumption. Energy conservation measures to reduce gasoline consumption would include carpooling and lowering driving speeds, whereas energy efficiency measures would include designing and manufacturing automobiles that travel farther on a gallon of fuel. Conservation and efficiency accomplish the same goal—saving energy.

Many energy experts consider energy conservation and energy efficiency the most promising energy "sources" available because they save energy for future use and buy us time to explore new energy alternatives. Developing technologies for energy conservation and efficiency costs less than developing new sources or supplies of energy; the technologies also improve the economy's productivity. The adoption of energy-efficient technologies generates new business opportunities, including the research, development, manufacture, and marketing of those technologies.

Energy-efficient technologies and greater efforts at conservation also provide important environmental benefits by reducing air pollution. Carbon dioxide emissions that contribute to global climate change, acid precipitation, and other environmental problems are related to large quantities of energy production and consumption.

> **energy conservation**
> Using less energy—by reducing energy use and waste, for example.
>
> **energy efficiency**
> Using less energy to accomplish a given task—by using new technology, for example.

Energy Consumption Trends and Economics

Although the U.S. economy has become more energy efficient, total energy consumption continues to slowly increase, partly due to an increase in population. The per person energy consumption in developing nations is substantially less than that in industrialized countries (see Figure 17.1a), but the greatest per person increase in energy consumption today is occurring in these developing nations, particularly in China and India. The rising energy demand in developing nations is caused by increases in economic development and population, as well as the use of older, less expensive, and less energy-efficient technologies.

Developing countries are forced to balance developing their economies with controlling environmental degradation. At first glance, these two goals appear mutually exclusive. However, both can be realized by adopting the new technologies now being developed in industrialized nations to achieve greater energy efficiency.

Energy-Efficient Technologies

The development of more efficient appliances, automobiles, buildings, and industrial processes has helped reduce energy consumption in highly developed countries. Compact fluorescent light bulbs produce light of comparable quality to that of incandescent light bulbs but require only 25 percent of the energy and last up to 15 times longer (see chapter opener). Although relatively expensive, the energy-efficient bulbs more than pay for themselves in energy savings. New condensing furnaces require approximately 30 percent less fuel than conventional gas furnaces. "Superinsulated" buildings use 70 to 90 percent less heat energy than buildings insulated using standard methods (**Figure 18.13**).

Superinsulated buildings • Figure 18.13

a. A superinsulated home is so well insulated and airtight that it doesn't require a furnace in winter. Heat from the inhabitants, light bulbs, and appliances provides most or all the necessary heat.

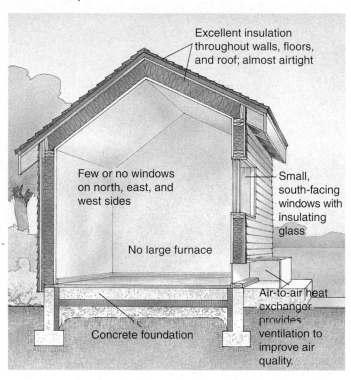

Excellent insulation throughout walls, floors, and roof; almost airtight

Few or no windows on north, east, and west sides

Small, south-facing windows with insulating glass

No large furnace

Concrete foundation

Air-to-air heat exchanger provides ventilation to improve air quality.

b. A superinsulated office building in Toronto, Canada, has south-facing windows with insulating glass. The building is so well insulated it uses no furnace.

Ontario Power Generation

The **National Appliance Energy Conservation Act (NAECA)** sets national appliance efficiency standards for refrigerators, freezers, washing machines, clothes dryers, dishwashers, room air conditioners, and ranges and ovens (including microwaves). For example, refrigerators built today consume 80 percent less energy than comparable models built in the early 1970s (**Figure 18.14**).

Energy costs often account for 30 percent of a company's operating budget. It makes good economic sense for businesses in older buildings to invest in energy improvements, which often pay for themselves in a few years. Implementing energy improvements may be as simple as fine-tuning existing heating, ventilation, and air conditioning systems or as major as replacing all the windows and lights. Both the environment and a company's bottom line can benefit from any energy improvements.

Automobile efficiency has improved dramatically since the mid-1970s as a result of the use of lighter materials and designs that reduce air drag. The average fuel efficiency of new passenger cars doubled between the mid-1970s and the mid-1980s. It declined after

Energy use standard for new refrigerators, 1972–2010 • Figure 18.14

Refrigerators sold today are four times as efficient as those sold in the early 1970s. Advances in insulation, coolant, and compressor technologies have all contributed to this improvement. Refrigerators today are also less expensive and more reliable.

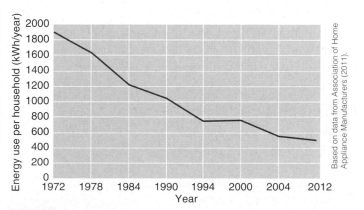

Based on data from Association of Home Appliance Manufacturers (2011).

Think Critically How old is the refrigerator that you most often use? How much energy would you save each year by replacing it with a recent model?

The Daihatsu prototype UFE (ultra fuel economy) III hybrid can achieve 169 miles per gallon.

Iain Materton/Alamy

that, as larger vehicles became more popular; light trucks and sport utility vehicles (SUVs) are now a substantial fraction of the passenger vehicles in the United States. Further, significant gains could easily be made using current technology. Automobiles with fuel efficiencies of 60 to 65 mpg could be routinely manufactured within the next decade or so, and manufacturers are developing even more efficient models for the future (**Figure 18.15**).

Cogeneration One energy technology that has a bright future is cogeneration, or **combined heat and power (CHP)**. Cogeneration involves the production of two useful forms of energy from the same fuel. A CHP system

> **cogeneration (CHP)** An energy technology that involves recycling "waste" heat.

generates electricity, and then the steam produced during this process is used rather than wasted. The system's overall conversion efficiency (that is, the ratio of useful energy produced to fuel energy used) is high because some of what is usually waste heat is incorporated into the process.

Cogeneration can be cost-effective on both small and large scales. Modular CHP systems enable hospitals, hotels, restaurants, factories, and other businesses to harness steam that would otherwise be wasted to heat buildings, cook food, or operate machinery before it cools and gets pumped back into the boiler as water (**Figure 18.16**). Larger CHP systems can produce electricity for local utilities.

Cogeneration • Figure 18.16

In this example of a cogeneration system, fuel combustion generates electricity in a generator. The electricity produced is used in-house or sold to a local utility. The waste heat (leftover hot gases or steam) is recovered for useful purposes, such as industrial processes, heating of buildings, hot water heating, and generation of additional electricity.

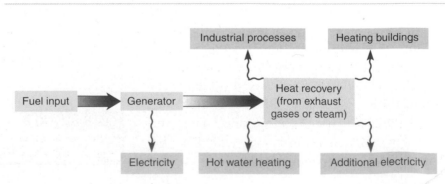

EnviroDiscovery
Netting the Benefits of Home Energy Production

Environmentally aware homeowners have often paid a large economic price when installing renewable energy sources such as solar panels or wind-driven generators. Such systems can be expensive to purchase and install, and they provide electricity only intermittently, when weather conditions permit. During unfavorable conditions, residents must purchase electricity from their local utility.

In a growing trend, utility companies are permitting homeowners who produce their own energy to "net meter" their electricity. Any excess energy homeowners generate is supplied to the utility's power grid, and the homeowners' electric meters run backward. Net metering offsets energy costs over a billing period by essentially providing individuals the full retail price for the electricity they generate. By itself, net metering doesn't turn residential energy generation into an inexpensive prospect, but it certainly makes it more affordable, encouraging the additional development of renewable energy for homes. Some utilities and governments have made it easier for consumers to finance installations, for example, by providing tax incentives or promoting long-term loans.

Electric meters can run backward for homeowners participating in net metering.

Electric Power Companies and Energy Efficiency

Changes in the regulations that govern electric utilities allow these companies to make more money by generating less electricity. Such programs provide incentives for energy conservation and thereby reduce power plant emissions that contribute to environmental problems.

Electric utilities can often avoid the massive expenses of building new power plants or purchasing additional power by helping electricity consumers save energy. Some utilities support energy conservation and efficiency by offering cash awards to consumers who install energy-efficient technologies. Other utilities give customers energy-efficient compact fluorescent light bulbs, air conditioners, or other appliances. They then charge slightly higher rates or a small leasing fee, but the greater efficiency results in savings for both the utility company and the consumer.

The utility company makes more money from selling less electricity because it does not have to invest in additional power generation to meet increased demand. The consumer saves because the efficient light bulbs or appliances use less energy, which more than offsets the higher rates.

New technologies also allow electric utilities to monitor when energy is used during the day, and even make changes to customer energy use to reduce peak demand. Peak demand is when customers use the greatest amounts of energy, such as on very hot afternoons when every air conditioner is running at maximum. Customers who are charged higher rates at these times can shift some of their activities to later at night or early mornings. Alternatively, they can get lower rates by allowing the utility to remotely shut off air conditioning for short periods of time. According to the American Council for an Energy-Efficient Economy, U.S. electric power plants are themselves an important target for improved energy efficiency. Much heat is lost during the generation of electricity. If all that wasted energy were harnessed—by cogeneration, for example—it could be used productively, thereby conserving energy.

Another way to increase energy efficiency would be to improve our electric grids because about 10 percent of electricity is lost during transmission. To accomplish

Some energy-saving measures for the home • Figure 18.17

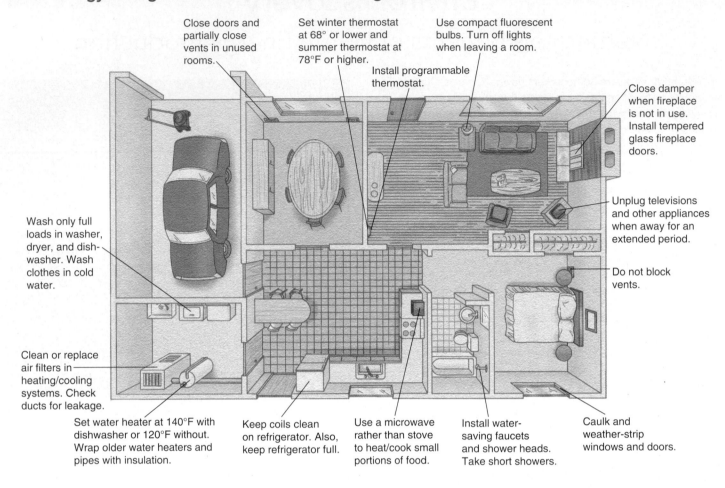

Close doors and partially close vents in unused rooms.

Set winter thermostat at 68° or lower and summer thermostat at 78°F or higher.

Install programmable thermostat.

Use compact fluorescent bulbs. Turn off lights when leaving a room.

Close damper when fireplace is not in use. Install tempered glass fireplace doors.

Wash only full loads in washer, dryer, and dish-washer. Wash clothes in cold water.

Unplug televisions and other appliances when away for an extended period.

Do not block vents.

Clean or replace air filters in heating/cooling systems. Check ducts for leakage.

Set water heater at 140°F with dishwasher or 120°F without. Wrap older water heaters and pipes with insulation.

Keep coils clean on refrigerator. Also, keep refrigerator full.

Use a microwave rather than stove to heat/cook small portions of food.

Install water-saving faucets and shower heads. Take short showers.

Caulk and weather-strip windows and doors.

this, some energy experts envision that future electricity will be generated far from population centers, converted to supercooled hydrogen, and transported through underground superconducting pipelines. The technology to build such conduits has not yet been developed.

Energy Conservation at Home

The average U.S. household spends several thousand dollars each year on utility bills. This cost could be reduced considerably with investment in energy-efficient technologies (**Figure 18.17**). Although a more energy-efficient house might cost more up front, depending on the technologies employed, the improvements usually pay for themselves in two or three years. Energy efficiency has become an essential element of design codes nationwide.

Some energy-saving improvements, such as thicker wall insulation, are easiest to install while a home is being built. Other improvements can be made to older homes to reduce heating and cooling costs and enhance energy efficiency. Examples include installing thicker attic insulation, installing storm windows and doors, caulking cracks around windows and doors, replacing inefficient furnaces and refrigerators, and adding heat pumps. Most local utility companies will also perform an energy audit on a home for little or no charge.

CONCEPT CHECK **STOP**

1. **What** is the difference between energy conservation and energy efficiency?

2. **How** can you conserve energy at home?

Green Architecture

A recent addition to the Midtown Manhattan skyline represents a new standard in green architecture in New York City (see photograph). Completed in 2006, the Hearst Tower, home of Hearst Publishing, was designed to achieve a degree of energy efficiency 26 percent higher than that of standard office buildings. The building's innovations earned it the city's first Gold LEED (Leadership in Energy and Environmental Design) certification from the U.S. Green Building Council.

The Hearst Tower's striking "diagrid" frame, which incorporates a system of steel and glass triangles, floods the interior with natural light while using approximately 2000 tons less steel than a conventional frame. Most of the steel (90 percent) used in the 46-story, 80,000 m² (856,000-square-ft) building is recycled.

Other energy-saving design features in the Hearst Tower boost the efficiency of the building's heating and cooling systems. Windows are coated to reduce solar radiation, and heating and air conditioning equipment cools and ventilates using only outside air for three-quarters of the year. In addition, embedded polyethylene tubes within the atrium's limestone floor circulates water for both cooling and heating. Unique to the building's design, a 10-story waterfall called the "Icefall" chills the atrium. The Icefall's water supply comes from rainwater collected at the roof and drawn into a 53,000-L (14,000-gal) basement reclamation tank. This water is also used to irrigate the building's plants.

Natural light is enhanced by the placement of few internal walls and only low partitions. Sensors turn off lights in empty rooms and control the amount of artificial light provided based on the natural light being received.

Finally, building health is protected by several interior features. Low-vapor paints and low-toxicity sealants coat surfaces, while furniture and carpet make use of low-toxicity recycled content and materials obtained from sustainable forests.

The Hearst Tower in Manhattan offers important contributions to green architecture.

Nick Wood/Alamy

Summary

1 Direct Solar Energy 444

1. An **active solar heating** system collects solar energy and then relies on pumps or fans to distribute heat. **Passive solar heating** systems distribute heat using mechanical devices.

2. A **photovoltaic (PV) solar cell** is a wafer or thin film of solid-state materials, such as silicon or gallium arsenide, that is treated with certain metals in such a way that it generates electricity when solar energy is absorbed. Manufacturing PVs requires toxic industrial chemicals, but PVs generate electricity with no pollution and minimal maintenance. PVs are limited by their low efficiency and by the amount of land needed for their large-scale use. The cost of PV's is much lower than it was three decades ago and continues to fall. **Solar thermal electric generation** is a means of producing electricity in which the sun's energy is concentrated by mirrors or lenses onto a fluid-filled pipe; the heated fluid is used to generate electricity. Solar thermal energy systems are efficient and provide significant environmental benefits; but they are only now becoming more cost-competitive with fossil fuels.

3. **Fuel cells** convert chemical energy into electricity; a fuel cell requires hydrogen fuel and oxygen (from the air).

James P. Blair/NG Image Collection

2 Indirect Solar Energy 450

1. **Biomass** is plant and animal material used as fuel. Biomass fuel—materials such as wood, fast-growing plant and algal crops, crop wastes, sawdust and wood chips, and animal wastes—is burned to release energy. Biomass fuels can be solid, liquid, or gas.

2. **Wind energy** is electric energy obtained from surface air currents caused by the solar warming of air. Restricted primarily to areas with consistent winds, wind power is a clean and cost-effective source of energy. Wind turbines can kill birds and bats, and some people find their appearance unpleasant. **Hydropower** is a form of renewable energy that relies on flowing or falling water to generate electricity. Hydropower is highly efficient, but the dams built in traditional hydropower projects can greatly disrupt the natural environment and displace local residents.

3 Other Renewable Energy Sources 456

1. **Geothermal energy** is the use of energy from Earth's interior for either space heating or generation of electricity. Deep geothermal energy uses high-temperature heat from Earth's interior. **Ground-effect heat pumps** take advantage of constant but relatively low temperatures just below Earth's surface. **Tidal energy** is a form of renewable energy that relies on the ebb and flow of the tides to generate electricity.

4 Energy Solutions: Conservation and Efficiency 458

1. **Energy conservation** is using less energy—for example, by reducing energy use and waste. **Energy efficiency** is using less energy to accomplish a given task—for example, with new technology. Examples of energy conservation measures that reduce gasoline consumption include carpooling and lowering driving speeds; energy efficiency measures include designing and manufacturing more fuel-efficient automobiles.

2. Households can conserve energy in many ways, including by keeping vents and refrigerator coils clean, replacing inefficient light bulbs and appliances, and adjusting thermostats.

Key Terms

- active solar heating 444
- biomass 450
- cogeneration (CHP) 460
- energy conservation 458
- energy efficiency 458
- fuel cell 449
- geothermal energy 456
- hydropower 454
- passive solar heating 445
- photovoltaic (PV) solar cell 445
- solar thermal electric generation 447
- wind energy 452

What is happening in this picture?

These hoses suck methane, which is used as fuel, from decomposing trash in California.

- Is this a renewable energy source? a form of conservation? Explain.

- What would be some advantages and disadvantages of this type of energy production?

Critical and Creative Thinking Questions

1. Do you think that a vehicle fuel efficiency of 60 miles per gallon could be reasonably achieved at present? Explain your answer.

2. In response to the Fukushima Daiichi disaster, Japan is considering eliminating all of its nuclear power plants by 2040. What combination of fossil fuels, renewable energy resources, efficient technologies and conservation measures could best achieve this?

3. Explain the following statement: Unlike fossil fuels, solar energy is not resource limited but is technology limited.

4. One advantage of the various forms of renewable energy, such as solar thermal and wind energy, is that they cause no net increase in atmospheric carbon dioxide. Is this true for biomass? Why or why not?

5. Give an example of how one or more of the renewable energy sources discussed in this chapter could have a negative effect on each of these aspects of ecosystems: soil preservation, natural water flow, foods used by wild plant and animal populations, and preservation of the diversity of organisms found in an area.

6. Explain how energy conservation and efficiency are major "sources" of energy.

7. Evaluate which forms of energy, other than fossil fuels and nuclear power, have the greatest potential where you live.

Sustainable Citizen Question

8. Consider Figure 18.14, which shows dramatic improvements in the efficiency of refrigerators over time. Have there been similar improvements in other technologies? How much can we rely on technological improvements to reduce our energy demand if energy prices continue to rise?

---

9. List energy conservation measures you could adopt in each of the following aspects of your life: washing laundry, lighting, bathing, cooking, buying a car, and driving a car.

The map below shows the average daily total of solar energy (on an annual basis) received on a solar collector that tilts to compensate for latitude.

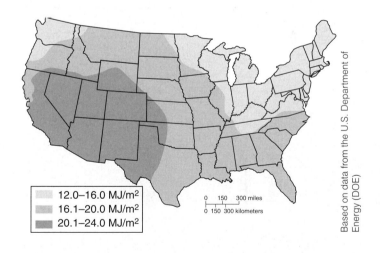

░	12.0–16.0 MJ/m²
▒	16.1–20.0 MJ/m²
▓	20.1–24.0 MJ/m²

0   150   300 miles
0 150 300 kilometers

Based on data from the U.S. Department of Energy (DOE)

10. Based on the map, is Nevada or Maine better suited for producing electricity from solar energy?

11. Which of the following do you think would be best for reducing transportation energy use in the United States: more efficient cars, fuel cell cars, or public transportation? Explain.

✓ THE PLANNER

Graphing Appendix

Introduction

LEARNING OBJECTIVES

1. **Define** *graph*.
2. **Explain** why graphs are so important in science.

A **graph** is a picture that expresses the relationship between two or more ideas or quantities. Graphs make this relationship easier to see and analyze because our brains are much better at interpreting pictures than words or rows and columns of numbers. For example, suppose we wanted to investigate how the eruption of the 1991 Philippine volcano Mount Pinatubo affected average global air temperatures. We could collect a series of temperature measurements before and after this eruption and put this information in a table like this:

Year	Global average temperature (°C)
1987	14.28
1988	14.34
1989	14.23
1990	14.39
1991	14.37
1992	14.18
1993	14.19
1994	14.28
1995	14.42
1996	14.32
1997	14.45
1998	14.61

However, it is much easier to interpret these numbers when they are arranged into a graph like this (**Figure A.1**):

Figure A.1

Global Average Temperature, 1987 to 1998. Climate scientists observed that the years following Mount Pinatubo's eruption were cooler than previous and subsequent years. This brief cooling period temporarily interrupted a longer-term warming trend (Chapter 9).

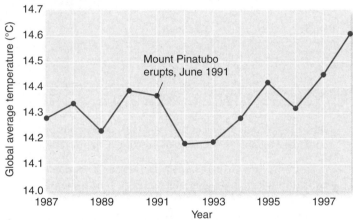

Data from NASA Global Land-Ocean Temperature Index.

For instance, we can tell just by glancing at this graph that there was a substantial drop in the average global temperature immediately following this eruption, and that it took four years for this temperature to reach its pre-eruption level. Consequently, this and most other science textbooks are filled with graphs to help you comprehend the ideas and concepts they present. In order to understand this material and think like a scientist, you therefore must become proficient at reading and interpreting these graphs.

This appendix will help you accomplish these goals by providing a general overview of the nuts and bolts of graphing, discussing the different kinds of graphs used in this textbook, and explaining how to use these graphs to answer scientific questions. It will also give you an opportunity to test your understanding of this material by taking a graphing quiz, formulating your own questions and hypotheses, and creating your own graphs to help you answer your questions and test your hypotheses.

CONCEPT CHECK 🛑 STOP

1. **What** is a graph?
2. **Why** are graphs so important in science?

Graphing Nuts and Bolts

LEARNING OBJECTIVES

1. **Define** *data* and *variable*.
2. **Distinguish** between independent and dependent variables.
3. **Describe** the following components of a graph: x-axis, y-axis, title, caption, legend, future projection, and measurement unit.

In science, **data** simply refers to factual information. Note that because one piece of factual information is called a "datum," even though it may sound funny, scientists are grammatically correct when they make statements such as "These temperature data are important because ..." or "This temperature datum is important because ..."

A **variable** is a factor whose values can change. Many graphs illustrate the relationship between values of a **dependent variable** and values of an **independent variable**. As the names imply, the values of a dependent variable are assumed to depend on the values of an independent variable. To investigate complex interactions and test a series of competing hypotheses, scientists often examine how the values of one independent variable may affect the values of one or more other dependent variables.

For example, suppose an ecologist hypothesizes that the complexity of plant communities directly affects the number of bird species present in a given area. This ecologist could test this hypothesis by recording the number of bird species in different areas and quantifying the complexity of the vegetation in each of these areas. These data could then be presented in a graph like this (**Figure A.2**):

Figure A.2

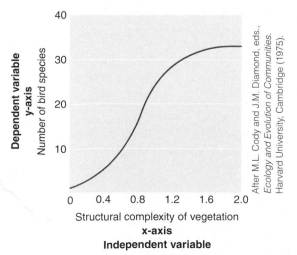

After M.L. Cody and J.M. Diamond, eds., *Ecology and Evolution of Communities.* Harvard University, Cambridge (1975).

By convention, the independent variable is always presented on the horizontal or **x-axis** and the dependent variable is always presented on the vertical or **y-axis**. Thus in this case, the ecologist assumes that the number of bird species *depends* on the structural complexity of the vegetation.

In addition to the x- and y-axes, it is important to recognize and understand the other components that are typically used to construct scientific graphs. As illustrated in the figure below (**Figure A.3**), most graphs start with a **title**, which succinctly describes the information they present. There is often a **caption** beneath this title that states the major implication of the data presented in the graph. Graphs that portray more than one series of data such as this one may use a **legend** that distinguishes and explains each of these data series. Reading a graph's title, caption, and legend before examining its actual data will often help you better understand and interpret its purpose and results.

Figure A.3 • Fertility changes in selected developing countries

Since the 1960s, fertility levels have dropped dramatically in many developing countries.

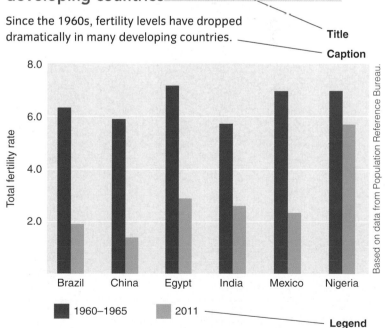

Based on data from Population Reference Bureau.

Some graphs illustrate how the relationship between two variables may change in the future. These **future projections** are usually represented by a dashed line (**Figure A.4**, next page). In this case, the graph projects the rate of future human population growth as a function of three different estimated fertility rates. Also note that because it is not clear without further explanation, the y-axis title on this graph includes the **measurement unit** (billions) portrayed along this axis.

Figure A.4

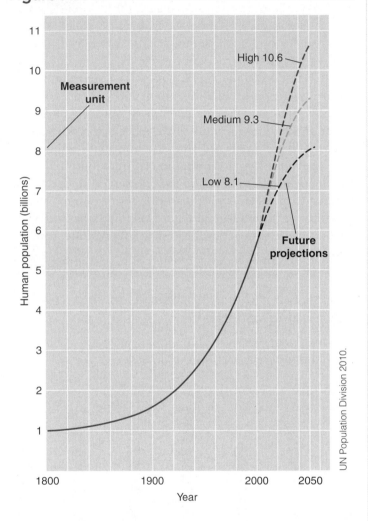

Human population (billions)

Measurement unit

High 10.6

Medium 9.3

Low 8.1

Future projections

UN Population Division 2010.

Year

Graph Types

LEARNING OBJECTIVES

1. **Explain** the difference between continuous and discrete data.
2. **Distinguish** among the following types of graphs: line graph, bar graph, and pie chart.

Scientists use different types of graphs to visualize and interpret different kinds of data. **Continuous data** can take on an infinite number of values within a given range, and are usually associated with measurable variables such as temperature or length. For example, there are an infinite number of measureable temperatures between 45 and 46 degrees Centigrade and lengths between 45 and 46 meters (45.12, 45.593, etc.). In contrast, **discrete data** may only assume a finite number of distinct values, and are usually associated with variables that are counted, such as the number of people within a given area or the number of trees in the forest (you can't count 45.12 people or 45.593 trees).

One of the most common graph types is a **line graph** in which one or more lines connect a series of data points together. Line graphs are used when the y-axis portrays continuous numeric data and the x-axis portrays either continuous numeric data or discrete categorical data that form a sequential series such as months or years. For example, the line graph below (**Figure A.5**) shows how the number of Earths needed to absorb humanity's ecological footprint has increased over the past five decades. Note that we can determine the time our collective footprint overshot Earth's capacity by finding the point on the x-axis that corresponds to the point at which this line exceeds the 1.0 value on the y-axis (around 1988).

Figure A.5 • Line Graph

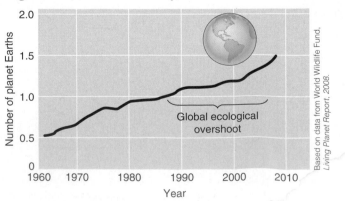

Number of planet Earths

Global ecological overshoot

Year

Based on data from World Wildlife Fund. *Living Planet Report, 2008.*

| CONCEPT CHECK | STOP |

1. **What** are data? What is a variable?
2. **Explain** the difference between a dependent and an independent variable.
3. **Draw** your own graph and include and label each of the following components: x-axis, y-axis, title, caption, legend, future projection, and the measurement units of the x- and y-axes.

A **bar graph** is often used when one variable is a number and the other variable is a category. For example, the bar graph below shows how a numeric variable (per capita ecological footprint) varies across a categorical variable (country). In such cases the height of each bar represents the numeric value for each category. Thus in this case (**Figure A.6**) we can see that the per capita ecological footprint in the United States is more than twice as great as France's and more than five times greater than India's.

Figure A.6 • Bar Graph

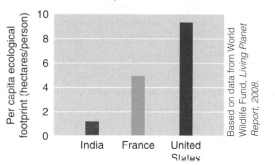

Figure A.7

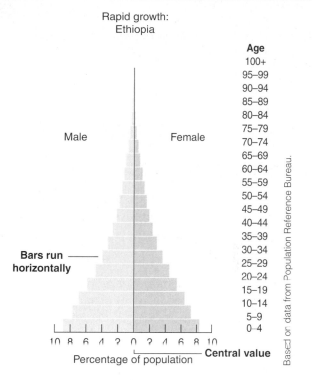

Rapid growth: Ethiopia

Male   Female

Bars run horizontally

Percentage of population   **Central value**

Based on data from Population Reference Bureau.

Some bar graphs are constructed so that their bars run horizontally instead of vertically (**Figure A.7**). Note that in contrast to **Figure A.6**, in this case the x-axis portrays the numeric variable (population percentage) and the y-axis portrays the categorical variable (discrete age ranges). Also note that bars extend from their central value of zero on the x-axis to the right to display the numeric values for females, and to the left to display the numeric values for males. In this horizontal orientation, the numeric value of each bar is determined by its length rather than its height. By examining this graph we can infer that Ethiopia's population is likely to keep growing rapidly because the majority of its males and females are in the younger, pre-reproductive age categories.

A pie chart shows the proportion of some total value taken up by two or more different categories. Each category is graphically portrayed as a "slice" of the total (the entire "pie"), and the size of each of these slices corresponds to its proportion

of the whole. In the pie chart below (**Figure A.8**), we can readily see from the size of the different slices that the mining industry generates over three-quarters of the total amount of solid waste we produce, and that less than 2 percent of this waste comes from municipal sources. By examining the second pie chart that illustrates the proportional contribution of the different materials within the municipal solid waste slice, we can also see that paper and paperboard make up almost a third of the entire municipal waste category.

Figure A.8 • Pie Chart

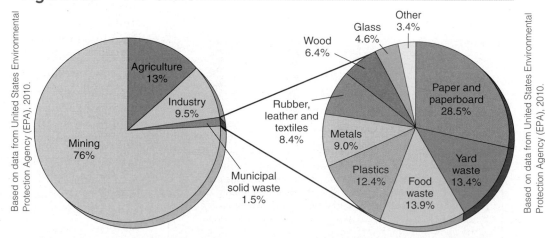

a. Composition of Total Solid Waste, 2010   b. Composition of Municipal Solid Waste, 2010

Based on data from United States Environmental Protection Agency (EPA), 2010.

1. **What** is the difference between continuous and discrete data?

2. **What** are line graphs, bar graphs, and pie charts? Under what circumstances would you use each of these different types of graphs?

Interpreting and Using Graphs

LEARNING OBJECTIVES

1. **Explain** how graphs can be used to answer questions and test hypotheses.

2. **Use** graphs to generate your own questions and hypotheses.

Much of the study and practice of environmental science involves asking questions and testing hypotheses related to these questions. As we have seen, graphs are a particularly effective tool for visualizing and interpreting the kinds of data scientists collect to investigate their questions and test their hypotheses. In fact, some of the *Critical and Creative Thinking Questions* at the end of each chapter in this book can only be answered by interpreting one or more of the graphs included in these sections. For example, consider the question below and its accompanying bar graph (**Figure A.9**), from the end of Chapter 5 (How Ecosystems Work).

Figure A.9

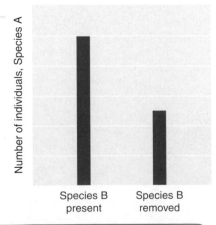

Number of individuals, Species A

Species B present   Species B removed

Think Critically Ecologists investigating interactions of two species at a study site first counted individuals of Species A and then removed all Species B individuals. Six months later, the ecologists again counted individuals of Species A. Viewing their results as graphed below, what is the likely ecological interaction between Species A and Species B? Explain your answer.

Answering this question requires us to first correctly interpret what this graph is showing us. In this case, we can conclude that the number of individuals of Species A declined after the ecologists removed Species B because the height of the bar over the "Species B removed" label is lower than the height over the "Species B present" label. Thus this graph is showing us that the presence of Species B appears to have a positive effect on the abundance of Species A. However, this graph does not tell us whether the relationship between these two species is an example of a mutualism (both species benefit from their interaction) or a commensalism (Species A benefits from its interaction with Species B but Species B is neither helped nor harmed by this interaction), as explained earlier in the chapter, or even if Species A is a predator of Species B.

These ecologists might therefore ask this follow-up question: "Do Species A and Species B have a mutualistic relationship in this study system?" If they had initially observed that there appeared to be more individuals of Species B in areas where there were also more individuals of Species A, they might hypothesize that the relationship between these two species is in fact mutualistic. They could investigate this question and test their hypothesis by recording the number of individuals of Species B in this study site before and after removing Species A. They could then answer their question and support or reject their hypothesis by using these data to produce another bar graph similar to their original one, except this time the y-axis would represent the number of individuals of Species B, and the x-axis would show Species A present and Species A removed.

Figure A.10 depicts another question and its accompanying graphs from the end of Chapter 7 (Human Population Change and the Environment).

We can tell just by glancing at these pie charts that France is the most urbanized of these three countries and Ethiopia is the most rural. Yet we can also use the patterns revealed by these simple charts to generate more complex and interesting questions and hypotheses

Figure A.10

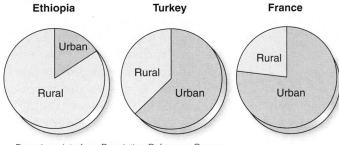

Based on data from Population Reference Bureau.

Think Critically Urbanization varies from one country to another. Which countries are mostly urban? Which countries are mostly rural?

related to the material presented in this chapter. For instance, we could ask whether the amount of urbanization depends on the rate of human population growth. Some might hypothesize that high population growth rates are associated with high urbanization rates because there is nowhere else for all those people to go. However, others might hypothesize that high population growth rates are associated with low urbanization rates because people in urban areas tend to have fewer children than people in rural areas.

To investigate this question, we could collect urbanization and population growth data for different areas within a given country, across many different countries, or both. In each case, we could visualize these data by constructing a line graph with "population growth rate" (the independent variable) on the x-axis and "urbanization (the dependent variable) on the y-axis. If the line connecting the data points on this graph ran uphill, we would conclude that urbanization appears to increase with population. Conversely, if this line ran downhill, we would conclude that urbanization appears to decrease with population growth. If the line was flat, we would conclude that there is no consistent relationship between population growth and urbanization.

CONCEPT CHECK **STOP**

1. **Which** of the four pollutants illustrated in the bar graphs below shows the greatest decline from Location 1 to Location 2? Suppose a scientist hypothesized that the overall level of air pollution at Location 2 will be lower than at Location 1. Do the data presented in these bar graphs support this hypothesis? Why or why not?

2. **Study** the line graph below, then devise two original questions that further explore some aspect of the patterns revealed by this graph. What would you hypothesize the answer to each of your questions will be, and why? What data would you collect to investigate your questions and test your hypotheses, and how would you graph each of these data sets?

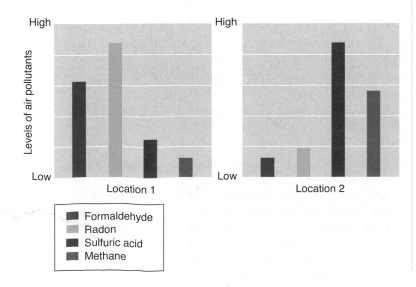

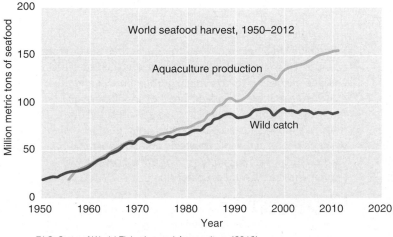

FAO *State of World Fisheries and Aquaculture (2012).*

Review, Test, and Apply Your Graphing Knowledge

LEARNING OBJECTIVES

1. **Review** and synthesize the material presented in this Graphing Appendix.

2. **Assess** your overall understanding of this material and review any sections that remain unclear.

We have seen how useful graphs can be for visualizing and interpreting data, answering scientific questions and testing hypotheses, and generating new questions and hypotheses that may increase our knowledge and understanding of environmental science. The diagram below (**Figure A.11**) presents a visual overview and synthesis of the material presented in this Graphing Appendix.

Dissecting and interpreting graphs • Figure A.11

 THE PLANNER

Graphs help scientists answer questions, test hypotheses, and generate new questions and hypotheses that can increase our basic knowledge and help us solve important real-world problems. Analyzing graphs systematically can help you better understand and interpret them and create your own scientific questions and hypotheses.

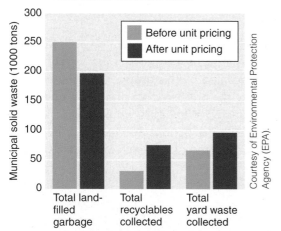

Effects of Unit Pricing on Garbage Collection and Waste Diversion Rates

Courtesy of Environmental Protection Agency (EPA).

Some municipalities have tried to reduce their solid waste by employing unit pricing, a system in which citizens are charged according to how much garbage they generate, but wastes that are diverted from the landfill through recycling and composting programs are collected for free. This graph shows how the total amount of landfilled garbage decreased while recycling and compostable yard waste collections increased after unit pricing was implemented in San Jose, California.

Think Critically Would you hypothesize that unit pricing would be an effective strategy for decreasing the amount of landfilled garbage in poor countries? Explain the reasoning behind your hypothesis, and what you would do to rigorously test it.

① Start by reading the graph's title and caption. Note that the title succinctly describes the subject matter of this graph, and that the caption provides important background information and summarizes the most important result portrayed in this graph.

② Examine the graph's descriptive components. The x-axis shows the three categories of wastes described in the caption—landfilled garbage, recyclables, and yard wastes—and the y-axis shows the amount of waste collected. Note that the unit on the y-axis scale is "1000 tons," so the amount of landfilled garbage collected before unit pricing is 250,000 tons, not 250 tons. The legend explains that the green bars represent the "before unit pricing" data and the red bars represent the "after unit pricing" data.

③ Study the graph's data. Note how the amount of landfilled garbage decreased after unit pricing was implemented while the amount of wastes diverted from the landfill through recyclable and yard waste collections increased after unit pricing.

④ Analyze the graph's overall results and think about their larger implications. For instance, these data suggest that unit pricing may be an effective strategy for reducing the amount of landfilled solid waste and increasing citizen participation in recycling and composting programs.

⑤ Build on the graph's results and implications by thinking of additional questions to investigate and hypotheses to test. For example, "Does the amount of waste diverted from landfills after unit pricing is implemented decrease over time?" We might hypothesize that some people may eventually decide that the money they save on their garbage bill by recycling and composting is not worth all that extra effort. We could investigate this question and support or refute our hypothesis by quantifying the amount of garbage collected at regular time intervals after a unit pricing system is first implemented.

1. **Test** your understanding of this Graphing Appendix by taking the following quiz:

   1. Label each of the following terms on the below graph: title, caption, legend, x-axis, y-axis.

   ## Fertility changes in selected developing countries
   Since the 1960s, fertility levels have dropped dramatically in many developing countries.

   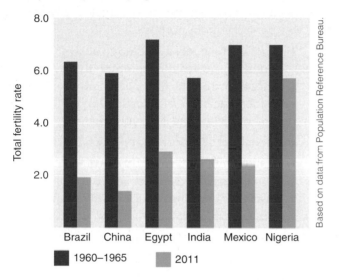

   Based on data from Population Reference Bureau.

   2. This graph is an example of a
      a. Scatter plot   c. Bar graph
      b. Line graph   d. Pie chart

   3. Which of the following countries had the greatest decline in their total fertility rate from the 1960's to 2011?
      a. India   c. Nigeria
      b. Mexico   d. Egypt

   4. The values of a dependent variable
      a. Are assumed to depend on the values of an independent variable
      b. Are assumed to affect the values of an independent variable
      c. May take on an infinite number of possibilities between a specified range
      d. May take on only a finite number of possibilities between a specified range

   5. Discrete data
      a. Are assumed to depend on the values of an independent variable
      b. Are assumed to affect the values of an independent variable
      c. May take on an infinite number of possibilities between a specified range
      d. May take on only a finite number of possibilities between a specified range

   6. Which of the following graphs is used to show the proportions of some total value taken up by two or more categories?
      a. Scatter plot
      b. Line graph
      c. Bar graph
      d. Pie chart

2. **Demonstrate** your ability to apply the material presented in this Graphing Appendix by performing the following tasks:

   1. Suppose you are a scientist studying the effects of a landfill on the diversity of organisms living in a nearby stream. To investigate this topic, you count the number of different species in a series of water samples collected at successively greater distances from this landfill. When you have finished collecting all of these data, you organize them into a table like this:

Distance of water sample from landfill (in meters)	Number of different species in sample
10	2
25	3
50	3
70	5
100	8
135	8
175	7
200	13
250	15
300	17

To visualize and interpret these numbers, convert this table into a line graph. The x-axis of this graph will be the "Distance of water sample from landfill" and the y-axis will be the "Number of different species in sample." Label both of these axes, provide the measurement unit for the x-axis, plot all of the above data, and include a title and caption for your graph. When you are finished, interpret your results. Do you think the landfill is significantly affecting the diversity of species living in this stream? Why or why not?

2. Suppose you are a scientist studying the biological effectiveness and economic costs of three different pesticides designed to keep insects from eating corn. To test these pesticides, you conduct an experiment in which you count the number of kernels damaged by insects on 1000 ears of corn sprayed with pesticide A, 1000 ears of corn sprayed with pesticide B, and 1000 ears of corn sprayed with pesticide C. You also quantify the cost of purchasing and applying each of these pesticides to their respective 1000 ears of corn. When this experiment is over and you have collected all of your data, you organize them into a table like this:

Pesticide	Average number of kernels damaged by insects	Average cost of pesticide
A	32	$1.87
B	41	$2.09
C	35	$1.11

Convert these data into whatever type of graph you think will best help you visualize and interpret the results of this experiment. Be sure to include all of the components that normally accompany the type of graph you selected. When you are finished, interpret your results. Which pesticide would you recommend, and why?

3. Ask your own original scientific question that is related to the material covered in one or more of the chapters in this book, then:

a. Create your own hypothesis that predicts the answer to your question. Explain the reasoning behind your hypothesis.

b. Gather the necessary data to answer your question. You may accomplish this by finding these data on the internet, in print, or by conducting an original investigation that enables you to collect the data yourself.

c. Construct an appropriate graph that helps you visualize and interpret these data.

d. Use this graph to answer your original question. Was your hypothesis supported? Why or why not?

e. Ask another original question that builds on the results and implications of the data presented in your graph. What would you hypothesize the answer to this question will be, and why? Explain how you would investigate this question and test your hypothesis.

Glossary

acid deposition A type of air pollution that includes sulfuric and nitric acids in precipitation, as well as dry acid particles that settle out of the air.

acid mine drainage Pollution caused when sulfuric acid and dangerous dissolved materials such as lead, arsenic, and cadmium wash from mines into nearby lakes and streams.

active solar heating A system of putting the sun's energy to use in which collectors absorb solar energy and pumps or fans distribute the collected heat.

acute toxicity Adverse effects that occur within a short period after high-level exposure to a toxicant.

age structure The number and proportion of people at each age in a population.

air pollution Various chemicals (gases, liquids, or solids) present in the atmosphere in high enough levels to harm humans, other organisms, or materials.

aquaculture The growing of aquatic organisms (fishes, shellfish, and seaweeds) for human consumption.

aquifer depletion The removal of groundwater faster than it can be recharged by precipitation or melting snow.

artificial eutrophication Overnourishment of an aquatic ecosystem by nutrients such as nitrates and phosphates due to human activities such as agriculture and discharge from sewage treatment plants.

atmosphere The gaseous envelope surrounding Earth.

benthic environment The ocean floor, which extends from the intertidal zone to the deep-ocean trenches.

biocentric preservationist A person who believes in protecting nature from human interference because all forms of life deserve respect and consideration.

biochemical oxygen demand (BOD) The amount of oxygen that microorganisms need to decompose biological wastes into carbon dioxide, water, and minerals.

biodiversity hotspots Relatively small areas of land that contain an exceptional number of endemic species and are at high risk from human activities.

biological diversity The number and variety of Earth's organisms; consists of three components: genetic diversity, species richness, and ecosystem diversity.

biological diversity The number and variety of Earth's organisms.

biological magnification The increase in toxicant concentrations as a toxicant passes through successive levels of the food chain.

biomass Plant and animal material used as fuel.

biome A large, relatively distinct terrestrial region with similar climate, soil, plants, and animals, regardless of where it occurs in the world.

biosphere The layer of Earth that contains all living organisms.

biotic potential The maximum rate at which a population could increase under ideal conditions.

boreal forest A region of coniferous forest (such as pine, spruce, and fir) in the Northern Hemisphere; located just south of the tundra. Also called *taiga*.

broad-spectrum pesticide A pesticide that kills a variety of organisms, including beneficial organisms, in addition to the target pest.

bycatch The fishes, marine mammals, sea turtles, seabirds, and other animals caught unintentionally in a commercial fishing catch.

carbon management Ways to separate and capture the CO_2 produced during the combustion of fossil fuels and then sequester (store) it.

carcinogen Any substance (for example, chemical, radiation, virus) that causes cancer.

carrying capacity (K) The largest population a particular environment can support sustainably (long term), if there are no changes in that environment.

chaparral A biome with mild, moist winters and hot, dry summers; vegetation is typically small-leaved evergreen shrubs and small trees.

chlorofluoro-carbons (CFCs) Human-made organic compounds that contain chlorine and fluorine; now banned because they attack the stratospheric ozone layer.

chronic toxicity Adverse effects that occur after a long period of low-level exposure to a toxicant.

clear-cutting A logging practice in which all the trees in a stand of forest are cut, leaving just the stumps.

climate The typical weather patterns that occur in a place over a period of years.

cogeneration (CHP) An energy technology that involves recycling "waste" heat.

command and control regulation Pollution control laws that work by setting limits on levels of pollution.

community A natural association that consists of all the populations of different species that live and interact together within an area at the same time.

compact development Design of cities in which tall, multiple-unit residential buildings are close to shopping and jobs, and all are connected by public transportation.

competition The interaction among organisms that vie for the same resources in an ecosystem (such as food or living space).

conservation biology The scientific study of how humans affect organisms and of the development of ways to protect biological diversity.

conservation easement A legal agreement that protects privately owned forest, rangeland, or other property from development for a specified number of years.

conservation tillage A method of cultivation in which residues from previous crops are left in the soil, partially covering it and helping to hold it in place until the newly planted seeds are established.

contour plowing Plowing that matches the natural contour of the land.

Coriolis effect The tendency of moving air or water to be deflected from its path and swerve to the right in the Northern Hemisphere and to the left in the Southern Hemisphere.

cost–benefit diagram A diagram that helps policy makers make decisions about costs of a particular action and benefits that would occur if that action were implemented.

crop rotation The planting of a series of different crops in the same field over a period of years.

deep ecology worldview A worldview based on harmony with nature, a spiritual respect for life, and the belief that humans and all other species have an equal worth.

deforestation The temporary or permanent clearance of large expanses of forest for agriculture or other uses.

degradation (of land) Natural or human-induced reduction in the potential ability of the land to support crops or livestock.

demographics The applied branch of sociology that deals with population statistics.

demographic transition The process whereby a country moves from relatively high birth and death rates to relatively low birth and death rates.

desert A biome in which the lack of precipitation limits plant growth; deserts are found in both temperate and tropical regions.

desertification Degradation of once-fertile rangeland or tropical dry forest into nonproductive desert.

dose–response curve In toxicology, a graph that shows the effects of different doses on a population of test organisms.

dust dome A dome of heated air that surrounds an urban area and contains a lot of air pollution.

ecological niche The totality of an organism's adaptations, its use of resources, and the lifestyle to which it is fitted.

ecological succession The process of community development over time, which involves species in one stage being replaced by different species.

ecology The study of the interactions among organisms and between organisms and their abiotic environment.

economic development An expansion in a region's or country's economy, viewed by many as the best way to raise the standard of living.

ecosystem A community and its physical environment.

ecosystem services Important environmental benefits such as clean air, clean water, and fertile soil, that the natural environment provides.

El Niño–Southern Oscillation (ENSO) A periodic, large-scale warming of surface waters of the tropical eastern Pacific Ocean that temporarily alters both ocean and atmospheric circulation patterns.

endangered species A species that faces threats that may cause it to become extinct within a short period.

endemic species Organisms that are native to or confined to a particular region.

energy conservation Using less energy—by reducing energy use and waste, for example.

energy efficiency Using less energy to accomplish a given task—by using new technology, for example.

energy flow The passage of energy in a one-way direction through an ecosystem.

enhanced greenhouse effect Additional atmospheric warming produced as human activities increase atmospheric concentrations of greenhouse gases.

enrichment The process by which uranium ore is refined after mining to increase the concentration of fissionable U-235.

environmental ethics A field of applied ethics that considers the moral basis of environmental responsibility.

environmental justice The right of every citizen to adequate protection from environmental hazards.

environmental science The interdisciplinary study of humanity's relationship with other organisms and the physical environment.

environmental worldview A worldview based on how the environment works, our place in the environment, and right and wrong environmental behaviors.

epidemiology The study of the effects of chemical, biological, and physical agents on the health of human populations.

estuary A coastal body of water, partly surrounded by land, with access to the open ocean and a large supply of fresh water from a river.

evolution The cumulative genetic changes in populations that occur during successive generations.

exponential population growth The accelerating population growth that occurs when optimal conditions allow a constant reproductive rate.

external cost A harmful environmental or social cost that is borne by people not directly involved in selling or buying a product.

extinction The elimination of a species from Earth.

first law of thermodynamics A physical law which states that energy cannot be created or destroyed, although it can change from one form to another.

fission The splitting of an atomic nucleus into two smaller fragments, accompanied by the release of a large amount of energy.

flowing-water ecosystem A freshwater ecosystem such as a river or stream in which water flows in a current.

fluidized-bed combustion A clean-coal technology in which crushed coal is mixed with limestone to neutralize acidic compounds produced during combustion.

food insecurity The condition in which people live with chronic hunger and malnutrition.

forest decline A gradual deterioration and eventual death of many trees in a forest.

freshwater wetlands Lands that shallow fresh water covers for at least part of the year; wetlands have a characteristic soil and water-tolerant vegetation.

fuel cell A device that directly converts chemical energy into electricity. A fuel cell requires hydrogen and oxygen from the air.

genetic engineering The manipulation of genes (for example, taking a specific gene from one species and placing it into an unrelated species) to produce a particular trait.

genetic resistance An inherited characteristic that decreases the effect of a given agent (such as a pesticide) on an organism (such as a pest).

geothermal energy Energy from Earth's hot interior, used for space heating or generation of electricity.

germplasm Any plant or animal material that may be used in breeding.

green chemistry A subdiscipline of chemistry in which commercially important chemical processes are redesigned to significantly reduce environmental harm.

greenhouse gases Gases—including water vapor, carbon dioxide, methane, and certain other gases—that absorb infrared radiation.

groundwater The supply of fresh water under Earth's surface that is stored in underground aquifers.

growth rate (r) The rate of change (increase or decrease) of a population's size, expressed in percentage per year.

gyres Large, circular ocean current systems that often encompass an entire ocean basin.

habitat corridor A protected zone that connects isolated unlogged or undeveloped areas.

habitat fragmentation The breakup of large areas of habitat into small, isolated patches.

hazardous waste A discarded chemical that threatens human health or the environment.

high-level radioactive wastes Radioactive solids, liquids, or gases that initially give off large amounts of ionizing radiation.

highly developed countries Countries with complex industrialized bases, low rates of population growth, and high per person incomes.

hydraulic fracturing The use of pressurized water and chemicals to extract natural gas from deep layers of shale.

hydropower A form of renewable energy that relies on flowing or falling water to generate electricity.

incentive-based regulation Pollution control laws that work by establishing emission targets and providing industries with incentives to reduce emissions.

industrialized agriculture Modern agricultural methods that require large capital inputs and less land and labor than traditional methods.

infant mortality rate The number of deaths of infants under age 1 per 1000 live births.

infrared radiation Electromagnetic radiation with wavelengths longer than those of visible light but shorter than microwaves; perceived as invisible waves of heat.

integrated waste management A combination of the best waste management techniques into a consolidated program to deal effectively with solid waste.

intertidal zone The area of shoreline between low and high tides.

invasive species Foreign species that spread rapidly in a new area if free of predators, parasites, or resource limitations that may have controlled their population in their native habitat.

landscape A region that includes several interacting ecosystems.

less developed countries Countries with low levels of industrialization, very high rates of population growth, very high infant mortality rates, and very low per person incomes relative to highly developed countries.

low-level radioactive wastes Solids, liquids, or gases that give off small amounts of ionizing radiation.

marginal cost of pollution abatement The added cost of reducing one unit of a given type of pollution.

marginal cost of pollution The added cost of an additional unit of pollution.

mass burn incinerator A large furnace that burns all solid waste except for unburnable items such as refrigerators.

microirrigation A type of irrigation that conserves water by piping it to crops through sealed systems.

minerals Elements or compounds of elements that occur naturally in Earth's crust.

moderately developed countries Countries with medium levels of industrialization and per person incomes lower than those of highly developed countries.

monoculture Ecological simplification in which only one type of plant is cultivated over a large area.

municipal solid waste Solid materials discarded by homes, offices, stores, restaurants, schools, hospitals, prisons, libraries, and other commercial and institutional facilities.

national income accounts Measures of the total income of a nation's goods and services for a given year.

natural capital Earth's resources and processes that sustain living organisms, including humans; includes minerals, forests, soils, water, clean air, wildlife, and fisheries.

natural selection The tendency of better-adapted individuals—those with a combination of genetic traits best suited to environmental conditions—to survive and reproduce, increasing their proportion in the population.

neritic province The part of the pelagic environment that overlies the ocean floor from the shoreline to a depth of 200 m (650 ft).

nonmunicipal solid waste Solid waste generated by industry, agriculture, and mining.

nonpoint source pollution Pollution that enters bodies of water over large areas rather than being concentrated at a single point of entry.

nonrenewable resources Natural resources that are present in limited supplies and are depleted as they are used.

nuclear energy The energy released by nuclear fission or fusion.

nuclear reactor A device that initiates and maintains a controlled nuclear fission chain reaction to produce energy for electricity.

nutrient cycling The pathway of various nutrient minerals or elements from the environment through organisms and back to the environment.

oceanic province The part of the pelagic environment that overlies the ocean floor at depths greater than 200 m (650 ft).

optimum amount of pollution The amount of pollution that is economically most desirable.

overburden Soil and rock overlying a useful mineral deposit.

overgrazing A situation that occurs when too many grazing animals consume the plants in a particular area, leaving the vegetation destroyed and unable to recover.

overnutrition A type of malnutrition in which an overconsumption of calories leaves the body susceptible to disease.

ozone thinning The removal of ozone from the stratosphere by human-produced chemicals or natural processes.

passive solar heating A system of putting the sun's energy to use that does not require mechanical devices to distribute the collected heat.

pathogen An agent (usually a microorganism) that causes disease.

persistent organic pollutants (POPs) Persistent toxicants that bioaccumulate in organisms and travel through air and water to contaminate sites far from their source.

pesticide A toxic chemical used to kill pests.

photochemical smog A brownish-orange haze formed by chemical reactions involving sunlight, nitrogen oxides, and hydrocarbons.

photosynthesis The biological process that captures light energy and transforms it into the chemical energy of organic molecules, which are manufactured from carbon dioxide and water.

photovoltaic (PV) solar cell A wafer or thin film of solid-state materials, such as silicon or gallium arsenide, that is treated with certain metals in such a way that the film generates electricity when solar energy is absorbed.

plate tectonics The study of the processes by which the lithospheric plates move over the asthenosphere.

point source pollution Water pollution that can be traced to a specific point of entry.

population A group of organisms of the same species that live together in the same area at the same time.

population ecology The branch of biology that deals with the number of individuals of a particular species found in an area and how and why those numbers increase or decrease over time.

poverty A condition in which people are unable to meet their basic needs for food, clothing, shelter, education, or health.

precautionary principle The idea that new technologies, practices, or materials should not be adopted until there is strong evidence that they will not adversely affect human or environmental health.

predation The consumption of one species (the prey) by another (the predator).

primary air pollutants Harmful chemicals that enter directly into the atmosphere due to either human activities or natural processes.

primary treatment Treatment of wastewater that involves removing suspended and floating particles through mechanical processes.

radiative forcing For greenhouse gases, the capacity to retain heat in Earth's atmosphere.

rangeland Grassland area that is not intensively managed and is used for grazing livestock.

renewable resources Resources that are replaced by natural processes and that can be used forever, provided they are not overexploited in the short term.

replacement-level fertility The number of children a couple must produce to "replace" themselves.

restoration ecology The study of the historical condition of a human-damaged ecosystem, with the goal of returning it as closely as possible to its former state.

risk assessment The quantitative and qualitative characterization of risks so that they can be compared, contrasted, and managed.

risk The probability of harm (such as injury, disease, death, or environmental damage) occurring under certain circumstances.

runoff The movement of fresh water from precipitation and snowmelt to rivers, lakes, wetlands, and the ocean.

salinization The gradual accumulation of salt in soil, often as a result of improper irrigation methods.

saltwater intrusion The movement of seawater into a freshwater aquifer near the coast.

sanitary landfill The most common disposal site for solid waste, where waste is compacted and buried under a shallow layer of soil.

savanna A tropical grassland with widely scattered trees or clumps of trees.

scientific method The way a scientist approaches a problem, by formulating a hypothesis and then testing it.

secondary air pollutants Harmful chemicals that form in the atmosphere when primary air pollutants react chemically with one another or with natural components of the atmosphere.

secondary treatment Biological treatment of wastewater to decompose suspended organic material; secondary treatment reduces the water's biochemical oxygen demand.

second law of thermodynamics A physical law which states that when energy is converted from one form to another, some of it is degraded into heat, a less usable form that disperses into the environment.

sewage Wastewater from drains or sewers (from toilets, washing machines, and showers); includes human wastes, soaps, and detergents.

shelterbelt A row of trees planted as a windbreak to reduce soil erosion of agricultural land.

sick building syndrome Eye irritations, nausea, headaches, respiratory infections, depression, and fatigue caused by indoor air pollution.

smelting The process in which ore is melted at high temperatures to separate impurities from the molten metal.

soil erosion The wearing away or removal of soil from the land.

soil horizons Horizontal layers into which many soils are organized, from the surface to the underlying parent material.

soil The uppermost layer of Earth's crust, which supports terrestrial plants, animals, and microorganisms.

solar thermal electric generation A means of producing electricity in which the sun's energy is concentrated using mirrors or lenses onto a fluid-filled pipe; the heated fluid is used to generate electricity.

source reduction An aspect of waste management in which products are designed and manufactured in ways that decrease the amount of solid and hazardous waste in the solid waste stream.

species richness The number of different species in a community.

spent fuel Used fuel elements that were irradiated in a nuclear reactor.

spoil bank A hill of loose rock created when the overburden from a new trench is put into the already excavated trench during strip mining.

standing-water ecosystem A body of fresh water surrounded by land and whose water does not flow; a lake or a pond.

subsistence agriculture Traditional agricultural methods that are dependent on labor and a large amount of land to produce enough food to feed oneself and one's family.

subsurface mining The extraction of mineral and energy resources from deep underground deposits.

surface mining The extraction of mineral and energy resources near Earth's surface by first removing the soil, subsoil, and overlying rock strata.

surface water Precipitation that remains on the surface of the land and does not seep down through the soil.

sustainability The ability to meet humanity's current needs without compromising the ability of future generations to meet their needs.

sustainable agriculture Agricultural methods that maintain soil productivity and a healthy ecological balance while having minimal long-term impacts.

sustainable consumption The use of goods and services that satisfy basic human needs and improve the quality of life but that also minimize resource use.

sustainable forestry The use and management of forest ecosystems in an environmentally balanced and enduring way.

sustainable soil use The wise use of soil resources, without a reduction in the amount or fertility of soil, so it is productive for future generations.

sustainable water use The wise use of water resources, without harming the essential functioning of the hydrologic cycle or the ecosystems on which present and future humans depend.

sustainable development Economic growth that meets the needs of the present without compromising the ability of future generations to meet their needs.

symbiosis An intimate relationship or association between members of two or more species; includes mutualism, commensalism, and parasitism.

systems perspective A perspective that considers not just immediate or intended effects of activities, but all of the impacts of those activities in other places or at other times.

temperate deciduous forest A forest biome that occurs in temperate areas where annual precipitation ranges from about 75 cm to 150 cm (30 to 60 in).

temperate grassland A grassland with hot summers, cold winters, and less rainfall than is found in the temperate deciduous forest biome.

temperate rain forest A coniferous biome with cool weather, dense fog, and high precipitation.

temperature inversion A layer of cold air temporarily trapped near the ground by a warmer upper layer.

tertiary treatment Advanced wastewater treatment methods that are sometimes employed after primary and secondary treatments.

threatened species A species whose population has declined to the point that it may be at risk of extinction.

total fertility rate (TFR) The average number of children born to each woman.

toxicology The study of toxicants, chemicals with adverse effects on health.

tropical rain forest A lush, species-rich forest biome that occurs where the climate is warm and moist throughout the year.

tundra The treeless biome in the far north that consists of boggy plains covered by lichens and mosses; it has harsh, cold winters and extremely short summers.

ultraviolet (UV) radiation Radiation from the part of the electromagnetic spectrum with wavelengths just shorter than visible light; can be lethal to organisms at high levels of exposure.

undernutrition A type of malnutrition in which an underconsumption of calories or nutrients leaves the body weakened and susceptible to disease.

urban heat island Local heat buildup in an area of high population.

urbanization A process whereby people move from rural areas to densely populated cities.

utilitarian conservationist A person who values natural resources because of their usefulness to humans but uses them sensibly and carefully.

water pollution A physical, biological, or chemical change in water that adversely affects the health of humans and other organisms.

Western worldview A worldview based on human superiority over nature, the unrestricted use of natural resources, and economic growth to manage an expanding industrial base.

wilderness A protected area of land in which no human development is permitted.

wind energy Electric energy obtained from surface air currents caused by the solar warming of air.

zero population growth The state in which the population remains the same size because the birth rate equals the death rate.

Index

Note: A page number followed by f indicates figures, illustrations, or maps; a page number followed by t indicates tables, charts, or graphs.

A

Abiotic environment, 98
Acid deposition, 110, 111, 234–238, 234f, 236f, 422
   effects of, 234f, 235, 236f, 423f
   and mountaintop removal, 422
   politics of, 235
   recovery from, 235, 273–238
Acid mine drainage, 306, 306f, 422
Acid rain, 61, 234, 237–238
Active solar heating, 444–445, 444f
Acute toxicity, 77, 85
Adaptation, 147, 229
   to climate change, 229–230
   and predator–prey interactions, 119
Additivity, 87, 88
Adirondack Mountains, New York State, 237, 237f
Aerosols, 108, 224
Africa. See also Individual countries
   food insecurity in, 350, 352
   gender roles, 175f
   GM crop opposition, 363
   wildlife reserves in, 343f
Age structure, 172–174, 173f, 173t, 174f
Age structure diagrams, 172, 173f
Agricultural land, 230, 355
Agriculture, 353–368
   challenges for, 355–358
   and deforestation, 330, 331f
   genetic engineering, 362–363
   and global climate change, 226
   improving, 40–41
   industrialized, 353, 353f, 358, 359f
   organic, 361, 368, 368t
   pest control, 364–367
   reducing water waste, 256
   resource conservation in, 50
   soil conservation and regeneration, 314–316, 315f
   soil erosion reduction, 61
   sustainable, 360–362
   types of, 353–354
   as water pollution source, 262
   water use for, 247, 248, 248f
Agroecosystem, 361
Air pollutants, 108, 198, 206–209, 212
Air pollution, 196, 196f, 199f
   in developing countries, 10
   effects of, 201–205
   emissions reduction, 61
   indoor, 209–211
   outdoor, 198–200
   and respiratory disease, 89f
   sources of, 198–200
   types of, 196–198

Air toxics, 198
Alaska
   Aleutian Islands, 282, 282f
   Arctic National Wildlife Refuge, 439, 439f
   boreal forest, 329
   Exxon Valdez oil spill, 428, 429f
   Fort Yukon, 132t
   Muir Glacier, 216, 216f–217f
   Pribilof Islands, 164, 164f
   Tongass National Forest, 344, 344f
   tundra, 129f
   Wrangell–Saint Elias Wilderness, 339f
Aleutian Islands, 282, 282f
Alpine tundra, 132
Alternative agriculture, 360
Alternative energy, 48. See also Renewable energy resources
Amazonas State, Brazil, 167f
Amazon Basin, Brazil, 40f, 330, 342f
Amazon rain forest, 79f
American Dust Bowl, 54
Ammonification, 110
Amphibian species, 15f
Amsterdam, Netherlands, 37f
Animals
   commercial harvest of, 385–386
   decline in varieties, 356
   diseases carried by, 80, 81f
   habitat corridors for, 326
   increasing yields, 358
   in national parks, 336, 338
Antagonism, 88
Antarctica
   climate change in, 228f
   ozone thinning over, 231
   protected lands in, 343f
   water temperatures around, 227
Anthropocentric worldview, 34
Antibiotics, given to livestock, 358
Antiquities Act (1906), 53
Aquaculture, 41, 286f, 287, 287f, 287t
Aquatic ecosystems, 142–147
   brackish, 146–147, 146f
   freshwater, 142–145, 143f–145f
Aquifers, 246
Aquifer depletion, 248–249, 249f
Aral Sea, 251, 251f
Arctic
   ozone thinning over, 231
   protected ecosystems in, 342f
   tundra, 129f, 132, 132t
   warming of, 15f
Arctic National Wildlife Refuge, Alaska, 439, 439f
Arctic sea ice. See Polar ice caps
Arid lands, 247
Ariquemes, Brazil, 227f
Arizona
   Cabeza Prieta Wilderness, 339f
   desert, 129f, 139f
   Grand Canyon National Park, 337f

Artificial eutrophication, 261
Assimilation, 110
Association of Environmental Studies and Sciences, 58
Asthenosphere, 298
Aswan High Dam, Egypt, 455, 455f
Atchafalaya National Wildlife Refuge, Louisiana, 145f
Athens, Greece, 100f
Atlantic forest, Brazil, 32f
Atlantic Ocean
   growth of, 298, 299
   Hibernia oil platform, 62f
Atmosphere, 99, 192–195, 192f, 193f
   and biogeochemical cycles, 108, 108f, 109, 109f, 111, 111f
   carbon dioxide in, 122
   circulation in, 194–195
   and climate, 218–221
   composition of, 192, 192f, 192t
   global distillation effect, 190
   layers of, 193f
   ocean–atmosphere interaction, 276–278
Atmospheric science, 16
Audubon, John James, 51
Auroras, 193f
Australia
   organic agriculture in, 368
   protected lands in, 343f
   Queensland flooding, 247f

B

Background extinction, 378
Bailey Elementary School, Falls Church, Virginia, 31f
Bangladesh
   Dhaka open-air market, 80f
   infant mortality in, 171f
   microcredit in, 180f
   polio vaccinations in, 166f
   population and population density of, 170t
Barren (polar ice) areas, 131f
Battery Park City, New York City, 37f
Bayview Hunters Point, San Francisco, California, 35
Bear Lake, Rocky Mountain National Park, Colorado, 143f
Bee colonies, 118, 118f
Belem, Brazil, 141t
Belize, 19f
Benthic environment, 278–281, 279f
Benthos, 142
Bioaccumulation, 82f, 83, 365
Biocentric preservationists, 53
Biocentric worldview, 34
Biochemical oxygen demand (BOD), 258, 258t
Biodegradable plastic, 401
Biodiesel, 450

Biodiversity. *See* Biological diversity
Biodiversity hotspots, 380, 381f
Biofuels, 357
Biogas, 450
Biogas digesters, 450, 451f
Biogeochemical cycles, 106
    carbon cycle, 106–107, 107f
    forests' role in, 325
    hydrologic cycle, 107–109, 108f
    nitrogen cycle, 109–110, 109f
    phosphorus cycle, 111–113, 112f
    sulfur cycle, 110–111, 111f
Biological controls, 366
Biological diversity (biodiversity), 39, 375,
    375f–376f. *See also* Species
    declining, 379–381, 381f
    hotspots, 380
    in land cover, 131f
    levels of, 375f
    loss of, 39–40
    need for, 375–376
    threats to, 383f
Biological magnification, 82t, 83, 365,
    366f
Biological nitrogen fixation, 361
Biological oxygen demand, 258
Biomagnification, 82f
Biomass energy, 450–451, 450f, 451f
Biomes, 128–142, 128f, 129f
    boreal forest, 133, 133f, 133t
    chaparral, 137, 137f, 137t, 138
    climate factors shaping, 129f
    desert, 138–139, 139f, 139t
    savanna, 140, 140f, 140t
    temperate deciduous forest, 135, 135f,
      135t
    temperate grassland, 136, 136f, 136t
    temperate rain forest, 134, 134f, 134t
    terrestrial, 128f
    tropical rain forest, 140–142, 141f, 141t
    tundra, 132, 132f, 132t
    vegetation types, 130f–131f
Biosphere, 99
Biotic environment, 98
Biotic potentials, 161
Birmensdorf, Switzerland, 198f
Birth rates, 165, 170, 171f, 172
Black Forest, Germany, 235, 236f
Blast furnaces, 305f
Bleaching, 42f, 289, 289f
Bluefin tuna, 272–273, 273t
Bolivia
    Eduardo Avaroa Andean Fauna
      National Reserve, 50f
    Madidi National Park, 342f
Bonneville Dam, Columbia River, 255f
Boreal forests, 133, 133f, 133t, 329, 329f
Borneo, Malaysia, 331f
Boston, Massachusetts, 450
Botswana, 119f
Bottom ash, 402
Brackish ecosystems, 146–147, 146f

Brandt, Willy, 38
Brazil
    Amazonas State, 167f
    Amazon Basin, 342f
    Ariquemes, 227f
    Atlantic forest, 32f
    Curitiba, 186f
    deforestation, 331f
    deforestation in, 330
    ethanol consumption, 451
    population and population density of,
      170t
    poverty in, 5f
    Roraima State, 40f
Bridalveil Fall, Yosemite National Park,
    323f
British Columbia, Canada, 13f
Broad-spectrum pesticides, 364–365
Brown, Lester R., 36
Brownfields, 182, 182f, 183f
Built-up areas, 131f
Burma (Myanmar), 392f
Bush, George W., 292
Bycatch, 286

C

Cabeza Prieta Wilderness, Arizona, 339f
Cabin Creek, West Virginia, 422f
Cades Cove, Great Smoky Mountains
    National Park, Tennessee, 337f
Cahn, Robert, 34
California
    Bayview Hunters Point, San Francisco,
      35
    chaparral, 137f
    Culver City, 137t
    goats and wildfire prevention, 138
    Hetch Hetchy Valley, Yosemite
      National Park, 53f
    Joshua Tree National Park, 60f
    Kern County, 428
    Los Angeles, 204, 204f
    Sierra Nevada Mountains, 372,
      372f–373f
    solar thermal electric generation, 448f
    Yosemite National Park, 323f
Cambodia, 38f
Cameroon, 37f
Canada
    boreal forest, 329, 329f
    British Columbia, 13f
    Fort Smith, 133t
    Ontario, 68f, 77, 247t
    Saskatchewan, 133f
    Yukon, 190
Cancer, 76
    lung, 76, 76f, 88f, 210
    malignant melanoma, 233
Cancer-causing substances, 86–87
Cap and trade, 67, 68
Captive breeding, 389f

Carbon cycle, 106–107, 107f, 122
Carbon dioxide, 197
    in the atmosphere, 223t
    country emission estimates, 238, 238t
    effects of increase in, 42, 42f
Carbon management, 229
Carbon monoxide, 197
Carbon oxides, 197
Carcinogen, 86
Carnivores, 102, 102f
Carrying capacity, 39, 163
    and population growth, 162–164, 163f,
      163t, 167
    of rangeland, 333, 334
Carson, Rachel, 26, 54–55, 55f
Cascade Range, Washington State, 340f
Catskill Mountains, New York, 260f
Cellular respiration, 101
Certified organic, 368
Chad, 251
Chandi Chowk bazaar (Delhi, India), 420f
Chaparral, 137, 137f, 137t, 138
Chattanooga, Tennessee, 212, 212f
Chemical accidents, 412
Chernobyl, Ukraine, 434, 435f
Chiapas, Mexico, 234f
Chicago, Illinois, 8, 8f
Children
    child labor, 38f, 176f
    pollution sensitivity of, 89
    value on gender of, 176, 177
China
    acid deposition in, 234
    air pollution in, 208–209, 208f
    aquaculture in, 41
    biogas digesters in, 450
    endangered species in, 3t
    environmental education in, 58f
    high-sulfur coal in, 234f
    hydropower in, 454
    Jiangxi, 81f
    Lianoning Province, 208f
    Lingwu, 257f
    Liuzhou, Guangxi, 268f
    Loess Plateau, 44, 44f
    Manchuria, 190
    Mount Hua, 203, 203f
    population and population density of,
      170t
    rare earth metals in, 304
    renewable energy in, 48t
    sandstorm in, 194f
    scrap imports by, 404
    Shanxi Province, 421f
    Sichuan Province landslide, 300
    Three Gorges Dam, 269, 269f, 454f, 455
    water pollution in, 268, 268f
    Wolong Nature Reserve, 343f
    Yellow River, 251
    Yongchuan, Chongqing, 118f
Chlorofluorocarbons (CFCs), 231, 233
Chronic bronchitis, 201

Chronic toxicity, 77
Circulation, ocean, 274–276, 275f
Cities. *See also* Urbanization
    dust domes over, 205, 205f
    as dynamic ecosystems, 181, 181f
    largest, 184, 185f
    reducing water waste in, 256
    sustainable, 43
Civilian Conservation Corps, 54
Clallam Bay, Sekiu, Washington, 99f
Clean Air Act (CAA, 1970), 207–208
Clean Air Act Amendments (1990), 67,
        68, 198, 207, 423
Clean Water Act, 266–267
Clean Water Act amendments (1987),
        266–267
Clear-cutting, 326, 327f, 344
Cleveland, Grover, 52
Cleveland, Ohio, 61, 64f
Climate(s), 129, 218–221
    biomes shaped by, 129f
    factors determining, 218
    and volcanic eruptions, 199f
Climate Adaptation and Mitigation
        e-Learning portal, 58
Climate change, 222–230. *See also* Global
        warming
    adaptation to, 229–230
    and arctic tundra changes, 132
    and carbon cycle, 122
    causes of, 223–225
    effects of, 226–227, 228f
    government policies related to, 48
    human-caused, 216
    international implications of, 238,
        238t
    mitigation of, 41–43, 229
    and ocean temperatures, 290
    and ozone thinning, 231–233
    and precautionary principle, 91, 92
    and water problems, 249
Climate Commitment (American College
        and University Presidents), 58
Clinton, Bill, 35, 36
Closed shrublands, 130f
Coal, 418, 419t, 421–423, 421f
Coal-fired power plants, 235, 237
    cleaner, 422, 423
    and climate change, 122
    nuclear plants *vs.*, 433t
Coastal development, 288
Coevolution, 116, 116f
Coffee cultivation, 380
Cogeneration, 424, 460, 460f
Colorado
    Bear Lake, Rocky Mountain National
        Park, 143f
    Great Sand Dunes National Park, 320,
        320f–321f
Colorado River Compact (1922), 250
Colorado River Delta, Mexico, 250f
Coltan, 317, 317f

Columbia River
    Bonneville Dam, 255f
    dams, 255–256, 255f
    Grand Coulee Dam, 255f
    hydropower from, 454
Columbia River Basin, 250
Combined heat and power (CHP), 424,
        460, 460f
Combined sewer overflow, 262
Combined sewer systems, 262
Combustion, 106, 107
Command and control regulations, 67
Commensalism, 116, 117f
Commercial harvest, 385–386
Communities, 98, 99f
    succession in, 151–153
Compact development, 183, 186
Competition, 120–121
Composting, 403, 403f, 404
Comprehensive Environmental Response,
        Compensation, and Liability Act
        (CERCLA), 409, 412
Condenser (nuclear reactors), 432
Congaree National Park, South Carolina,
        320, 321f
Conservation, 50
    of land resources, 341
    in mid-20th century, 54–55
    policies and laws, 390–392
    soil, 314–316
    of species, 388–390, 389f
    water, 256, 257f
Conservation biology, 386–390
Conservation easements, 335, 355
Conservation Reserve Program (CRP),
        316
Conservation tillage, 41, 314, 315f
Consumers, 101–102, 102f
Consumption, 4
    in highly-developed countries, 10
    oil, 13t
    and population growth, 6–7, 9
    sustainable, 28–30
Contamination
    of groundwater, 263f
    from mining, 308
    reduced human exposures to, 61
Continental shelves, 425
Contour plowing, 314
Control group, 18
Convection
    atmospheric, 194f
    of building heat, 445
Convention on Biological Diversity, 392
Convention on International Trade in En-
        dangered Species of Wild Flora and
        Fauna (CITES), 392
Convergent plate boundaries, 299, 299f
Copper Basin, Tennessee, 296, 296f–297f,
        308
Corals, 280
Coral bleaching, 42f, 289f

Coral reefs, 280–281, 280f, 281f, 288–289
Coriolis effect, 195, 195f, 274
Cornell University, New York, 17f
Costa Rica, 117f, 129f
Cost–benefit diagram, 66–67, 67f
Costco Busan oil spill (San Francisco Bay,
        2007), 441, 441f
Council on Environmental Quality, 59
Cowles, Henry, 152–153
Crater Lake, Oregon, 260f
Crops
    decline in varieties, 356
    genetically modified, 357f, 363, 363f
    genetic engineering of, 362–363, 362f
    increasing yields, 356–357, 357f
    world production of, 363f
Cropland, 131f
Crop rotation, 314
Crude oil, 423
Cullet, 406
Cultural diversity, 40
Culture, fertility rates and, 175–177
Culver City, California, 137t
Curitiba, Brazil, 186, 186f
Custer State Park, South Dakota, 136f
Cuyahoga River, Cleveland, Ohio, 61

D

Dams, 255–256, 255f, 454–455, 454f, 455f
Darwin, Charles, 147, 147f
Data, 17
DDT (dichlorodiphenyltrichloroethane),
        81–84, 82f, 84f, 366
Dead zones, 110, 293
Death, causes of, 74f
Death rate, 165, 166
Deciduous broad-leaved forest, 130f, 135f
Deciduous needle-leaved forest, 130f
Decommissioning (nuclear power plants),
        438
Decomposers, 102, 102f
Deep ecology worldview, 32–34
Deep geothermal energy, 456, 457f
Deepwater Horizon oil spill, 428, 429f, 430
Deep-well injection, 414
Deforestation, 39, 328–331, 331f
    and biodiversity decline, 380
    boreal forests, 329
    and climate change, 122
    in early United States, 51
    and evolution, 380
    as global issue, 14f
    results of, 329
    tropical forests, 330
Delhi, India, 420f
Dematerialization, 405
Democratic Republic of the Congo
        (DRC), 317
Demographics, 170–174, 171f
Demographic transition, 171f, 172
Denitrification, 110

Density (ocean water), 274
Derelict lands, 308
Desalinization, 254
Desert, 129f, 131f, 138–139, 139f, 139t
Desertification, 14f, 334–335, 334f
Detritus, 102, 132
Deval, Bill, 33
Dhaka, Bangladesh, 80f
Differential reproductive success, 148
Dimakay Island, Philippines, 148f
Dinosaur National Monument, 53
Dioxins, 87, 87f, 410–411, 410f
Direct solar energy, 444–449
    active solar heating, 444–445, 444f
    passive solar heating, 445, 445f
    photovoltaic solar cells, 445–447, 446f,
        447f
    solar-generated hydrogen, 448–449,
        449f
    solar thermal electric generation,
        447–448, 448f
Disease(s). See also specific diseases, e.g.:
        Cancer
    emerging, 78–81
    and global climate change, 227, 227f
    pandemics, 80, 81, 278
    waterborne, 455
Disease-causing agents, 77–78, 78t
Dispersal, 161, 174–175
Divergent plate boundaries, 298, 299, 299f
Dose, 85
Dose–response curve, 86, 86t
Drainage basin, 245
Drinking water
    improvement in, 61
    purification of, 264–265, 264f
    waste from bottled water, 398f
Drip irrigation, 256
Dust dome, 205, 205f

E

Earth, layers and surface structure of, 298,
    298f
Earth Charter, 26
Earth Day, 55–56, 56f
Earthquakes, 300, 300f, 301
Eco-justice, 35. See also Environmental
        justice
Ecological footprint, 10
Ecologically certified wood, 328
Ecologically sustainable forest manage-
    ment, 325–326
Ecological niches, 113–115
Ecological overshoot, 10, 11t
Ecological pyramids, 105, 105f
Ecological succession, 151–153, 152f
Ecology, 16, 98
    landscape, 98–99
    population, 160–164
    restoration, 387, 388f

Ecology, Community and Lifestyle (Arne
        Naess), 33
Economics
    defined, 62
    environmental, 62–68
Economic development, 352
Economic growth, 4
Eco-roofs, 8
Ecosystems, 96–124
    aquatic, 142–147
    biogeochemical cycles, 106–113
    ecological niches, 113–115
    and ecology, 98–99
    energy flow through, 100–106
    interactions among organisms, 116–122
    most-endangered, 341
    protected, 342f–343f
    at risk of climate-change loss, 227, 228f
    worldwide protected ecosystems,
        342f–343f
Ecosystem diversity, 375, 375f
Ecosystem services, 145, 376, 376f
    of the atmosphere, 192
    biodiversity and, 39
    from government-owned lands and
        rural areas, 322–323
    soil organisms and, 311–312
    and species richness, 376, 376f, 376t
    wetlands and, 145
Ecotone, 374
Ecuador, 3t, 349f
E-cycling, 415
Edge effect, 375
Edison, Thomas, 30
Eduardo Avaroa Andean Fauna National
        Reserve, Bolivia, 50f
Education, fertility rates and, 178, 178f,
        179f
Effective dose-50 percent (ED_{50}), 86
Egypt
    Aswan High Dam, 455, 455f
    family planning in, 37f, 179f
    Sinnuris, 179f
Ehrlich, Paul R., 10, 55
El Capitan, Yosemite National Park, 323f
Electrostatic precipitators, 206, 206f
El Niño–Southern Oscillation (ENSO),
        276–277, 276f, 290
El Yunque National Forest, Puerto Rico,
        141f
Emerging diseases, 78–81
Emigration, 161
Emphysema, 201
Endangered ecosystems, 341
Endangered species, 3t, 378, 379f, 390f,
        391f
    captive breeding of, 389f
    human causes of, 382, 384–386
    protecting, 393, 393f
    recovery of, 61
    reintroducing to nature, 388
    seed banks, 388, 389f, 390

    trade in products made from, 392f
    in the U.S., 390t
Endangered Species Act (1973), 390, 391
Endemic species, 379
Endocrine disrupters, 92
Energy, 100, 458–459, 462
    alternative sources of, 48. See also
        Renewable energy resources
    and climate change, 42–43
    commercial sources of, 423f
    conservation of, 458–459, 462
    consumption of, 420, 420f, 421f,
        426f–427f
    efficiency, energy, 458–462
    global supply of, 426f–427f
    light bulb use of, 30f
    nonrenewable sources of,
    potential and kinetic, 100, 100f
Energy conservation, 458–459, 462, 462f,
        463
Energy efficiency, 458–462, 459f
Energy flow, 100–106, 103f
England, 20f
Enhanced greenhouse effect, 41, 224,
        224f
Enrichment
    uranium, 431
    water, 258
Entombment (nuclear power plants), 438
Entropy, 101
Environmental challenges, 2
    global, 14f–15f
    handling, 20–23, 21f
Environmental chemistry, 16
Environmental economics, 62–68
Environmental education, 58
Environmental ethics, 31
Environmental health hazards, 72–94
    air pollution, 201
    determining health effects of
        pollutants, 85–89
    disease-causing agents, 77–78
    emerging diseases with environmental
        change, 78–81
    movement and persistence of
        toxicants, 81–83
    from ozone depletion, 231, 233
    precautionary principle, 90–92
    risk management with, 74–76
Environmental history, 48, 51–57
    environmental movement, 55–56
    forest protection, 51–52
    mid-20th century conservation, 54–55
    national parks and monuments,
        52–54
    timeline of, 57
Environmental impacts
    of coal, 422
    of industrialized agriculture, 358
    of oil and natural gas, 428–430
Environmental impact statements (EISs),
        59–60, 59f

Environmentalists, 55
Environmental justice, 35–36
Environmental legislation, 52–54, 59–61
Environmental movement, 55–56, 340
Environmental Protection Agency (EPA),
    60, 207, 208
Environmental regulations, 60
Environmental resistance, 162–164
Environmental science, 16–20
    goals of, 16–17
    as a process, 17–20
Environmental taxes, 67
Environmental Working Group, 84
Environmental worldview, 32
Epicenter, 300
Epidemiology, 77
Epiphytes, 116
Equatorial uplift, 220
Erosion, soil, 313, 313f
Estracada, Oregon, 134t
Estuaries, 146, 146f
    as brackish ecosystems, 146–147, 146f
    and overdrawing of surface waters,
        249
    tidal power problems, 457
Ethanol, 450, 451
Ethics, 31, 36
Euphotic zone, 282
Europe
    ethanol consumption, 451
    GM crop opposition, 363
European Union
    automobile recycling, 396
    GM crop moratorium, 363
    precautionary principle in, 91
Eutrophication, 143, 144, 260, 261
Eutrophic lakes, 260, 260f
Everglades, Florida, 126, 126f–127f, 127t
Evergreen broad-leaved forest, 130f
Evergreen needle-leaved forest, 130f
Evolution, 147–150, 150f, 377
Expansionist worldview, 32
Experimental group, 18
Exponential population growth, 162,
    162f, 162t
Ex situ conservation, 386
External costs, 65, 306
Extinction, 378, 379f
Exxon Valdez oil spill, 428, 429f, 430
Eyjafjallajokull Volcano, Iceland, 199f

F

Falls Church, Virginia, 31f
Family planning, 37f, 39, 178–180, 179f
Faults, 300
Fecal coliform test, 78, 79f
Federal Land Policy and Management Act
    (1976), 335
Federal lands, 322f
    management of, 322–323, 323t, 340

national parks, 336–338
wilderness areas, 338–339
Fertility rates, 172t, 174
    and culture, 175–177
    and education, 178, 178f, 179f
    and government policies, 180
    replacement-level, 170
Fibrous root systems, plants with, 333
Fiji, 280f
First law of thermodynamics, 100
Fisheries, 284, 286–287
Fishing, 272
    commercial fishing methods, 286f
    entanglement in gear, 15f
    overfishing, 272
Fish ladders, 255f
Fission, 431–432, 431f
Flooding, 246–247
Floodplain, 246–247
Florida
    Everglades, 126, 126f–127f, 127t
    green/brown anoles, 114, 114f
    Lake Apopka, 92f
    Orlando Wetlands Park, 266f
    Pelican Island National Wildlife
        Refuge, 339f
Flowing-water ecosystems, 144, 144f
Fluidized-bed combustion, 423
Fly ash, 402, 419t
Focus (earthquakes), 300
Food chains, 103, 103f, 105
Food insecurity, 40–41, 350
Food Quality Protection Act amendments
    (1996), 92
Food resources, 350–352. *See also*
        Agriculture
    decline in crop varieties, 356
    genetically modified, 357f, 363, 363f
    increasing yields, 356–357, 357f
    world grain stocks, 348, 348f–349f
Food security, 40–41, 348
Food Security Act (Farm Bill, 1985), 316
Food webs, 103, 104f, 113
Forbs, 333
Ford Motor Company, 8
Forest decline, 235, 236f
Forest management, 325–327
Forest Reserve Act (1891), 52
Forests, 324–332. *See also* Deforestation
    boreal, 133, 133t
    and climate change, 229
    in hydrologic cycle, 324f
    management of, 325–327
    protecting and restoring, 39, 51–52
    temperate deciduous, 135, 135t
    temperate rain forest, 134, 134t
    threats to, 131f
    tropical rain forest, 140–142, 141t
    in the U.S., 332
Forest Stewardship Council (FSC), 328
Fort Smith, Northwest Territories,
        Canada, 133t

Fort Yukon, Alaska, 132t
Fossil fuels, 106, 107, 111
    air pollutants from, 200
    greenhouse gases from, 229
    and nuclear energy, 432, 434
    reducing reliance on, 442
    from solar energy, 219
Fossil record, 150f
Foundation for International Community
        Assistance (FINCA), 180
Fracking (hydraulic fracturing), 428
France, 11t
Freshwater ecosystems, 142–145,
        143f–145f, 230
Freshwater resources, 242–270
    aquifer depletion, 248–249, 249f
    global water issues, 251–254,
        252f–254f
    importance of, 244
    improving water quality, 264–269
    limited supply of, 242
    management of, 254–257
    overdrawing surface waters, 249–250
    pollution of, 258–263
    problems with, 246–253
    salinization of irrigated soil, 250
    shared, 251, 252f–253f, 254
Freshwater swamps, 145f
Freshwater wetlands, 145
Frog die-offs, 372, 372f–373f, 373t
Frontier attitude, 32, 51
Fuel assemblies (nuclear reactors), 432
Fuel cells, 449, 449f
Fuel consumption, climate change and,
        122
Fuel efficiency, 459–460, 460f
Fuel rods, 432
Fukushima Daiichi nuclear disaster
        (Japan, 2011), 435, 435f
Fundamental niche, 113
Fungicides, 364, 364f
Fusion, 431

G

Galápagos Islands, 102f, 149f
Galápagos National Park, Ecuador,
        342f
Game Management (Leopold), 54
Ganges River, India, 268, 268f
Gause, G. F., 163
Gender roles
    and status of women, 177–178, 177f
    and total fertility rate, 175, 175f
Gene banks, 388, 390
General Circulation Models (GCMs), 226
Genetically modified (GM) crops, 357f,
        362f, 363, 363f
Genetic diversity, 375–377, 375f
Genetic engineering, 362–363, 362f, 377
Genetic resistance, 365, 365t
Genuine progress indicator (GPI), 63

Georgetown University Intercultural Center, Washington, D.C., 446f
Geosciences, 16
Geothermal energy, 456, 457f
   deep, 456, 457f
   shallow, 456, 457
Germany
   Black Forest, 235, 236f
   Messel Pit, 150f
   renewable energy in, 48, 48t
Germany Valley, West Virginia, 135f
Germplasm, 356
Gifford Pinchot National Forest, Cascade Range, Washington, 340f
Glaciers, 216, 217t
Glass recycling, 405, 406
Global distillation effect, 190
Global environmental issues, 14f–15f
Global warming, 14f, 15f, 216. See also Climate change
Global water issues, 251–254, 252f–254f
Goats, 138, 138f
Gold, 307
Goodall, Jane, 58f
Gotland Island, Sweden, 34f
Government-owned lands, 322. See also Federal lands
Grand Banks, Atlantic Ocean, 62f
Grand Canyon National Park, Arizona, 337f
Grand Coulee Dam, Columbia River, Washington State, 255f
Grangemouth, United Kingdom, 13f
Grasslands, 131f
   rangelands, 333–335, 333f
   savanna, 140, 140f, 140t
   temperate, 136, 136f, 136t
   wooded, 130f
   Gray water, 256
Great Sand Dunes National Park, Colorado, 320, 320f–321f
Great Smoky Mountains National Park, 65f, 337f
Green architecture, 457, 463
Greenbelt Movement (Kenya), 56
Green chemistry, 413
"Green" forestry, 328
Greenhouse effect, enhanced, 41, 224
Greenhouse gases, 223, 224, 224f, 225f
   and climate warming, 233
   dealing with, 229
Green revolution, 356–357, 361
Green roofs, 8, 8f
Gross domestic product (GDP), 63
Gross primary productivity (GPP), 105
Groundwater, 245, 246f
Groundwater pollution, 263, 263f
Groundwater recharge, 245
Growth rate, 161
Guanajuato, Mexico, 251
Gulf of California, Mexico, 250f

Gulf of Mexico, 293, 293f, 293t, 428–430, 429f
Gulf Stream, 275, 275f
Gunung Palung National Park, Indonesia, 343f
Guwahati, India, 331f
Gyres, 274

H

Habitat corridors, 326
Habitat fragmentation, 358, 382
Habitats, 113
   bird, 380
   destruction/fragmentation/degradation of, 382
   marine, 284
   protecting, 386–387
   restoring, 387–388
Haeckel, Ernst, 98
Haiti earthquake (2010), 300
Hanging Rock State Park, North Carolina, 101f
Harrison, Benjamin, 52
Hawaii, 218f, 342f
Hawaiian Islands, formation of, 299
Hawaii Volcanoes National Park, Hawaii, 342f
Hayes, Denis, 55
Hazardous air pollutants (HAPs), 198
Hazardous waste, 409–415
   cleanup of, 413f
   high-tech, 415
   managing, 412–414
   production of, 413–414
   types of, 409–411
Health hazards. See Environmental health hazards
Hearst Tower, Manhattan, 463, 463f
Herbicides, 364
Herbivores, 102, 102f
Heritable variation, 148
Hetch Hetchy Valley, California, 53, 53f
Hibernia oil platform, Grand Banks, Atlantic Ocean, 62f
High-grade ores, 302
High-input agriculture, 353. See also Industrialized agriculture
High-level radioactive wastes, 436
Highly developed countries, 4, 6
   birth and infant mortality rates in, 170, 171f
   consumption in, 9
   population stabilization in, 172
High-tech waste, 415
Hill, Julia "Butterfly," 346, 346f
Hindman, Kentucky, 308
Hoh Rain Forest, Washington State, 134f
Holdren, John P., 10
Honduras, 89f
Hormones, 92, 358

Hotspots
   biodiversity, 380, 381f
   volcanic, 299
Houses
   active solar heating, 444–445, 444f
   energy conservation for, 462, 462f
   net metering of electricity, 461, 461f
   passive solar heating, 445f
   preventing water pollution in, 267
   superinsulated, 459f
   thermal images of, 442, 442f–443f
Houston, Texas, 35
Hubbard Brook Experimental Forest, New Hampshire, 237
Human activities
   and climate change, 122, 216
   environmental health hazards from, 78–80
   and extinctions, 378
   and greenhouse gas accumulation, 223, 224
   and nitrogen cycle, 110
   and phosphorus cycle, 113
   and tropical rain forest destruction, 142
Humanely raised and handled animals, 368
Human impacts, 2, 4–11
   IPAT equation assessment of, 10–11
   on natural flow of water, 247, 247f
   on oceans, 284–290
   population growth and resource consumption, 6–7, 9–11
   rich–poor country gap, 4–6
   on rivers and streams, 144
   on wildfires, 154
Human population patterns, 165–167, 168f–169f, 169t. See also Population growth
Humus, 310, 310f
Hunting, 385
Hydraulic fracturing (fracking), 428
Hydrocarbons, 197, 198
Hydrogen, solar-generated, 448–449, 449f
Hydrogen bonds, 244, 245
Hydrologic cycle, 107–109, 108f, 245, 245f, 246f, 324f
Hydropower, 454–455, 454f, 455f
Hydrosphere, 99
Hydrothemal reservoirs, 456
Hypotheses, 18
Hypoxia, 293

I

Iceland
   Eyjafjallajokull volcano, 199f
   geothermal power plant, 457f
   as volcanic island, 299
Igneous rocks, 301, 301f
Illinois, 8, 8f
Immigration, 161
Incentive-based regulations, 67

Incineration
    hazardous waste, 413
    solid waste, 402–403, 402f, 403f
India
    biogas digesters in, 450, 451f
    Delhi, 420f
    ecological footprint of, 11t
    endangered species in, 3t
    Ganges River, 268, 268f
    Guwahati, 331f
    Kolkata (Calcutta), 29f
    population and population density of,
      170t
    population pressures in, 158, 158t
    water resources in, 244f
Indian Ocean, 300
Indirect solar energy, 450–455
    biomass energy, 450–451, 450f, 451f
    hydropower, 454–455, 454f, 455f
    wind energy, 452, 453f
Indonesia
    endangered species in, 3t
    Gunung Palung National Park, 343f
    mangrove restoration, 37f
    population and population density of,
      170t
Indoor air pollution, 209–211
Industrial ecosystem, 212
Industrial facilities
    air pollutants from, 200
    reducing water waste in, 256, 257f
Industrialized agriculture, 353
    energy inputs in, 353f
    environmental impacts of, 358, 359f
Industrial smog, 201
Infant mortality rates, 170, 171f
Influenza pandemics, 80, 81, 278
Infrared radiation, 220
Inherent safety, principle of, 412
Insecticides, 364
In situ conservation, 386
Insolation, 219–220, 219f, 220f
Instrumental value, 33f
Integrated pest management (IPM), 361,
    366, 367, 367f, 367t
Integrated waste management, 408, 408f
Intercropping, 354
Interspecific competition, 120
Intertidal zone, 278, 279f
Intraspecific competition, 120
Intrinsic rates of increase, 161
Intrinsic value, 33, 33f
Invasive species, 338, 384–385, 384f
IPAT equation, 10–11
Irrigation, 247, 248
    reducing water waste, 256
    and soil salinization, 250, 313, 313f,
      358
    water withdrawals for, 253f
Israel, 171f, 257f

J
Japan
    automobile recycling, 396
    earthquake and tsunami (2011), 300,
      301
    Fukushima Daiichi nuclear disaster
      (2011), 435, 435f
    population and population density of,
      170t
    renewable energy in, 48t
    Tokyo, 13f
    Tokyo-Yokohama-Osaka-Kobe agglom-
      eration, 184
    typical family from, 6f
Jiangxi, China, 81f
Joshua Tree National Park, California, 60f

K
Kalahari Desert, South Africa, 120f
Kamchatka, Russia, 343f
Kansas
    center-pivot irrigation, 248f
    Lawrence, 136t
    prairie soil, 309f
    wind farm, 453f
Kelps, 281, 281f
Kentucky, 308
Kenya
    Greenbelt Movement, 56
    Nairobi, 185f
    nomadic sheep herders, 354f
    West Pokot, 41f
Kern County, California, 428
Keystone species, 121, 121f, 122
Kilimanjaro National Park, Tanzania,
    140f
Kinetic energy, 100, 100f
Kingsport, Tennessee, 35f
Kingston, Tennessee, 418, 418f–419f
Kolkata (Calcutta), India, 29f
Kouakourou, Mali, 7f
Kuwait, 430
Kyoto Protocol, 238

L
Lagos, Nigeria, 242, 242f–243f
Laguna Colorada, Bolivia, 50f
Lakes, 142–144, 143f
    acidic, 237
    eutrophic, 260–261, 260f
    oligotrophic, 260, 260f
    UV radiation damage to, 233
Lake Apopka, Florida, 92f
Lake Michigan, 152
Lake Washington, Washington State,
    96–97, 96f, 97t
Land degradation, 334, 358
Landfills
    hazardous waste, 414, 414f
    sanitary, 400–403, 400f–402f

Land resources, 320–346
    clear-cutting, 344
    conservation of, 341
    forests, 324–332
    land use in the U.S., 322–323
    management of federal lands, 340
    national parks, 336–338
    protected ecosystems, 342f–343f
    rangelands, 333–335
    wilderness areas, 338–339
Landscape, 98
Landscape ecology, 98–99
Landslide, 300
Land use (U.S.), 322–323
La Niña, 277–278, 290
Las Conchas fire (New Mexico, 2011),
    154f
Latin America, 175f
Latitude
    and precipitation, 221
    and seasons, 220
    and solar intensity, 220f
Lava, 299
Lawrence, Kansas, 136t
Leachate, 400
Leaching, 311
Lead pollution, 209
Leopold, Aldo, 54, 54f
Less developed countries (LDCs), 6, 170,
    171f
Lethal dose, 85
Lethal dose-50 percent (LD_{50}), 85–86,
    85t
Lianoning Province, China, 208f
Life-cycle assessment, 22
Life history characteristics, 161
Life zones (oceans), 278–283
Light bulbs, 30, 30f
Limnetic zone, 142, 143, 143f
Lincoln High School, Nebraska, 179f
Lingwu, China, 257f
Liquefied petroleum gas, 424
Literacy, environmental, 58
Lithosphere, 99, 298
Littoral zone, 142, 143f
Liuzhou, Guangxi, China, 268f
Locusts, 361, 361f
Loess Plateau, China, 44, 44f
Los Angeles, California, 204, 204f
Louisiana
    Atchafalaya National Wildlife Refuge,
      145f
    New Orleans, 23f
Love Canal, New York State, 409, 409f
Low-grade ores, 302
Low-input agriculture, 360. *See also*
    Sustainable agriculture
Low-level radioactive wastes, 436
Lung cancer, 76, 76f, 88f, 210
Lusaka, Zambia, 140t

M

Maathai, Wangari, 56, 56f
MacArthur, Robert, 115
McKinley, William, 52
Madagascar, 381f
Madidi National Park, Bolivia, 342f
Magma, 299, 301f
Magnuson Fishery Conservation Act, 291
Magnuson-Stevens Fishery Conservation and Management Act, 291
Malaria, 227
Malaysia, 331f
Maldives, 226f, 228f
Mali, 7f, 334f
Malignant melanoma, 233
Malthus, Thomas, 165
Man and Nature (Marsh), 52
Manchuria, China, 190
Manganese nodules, 289, 290
Mangrove forests, 146–147, 146f
Manhattan, New York City, 463, 463f
Manila, Philippines, 43f
Marginal cost, 66
Marginal cost of pollution, 66, 66f
Marginal cost of pollution abatement, 66, 66f
Mariculture, 287
Marine pollution, 284, 287–289
Marine snow, 283
Marsh, George Perkins, 52
Marshes
    freshwater, 142, 145
    salt, 146, 146f, 147
Massachusetts, 450
Mass burn incinerators, 402–403, 403f
Mass extinction, 378
Matter, cycles of. *See* Biogeochemical cycles
Mauritania, 251
Maximum containment level, 266
Maximum population growth, 161, 162
Mayan Palace (Palenque, Chiapas, Mexico), 234f
Mediterranean climates, 137
Megacities, 184
Meltdown, 434
Merbold, Ulf, 192
Mesosphere, 193f
Messel Pit, Germany, 150f
Metals, 302, 405, 406
Metamorphic rocks, 301, 301f
Methanol, 450
Mexico
    Colorado River Delta, 250f
    drought, 251
    Mayan Palace at Palenque (Chiapas), 234f
    sustainable forestry in, 326
    typical family from, 6f
    Yaqui Indian pesticide exposure, 72
Mexico City, Mexico, 43, 251
Microcredit, 180

Microirrigation, 256, 257f
Micronesia, 146f
Minerals, 301–308, 303f
    "conflict," 317
    economic impact of mining, 302, 304
    environmental implications of mining, 306–308
    extraction of, 304–305, 305f
    nonfuel, 298t
    offshore extraction of, 289, 290
    processing of, 305, 305f
    in soil, 309
    in soil nutrient cycling, 312
Mining, 304–305, 305f
    coal, 422
    economic impact of, 302, 304
    environmental impacts of, 306–308
    types of, 304f
Minneapolis, Minnesota, 413f
Minnesota
    Minneapolis, 413f
    Taylors Falls, 32f
Mitigation
    of climate change, 41–43, 229
    defined, 229
Mixed forest, 130f
Moderately developed countries (MDCs), 6, 170
Modern synthesis, 148
Modular incinerators, 403
Molecular biology, 151f
Moment magnitude scale, 300
Monocultures, 325, 354, 364
Montreal Protocol, 233
Mosquitos, 80
Mountaintop removal (coal), 422
Mount Hua, China, 203, 203f
Mount Mitchell, North Carolina, 423f
Mount Pinatubo, Philippines, 199f, 225f, 299
Muir Glacier, Alaska, 216, 216f–217f
Muir, John, 52–53, 52f
Multicropping, 40, 41
Municipal sewage treatment, 265–266, 265f
Municipal solid waste, 398, 399, 399f, 400f
    composition of, 399f
    disposal of, 400f
    reuse/recycling/recovery of, 61
Municipal water supplies
    drinking water, 264–265, 264f
    reducing waste in, 256
Murmansk, Russia, 438, 438f
Mutation, 148, 150
Mutualism, 116, 117f
Myanmar (Burma), 392f
Myers, Norman, 380

N

Naess, Arne, 32, 33
Nairobi, Kenya, 185f

Nanomaterials, 410, 410f
Nanotechnology, 410
Na Pali Coast, Kauai, Hawaii, 218f
Narrow-spectrum pesticides, 364
Nashville, Tennessee, 135t
National Appliance Energy Conservation Act (NAECA), 459
National conservation strategy, 392
National Council for Science and the Environment, 58
National emission limitations, 267
National Environmental Education Act (1990), 58
National Environmental Policy Act (NEPA, 1970), 59, 60
National forests, 332
National income accounts, 63–65
National marine sanctuaries, 292, 292f
National monuments, 52–54, 60
National Oceanic and Atmospheric Administration (NOAA), 292
National parks, 52–54, 60, 320, 322f, 336–338
National Park Service (NPS), 53, 336
National Wilderness Preservation System (NWPS), 60, 61, 338, 339, 339f, 387
National Wildlife Refuge System, 387
Natural capital, 62, 63, 63f, 302
Natural gas, 423–430
    environmental impacts of, 428–430
    global supply of, 426f–427f, 428
    reserves of, 425–428
Natural increase, 161
Natural selection, 148–150
Nebraska, 179f
Negev Desert, Israel, 257f
Nekton, 142, 282
Nelson, Gaylord, 55
Neritic province, 278, 279f, 281, 282f
Net domestic product (NDP), 63
Netherlands, 37f
Net primary productivity (NPP), 105, 105t
Nevada
    Reno, 139t
    Yucca Mountain, 22, 437, 437f
New Hampshire, 237
New Haven, Connecticut, 442, 442f–443f
New Mexico
    Las Conchas fire, 154f
    Santa Fe, 445f
    thunderstorm in, 193f
New Orleans, Louisiana, 23, 23f
New York City
    Battery Park City, 37f
    Hearst Tower, 463, 463f
    South Bronx, 90f
    Times Square, 9f
New York State
    Adirondack Mountains, 237, 237f
    Catskill Mountains, 260f
    Cornell University, 17f
    Love Canal, 409, 409f

New Zealand
  North Island, 316f
  ozone thinning over, 232t
  protected lands in, 343f
Niches, 113–115
Nigeria
  DDT-treated mosquito netting, 84f
  gender discrimination in education,
    177f
  Lagos, 242, 242f–243f
  oil extraction and health in, 36
  population and population density of,
    170t
Nile River, Egypt, 455, 455f
NIMBY ("not in my back yard") response,
    22
Nitrification, 110
Nitrogen cycle, 109–110, 109f
Nitrogen fixation, 110, 110f
Nitrogen oxides, 197, 235
Nomadic herding, 354, 354f
Nonmetallic minerals, 302
Nonmunicipal solid waste, 399, 399t
Nonpoint source pollution, 261, 261f, 262
Nonrenewable energy resources, 418–440.
    See also Fossil fuels
  coal, 122, 235, 237, 418, 419t, 421–423
  natural gas, 423–430
  nuclear, 22f, 430–438
  oil, 13t, 290, 423–430, 439
Nonrenewable resources, 7
North America
  Mount Mitchell, 423f
  satellite view at night, 5f
  tidal power in, 457
North American Association for Environ-
    mental Education, 58
North Carolina
  Great Smoky Mountain National Park,
    65f
  Hanging Rock State Park, 101f
  Okracoke Harbor, Outer Banks, 146f
North Island, New Zealand, 316f
No-tillage, 314
Nuclear energy, 430–438, 430f
  conventional nuclear fission, 431–432,
    431f, 432f
  decommissioning nuclear power
    plants, 438
  and fossil fuels, 432, 434
  and nuclear weapons, 435
  radioactive wastes from, 436–438
  safety of nuclear power plants,
    434–435
Nuclear fission, 431–432, 431f, 432f
Nuclear fusion, 431
Nuclear power plants, 22f
  coal-fired plants *vs.*, 433t
  decommissioning, 438
  safety of, 434–435
Nuclear reactors, 432, 433f
Nuclear Waste Policy Act (1982), 436–437

Nuclear weapons, 435
Nutrient cycling, 312, 312f

O

Obama, Barack, and administration, 339,
    437
Ocean(s), 272–294
  acidifying, 15f
  addressing problems of, 291–292
  and biogeochemical cycles, 108, 108f,
    109, 109f, 111–113, 111f, 112f
  and climate change, 227, 228f
  Coriolis effect, 195f
  dead zones, 293
  human impacts on, 284–290
  major life zones in, 278–283
  ocean–atmosphere interaction,
    276–278
  patterns of circulation in, 274–276,
    275f
  threatened, 14f
Ocean–atmosphere interaction, 276–278
Ocean conveyor belt, 274–276, 275f
Ocean Dumping Ban Act, 288
Oceanic province, 278, 279f, 283, 283f
Ocean meridional circulation, 274
Ocean ranches, 287, 287f
Ogallala Aquifer, 249, 249f
Ohio
  Cleveland, 61, 64f
  Cuyahoga River fire, 61
Ohio Valley, 237
Oil, 423–430
  consumption of, 13t
  drilling on wildlife refuges, 439
  environmental impacts of, 428–430
  global supply of, 426f–427f, 428
  offshore extraction of, 290, 290f
  reserves of, 425–428, 426f–427f
Oil pollution, 15f
Oil Pollution Act (1990), 428
Oil spills, 428–430, 429f, 441, 441f
Okavango Delta, Botswana, 119f
Okracoke Harbor, Outer Banks, North
    Carolina, 146f
Oligotrophic lakes, 260, 260f
Ontario, Canada
  acid-rain-damaged buildings, 68f
  E. coli outbreak in, 77
  runoff before and after urbanization,
    247t
Open dumps, 400
Open management (fishing), 286–287
Open-pit surface mining, 304, 304f
Open shrublands, 131f
Optimum amount of pollution, 67
Ore, 302
Oregon
  Bonneville Dam, Columbia River, 255f
  Crater Lake, 260f
  Estracada, 134t

Portland, 184
  Proxy Falls, Cascade Range, 221f
Organic agriculture, 361, 368, 368f, 368t
Organic foods, 368
Organic Food Production Act (1990), 368
Organisms
  ecosystem interactions among,
    116–122
  soil, 311–312, 311f
  value/importance of, 377
*The Origin of Species by Means of Natural
    Selection* (Darwin), 147
Orlando Wetlands Park, Florida, 266f
Ottawa, Ontario, Canada, 68f
Otters, 282f
Our Common Future (U.N. World
    Commission on Environment and
    Development), 28
Overburden, 304
Overdrawing surface waters, 249–250
Overfishing, 272
Overgrazing, 334, 335
Overhunting, 385
Overnutrition, 350
Oxides, 302
Ozone, 198
  in California, 204t
  peak and average daily, 204t
Ozone layer, 193f
Ozone thinning, 15f, 231–233, 232f, 232t
  causes of, 231
  effects of, 231, 233
  reversing, 233

P

Pacala, Stephen, 122
Pakistan, 170t, 176f
Palau, Micronesia, 146f
Pandemic disease, 80
Parasitism, 116, 117f, 118
Parent material, 309
Particulate matter, 196–197, 212t
Passive solar heating, 445, 445f
Pathogens, 78, 364
Pelagic environment, 279f
Pelican Island National Wildlife Refuge,
    Florida, 339f
Pennsylvania
  Three Mile Island nuclear disaster,
    434
  Washington's Landing, Herr's Island,
    Pittsburgh, 183f
  York County, 355f
Permafrost, 132
Per person GNI PPP, 170
Persian Gulf oil spill, 430
Persistence (of substances), 83
Persistent organic pollutants (POPs), 84,
    84t, 191t
Peru, 144f
Pest control, 364–367

Pesticides, 72, 358, 364, 366f
  alternatives to, 366–367, 367t
  benefits of, 364
  concentrations of, 191t
  contamination by, 358
  mobility of, 83f
  problems with, 364–366
Pests, 364
Petrochemicals, 424
Petroleum, 423. *See also* Oil
Pheromones, 366
Philippines
  Dimakay Island, 148f
  Manila, 43f
  Mount Pinatubo, 199f, 225f, 299
  smoking in, 88f
Phosphorus cycle, 111–113, 112f
Photochemical smog, 110, 201, 202, 202f, 204f
Photodegradable plastic, 401
Photosynthesis, 100–101, 101f, 106
Photovoltaic (PV) solar cells, 445–447, 446f, 447f
Phytoplankton, 142, 282
Pinchot, Gifford, 52
Pioneer community, 152
Pittsburgh, Pennsylvania, 183f
Plan B 2.0 (Lester R. Brown), 36
Plankton, 142, 280
Plasma torch, 413
Plastic, in sanitary landfills, 398f, 401
Plate boundaries, 298, 298f, 299f
Plate tectonics, 298–301
Poaching, 385
Point source pollution, 261, 261f
Polar easterlies, 195
Polar ice caps, 15f, 216, 217t, 226t
Polar molecules, 244
Pollutants. *See also* Hazardous waste
  air, 108, 206–209, 212
  determining health effects of, 85–89
  global distillation effect, 190
  persistent organic, 84, 84t
  primary and secondary, 196
Polluted runoff, 261, 262
Pollution, 63. *See also specific types, e.g.,*
    Water pollution
  acceptable amount of, 5, 66
  children's sensitivity to, 89
  economic view of, 65–67
  and environmental justice, 35–36
  as global issue, 15f
  from incinerators, 402
  marginal cost of, 66, 66f
  optimum amount of, 67
  and species endangerment, 382
Pollution control
  costs and benefits of, 64–65
  economic strategies for, 67
  through legislation, 61
Polychlorinated biphenyls (PCBs), 411, 411f

Polyculture, 354
Ponds, 142–144
Population(s), 4, 98, 160, 160f
  characteristics of, 160
  evolution of, 147–150
  size changes in, 161. *See also* Population growth
  stabilizing, 38–39, 174–180
  world, 165
The Population Bomb (Ehrlich), 55
Population crash, 164, 164t
Population ecology, 160–164
Population growth, 4, 5t, 158–188
  and carrying capacity, 162–164, 167
  and consumption, 6–7, 9–11
  demographics of countries, 170–174
  exponential, 162, 162f, 162t
  human population patterns, 165–167, 168f–169f, 169t
  limits on, 148
  maximum, 161, 162
  population ecology, 160–164
  and poverty, 5f, 9, 38, 39
  projecting, 166–167, 166f
  stabilizing world populations, 174–180
  and technological advances, 55
  and urbanization, 181–186
  and water shortage, 248, 250
  and world hunger, 350–352, 351f
  zero, 166
Population growth momentum, 173
Portland, Oregon, 184
Potential energy, 100, 100f
Poverty, 4
  eliminating, 38
  and food insecurity, 40
  and food problems, 352
  and population growth, 5f, 9, 38, 39
Power plants. *See also specific types, e.g.,*
    Coal-fired power plants
  air pollutants from, 200
  carbon capture, 229, 230f
Precautionary principle, 90–92
Precipitation, 220, 221
  acid rain, 61, 234, 237–238
  and air pollution, 203f
  average monthly, 132t–137t, 139t–141t
  and climate, 218
  rain shadow, 221f
Predation, 119, 119f
Predator–prey interactions, 119–120
Prescribed burning, 154
Preservation, 50
Preston County, West Virginia, 306f
Prevailing winds, 195
Pribilof Islands, Alaska, 164, 164f
Primary air pollutants, 196, 196f, 200, 200f
Primary consumers, 101–102, 102f
Primary productivity, 132
Primary sludge, 265
Primary succession, 152–153, 152f

Primary treatment (sewage), 265, 265f
Prime farmland, 355
Prince William Sound, Alaska, 428, 429f
Principle of inherent safety, 412
Producers, 101
Productivity
  in ecosystems, 105–106, 105f
  primary, 132
Profundal zone, 143
Proxy Falls, Cascade Range, 221f
Public Rangelands Improvement Act (1978), 335
Puerto Rico
  aquaculture, 287f
  climate change in, 228f
  El Yunque National Forest, 141f
Pyramids of energy, 105
Pyramids of numbers, 105, 105f

Q

Queensland, Australia, 247f

R

Radiative forcing, 224, 225t
Radioactive waste disposal, 22
Radioactive wastes, 436–438
Radon, 210–211, 210f
Rain forests
  as global issue, 15f
  temperate, 134, 134t
  tropical, 39, 129f, 140–142, 141f, 329, 330, 379, 380
Rain shadow, 221, 221f
Range (of species), 378
Rangelands, 333–335, 333f
Rare earth metals, 304
Reactor core, 432
Realized niche, 113, 114f
Reclamation, soil, 296, 314, 316
Recycling
  exports of scrap, 404
  of old cars, 396
  of paper, 406
  solid waste, 396, 396f–397f, 405–408, 406f
  water, 256
Refining
  environmental impacts of, 307–308
  petroleum, 424f
Reforestation, 389f
Refuse-derived fuel incinerators, 403
Religious values, population growth and, 176, 177
Renewable energy resources, 442–464
  deep geothermal energy, 456, 457f
  direct solar, 444–449
  geothermal energy, 456–457
  indirect solar, 450–455
  tidal energy, 457
Renewable resources, 7

Reno, Nevada, 139t
Replacement-level fertility, 170
Reproductive capacity, 148
Reservoirs, 255–256
Resources, 7, 9, 48, 50. *See also* Consumption
    alternative energy, 48
    depletion of, 63–64
    nonrenewable, 7.
    protecting and restoring, 39–40
    renewable, 7.
Resource conservation, 50
Resource Conservation and Recovery Act
        (RCRA, 1976), 412
Resource degradation, 63
Resource partitioning, 114–115, 115f
Resource preservation, 50
Response, 85
Restoration
    of habitats, 387–388, 388f
    of mining lands, 308, 308f
    of rangelands, 335
Restoration ecology, 387, 388f
Reuse
    of old cars, 396
    of solid waste, 405
    water, 256
Rhine River Basin, 254, 254f
Rhizobium, 110
Richter scale, 300
Risk, 74
Risk assessment, 75–76
    of cancer-causing substances, 87
    of chemical mixtures, 87–88
    steps in, 75f
Risk management, 74–76
Risong Bay, Palau, Micronesia, 146f
Rivers, 142, 144, 144f
Rock cycle, 301, 301f
Rocky Mountain National Park, Colorado,
        143f
Rodenticides, 364
Roosevelt, Franklin Delano, 38, 54
Roosevelt, Theodore, 52, 52f, 387
Roraima State, Brazil, 40f
Runoff, 108, 245
    polluted, 261, 262
    and sea level rise, 230
Russia
    boreal forest, 329
    Kamchatka Peninsula, 343f
    Murmansk nuclear waste site, 438f
    population and population density of,
        170t

S

Safe Drinking Water Act (1974), 266
Safe Drinking Water Act amendments
        (1996), 92
Sahel region, Mali, 334f
St. Croix River, Taylors Falls, Minnesota,
        32f

Salinization, 250, 313, 313f
    of irrigated soil, 250, 313, 313f, 358
    and land degradation, 358
    of surface water, 250
Salt marshes, 146, 146f, 147
Saltwater intrusion, 230, 248
A Sand County Almanac (Leopold), 54, 54f
San Francisco, California
    Bayview Hunters Point, 35
    plate boundaries, 299
San Francisco Bay, 441, 441f
Sanitary landfills, 400–403, 400f–402f
Santa Fe, New Mexico, 445f
Sao Paulo, Brazil, 5f
Sareks National Park, Sweden, 343f
Saskatchewan, Canada, 133f
Savanna, 140, 140f, 140t
Scenarios, 229
Schistosomiasis, 455
Science, 17
    prediction in, 18, 20
    as process, 17–18
Scientific method, 18, 19f
Scrubbers (power plant), 422
Sea grasses, 281, 281f
Sea level rise, 226, 226f
    and ocean temperatures, 290
    and runoff, 230
Seasons, latitude and, 220
Secondary air pollutants, 196, 196f
Secondary consumers, 102, 102f
Secondary sludge, 266
Secondary succession, 153, 153f, 332
Secondary treatment (sewage), 265–266,
        265f
Second green revolution, 361
Second law of thermodynamics, 101
Sedimentary rocks, 301, 301f
Sedona, Arizona, 139f
Seed banks, 388, 389f, 390
Seed tree cutting, 326, 327f
Seismic waves, 300
Sekiu, Washington, 99f
Selective cutting, 326, 327f
Semiarid lands, 247
Senegal, 361f
Sessions, George, 33
Severn River, Shrewsbury, England, 20f
Sewage, 258
    in developing countries, 268
    water pollution from, 258, 259f
    in water systems, 77–78
Shade plantations, 380
Shaft mines, 305, 305f
Shanxi Province, China, 421f
Shelterbelts, 316, 316f
Shelterwood cutting, 326, 327f
Shifting cultivation, 354
Shortgrass prairies, 136
Shrewsbury, England, 20f
Siberia, 329
Sick building syndrome, 209, 210

Sierra Nevada Mountains, California, 372,
        372f–373f
Silent Spring (Carson), 26, 54–55
Simon, Julian, 55
Sinks, environmental, 62, 63
Sinnuris, Egypt, 179f
Slag, 305
Slash-and-burn agriculture, 330, 354
Slope mines, 305, 305f
Sludge, 265, 266
Smelting, 305
Smith, Robert Angus, 234
Smog, 201, 202
Smoking, 76, 88
Socolow, Robert, 122
Soil, 309–316, 309f
    agricultural, 361
    conservation and regeneration of,
        314–316
    erosion of, 313
    formation/composition of, 309–311
    organisms in, 311–312, 311f
    pollution, 313
    reclamation, 296
Soil Conservation Service, 54
Soil erosion, 313, 313f
Soil horizons, 311
Soil organisms, 311–312, 311f
Soil pollution, 313
Soil profile, 310, 310f, 311
Solar energy
    and atmospheric circulation, 194f
    direct, 444–449
    finding locations for installations, 48
    indirect, 450–455
    insolation, 219–220, 219f, 220f
Solar-generated hydrogen, 448–449, 449f
Solar thermal electric generation, 447–
        448, 448f
Solid waste, 398–408, 405f
    Curitiba garbage purchase program,
        186
    disposal of, 400–404
    reducing, 404–408
    types of, 398–399, 399t
Solvent, water as, 245
Solving environmental problems, 20–23
Source reduction (solid waste), 405
South Africa, 120f
South Asia, 350
South Bronx, New York City, 90f
South Carolina, 320, 321f
South Dakota, 136f
Soviet Union, 251
Spain, 48t
Species, 98, 374. *See also* Biological
        diversity (biodiversity)
    amphibians, 372, 372f–373f, 373t
    conserving, 388–390, 389f
    endangered and threatened, 378, 379f,
        382, 384–386
    endemic, 379

extinction of, 378, 379f
invasive, 338, 384–385, 384f
keystone, 121, 122
number of, 374
in tropical rain forests, 141
Species richness, 374–376, 374f–376f,
    376t
Spent fuel, 434, 436, 436f
Spoil bank, 304
Standing-water ecosystems, 142
Stationary sources (pollutants), 200
Steam generators, 432
Stegner, Wallace, 54
Stockholm Convention on Persistent
    Organic Pollutants, 84
Storage (decommissioned nuclear
    plants), 438
Stratosphere, 193f
Streams, 142, 144, 237, 237f
Strip cropping, 314
Strip mining, 304, 304f, 422
Subduction, 299
Sublethal dose, 85
Subsidence, 248
Subsidies, 291
    for ethanol, 451
    for fishing industry, 291
Subsistence agriculture, 354
Subsurface mining, 304, 305, 305f, 422
Succession, 151–153
    primary, 152–153, 152f
    secondary, 153, 153f
Sulfides, 302
Sulfur cycle, 110–111, 111f
Sulfur dioxide, 235
Sulfur emissions, 68
Sulfur oxides, 197
Summer Olympics (2004, Athens,
    Greece), 100f
Sun plantations, 380
Superfund Act. See Comprehensive Envi-
    ronmental Response, Compensa-
    tion, and Liability Act (CERCLA)
Superfund National Priorities List,
    412–413
Surface mining, 304, 304f, 422, 422f
Surface Mining Control and Reclamation
    Act (1977), 308, 422
Surface waters, 245
    overdrawing, 249–250
    seepage of, 246, 246f
Sustainability, 12–15, 26–45
    and ecosystem changes, 98
    and environmental justice, 35–36
    overall plan for, 36–43
    values in solving problems of, 31–34
Sustainable agriculture, 360–362, 360f,
    367
Sustainable cities, 43
Sustainable consumption, 28–30
Sustainable development, 28, 28f
Sustainable forestry, 325–326

Sustainable soil use, 312
Sustainable water usc, 254
Swamps, 142, 145, 145f
Sweden, 34f, 343f
Switzerland, 198f
Symbiosis, 116, 117f
Synergy, 87–88
Systems perspective, 55

T

Taiga, 133. See also Boreal forests
Tailings, 307, 307f
Tallgrass prairies, 136
Tambopata River, Peru, 144f
Tantalum, 317, 317f
Tanzania
    commensalism, 117f
    Kilimanjaro National Park, 140f
    locust swarms in, 361
    used clothing market, 417f
TAO/TRITON array, 277
Taxes, environmental, 67
Taylor Grazing Act (1934), 335
Taylors Falls, Minnesota, 32f
Temperate deciduous forest, 130f, 135,
    135f, 135t
Temperate grassland, 136, 136f, 136t
Temperate rain forests, 134, 134f, 134t
Temperature(s)
    average monthly, 132t–137t,
        139t–141t
    and climate, 218
    mean annual global, 222t. See also
        Climate change
    ocean, 290
Temperature inversion, 202, 204
Tennessee
    Cades Cove, Great Smoky Mountains
        National Park, 337f
    Chattanooga, 212, 212f
    Copper Basin, 296, 296f–297f, 308
    Kingsport, 35f
    Kingston, 418, 418f–419f
    Nashville, 135t
Terracing, 314, 315f
Terrestrial biomes, 128f
Tertiary consumers, 102, 102f
Tertiary treatment (sewage), 266
Texas, 35
Thailand, 33f
Theory(ies), 18
Thermal stratification, 143
Thermodynamics, 100
    first law of, 100
    second law of, 101
Thermosphere, 193f
Thoreau, Henry David, 51, 52
Threatened species, 378, 390f
    human causes of, 382, 384–386
    protecting, 393, 393f
    in the U.S., 390t

Three Gorges Dam, China, 269, 269f,
    454f, 455
Three Mile Island, Pennsylvania, 434
Threshold level, 86
Tidal energy, 456, 457
Tidal pool communities, 99f
Tides, 457
Tillage, 314
Times Square, New York, 9f
Tires, recycling, 402f, 406, 408
Tokyo, Japan, 13f
Tongass National Forest, Alaska, 344,
    344f
Topography, 309
    and air pollution, 202–204, 203f
    and precipitation, 221
    and soil formation, 309
Total fertility rate (TFR), 170, 175–177
Toxicants, 77, 81–83
Toxicity, 77
Toxicology, 77
Toxic waste. See Hazardous waste
Tradable permits, 67, 68
Trade winds, 195
Transform plate boundaries, 299, 299f
Transpiration, 324
Transportation, urban, 43, 184
Tree harvesting, 326, 327f, 344
Trickle irrigation, 256
Trophic level, 103, 103f
Tropical climates, 218f
Tropical dry forest, 140, 141, 329, 330
Tropical forests, 330
Tropical rain forests, 129f, 140–142, 141f,
    330
    declining biodiversity in, 379, 380
    deforestation, 329
    loss of, 39
Troposphere, 193f
Tsunamis, 300
Tundra, 129f, 132, 132t
Turneffe Atoll, Belize, 19f
Turnovers, 143

U

Ukraine, 434, 435f
Ultraviolet (UV) radiation, 231
Undernutrition, 350
United Kingdom, 13f, 26t
U.N. Convention on the Law of the Sea
    (UNCLOS), 291
U.N. Fish Stocks Agreement, 291
U.N. International Maritime Organization
    International Convention for the
    Prevention of Pollution from Ships
    (MAR-POL), 287, 288
U.N. World Commission on Environment
    and Development, 28
United Nations Intergovernmental Panel
    on Climate Change (IPCC), 222,
    223

United States
  acid deposition in, 235, 237
  ecological footprint of, 10, 11t
  endangered ecosystems in, 341, 341t
  endangered species in, 3t
  ethanol consumption, 451
  export of scrap to China, 404
  forests in, 332
  GM crops in, 363
  hydropower in, 454–455
  land use in, 322–323.
  mineral stockpiling, 304
  organic agriculture growth, 368t
  organic food demand, 368
  overnutrition in, 350
  politics of acid deposition in, 235
  population and population density of, 170t
  precautionary principle in, 91
  public lands in West, 342f
  renewable energy in, 48t
U.S. Commission on Ocean Policy, 291
University of Wisconsin-Madison Arboretum, 388, 388f
Upwelling, 276–277, 277f
Urban agglomerations, 184
Urban areas
  air pollution in, 201, 202
  disease transmission in, 80
  environmental problems of, 182–183, 182f, 183f
  shift from rural areas to, 184t
Urban heat islands, 204–205, 205f, 205t
Urbanization, 181, 184t
  environmental benefits of, 183–184
  environmental problems with, 182–183, 182f, 183f
  and population growth, 181–186
  trends in, 184–185
Utah, 337f
Utilitarian conservationists, 52

V

Values, 31–34
Variable, 18
Vegetation, in biomes, 130f–131f
Virginia
  Falls Church, 31f
  salvage yard, 396f–397f
Volcanoes, 299
  and air pollution, 198, 199f
  and global temperatures, 225f
Voluntary simplicity, 29–30

W

Washington, D.C., 413f, 446f

Washington's Landing, Herr's Island, Pittsburgh, Pennsylvania, 183f
Washington State
  Clallam Bay, Sekiu, 99f
  Gifford Pinchot National Forest, 340f
  Grand Coulee Dam, Columbia River, 255f
  Hoh Rain Forest, 134f
  Lake Washington, 96–97, 96f–97f
  Proxy Falls, Cascade Range, 221f
  Seattle, 29f
Wastes
  hazardous. See Hazardous waste
  radioactive, 436–438
  solid. See Solid waste
Wastewater treatment, 265–266, 265f, 266f
Water. See also Freshwater resources; Ocean(s)
  aquatic ecosystems, 142–147
  and biogeochemical cycles, 108, 108f, 109, 109f, 111–113, 112f, 245, 245f
  distribution of, 242t
  for fast-growing cities, 185f
  hydrologic cycle, 245–246, 245f, 246f
  hydropower, 454–455, 454f, 455f
  importance of, 244
  limited supply of, 242
  properties of, 244–245
  soil erosion caused by, 313f
  tidal energy, 457
Waterborne diseases, 77–78, 78t
Water conservation, 256, 257f
Water management, 254–257
Water pollution, 258–269
  diseases transmitted by, 78t
  groundwater, 263
  improving water quality, 264–269
  from mining, 306
  monitoring, 20f
  sources of, 261–263
  types of, 258–261
Water quality, improving, 264–269
Water scarcity
  in cities, 43
  and population growth, 248
  projected, 244
Watersheds, 108, 245, 252t
Water table, 248
Wat Mahathat, Thailand, 33f
Weather, 218
  and air pollution, 202–204, 203f
  and ocean–atmosphere interaction, 276–278
Weathering processes, 309
West Africa, 251
Westerlies, 195

Western worldview, 32–34, 32f, 33f
West Pokot, Kenya, 41f
West Virginia
  Cabin Creek, 422f
  coal-burning power plants, 68f
  Germany Valley, 135f
  Preston County, 306f
Wetlands, 245
  constructed, 266, 266f
  freshwater, 145, 145f
Wet sulfate, 61
Whitman, Christine Todd, 91
Wilderness, 338
Wilderness Act (1964), 338, 339
Wilderness areas, 338–339
"Wilderness Essay" (Stegner), 54
Wildfires, 138, 138f, 154, 154t
Wildlife refuges, 322f, 387, 439
Winds, 195, 195f
Wind energy, 452, 453f, 453t
Wisconsin, 388, 388f
Wise-use movement, 340
Wolong Nature Reserve, China, 343f
Women, status of, 177–178, 177f
Wooded grasslands, 130f
Woodlands, 130f
World Conservation Strategy, 391–392
World hunger, 350–352, 351f
Worldviews, 31–34
Wrangell–Saint Elias Wilderness, Alaska, 339f

Y

Yangtze River, China, 454f, 455
Yellow River, China, 251
Yellowstone National Park, 153, 336, 338, 338f
Yields (agricultural), 353, 356–357, 357f
Yongchuan, Chongqing, China, 118f
York County, Pennsylvania, 355f
Yosemite National Park, California, 53f, 323f
Yosemite National Park Bill (1890), 52
Yucca Mountain, Nevada, 22, 437, 437f
Yukon, Canada, 132f, 190

Z

Zero population growth, 166
Zimbabwe, 79
Zion National Park, Utah, 337f
Zooplankton, 142, 282
Zooxanthellae, 280